American Institute of Chemical Engineers

8th Topical Conference on Microreaction Technology 2005

Held at the 2005 AIChE Spring National Meeting

April 10-14, 2005
Atlanta, Georgia, USA

Printed from e-media with permission by:

Curran Associates, Inc.
57 Morehouse Lane
Red Hook, NY 12571
www.proceedings.com

ISBN: 978-1-60423-468-8

Some format issues inherent in the e-media version may also appear in this print version.

TK - IMRET - 8th International Conference on Microreaction Technology

Session 128 - Plenary Talk - Mixing of Miscible Fluids in Microstructured Mixers

Chair: Yong Wang

Session 129 - Unit Operations

Chair: Asterios Gavriilidis

Session 130 - Characterization and Simulation of Microstructured Devices

Chair: MichaelMatlosz

Session 133 - Modularization and Multiscale Design

Chair: Claude de Bellefon
Vice Chair: Shinji Hasebe

Session 134 - Microstructured Devices as Tools in Chemical Research

Chair: Thomas Schwalbe
Vice Chair: Masaaki Sato

Session 137 - Nanostructures and Nanoparticles

Chair: Steve J. Haswell
Vice Chair: Elias J. Klemm

Session 138 - Single Reactors II

Chair: Albert Renken
Vice Chair: Jun-ichi Yoshida

128a Mixing of Miscible Fluids in Microstructured Mixers

Friedhelm Schoenfeld, Holger Loewe, Patrick Loeb, Volker Hessel

Mixing in minute volumes is probably the by far most frequently investigated operation for chemical micro process engineering. It is the essential step to improve many single-phase liquid organic reactions with regard to selectivity and is of even more importance for gas/liquid and liquid/liquid reactions to provide sufficiently large interface for mass transport. Mixing is even applied in some cases for micro-space gas-phase reactions, e.g. when contacting gases in an explosive regime. Many mixing principles and devices were developed also in the framework of miniature analytical chip systems; some of them are amenable for performing chemical reactions as well. Consequently, there have been several tens, if not hundreds, of different microstructured mixers developed and today's commercial offer is large, including even pilot- and production-type apparatus. The scientific literature on mixing with such tools is extraordinarily broad and many in-depth investigations provide detailed insight. In this context, this review wants to give a comprehensive overview about mixing principles and realised respective microstructured devices. Their characterisation will be exemplarily given; applications will be shown as the case arises, but the focus is on the microstructured mixers themselves.

Mixing in minute spaces is possible by active and passive means. Active mixing involves the use of external energy, whereas passive mixing simply uses part of the flow energy, often provided anyway. Active mixing will be done, if the latter is not sufficient, e.g. for slowly moving streams or standing volumes, which is often the case in analytical and biological applications. Often chemical uses are concerned with achieving reasonable throughputs and thus can rely on passive mixing solely. Both generic principles follow the two same intentions to carry out efficient mixing. The reduction of diffusion distances serves to come to higher concentration gradients. A second means is the enlargement of internal interfaces for mixing which is done by convection, e.g. via recirculation flows. By superposition of convective eddy structures, chaotic patterns with large stretching factors can be formed, expanding the interfaces in an exponential manner. Besides diffusion and convection, there remains turbulence for specialty mixers serving a small field of selected applications.

Passive micro mixers use all kinds of external energy which generates convection in a fluid. Among these are the application of electrical and magnetic fields (and their combination), ultrasound, surface acoustic waves, bubble induced micro-streaming, pumping by integrated valves and micro pumps, stirring by miniature impellers, and a good deal more. Several examples will be given here, in particular referring to electro-hydrodynamic mixing, but shortly covering the other means as well.

With even higher relevance for chemical processing, the presentation will unroll the different passive types of microstructured mixers. Basically, two major classes with a large number of sub-versions can be distinguished. Multi-laminating mixers reduce the diffusion distances by highly regular concentration patterns of many lamellae with typical thickness of a few 10 μm. Recently, geometric focusing techniques allowed to further thin the lamellae down to a few μm or even thinner, which is equivalent to ms-mixing for most organic molecules in low-viscous liquid media. The second major class of passive micro mixers is based on creating convection; mostly this is done by posing internal structures into the flow which induce transverse circulatory motion.

Multi-laminating mixers comprise bifurcation or interdigital feed arrangements, splitting the flow by use of symmetry or pressure barriers. The flow patterns of these mixers are well documented in all their details meanwhile by optimising methods for direct flow visualization, e.g., how the lamellae may tilt when abruptly changing their direction as given in some focusing elements. Besides the focusing technique, the multi-lamellae may be guided into a large chamber where turbulence sets in, further diminishing the diffusion distances. As an indirect way, splitting and recombination (SAR) can principally generate multi-lamination patterns in a serial rather than in a parallel manner. Although this has often been claimed, the corresponding mixers actually induced circulatory motion for mixing, rather than performing ideal SAR action; at very small, Reynolds numbers, friction and inertia forces distort the multi-lamellae patterns at the expense of interface growth. Just recently, a properly functioning SAR micro mixer has been proposed and evaluated, which contained as essential splitting element a cutting sheet followed by separated flow guidance of the two cut streams (being recombined thereafter). Detailed investigations on hydrodyamics using CFD simulations and experimental flow visualisation were supplemented by semi-analytical descriptions for calculating mixing (which is not simply feasible for micro-channel flows due to "numerical diffusion") and in-line concentration monitoring.

One of the first and still instructive examples for convective micro mixers is the so-called staggered herring-bone (SHM) micro mixer, having alternately arranged oblique grooves. The recirculation patterns were visualised by confocal fluorescence microscopy, proving the superiority of the chaotic mixing over non-structured and symmetrically grooved channels. In particular attractive is the small dependence on the mixing length of the SHM mixer on the Peclet number. This pioneering design was later optimised and followed by many successor works which will be given. Recently, perfect replicas of Sulzer and Kenics mixers having intersecting and helical internal packing, respectively, were realised at the miniature level. The so-called barrier-embedded micro mixer with slanted grooves or helical elements uses principles of static mixers known from polymer extrusion. Redirected flow mixers, like the Coanda mixer with Tesla structures, split a small part of the flow and re-guide it to the main stream to generate interfaces. Small packs, structured or by orientation of particles, were applied for mixing, too.

Apart from all these more or less complex micro mixer principles and designs, there is just recently a recalling on the potential of simply using T-junction micro mixers, resembling mixing tees used at macroscopic level. While these tools exhibit stratified, i.e. bi-laminated flow, at small Reynolds number which provides only a low interface, there is interpenetration of the streams in the so-called vortex regime which leads to quite efficient mixers. Alternatively, deliberately set periodical pulsations of the volume flow can assist mixing in T-junctions, which has been exploited in the so-called time-pulsing mixing. In addition, posing simple square flow obstacles or integrating Venturi nozzles can enhance mixing in T-junctions considerably. Specialised T-junctions operated at very high flow rates allow turbulent mixing by the impinging jet principle, based on high-energetic frontal collision of the streams.

Reviews of the authors in the field:

[1] Hessel, V., Hardt, S., Löwe, H.; Chemical Micro Process Engineering - Fundamentals, Modelling and Reactions, Wiley-VCH, Weinheim (2004).

[2] Hessel, V., Löwe, H., Müller, A., Kolb, G.; Chemical Micro Process Engineering - Processing and Plants, Wiley-VCH, Weinheim (2004).

[3] Hessel, V., Löwe, H.; "Micro mixers - a review on passive and active mixing priciples", Chem. Eng. Sci. (2005) accepted for publication.

[4] Hessel, V., Löwe, H.; "Mixing principles for microstructured mixers", chapter in ACS book "Microreactor technology and process intensification", accepted for publication, (2005).

[5] Jähnisch, K., Hessel, V., Löwe, H., Baerns, M.; "Chemistry in Microstructured Reactors", Angew. Chem. Int. Ed. 43, 4 (2004) 406-446.

[6] Pennemann, H., Watts, P., Haswell, S., Hessel, V., Löwe, H.; "Benchmarking of microreactor applications", Org. Proc. Res. Dev. 8, 3 (2004) 422-439.

[7] Hessel, V., Löwe, H.; "Micro chemical engineering: components - plant concepts - user acceptance: Part I", Chem. Eng. Technol. 26, 1 (2003) 13-24.

[8] Hessel, V., Löwe, H.; "Micro chemical engineering: components - plant concepts - user acceptance: Part II", Chem. Eng. Technol. 26, 4 (2003) 391-408.

[9] Hessel, V., Löwe, H.; "Micro chemical engineering: components - plant concepts - user acceptance: Part III", Chem. Eng. Technol. 26, 5 (2003) 531-544.

Selected technical papers of the authors in the field:

[10] Hessel, V., Hardt, S., Löwe, H., Schönfeld, F.; "Laminar mixing in different interdigital micromixers - Part I: Experimental characterization", AIChE J. 49, 3 (2003) 566-577.

[11] Hardt, S., Schönfeld, F.; "Laminar mixing in different interdigital micromixers - Part 2: Numerical simulations", AIChE J. 49, 3 (2003) 578-584.

[12] Schönfeld, F., Hessel, V., Hofmann, C.; "An optimised split-and-recombine micro mixer with uniform 'chaotic' mixing", Lab Chip 4, (2004) 65-69.

[13] Schönfeld, F., Hardt, S.; "Simulation of helical flows in microchannels", AIChE J. 50, 4 (2004) 771-778.

[14] Jiang, F., Drese, K. S., Hardt, S., Küpper, M., Schönfeld, F.; "Helical flows and chaotic mixing in curved micro channels", AIChE J. 50, 9 (2004) 2297-2305.

[15] Ehrfeld, W., Golbig, K., Hessel, V., Löwe, H., Richter, T.; "Characterization of mixing in micromixers by a test reaction: single mixing units and mixer arrays", Ind. Eng. Chem. Res. 38, 3 (1999) 1075-1082.

[16] Löb, P., Pennemann, H., Hessel, V.; "g/l-Dispersions in interdigital micromixers with different mixing chamber geometries", Chem. Eng. J. 101, 1-3 (2004) 75-85.

[17] Löb, P., Drese, K. S., Hessel, V., Hardt, S., Hofmann, C., Löwe, H., Schenk, R., Schönfeld, F., Werner, B.; "Steering of liquid mixing speed in interdigital micromixers - from very fast to deliberately slow mixing", Chem. Eng. Technol. 27, 3 (2003) 340-345.

128b Microreaction for Pharmaceutical Manufacturing

Adeniyi Lawal, Donald Kientzler, Raghunath Halder, Ronald S Besser, S Kiang, Sunitha Tadepalli, Woo Lee

The trend toward miniaturization of reaction systems for pharmaceutical applications gained strong acceptance in the quest for rapid production of libraries of candidate compounds for the drug discovery process. More slowly to mature, however, has been the implementation of microscale systems for the production of pharmaceutical compounds and intermediates. Microreactors seem like an obvious tool for this use since the most common reaction vessels are batch reactors which possess well-known limitations in heat and mass transport. Various reasons for delays in implementation exist, many of which are non-technical in nature, including regulatory issues.

Although the industry has been slow to embrace microreactors, there is presently great interest in many parts. One of the process steps which would appear to be ripe for implementation is hydrogenation, which accounts for a significant fraction of all pharmaceutical processes. As an example, we will describe the work now in progress in our center on this application.

The New Jersey Center for Microchemical Systems is situated in near proximity to its collaborators at Bristol-Myers Squibb, and a large cadre of other pharmaceutical researchers and manufacturers. A brief introduction will be given of the two organizations involved in this activity whose relationship could be considered as one model for collaborative interactions between academic and industrial groups. Past work reported by the research community for the use of microreaction in pharmaceutical synthesis applications will also be surveyed.

128c A CFD model for characterising axial mixing in microreactors operating under Taylor flow

Asterios Gavriilidis, Panagiota Angeli, Wael Salman

Keywords: Taylor flow, Axial mixing, Dispersion, Microchannels.

Taylor flow is one of the dominant gas/liquid flow patterns in microscale diameter channels. It consists of elongated gas bubbles of equivalent diameter larger than the tube diameter separated by liquid slugs. The bubbles adopt a characteristic capsular shape and completely fill the channel cross section, separated from the wall only by a thin film of liquid. Because of the presence of bubbles in front and at the back of the slugs, the flow field in the liquid is modified compared to single phase flow, and toroidal vortices are formed [1]. As the only means of communication between two successive slugs is the thin film surrounding the bubbles, reduced axial mixing is achieved. Also within the slugs liquid recirculation improves radial mass transfer, i.e. from liquid to wall and from gas to liquid [2, 3, 4].

Taylor flow can be important in two-phase applications that involve mass transfer, fluid/fluid or wall/fluid (such as in tubular reactors or tubes with absorbing walls for liquid chromatography), or in single phase liquid applications which suffer from large backmixing (for example tubular dialysers can be improved by operating in Taylor flow rather than in single phase). Much work on Taylor flow has been done in the context of monolith froth reactors (MFR) where the two fluid phases are introduced into the reactor as a froth and flow through the catalyst coated channels in a Taylor flow pattern [5, 6]. Residence time of the liquid in the reactor is important for determining conversion and selectivity in such systems. Another application of Taylor flow is in the field of combinatorial chemistry as a means of introducing sequentially different reactants/catalysts within a microchannel reactor/analyser.

For all the above applications, it is important to be able to predict the extent of axial mixing as a function of the hydrodynamics of the two phase flow. Previous attempts to model axial mixing in Taylor flow systems have assumed simplified models for the mass transfer within the liquid slug and from the slug to the liquid film at the wall. This results in models which predict satisfactorily experimental data for a limited range of operating conditions [5,6,7,8].

In this paper, a model is presented for the calculation of residence time distribution in a Taylor flow system that is not using any assumptions and therefore is not restricted to a particular range of operating conditions. To study periodic Taylor flow, a bubble and a slug, or a slug and two half bubbles, are often considered which are referred to as a unit cell. In our model, CFD simulations are used to obtain the exact residence time distribution in a single unit cell when a step change in the concentration of the liquid film entering the cell is applied. A convolution method is then developed to obtain from the residence time distribution of a single cell the residence time distribution in the whole microchannel. The model predictions compared well with experimental data by Thulasidas et. al. [5].

The effect on residence time distribution of various governing parameters such as slug length, Reynolds number, Peclet number and film thickness was investigated using this model. It was found that by decreasing slug length, film thickness and Peclet number axial mixing is reduced.

References

[1] Thulasidas, T. C., Abraham, M. A., Cerro, R. L., 1997, Flow patterns in liquid slugs during bubble train flow inside capillaries, Chemical Engineering Science 52[17], 2947-2962.

[2] Irandoust, S., Andersson, B. 1988, Mass transfer and liquid phase reactions in a segmented 2-phase flow monolithic catalyst reactor, Chemical Engineering Science 43[8], 1983-1988.

[3] Irandoust, S., Andersson, B., Simulation of flow and mass transfer in Taylor flow through a capillary, Computers & Chemical Engineering 13[45], 519-526.

[4] Bercic, G., Pintar,A., 1997, The role of gas bubbles and liquid slug lengths on mass transport in the Taylor flow through capillaries, Chemical Engineering Science 52[21, 22], 3709-3719.

[5] Thulasidas,T. C., Abraham,M. A., Cerro, R.L., 1999, Dispersion during bubble train flow in capillaries, Chemical Engineering Science 54, 61-76.

[6] Kreutzer, M., 2003, Hydrodynamics of Taylor flow in Capillaries and Monolith Reactors, PhD Thesis, Delft University Press.

[7] Pedersen, H., Hovarth, C., 1981, Axial dispersion in a segmented gas-liquid flow. Industrial & Engineering Chemistry Fundamentals 20, 181-186.

[8] Thiers, R. E., Reed, A.H., Delander, K., 1971, Origin of the lag phase of continuous-Flow Analysis Curves, Clinical Chemistry, Vol 17[1], 42-48.

Effective Mixing by the Use of Convective Micro Mixers

Michael Engler, Thomas Kiefer, Norbert Kockmann, Peter Woias
Laboratory "Design of Microsystems"
Institute of Microsystems Technology IMTEK, University of Freiburg

Prepared for Presentation at IMRET 8, 10th-14th April 2005, Plenary Talk – Mixing of Miscible Fluids in Microstructured Mixers, Unpublished

Effective Mixing by the Use of Convective Micro Mixers

Michael Engler, Thomas Kiefer, Norbert Kockmann, Peter Woias
Laboratory "Design of Microsystems"
Institute of Microsystems Technology IMTEK, University of Freiburg
Georges-Koehler-Allee 102, D-79108 Freiburg
Tel.: 0761/203-7504
Fax: 0761/203-7492
E-Mail: engler@imtek.de

Abstract

This paper shows the design, fabrication, and characterisation of micro mixers which operate by convective effects in relatively large microchannels with hydraulic diameters of about 300µm. Hydrodynamic effects at relatively high Reynolds numbers of up to 1000 lead to the creation of secondary vortices and to a very fast mixing by the generation of thin fluid lamellae. Due to the high flow velocities (around 1 m/s) in these channels, the risk of fouling or clogging is reduced whereas the throughput is enhanced. Simple mixing elements (T-mixer, L-mixer) are studied by CFD simulations to obtain their mixing performance. To improve the throughput of the mixers, 16 of these elements are parallelised on silicon chips sealed by Pyrex glass with dimensions of $20\times20\times1.5$ mm^3. These mixing manifolds are designed with the help of analytical calculations.

The mixers are experimentally characterised by measuring mass flow, pressure loss, mixing times and mixing quality. The color change of a pH indicator reaction (Bromothymole Blue) indicates the mixing length and time, while the mixing quality is obtained from the well-known iodide-iodate-reaction according to Villermaux and Dushman. Results for aqueous solutions give very short mixing times of about 500µs with mass flows up to 20kg/h at a pressure loss below 1 bar. To compare these mixers among each other and with micro mixers designed by other groups, a dimensionless number, the *Mixing Efficiency ME*, is proposed and calculated for several micro mixers.

Introduction

The successful implementation and operation of micro mixers has been shown by many authors in the past few years. Yet, micro mixers are still hardly used for chemical production purposes. Reasons are, besides the indecision to replace such a reliable and established technology like conventional batch reactors, micro reactor specific problems like low production rates or clogging of the micro channels.

On the other hand, micro mixers have several benefits. Due to their small dimensions, the mixing can be made very fast which is beneficial for a multitude of chemical reactions, especially when they are complex and consist of multiple steps. Furthermore, due to the high surface to volume ratio, the temperature of the mixing process can be controlled very easily and safety problems, e.g. when performing fast reactions in explosive regimes, can be reduced to a minimum. This altogether can significantly increase the yield of chemical reactions and save resources.

A good overview over already fabricated micro mixers is given by Nguyen and Wu (2005). The authors give a review over the different kinds of micro mixers which have been designed up to now, with most of these mixers still being on laboratory level. An even more detailed overview is given by Hessel et al. (2005). Here, also micro mixers are shown which have reached a certain point of maturity and are, in part, already commercially available. Most of the commercially available mixers presented up to now are mixing by molecular diffusion, where the diffusion happens in a strictly laminar flow regime. The slow diffusion process is supported by various design principles like split-and-recombine, multilamination or hydrodynamical focussing.

In this work, a novel approach for micro mixers is applied, which builds up on a well-known effect that is nevertheless new for an implementation in a micro mixer: Convective mixing, i.e. the generation of laminar secondary vortices leads to a very fast and effective mixing inside simple mixing elements. The first micro mixers, in which these effects have been discussed, were T-shaped micro mixers, so the principle is illustrated here using these geometries. For earlier results of convective mixing in T-shaped micro mixers, see e.g. Kockmann et al. (2003), Engler et al. (2004) and Wong et al. (2004).

The Convective Mixing Principle

Convective Mixing

When operating T-shaped micro mixers at relatively high Reynolds numbers above 50, one identifies the appearance of secondary vortices. These vortices arise because of the deflexion of the inlet flows at the T-junction. At Reynolds numbers higher than 150-250 (depending on the geometries), the vortices begin to engulf each other leading to an effective mixing. Fig. 1. shows illustrations of the flow situation inside a T-shaped micro mixer with cross sections of 300×300 μm^2 for the inlets and 600×300 μm^2 for the outlet. This pictures were obtained from CFD simulations with the commercial CFD-code CFD-ACE+ by ESI Group. The view is into the mixing channel, as illustrated on the left side. The figure in the middle shows a schematic sketch of the flow situation inside the cross section of the mixing channel (top) and a cutout of a plane inside the mixing channel seen from above (bottom). The mechanism for the creation of the thin fluid filaments is obviously a roll-up of the interface between the two inlet fluids. The right hand side figure shows the simulation of the flow field at two Reynolds numbers, Re_M = 179 (top) and Re_M = 249 (bottom) with 7000 streamlines, coloured black and white for the two inlet fluids. The effect of the rising Reynolds number on the intensity of the swirling and on the further thinning of the fluid lamellae can clearly be observed. Even at relatively low Reynolds numbers of 180, fluid filaments with a thickness of only 1 – 3 μm are created, as reported by Hoffmann et al. (2004). In these regions, the time for a full mixing by diffusion falls below milliseconds.

Experiments with dyed solutions have shown that the mixing becomes fully homogenous at distances of about 2 mm downstream in the mixing channel at Reynolds numbers larger than 350 for the mixer geometry given above. This results in mixing times of about 2 ms. The reason for this very fast mixing is the large amount of energy which is dissipated directly into the mixing process. This specific power dissipation ε inside the micro mixers is calculated by

$$\varepsilon := \frac{\Delta p_v \cdot \dot{V}}{m} = \frac{\Delta p_v \cdot \overline{u}_M}{\rho \cdot l_m}, \tag{1}$$

where Δp_v is the pressure loss originating from the vortices, $\dot{V}$ the volume flow, m the mass into which the energy is dissipated, $\overline{u}_M$ the mean velocity in the mixing channel, ρ the mean density of the fluids, and l_m the mixing length inside the mixing channel, after which the mixing is completed. Since the vortices relaminarisate in the range of the mixing length, ε can be calculated to a good approximation using the mixing length. First calculations indicate specific power dissipations of more than 3000 W/kg at $Re_M = 400$ for the above geometry. Detailed calculations and the correlation of mixing time to power dissipation are part of future research.

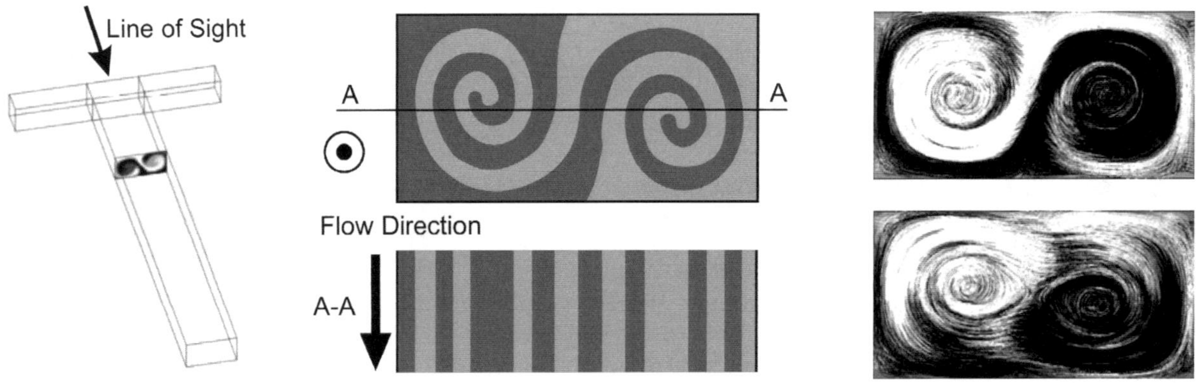

Fig. 1: Illustration of the mixer model (left), schematic sketch of the cross section of the mixing channel (top centre), corresponding view from above (bottom centre) and simulations of the flow situation, illustrated by 7000 streamlines each at two different Reynolds numbers, $Re_M = 179$ (top right) and $Re_M = 249$ (bottom right)

Characteristics

With the information given above, the following characteristics of the convective mixing principle can be given:

- Low pressure losses with high mass flows because of the wide channels,
- very fast mixing in the ms- or sub-ms-range for aqueous liquids,
- increased efficiency by raising the Reynolds number: Typical Reynolds numbers for convective mixing are between 250 and 1000,
- reduced danger of fouling and hardly any clogging potential due to the wide fluid channels (up to 600µm hydraulic diameter) and the high flow velocities of approximately 1 m/s for aqueous solutions.

Several micro mixers using the convective mixing principle have been fabricated and characterised. The core of the concept is the simple and very fast creation of thin fluid filaments by hydrodynamical effects only. The principle is very basic and easy to implement. There is no need for sophisticated internals or active items, nor is there any need to use very narrow channels. Its potential will be shown in the next paragraph, where 16 single convective mixing elements are parallelised on one silicon chip to increase the overall throughput.

Design of the Convective Micro Mixers

For the design, different single mixing elements are at first simulated by the above mentioned commercial CFD-code. The best of these elements are parallelised on on a single stacked chip by distributing them into two fluid layers. Here, an analytical approach to equalise the pressure losses over all 16 mixing elements is used.

Simulation of Single Mixing Elements

The single elements have been simulated using CFD-ACE+. The elements have been studied concerning the flow (velocity vector and pressure) and the molecular diffusion. A multitude of different convective mixing elements has been simulated, whereof the three best elements are presented here. These are the 3D-T-Mixer and the Tangential Mixer as elements combining the fluid and four L-bends in a row as an element guiding the fluid, shown in Fig. 2.

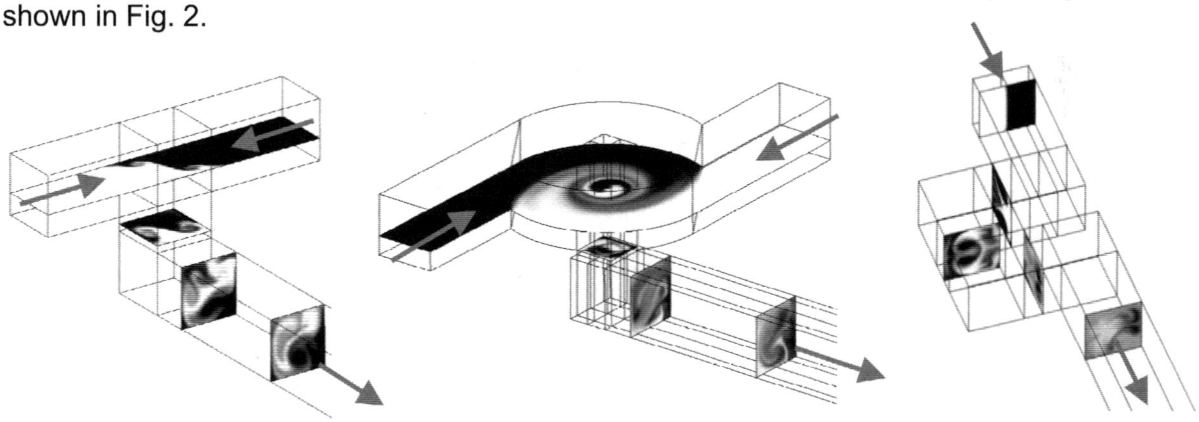

Fig. 2: Three of the simulated mixing elements; 3D-T-Mixer (left), Tangential Mixer (centre) and four L-bends in a row (right)

For a quantitative characterisation, the well-known mixing quality α after Zwietering and Dankwerts is used. The mixing quality α in a cross section of the mixing channel is calculated by

$$\alpha := 1 - \sqrt{\frac{\sigma_M^2}{\sigma_{\max}^2}}, \tag{2}$$

where σ^2 is the variance (with indices *max* standing for the maximum variance which equals 0.5 for symmetrical inlet conditions and *M* for the variance calculated on all nodes of the cross section, where the number of nodes amounts to 1200-1500, depending on the simulated geometry). A mixing quality of 0 means complete segregation, while 1 means homgeneous mixing. Details about the calculation can be found in Engler et al (2004). Results for the above mentioned mixing elements are shown in Fig. 3. A rising mixing quality occurs with increasing Reynolds number. The four L-mixers are solely used for a further mixing after the fluids are combined by the 3D-T-Mixer or the Tangential Mixer. The Tangential Mixer shows a high mixing quality at relatively low Reynolds numbers, while the 3D-T-Mixer is mixing well at higher Reynolds numbers. Unfortunately, the simulations become unstable at higher Reynolds numbers, so it was impossible to simulate the fully homogenous mixing (α = 1). However, this does not cause any limitations, since former experiments with T-shaped mixers (Engler et al. (2004)) have shown that fully homogeneous mixing is achieved at higher Reynolds numbers than those simulated.

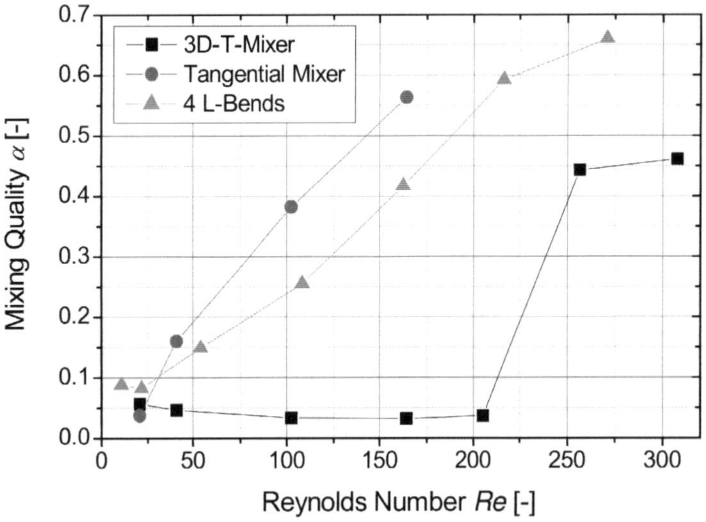

Fig. 3: Mixing qualities for the three different mixing elements shown in Fig. 2

Design of the Entire Mixers

The mixers demonstrated above are primarily optimised concerning the mixing quality, but not concerning the throughput. For this, the single mixing elements are parallelised on one chip with dimensions of 20×20 mm^2. These dimensions are arbitrarily chosen and can easily be modified for later designs, e.g. for an even higher throughput.

For topological reasons (the channels can not overlap on one layer), it is necessary to use two channel layers. This is implemented by using two silicon wafers, where one wafer holds the inlets and the primary micro mixers that combine the fluids. The other one incorporates fluid transportation in secondary mixers and the outlet. To decrease the pressure loss, the inlet and outlet holes have been widened compared to former designs. For the inlets, 3 holes for each inlet fluid with dimensions of 300×300 μm^2 have been used to achieve a better pressure homogeneity over the inlet channels. For the outlet, one hole with dimensions of 1200×1200 μm^2 has been located in the centre of the chip.

Two different mixers have been designed. Each mixer consists of 16 single mixing elements which are parallelised. The two mixers of Fig. 2 (3D-T Mixer and Tangential Mixer) have been implemented as primary units. Additionally, a meandering structure consisting of eight succesive L-elements has been attached as a secondary mixer directly after the mixing elements. These elements enhance the mixing after the combining mixing element and serve for a complete mixing at even low flow velocities.

To ensure that the flow properties are the same for each mixing element, it has been necessary to equalise the flow resistance for each inlet and outlet of the single mixing elements. For this, the inlet and the outlet channels have been partially widened to decrease the flow resistance. These different flow resistances have approximately been calculated using the Poiseuille equation together with the analytical equation for the pressure drop along a channel. Using a commercial computer algebra system (Mathematica®), the different channel lengths have then been calculated so that the pressure losses are the same over each of the inlet and the outlet channels. Fig. 4 gives the final mask layouts for the two entire mixers.

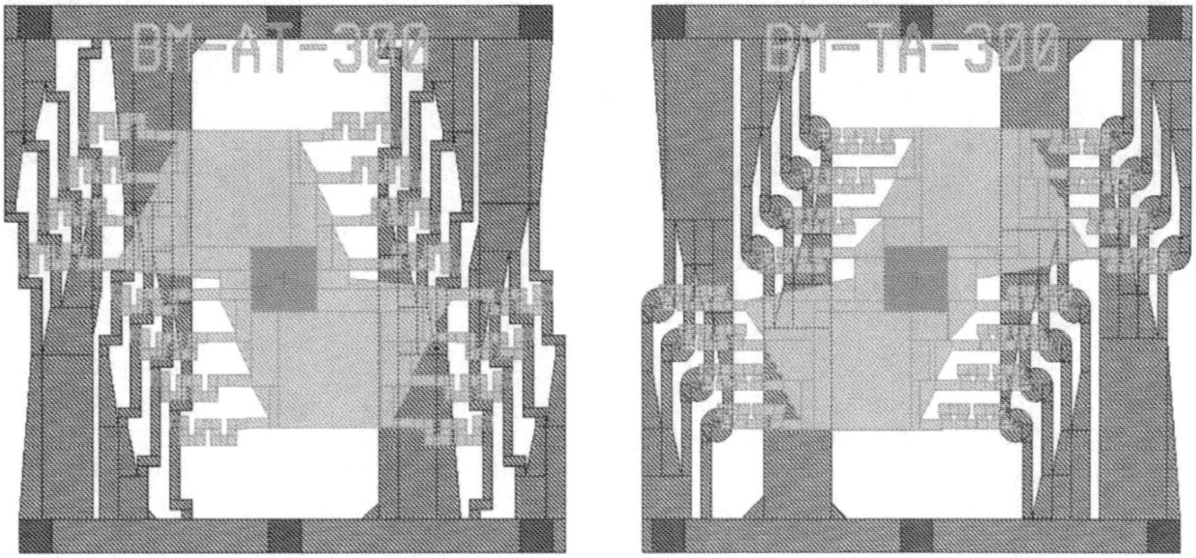

Fig. 4: Mask layouts for the 3D-T-Mixer (left) and the Tangential Mixer (right); the six inlet holes are positioned at the left and right side of the chip, while the outlet hole is in the centre; the lower layer (inlet layer) is drawn in green (or dark grey), while the upper layer (outlet layer, can later be observed by a microscope) is drawn in orange (or light grey).

Fabrication and Housing

Fabrication

The mixers are fabricated by standard silicon bulk micro machining. For the etching, the well established Advanced Silicon Etching (ASE) Process has been used. With this, rectangular channels with a high quality are fabricated which are very close to ideal rectangular channels, compare SEM pictures in Fig. 5. The two silicon layers are bonded by Silicon Fusion Bonding (SFB). During the bonding process, the breaking of the fragile silicon wafers with a multitude of channels has occured sometimes. For a further fabrication, a new low temperature bonding technique which was developed by Doll et al. (2004), shall be used. Finally, the two layers are sealed by Pyrex Glass using anodic bonding onto the upper layer. The lower layer contains the inlet and outlet holes. On a 4-inch silicon wafer, 12 different chips with dimensions of 20×20 mm^2 can be integrated. With this, 12 mixers can be produced in one fabrication step. After the fabrication, the chips are separated by dicing saws. An example of one of the fabricated chips (T-Cascade Mixer) is shown in Fig. 6, left side.

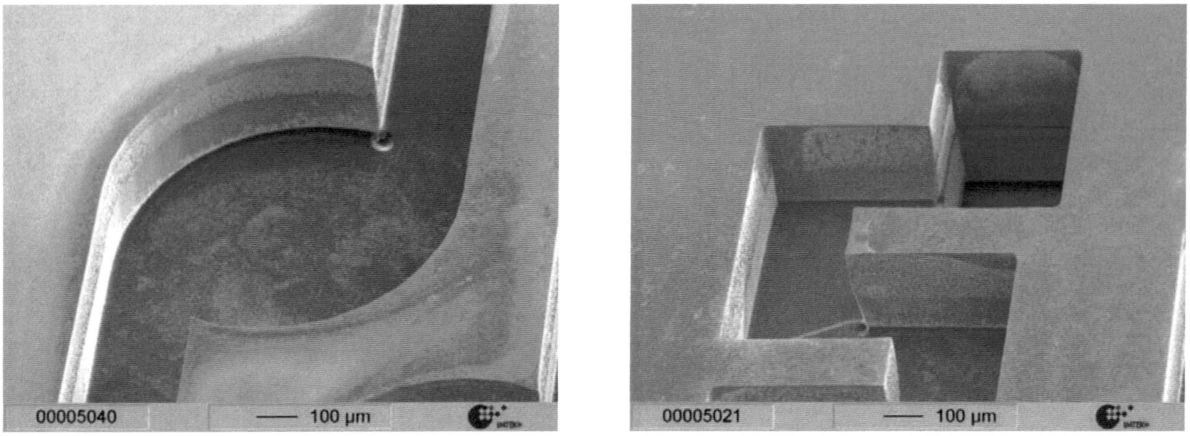

Fig. 5: SEM photos of the mixing element of the Tangential Mixer (left) and the mixing channel of the 3D-T-Mixer together with the hole combing the two fluid layers (right)

Housing

For the housing, a fluidic mount was made of aluminium by standard precision machining. Standard fluidic connectors for tubes with 4mm external diameter are used. To obtain the pressure loss over the whole mixer chip, piezo-resistive pressure sensors are attached to the side of the fluidic mount to measure the dynamic pressure of the moving fluid right before the inlets and after the outlet of the mixer, respectively. A thin layer of silicone (1 mm thickness), which was casted inside a mould, is laid between the fluidic mount and the mixer as a sealing. By pressing the mixer onto the fluidic mount, a leak-tight sealing for pressures up to 2 bar inlet pressure is attained. The housing is shown in Fig. 6, right side.

Fig. 6: The fabricated 3D-T-Mixer compared to a 1 Euro Cent coin (left) and the fluidic mount for the mixers together with fluidic connectors and piezo-resistive pressure sensors.

Experimental Characterisation

The mixers are characterised concerning their overall performance. This includes mass flow, pressure loss, mixing times, and mixing quality. The mixing time is measured by optical measurements of a colour reaction, while the mixing quality is determined using the well-known iodide-iodate-reaction according to Villermaux/Dushman.

Measurement Set-up and Pressure Losses

The zwo fluids are provided by pressurized tanks. The mass flow is adjusted by a pressure controller and mechanical valves, and measured using thermal flow meters (Bronkhorst LiquiFlow). For fast and effective measurements, a LabView program was created which controls the inlet pressure, and visualises and handles the measured data. To study the mixing optically, the fluidic mount containing the assembled mixer chip is placed under a microscope.

For the determination of the pressure loss, deionised water was used inside the mixers. The pressure was measured by the piezoelectric pressure sensors. The results for the two mixers are shown in Fig. 7. The Tangential Mixer has a pressure loss of nearly 1 bar at a mass flow of 10kg/h, while the 3D-T-Mixer exhibits a value of 0,5bar at a mass flow of about 14kg/h (10kg/h corresponds to *Re* = 600).

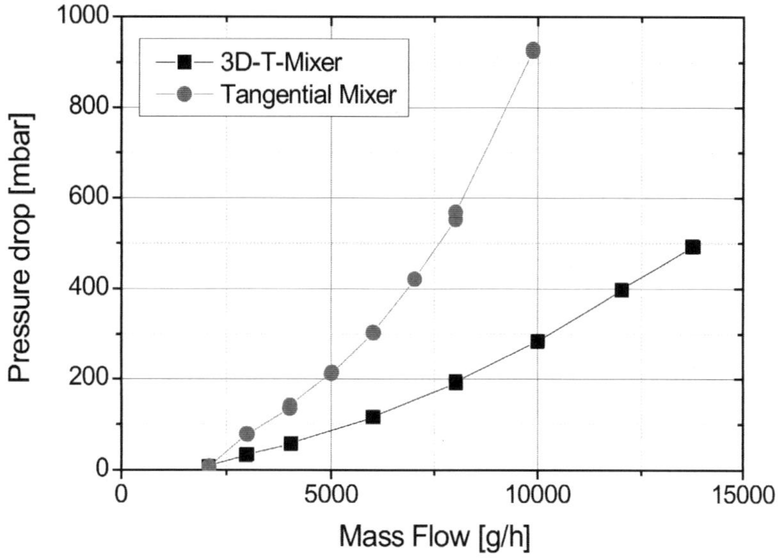

Fig. 7: Pressure loss of the 3D-T-Mixer and the Tangential Mixer over each entire chip

Mixing Times by Optical Measurements

For the determination of the mixing times, optical measurements using the pH-induced colour change of Bromothymole Blue (BTB) have been performed. One inlet fluid consists of a 1,5 mM solution of BTB with a pH value of 7, where BTB appears green. The other inlet fluid consists of a 67 mM solution of Na_2HPO_4 with a pH value of 8, which is colourless. When the fluids are mixed together, a pH value of approximately 7.5 adjusts. At this value, BTB appears blue. It is thus easy possible to optically distinguish the two unmixed inlet zones (green and colourless) and the mixed zone (blue).

With this, the mixing length l_m can be determined by image processing. It has to be mentioned, that these measurements are course up to now, and that they will be improved in later investigations. The order of magnitude is certainly reliable, but the numerical values may vary with a relative uncertainty of about 50%. The mean velocity was determined by measuring the mass flow:

$$\overline{u} = \frac{\dot{m}}{\rho_S \cdot A_M}, \tag{6}$$

where ρ_S is the density of the mixed solution (equivalent to water) and A_M is the area of the cross section of the mixing channel. The mixing time is calculated by the length until the mixture is fully homogeneous, the mixing length l_m, divided by the mean velocity:

$$t_m = \frac{l_m}{\overline{u}}. \tag{7}$$

Results are shown in Fig. 8. The mixing inside the Tangential Mixer happens at lower mass flows than that of the 3D-T-Mixer, as the simulations have indicated. Shorter mixing times in the Tangential Mixer have not been measured since partial mixing has no longer been observable at mass flows larger than 6kg/h. The mixing times for both mixers are very short, down to 300µs for the 3D-T-Mixer. Although the determination of the mixing time is a bit imprecise up to now, the order of magnitude is certainly correct and falls below microseconds.

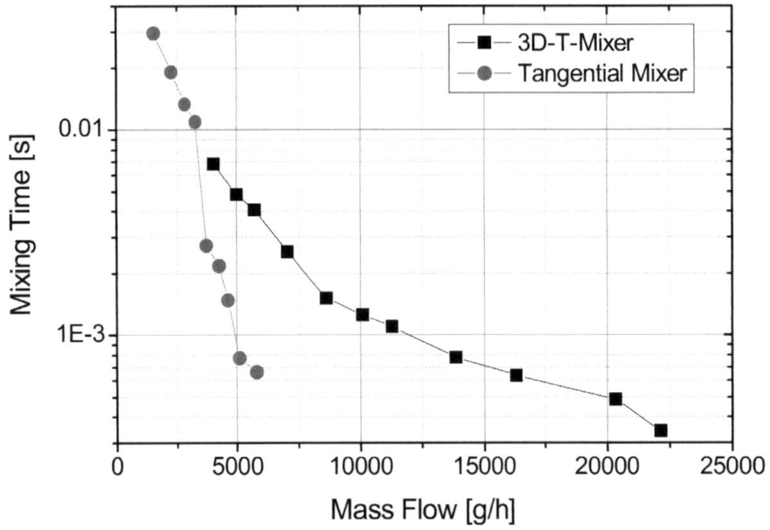

Fig. 8: Mixing Times for the 3D-T-Mixer and the Tangential Mixer with respect to the mass flow

Mixing Quality by the Iodide-Iodate-Reaction

To give a quantitative approach to the mixing quality, the well-known iodide-iodate-reaction by Villermaux/Dushman, a competitive parallel reaction scheme, is used. Details can be found in Guichardon et al. (2000a) and (2000b), while the implementation for micro mixers is shown in Panić et al. (2004). The scheme of the reaction is the following:

$$H_2BO_3^- + H^+ \xrightarrow{k_1} H_3BO_3\,, \tag{i}$$

$$5I^- + IO_3^- + 6H^+ \xrightarrow{k_2} 3I_2 + 3H_2O\,. \tag{ii}$$

Here, reaction (i) is nearly instantanous while reaction (ii) is about as fast as the mixing process. Since the H^+-ions are consumed by the first reaction, iodine I_2 is only created when there exists a local oversupply of H^+-ions, which corresponds to a "bad" mixing. The initial concentrations of the different species are crucial for the results. In this work, the initial concentrations shown in **Table 1** have been used.

Table 1: Initial concentrations of the involved species for the iodide-iodate-reaction

	$[H_2BO_3^-]_0$	$[H^+]_0$	$[I^-]_0$	$[IO_3^-]_0$
Inlet 1	0.0455 mol/l	0	0.0159 mol/l	0.0033 mol/l
Inlet 2	0	0.028 mol/l	0	0

The concentrations are given for one inlet only. Since the ratio of the mass flow of the two inlet streams is adjusted to 1:1, the overall concentrations in the mixed fluid would reduce to the half of the initial concentrations, if there wasn't any chemical reaction. When iodine I_2 is created inside the mixture, there is another reaction, an equilibrium reaction:

$$I_2 + I^- \leftrightarrow I_3^-\,. \tag{iii}$$

The amount of the created I_3^- can be measured spectroscopically with two characteristic absorption peaks in the UV range at 352 and 286 nm. According to Beer's Law, the concentration of I_3^- can be calculated by

$$[I_3^-] = \frac{\beta}{\varepsilon \cdot l_p}\,, \tag{8}$$

where β is the optical absorption measured by the spectrometer at 352 nm, l_p = 0.01 m the optical path length and ε the molar extinction coefficient of I_3^- at 352 nm, which amounts to 2550 m^2/mol (mean value of several references, see Guichardon et al. (2000a)). Using the mass balance and the equilibrium constant K_B of equation (iii), the concentration of iodine I_2 can be calculated by

$$-\frac{5}{3}[I_2]^2 + \left(\frac{[I^-]_0}{2} - \frac{8}{3}[I_3^-]\right) \cdot [I_2] - \frac{[I_3^-]}{K_B} = 0\,, \tag{9}$$

where $[I^-]_0$ is the concentration given in Tab.1 and K_B = 735 l/mol is the equilibrium constant. K_B stongly depends on the temperature with an approximation formula given by Guichardon et al. (2000a). The value used here corresponds to a temperature of 23°C. With this, a segregation index X_S can be calculated by

$$X_S = \frac{Y}{Y_{ST}}, \tag{10}$$

where Y is the ratio of the acid mole number consumed by reaction (ii) and the total acid mole number:

$$Y = \frac{2(n_{I_2} + n_{I_3^-})}{n_{H_0^+}} = \frac{4([I_2] + [I_3^-])}{[H^+]_0}, \tag{11}$$

where $[H^+]_0$ is the value given in Table 1. Finally, Y_{ST} is the value of Y in a total segregation case when mixing is considered to be infinitely slow. With this, reaction (i) and (ii) appear as quasi instantanous compared to mixing and acid is consumed in proportion to the local concentration of both borate and iodide/iodate ions. Y_{ST} is then related to the concentration of the formed iodine:

$$Y_{ST} = \frac{6[IO_3^-]_0}{6[IO_3^-]_0 + [H_2BO_3^-]_0}. \tag{12}$$

Results of measurements for the 3D-T-Mixer and the Tangential Mixer are shown in Fig. 9. The decrease of the segregation index X_S with rising Reynolds number is evident, since the rising mass flow leads to a stronger vortex creation, thinner fluid filaments and therefore shorter diffusion lengths. With this, the local oversupply of H^+-ions is faster equalised and less time is available for the creation of iodine. Altogether, the segregation index of the mixers is very small, indicating a very fast and effective mixing.

Unfortunately, there are different problems when analysing the results of the iodide-iodate-reaction. At first, the relation of the segregation index to the mixing time is highly non-linear, since the creation of iodine is only accomplished by a local oversupply of H^+-ions. As an example, imagine inlet 1 and 2 flowing alongside each other. The mixing mechanism would only be lateral diffusion and there would be no creation of iodine, no matter how large the diffusion lengths would be, since nearly all H^+-ions would immediately be consumed by $H_2BO_3^-$. In the case of the mixers given here, the creation of iodine is only accomplished since the fluid filaments have different thicknesses which leads to a local oversupply of H^+.

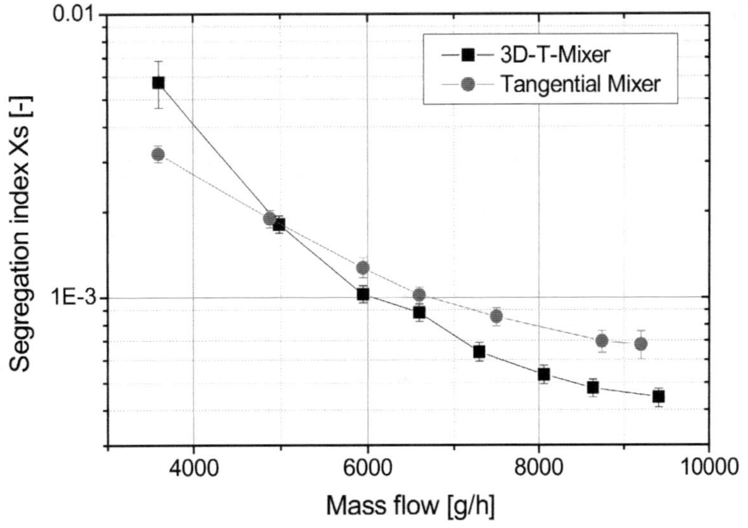

Fig. 9: Segregation Index X_S for the two micro mixers over the mass flow

The second problem is the comparison of the segregation indices measured here with values measured by other groups using different mixers. At first sight, this should easily be possible since the segregation index is calculated by the ratios of the initial concentrations and should be similar for a wide range of initial concentrations. However, this is not valid, since there is a strong dependance of the H^+-concentration from the pH-value. Commonly, H_2SO_4 is used as H^+-ion donor. Depending on the pH-value, sulphuric acid is capable of providing two H^+-ions. This means that the concentration of initial H^+-ions, which are provided inside the solutions, depends on the concentration of sulphuric acid. This complicates the comparability to other work groups. Even worse is the fact, that the concentration of H^+-ions changes dynamically during the reaction due to the change of the pH-value, which makes it even difficult to determine the "right" initial concentrations for the H^+-ions. Unfortunately, other groups have not mentioned this. There might be a need for more research concerning this fact. In this research, $[H^+]_0$ is obtained from pH-measurements. It has also been tried to avoid the problem by using HCl as donor of H^+-ions, since HCl is univalent. However, when doing so, strong precipitation has been observed and absorption measurements have been falsified. According to Guichardon et al. (2000a), the reason is the creation of chloride-iodide components.

A New Identification Number for the Comparison of Micro Mixers – the Mixing Efficiency

Approach

There is no possibility yet to compare different micro mixers concerning their performance in a simple manner. For this reason, a new identification number for micro mixers, a mixing efficiency, is proposed here, which shall serve as a simple tool to identify the quality of any arbitrary micro mixer. The mixing efficiency concentrates on micro mixers for chemical production purposes, not for applications of analytical chemistry. With these preconditions, three very important characteristics of micro mixers for chemical production purposes are identified:

- High mass flow,
- short mixing time,
- low pressure loss

Furthermore, the number should be dimensionless to be applied to different scales. For this, a dimensionless number has been searched combining the above attributes. The approach which has been chosen takes the ratio of three dimensionless numbers. The first two are the Euler and the Reynolds number,

$$Eu := \frac{\Delta p}{\rho \cdot \bar{u}^2} \quad \text{and} \quad Re = \frac{\bar{u} \cdot d_h}{\nu}. \tag{13}$$

The Euler number contains the influence of the pressure loss while the Reynolds number contains the influence of the mass flow by means of the mean velocity. The third dimensionless number is constituted by the ratio of the hydraulic diameter to the mixing length d_h/l_m. The higher this number is, the better is the performance of the mixer, since a high ratio indicates a short mixing length (compared to the hydraulic diameter) and therefore a good mixing performance. With the mixing time t_m, this number can further be written as

$$\frac{d_h}{l_m} = \frac{d_h}{\bar{u} \cdot t_m}. \tag{14}$$

These three identification numbers, Eu, Re and d_h/l_m are combined regarding the fact, that the resulting number should increase with increasing mass flow and decreasing pressure loss and mixing time:

$$\frac{d_h}{l_m} \cdot \frac{Re}{Eu} = \frac{d_h}{\bar{u} \cdot t_m} \cdot \frac{\rho \cdot \bar{u}^2}{\Delta p} \cdot \frac{\bar{u} \cdot d_h}{\nu} = \frac{\rho \cdot \bar{u}^2 \cdot d_h^2}{\nu \cdot \Delta p \cdot t_m} \overset{(6)}{=} \frac{d_h^2 \cdot \dot{m}^2}{\eta \cdot A_M^2 \cdot \Delta p \cdot t_m}, \tag{15}$$

where η is the dynamic viscosity. Regarding that A_M is equivalent to d_h^2 in a good approximation, the number is further simplified and gives the new identification number, the mixing efficiency *ME*:

$$ME := \frac{\dot{m}^2}{\eta \cdot d_h^2 \cdot \Delta p \cdot t_m}. \tag{16}$$

ME is built up by primary parameters and combines the requirements which were given above: It is dimensionless and increases with increasing mass flow, decreasing pressure loss and decreasing mixing time. It shall therefore be suitable to serve as a tool to compare different micro mixers for the use in chemical production processes.

Application

As a first evaluation, the mixing efficiency is applied to the mixers presented in this paper and to other micro mixers. At first, the single 3D-T-mixing and the Tangential mixing element is compared to the former T-Mixer shown by Engler et al. (2004). To compare mixing apparatuses, the fabricated 3D-T-Mixer and the Tangential Mixer are compared to three commercially available mixers from the Institut für Mikrotechnik Mainz IMM: the well-known interdigital mixer SSIMM, the CaterPillar Split-and-Recombine CPMM-R1200-V1.1., and the SuperFocus Mixer.

Values for mass flow, hydraulic diameter, pressure loss and mixing times (maximum and minimum values, resp.) as well as the calculated mixing efficiency are given in **Table 2** (For the dynamic viscosity, a value of 0.001 Pa·s has been used). When considering the single mixing elements, the 3D-T-mixing element shows the best performance, and even the simple T-element is better than the tangential element. When considering the apparatuses, the 3D-T-Mixer is the most efficient one with a value for *ME* of about 12,000. The value for the Tangential Mixer is about one fifth of the 3D-T-Mixer. The interdigital mixer has a value of 0,55. This is not really surprising, since this mixer was one of the first micro mixers ever investigated. The SuperFocus Mixer from IMM also has very good with a value of 3,000. However, the mixing efficiency introduced here gives a good possibility to compare different micro mixers among each other. It would be interesting to compare the values calculated here with values of micro mixers already implemented by other groups.

Table 2: Values of mass flow, hydraulic diameter, pressure loss, mixing time and mixing efficiency for several mixing elements and micro mixing apparatuses

	Max. Mass Flow $\dot{m}$ [kg/h]	Hydr. ∅ d_h [μm]	Max. Pressure Loss Δp [bar]	Min. Mixing Time t_m [ms]	**Mixing Eff. *ME* [-]**
Single Mixing Elements					
3D-T-Mixer Element (this work)	1.4	300	0.9	0.35	**48.0**
Tangential Mixer Element (this work)	0.9	300	1.8	0.4	**9.6**
T-Mixer (Engler et al. (2004))	1.0	400	0.2	2.0	**12.1**
Apparatuses					
3D-T-Mixer (this work)	22.0	300	1.0	0.35	**11856**
Tangential Mixer (this work)	15.0	300	2.0	0.4	**2411**
SuperFocus Micro Mixer (IMM (2004), Hessel et al. (2005))	350	909	10	4	**2860**
Caterpillar CPMM-R1200-V1.1 (IMM (2003))	90	1200	17	20	**12.8**
Interdigital Mixer SSIMM (IMM (2003))	2.0	71.0	2.8	400	**0.55**

Summary and Conclusion

Efficient Mixing has been shown by the use of convective micro mixers. The best of the mixers developed within this research, the 3D-T-mixer, allows a throughput of up to 22 l/h (see Fig. 8) at pressure losses below 1 bar. This performance is achieved by the use of convective flow, i.e. the hydrodynamical creation of secondary vortices, which serves for a very fast creation of thin fluid filaments and therefore a quick mixing. Simulations with the commercial CFD-code CFD-ACE+ from ESI Group have been used to design the single mixing elements. These single elements have been combined and parallelised on two silicon chips, whereupon the symmetrical flow conditions have been ensured by analytical calculations. The mixers have been fabricated by standard silicon bulk micromachining. Their characterisation has been carried out by a pH-induced color change of Bromothymole Blue and by spectroscopical measurements using the well known iodide-iodate-reaction according to Villermaux and Dushman.

To compare the different mixers among themselves and among other micro mixers, a new identification number has been derived, the mixing efficiency *ME*. This number combines important mixer characteristics like mass flow, mixing time and pressure loss. First evaluations for the fabricated mixers give very good values compared to a simple T-mixer and compared to the interdigital mixer SSIMM from IMM.

Acknowledgements

The authors wish to thank the German Research Association DFG for the financial support in the priority programm SPP1141 “Mischen”.

References

Doll, A., Goldschmidtböing, F., Woias, P. (2004)
Low-Temperature Plasma Assisted Wafer Bonding and Bond-Interface Stress Characterisation, 17th Proc. MEMS Conf., Maastricht, The Netherlands, 665-668

Engler, M., Kockmann, N., Kiefer, T., Woias, P., (2004)
Convective Mixing and its Application to Micro Reactors, Proc. 2nd Int. Conf. Micro-Minichannels, Rochester, USA

Guichardon, P., Falk, L., (2000a)
Characterisation of Micromixing Efficiency by the Iodide-Iodate Reaction System. Part I: Experimental Procedure, Chem. Eng. Sc. **55**, 4233-4243

Guichardon, P., Falk, L., Villermaux, J. (2000b)
Characterisation of Micromixing Efficiency by the Iodide-Iodate Reaction System. Part II: Kinetic Study, Chem. Eng. Sc. **55**, 4245-4253.

Hessel, V., Löwe, H., Müller, A., Kolb, G., (2005)
Chemical Micro Process Engineering, WILEY-VCH Verlag GmbH & Co. KGaA, Weinheim, Germany

Hoffmann, M., Schlüter, M., Räbiger, N., (2004)
Experimental Investigation of Mixing in a Rectangular Cross-Section Mixer Mixer, Proc. AIChE Annual Meeting, Austin, USA, 330b

IMM Institut für Miktotechnik Mainz GmbH (2003)
Process Technology of Tomorrow – The Catalogue, Mainz, Germany

IMM Institut für Miktotechnik Mainz GmbH (2004)
Hochdurchsatz bis in den m^3/h-Bereich, Mainz, Germany

Kockmann, N., Föll, C., Woias, P., (2003)
Flow Regimes and Mass Transfer Characteristics in Static Micro Mixers, Micromachining and Microfabrication, Photonics West, San Jose, USA

Nguyen, N.-T., Wu, Z., (2005)
Micromixers – a Review, J. Micromech. Microeng. **15**, R1-R16

Panić, S., Löbbecke, S., Türcke, T., Antes, J., Bošković, D. (2004)
Experimental Approaches to a Better Understanding of Mixing Performance of Microfluidic Devices, Chem. Eng. Journal **101**, 409-419

Wong, S.H., Ward, M.C.L., Wharton, W., (2004)
Micro T-mixer as a Rapid Mixing Micromixer, Sensors and Actuators B **100**, 359-379

Development of a New Micromixer for Rapid Mixing Utilizing Kinetic Energy and Diffusion

Hideharu Nagasawa[1), 2)], *Nobuaki Aoki*[2)], *Taisuke Maki*[2)], *Kazuhiro Mae*[2)]

1) Advanced Core Technology Laboratories, Fuji Photo Film CO., LTD.
210, Nakanuma, Minamiashigara-shi, Kanagawa 250-0193, Japan
2) Department of Chemical Engineering, Kyoto University
Katsura, Nishikyo-ku, Kyoto 625-8510, Japan

Abstract

A new type of micromixer for rapid mixing was developed based on mixing by utilizing both kinetic energy and molecular diffusion of fluids. The rapid mixing was realized by the collision of micro segments, which were divided into several streams in the radial, at the center of the mixer. From the results of conducting the Villermaux/ Dushman reaction and the precipitation of polymer particles by phase separation, it was clarified that the mixing performance of the developed micromixer was excellent, and the rapid mixing was driven by both the deformation of micro segments under high shear rate and instant molecular diffusion between micro segments. In addition, the mixer had a high throughput and corresponded to wide range of flow rate ratio of fluids. Thus, it is expected that the proposed design concept for the micro mixing will give us an efficient micro processing in future.

Keyword: Micromixer, Mixing performance, Kinetic energy

1 Introduction

Micromixer is an important device for micro chemical processing, because mixing is a key operation for heat transfer, mass transfer and reaction. Since flow in micromixer is generally within the laminar flow regime, the mixing is mainly driven by molecular diffusion. Therefore, there are some mixing methods to shorten diffusion time between fluids to be mixed. First is multi-lamination method to have been employed in several micromixers such as the inter-digital micromixer [1], static V-micro-jetmixer [2], and Super Focus [3]. The second is split and recombine method to shorten diffusion time [4–6]. The third is the method utilizing convection of fluids to assist molecular diffusion [7-10].

Thus, several distinct micromixers have been developed, but additional efforts should be made in order that micro mixing is applied to an actual process. The desired features as an industrial micromixer are as follows: rapid mixing, high throughput, wide range of mixing ratio, prevention of clogging in micro channel, and so on. In this study, we proposed a new micromixer having these desired functions by combining reasonably a kinetic energy with molecular diffusion.

2 Materials and Experiment

2.1 Proposed micromixer

The proposed micromixer, K-M mixer, was reasonably assembled by the functional elements required to effective micro mixing; the uniform feed to the micro segment, the instant mixing of the arranged micro segment, and the instant contraction flow just after mixing. Figure 1 shows the overview and Figure 2 shows the internal structure and schematic of flow in the micromixer in the typical one of the developed micromixer. The micromixer was consist of three functional parts: one is inlet plates to feed materials uniformly to the micro channel, second is mixing plates having several radial micro channels with the mixing zone at the center, the last is outlet plates for achieving instant contraction flow just after mixing. We fabricated several patterns of micro channels. The representative patterns used in this study are shown in Table 1.

The most distinguished point is to be aimed to qualify both the instant mixing of fluids and the high throughput by utilizing both kinetic energy and molecular diffusion of fluid in the center of mixing zone. In addition, since we flexibly selected the number of micro channel by the mixing plate having various pattern of inlet holes, the wide range of flow rate ratio of two fluids is available.

Figure 1 Overview of micromixer

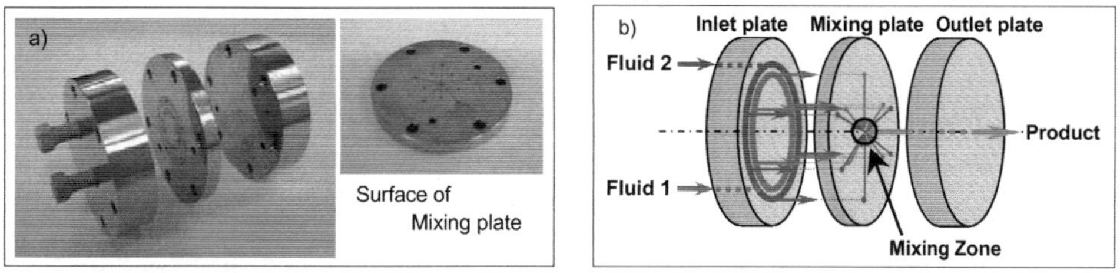

Figure 2 Details of micromixer.
a) The internal structure and b) schematic of flow in micromixer

Table 1 Dimension of micromixers fabricated

Type	K-M-CC-1	K-M-CC-2	K-M-CC-3	K-M-CC-4
Sketch of mixing zone	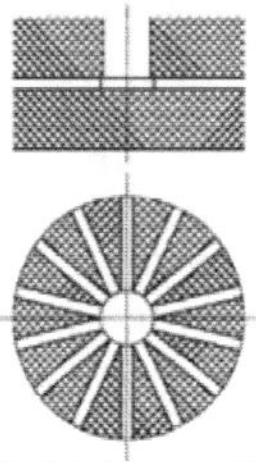	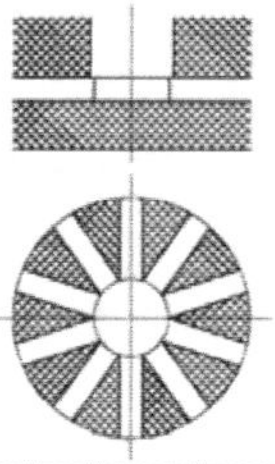	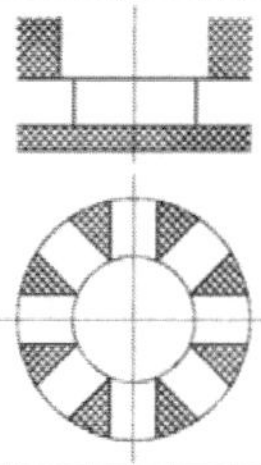	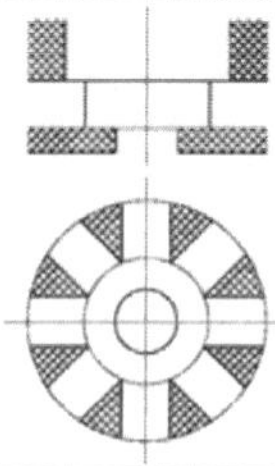
The number of channel, n_i	14	10	8	8
Channel width, W_i	50μm	100μm	200μm	200μm
Diameter of mixing zone, D_i	220μm	320μm	520μm	520μm
Diameter of outlet, R_i	200μm	360μm	640μm	690μm
Diameter of injection port, V_i	N/A	N/A	N/A	260μm

2.2 Mixing experiment using miscible fluids

The mixing performance of the proposed micromixer for a single phase system in two miscible fluids was evaluated by Ehrfeld's method [1] based on Villermaux / Dushman Reaction [11, 12]. The method is based on the competition of two parallel reactions in aqueous solution as follows:

$$CH_3COO^- + H^+ \leftrightarrow CH_3COOH \qquad (1)$$

$$5I^- + IO_3^- + 6H^+ \leftrightarrow 3I_2 + 3H_2O \qquad (2)$$

$$I_2 + I^- \leftrightarrow I_3^- \qquad (3)$$

In the experiments, the solution of 0.016 mol L^{-1} of KI, 0.0032 mol L^{-1} of KIO_3 and 1.33 mol L^{-1} of CH_3COONa was mixed with the solution of 0.138 mol L^{-1} of HCl in the micromixer at room temperature. The inlet volume flow rates changed in the range of 1.0 to 10 mL min^{-1} using two syringe pump (IC3210, Kd Scientific Co.). The product liquid just after mixing was immediately served a UV spectrometer (Shimadzu, Ltd., Multispec-1500), and then the absorbance of UV spectrometry at 352 nm was measured within 15 s. For comparison, we also performed a mixing experiments using T-shape device having 1.588 mm of inner diameter in order to clarify the effect of channel width on the mixing performance.

2.3 Mixing experiment using immiscible fluids

In this study, a precipitation of polystyrene particles was selected as a model mixing for evaluating the mixing performance of a multi phase system in two immiscible fluids [13]. The method is based on the immersion precipitation of polystyrene by contacting two immiscible fluids, that is, an aqueous solution containing di-iso-octyl sodium sulfosoccinate (AOT) and an organic solvent, tetrahydrofuran (THF), solution containing polystyrene.

In the experiments, the aqueous solution containing 0.05wt% of AOT and the THF solution was containing 0.5wt% of polystyrene. The volume ratio of the polystyrene solution to the aqueous solution was adjusted to 0.14. Total volume flow rate of polystyrene solution and the aqueous solution was 5, 10, 20 and 50 ml min^{-1} and the temperatures of all solutions were kept at 298 K. The product solution was sampled and served to Particle Size Distribution Analyzer (Horiba, Ltd., LA-920) to determine the particle size distributions. For comparison, other two types of micromixers were also used: One is an interdigital micromixer (IMM single mixer) provided by the Institut für Mikrotechnik Mainz GmbH (IMM) in Germany [1] and the other is YM-1 mixer developed by Kyoto University and Yamatake Co. [6]. The IMM single mixer and the YM-1 mixer are based on the multi-lamination mixing and Splitting/recombination mixing, respectively.

3 Results and Discussion

3.1 Mixing performance of the proposed micromixer for miscible fluid

To examine the mixing performance of K-M mixers between miscible fluids, we conducted the Villermaux/ Dushman Reaction under various conditions. Figure 3 shows the change in the intensity of absorbance at 352nm with the total flow rate. The intensity of absorbance at 352 nm decreased with the increase in the total flow rate for K-M-CC-3 and K-M-CC-4. For both mixers the intensities became near zero, which meant the mixing is sufficiently fast, beyond 3 mL min^{-1} of total flow rate. The K-M-CC-1 and K-M-CC-2 mixers showed extremely excellent mixing performance at even low flow rate. However, the K-M-CC-1 mixer having 50 μm of micro channel did not operate at the flow rate more than 1 mL min^{-1}, because the pressure drop was very high. On the other hand, the intensity of absorbance at 352 nm was large when using T-shape mixer. The intensity decreased gradually with the increase in the total flow rate, but the mixing was insufficient even at 10 mL min^{-1}. The mixing in the T-shape mixer is driven mainly by the kinetic energy because of long diffusion distance. The result clearly indicates that the feed of micro segment should be necessary for fast mixing. The change in the intensity using the IMM single mixer was also drawn for comparison [1]. The mixing performance of K-M-CC-3 and

K-M-CC-4 mixers were nearly equivalent to that of the IMM single mixer, whereas the size of micro channels of K-M-CC-3 and K-M-CC-4 were 200 μm whose value is 4 times larger than that of IMM single mixer. Furthermore, the K-M-CC-1 and K-M-CC-2 showed a mixing potential better than IMM single mixer. These results suggest that the contribution of kinetic energy on the mixing performance should be considered for micro mixing.

To make clear the effect of micro segment size feed and the kinetic energy on the mixing ability, we examined the relationship between the mixing performance and the shear rate at the outlet. Figure 4 shows the plot of the intensity of absorbance at 352 nm against the shear rate, u_{out}/D, respectively, at the outlet. From this figure, it was found that the mixing performance, which was indicated by the intensity of absorbance at 352 nm, correlated well with the shear rate at the outlet, u_{out}/D, irrespective of the channel size, the number of channel, and the size of mixing zone.

From this result, it was presumed that the mixing in K-M mixer proceeded as follows: The mixing in laminar flow is finally driven by molecular diffusion. If the fluid segment is small, the molecular diffusion between segments is easy to finish within a short time. On the other hand, it is well known that the fluid segment was transformed and deformed into small one by the shear stress in laminar flow. In K-M mixer the mixing having large micro channel, the segment size of fluid is fairly large, but the segment was deformed into small one at the mixing zone by the high shear rate. So, the mixing became to be sufficient beyond a certain flow rate. In the K-M mixer having small micro channel, the mixing is finished only by the molecular diffusion between segments even at low shear rate as soon as the segments encountered in the mixing zone. Thus, it was clarified that the proposed micromixer designed based on the mixing principle has an excellent mixing performance for miscible fluids under high throughput.

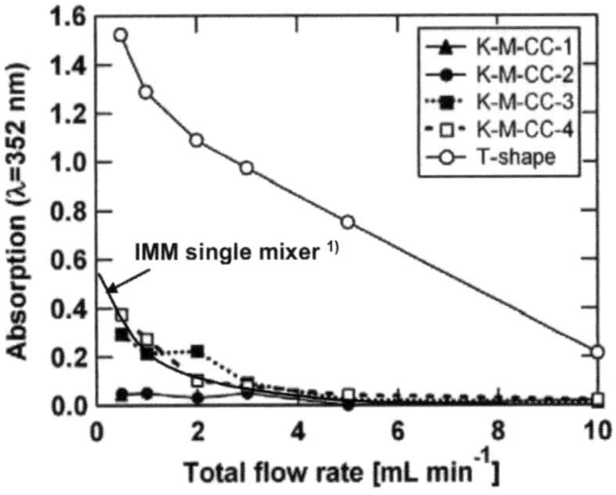

Figure 3 Change in the intensity of absorbance at 352 nm with the total flow rate

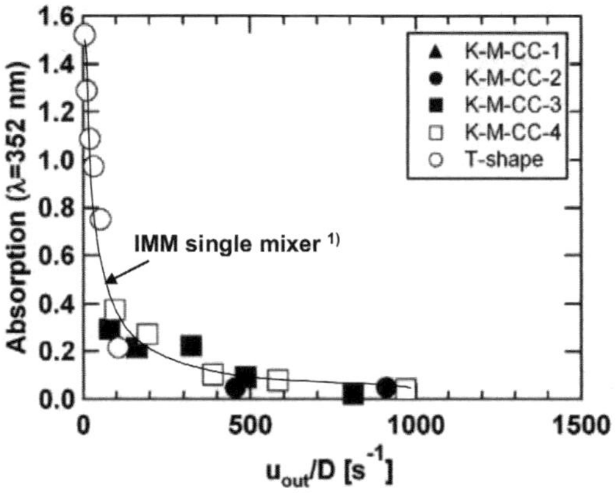

Figure 4 Change in the intensity of absorbance at 352 nm with the shear rate at the outlet

3.2 Mixing performance of the proposed micromixer for immiscible fluids

Next, the mixing performance of K-M mixer for immiscible fluids was investigated by the precipitation of polyethylene particles [13]. The K-M-CC-2 mixer having 100 μm of micro channel was used. Figure 5 shows the change in the particle size distribution with the total volume flow rate. In accordance with increasing total volume flow rate, a peak of distribution shifted toward small size and the distribution became narrow. But, the particle size distribution did not change little beyond 10 mL min^{-1} of the total flow rate. Comparing the result in Fig. 3, the sensitivity of the mixing was strong for the precipitation of polystyrene. This suggests that the precipitation method is useful to evaluate the mixing performance as addressed in the previous paper [13]. The median average size was 104nm and the coefficient of variation was 28.0 at the condition of 20mL/min of total volume flow rate. In this condition, the shear rate at the mixing zone was ca. 3200 s^{-1}, so the segment was deformed into small one instantly by the collision with the kinetic energy of fluid; as a result, the particle size became small.

To confirm the mixing performance, we compared the particle size distributions of polystyrene produced by K-M-CC-2 mixer, IMM single mixer, and YM-1 mixer, respectively, as shown in Figure 6. Total flow rates were 10 mL min^{-1} for both K-M mixer and IMM mixer. The flow rate of YM-1 mixer was 50 mL min^{-1}, which is the sufficient flow rate to minimize the particle size [13]. The median average particle size was 109nm and the coefficient of variation was ca. 33 for K-M-CC-2 mixer. These values were smaller than those by IMM mixer and YM-1 mixer. From these results, it was shown that the micromixer based on the mixing by both kinetic energy and molecular diffusion of fluids like the K-M mixer has advantage in the mixing performance rather than the multi-lamination type like the IMM mixer and the sprit-recombination type like the YM mixer for the mixing of two immiscible fluids.

On the other hand, productivity and operability is also important for an actual production. Using the K-M-CC-2 mixer, we successfully prepared fine polystyrene particles at 20 mL min^{-1} of the total volume flow rate. This is approximately corresponded to the production rate of 7 ton per year. By assembling the mixers, it is seemed to produce enough amounts of nano particles. Operability was also confirmed in the experiment. Although the volume flow rate ratio of two fluids was 14/86, the flow was stable without pulsation. In addition, the production was executed continuously without clogging in the micro channel. Thus, it was clarified that the K-M mixer had an attractive potential as a device of nano particle production.

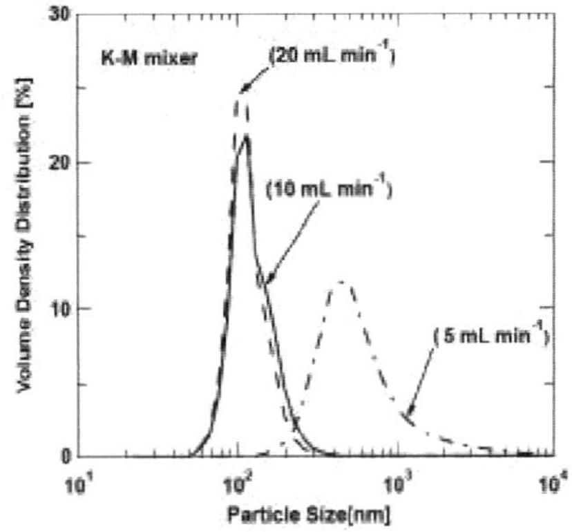

Figure 5 Change in the particle size distribution of polystyrene with flow rate using the K-M-CC-2 mixer

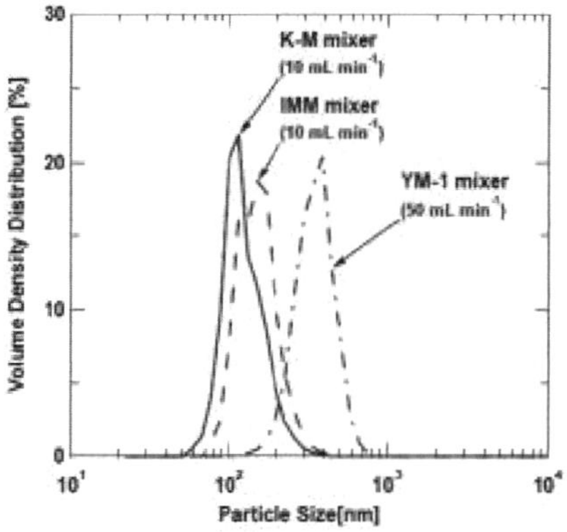

Figure 6 Comparison of the particle size distribution of polystyrene produced by several micromixers

4 Conclusions

We proposed a new concept of micromixer based on the essential of micro mixing. To confirm the design concept of micro mixing, we performed two mixing experiments, the Villermaux/ Dushman Reaction for a mixing of miscible fluids, and the precipitation of polystyrene particles by phase separation for a mixing of immiscible fluids using the developed micromixers. The mixing performance of the proposed mixer was superior to that of the IMM single mixer etc. In addition, the mixer has advantages of high throughput and stable operation in a wide range of flow rate ratio of two fluids. By examining the effect of the dimension of micro channel and flow rate on the mixing performance, it was clarified that the mixing was driven by both the deformation of micro segments under high shear rate and instant molecular diffusion between micro segments. Thus, the proposed micromixer is logically designed for actual processing, and it is expected to develop an efficient mixer for the production of valuable chemicals.

Acknowledgments

The research was conducted in the aid of The Research Association of Micro Chemical Process Technology (MCPT) for the Project of Micro-Chemical Technology for Production, Analysis, and Measurement Systems financially supported by New Energy and industrial Development Organization (NEDO). We appreciate to MCPT and NEDO for their assistance.

References

[1] Ehrfeld, W., K. Golbig, V. Hessel, H. Löwe and T. Richter; "Characterization of Mixing in Micromixers by a Test Reaction: Single Mixing Units and Mixer Arrays," *Ind. Eng. Chem. Res.*, **38**, 1075–1082 (1999).

[2] Ehlers, St., K. Elgeti, T. Menzel and G. Wießmeier; "Mixing in the Offstream of a Microchannel System," *Chem. Eng. Proc.*, **39**, 291–298 (2000).

[3] Hessel, V., S. Hardt, H. Löwe and F. Schönfeld; "Laminar Mixing in Different Interdigital Micromixers: I. Experimental Characterization," *AIChE J.*, **49**, 566–577 (2003).

[4] Ehrfeld, W., V. Hessel, H. Löwe; Microreactors, New Technology for Modern Chemistry, WILEY-VCH, Weinheim, Germany, 55–63 (2000).

[5] Kirner, T., J. Albert, M. Günther, G. Mayer, K. Reinhäckel and J. M. Köhler; "Static Micromixers for Modular Chip Reactor Arrangements in Two-step Reactions and Photochemical Activated Processes," *Chem. Eng. J.*, **101**, 65–74 (2004).

[6] Mae, K., T. Maki, I. Hasegawa, U. Eto, Y. Mizutani and N. Honda; "Development of a New Micromixer Based on Split/recombination for Mass Production and its Application to Soap Free Emulsifier," *Chem. Eng. J.*, **101**, 31–38 (2004).

[7] Mengeaud, V., J. Josserand and H. H. Girault; "Mixing Processes in a Zigzag Microchannel: Finite Element Simulations and Optical Study" *Anal. Chem.*, **74**, 4279–4286 (2002).

[8] Liu, Y. Z., B. J. Kim and H. J. Sung; "Two-fluid Mixing in a Microchannel," *Int. J. Heat Fluid Flow*, **25**, 986–995 (2004).

[9] Hong, C.-C., J.-W. Choi and C. H. Ahn; "A Novel In-plane Passive Microfluidic Mixer with Modified Tesla Structures," *Lab Chip*, **4**, 109–113 (2004).

[10] M. Engler, N. Kockmann, T. Kiefer and P. Woias, Numerical and Experimental Investigations on liquid mixing in static micromixers, *Chem. Eng. J.*, **101**, 2004, 315–322

[11] M.-A. Schneider, T. Maeder, P. Ryser and F. Stoessel, A Microreactor-based System for the Study of Fast Exothermic Reactions in Liquid Phase: Characterization of the System, *Chem. Eng. J.*, **101**, 2004, 241–250

[12] S. Pani´c, S. Loebbecke., T. Tuercke, J. Antes and D. Boškovi´, Experimental approaches to a better understanding of mixing performance of microfluidic, *Chem. Eng. J.*, **101**, 2004, 409-419

[13] T. Maki, Y. Omukai, H. Nagasawa and K. Mae, Production of Polymer Nanoparticles by Immersion Precipitation using Micromixer -Development of the Evaluation Method of the Mixing Performance of Micromixers-, *Proc. The 10th APCChE Congress*, 2004, 4B-02

128f Theoretical and numerical investigation of mixing processes in a T-shaped micro-mixer

Carsten Stemich, Dieter Bothe, H J Warnecke

In relation to the time-scale of chemical kinetics, diffusive transport in micro-devices is faster than in conventional mixers. To exploit the resulting potential for chemical process engineering, the mixing behavior of flow mixers at micro-scales needs to be further investigated. Based on numerical simulations, the mixing characteristics of a T-shaped micro-reactor with rectangular cross sections are investigated for three different flow regimes. For the description of the mixing efficiency the intensity of mixing as well as a measure of the scale of segregation which employs the concept of specific contact area are used. Furthermore, first results concerning experimental validation are given, showing a good qualitative agreement between simulation and experiment

[TK002 -129a]

SEPARATION PRINCIPLES IN MICRO PROCESS ENGINEERING

Norbert Kockmann, Peter Woias

Institute of Microsystem Technology, IMTEK
Laboratory for Design of Microsystems
Albert-Ludwig University of Freiburg
Georges-Koehler-Allee 102, 79110 Freiburg, Germany
{kockmann, woias}@imtek.de

Key words: micro process engineering, separation processes, micro cyclone, pressure diffusion, thermodiffusion, micro rectification

2005 AIChE Spring Meeting, Atlanta, GA, April 10-14,
Session [129] – Unit Operations
Asterios Gavriilidis, Chair

Separation Principles in Micro Process Engineering

N. Kockmann, P. Woias
Laboratory for Design of Microsystems, IMTEK,
Albert-Ludwig University of Freiburg, Germany

Abstract

Separation processes belong as well as chemical reaction technology to the large number of unit operations in process engineering and they are responsible for 60 to 80 % of plant investment costs. In micro process engineering, the separation techniques are very rare compared to the analytical applications, although the transport processes in microstructures show a good potential for separation processes, too. This presentation outlines the state of the art for separation unit operations in micro process engineering and the potential of further developments. A micro cyclone is presented, a mechanical separation process without moving parts and with an increasing separation factor for decreasing dimensions. The rectification as one of the main unit operations in the chemical industry is transferred to a microchip in a design study. The first results show the performance of the fluid flow and heat transfer within micro channels. The thermodiffusion as a coupled process of a temperature gradient with a species current shows only a weak effect, but with the possibility of parallel switching and cascading of many channels, the effect can be enhanced to an alternative separation process. The opportunities and possibilities of micro technology for process engineering in the separation field is demonstrated, limits and risks are shown.

Introduction

In the growing field of micro process engineering, micro reactors are belonging to the most predominant equipment and are a well established technology despite of many research needs in that area (Hessel et al. 2004). Due to their small dimensions the heat and mass transfer as well as the mixing of reactants is very fast and can be controlled in a very effective manner. The amount of reactants is very small and can be safely handled in explosive regimes, which is important for highly exothermic reactions or hazardous media. Nevertheless, chemical by-products, which have to be removed to gain a consumable product, accompany most of the products. Therefore, a broad variety of appropriate separation techniques is necessary for a complete plant in micro technology which is intended to be set up for the design of new chemical processes or for optimizing existing plants (Klemm et al. (2004)).
Separation processes in micro technology are often implemented in analytical applications, see Fang (1993) for a detailed overview. One of the first textbooks on micro reaction technology, “Microreactors” from Ehrfeld et al. (2000), comprises a sub-chapter on separation technology. A few examples of extraction and adsorption were shown with a short overview about separation technology. In their book chapter on micro process engineering, Klemm et al. (2004) are mentioning some separation processes like extraction, rectification, or membranes, but they also reclaim in their summary the urgent need for separation processes with microstructures for process development and homogeneous process set-up.
To categorize the huge variety of separation principles, one can start from the employed driving force. External force fields are the driving force in mechanical separation units like cyclones or centrifuges. Molecular processes are the driving forces for thermal separation units like evaporation or extraction. In adsorption or membrane separation also chemical processes may be involved for mass transfer processes in microstructures.
Another methodology to categorize the multiplicity of thermal processes is based on the employed phases for the separation, see Blass (1997). Single-phase separation includes thermodiffusion, pressure diffusion (see Schütte (1972)), or electrophoresis. Multiphase processes

are divided into processes with their own counter-phase like evaporation or rectification, with an additional counter-phase like extraction or absorption or with a structured membrane with various properties for osmotic separation, dialysis, or pervaporation, see Melin and Rautenbach (2003).
As a mechanical separation process without moving parts, cyclones are often used for separation of suspensions and dispersions. A detailed description can be found in Bockhardt et.al. (1997), chap. 6. The Little Things Factory ltf in Ilmenau offers a commercial micro cyclone in their product program, which is manufactured from glass by powder blasting. Unfortunately, no internal geometrical data are given, but the cyclone has a volume flow of about 200 ml/h with a pressure loss of 1 bar and exhibits separation factors z of about 100 to 1000 (ltf-gmbh (2005)).
Thermal separation processes represent the majority of conventional separation unit operations and some of them are already transferred into micro system technology. The distillation and rectification process is described in the textbooks of Geankopolis (2003) or Schönbucher (2003). On the IMRET7 conference, Sorata and Kusabe (2003) presented a rectification chip with one separation stage. The liquid and gaseous phase of a water-methanol mixture was mixed and guided through a residence channel. Behind the separation chamber ($8*8\ mm^2$) the outlet concentration of the liquid phase give a separation factor of 65% for one stage. A working group at the Technical University of Darmstadt, Prof. Hampe, investigated the reversible rectification with a microstructured device. Geometrical details were given in Klemm et al. (2004). A dechema presentation from 2004 gives more details with a volume flow of 20 ml/h and a separation factor of 2.5 (MicroChemTec (2004)).
The working group of Bob Wegeng investigated the absorption separation principle with the absorption micro heat pump for energy management purposes, see Wegeng et al. (1999). The stripping of toluene solved in water into nitrogen on a micro chip is described by Cypes and Engstrom (2003). They found an almost tenfold higher volumetric productivity in the microchip column compared to a conventional packed tower.
The extraction process is described by Ehrfeld et al. (2000) and Klemm et al. (2004). On recent conferences Okubo et al. (2003) describe the rapid extraction process on a chip and TeGrotenhuis et al. (1998) described micro devices for and transport processes in solvent extraction. Kusakabe et al. (2003) described the flow details and transport processes within a droplet in a micro contactor for solvent extraction. Their comparison with experimental data shows a good agreement for the extraction of ethyl acetate from n-hexane to water.
In this paper three examples of different separation principles are given to show the variety and the possibilities of micro system engineering. The mechanical separation process is represented by a micro cyclone, a feasibility study with first experimental results. A thermal separation process with phase change is represented by a design study of a micro rectification chip. Opportunities and limits will be explained, which can be transferred to other unit operations. The inherent problems of phase change, multiphase flow and fluid transport are described against the background of miniaturization. At last, a thermal separation process with single-phase flow is represented by the thermodiffusion, which has not yet been transferred to microstructures. With this example, some design issues and fabrication limits will be shown, which are very important for micro system technology. The paper ends with a discussion and outlook on further possibilities in separation technology, which can be treated in the near future.

Mechanical Separation – Micro Cyclone

For the separation of particulate systems the density difference between particles and carrier fluid is used within an applied force field like centrifugal force or acceleration forces. Centrifugal forces occur in a rotating system like a centrifuge or a cyclone (Bockhardt et al.

(1997) and separate larger and more dense particles from smaller and from the carrier fluid. The cyclone is a mechanical separation process of dispersions of solid – gas or solid – liquid mixtures without moving parts. The separation of immiscible liquids with different densities is possible, too, but can be done by other separation processes as well or better (settling, filtering, coalescing, see Bockhardt et al. (1997).
In a centrifuge the centrifugal acceleration scales proportional to the radius. Hence, a miniaturization will reduce the radial acceleration and therefore the separation efficiency, which is defined as the ratio of the acceleration by the force field to the gravity.

$$\text{separation factor } z = \frac{\text{centrifugal accelaration}}{\text{gravitation}} \tag{1}$$

Fluidic vortices possess a high velocity near the vortex center and a decreasing tangential velocity with increasing radius r. The tangential velocity of a real viscous vortex scales with r^{-n}, where n ranges from 0.5 to 0.8 (see Bockhardt et al. (1993)). The centrifugal acceleration is proportional to the square of the tangential velocity w_t divided by the radius r, which results for the separation factor.

$$z = \frac{{w_t}^2}{r\,g} = \frac{const.}{r^2 g} \tag{2}$$

Here, g is the acceleration of gravity. The separation factor increases with the inverse the square radius of the cyclone r^2, but also the pressure loss of the cyclone. This enhancement is very promising for miniaturization. Hence, small particles are more likely to be separated by smaller cyclones. In a feasibility study, a micro cyclone was realized by conventional precision machining in PMMA, see Fig. 1.

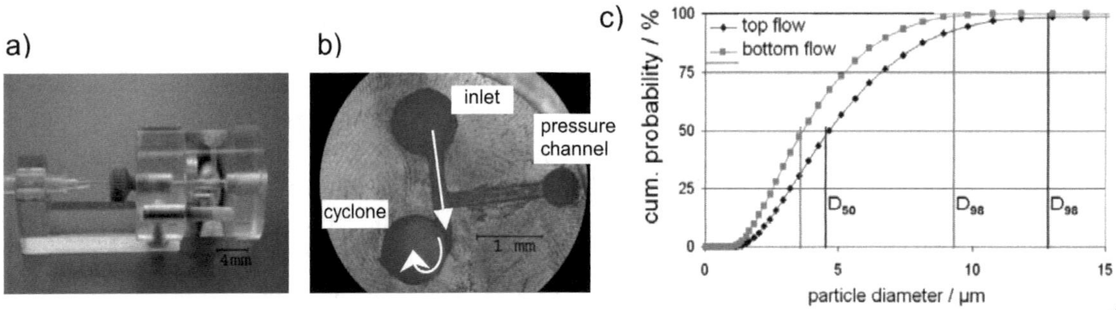

Fig. 1: a) Micro cyclone fabricated in PMMA, with HPLC connections; b) Cross section of the micro cyclone with inlet and channel for pressure measurement; c) cumulative probability of the particle diameter for the top and bottom flow of the cyclone, small particles are collected in the top flow.

The suspension (here: SiO_2 particles in water with a mean diameter of about 4.5 µm and a density of about 2500 kg/m^3) enters tangentially the cyclone hole. Driven by the centrifugal force the large particles are flowing to the outer radius; smaller particles are collected in the centre and leave the cyclone through the top flow. In Fig. 1 c) the separation efficiency of one cyclone can be seen at the D_{50}-value of the top and bottom flow, and more clearly at the D_{98} value (9 µm and 13 µm for the top and bottom flow, respectively).
For the next investigations, a fabrication in silicon with smaller diameter is intended to separate particles with diameter smaller than 5 µm. A cascading of many micro cyclones will give a higher separation factor but needs a higher pressure loss. The process can be extended to pressure diffusion, a coupled process described by the thermodynamics of irreversible processes and used for isotope separation, see Schütte (1972).

Thermal Separation – Micro Rectification

The separation of a liquid mixture with subsequent evaporation and condensation enriches the low boiling component in the vapor phase. This coupled heat and mass transfer is performed in small laboratory equipment like the micro distillation apparatus as well as in huge rectification towers like 60 m high cryogenic rectification columns of air separation units. The underlying properties and transfer processes in each stage or in packed beds should not be treated here in detail, other textbooks are more adequate, see Schönbucher (2003) or Geankopolis (2003). The Raoults law for an ideal mixture of two components describes the equilibrium correlation between the concentration of a component in the liquid and the partial pressure of the component in the vapor phase. The heat and mass transfer between the liquid and the gaseous phase can be calculated with the help of a two-film model.

With Raoults law the equilibrium curves of the saturation line and boiling point line can be drawn, and the separation steps can be visualized (McCabe-Thiele diagram). Sowata and Kusakabe (2003) showed this for the vapour-liquid separation on a chip. Two fluid streams, a vapour and a liquid mixture of water-methanol with both the same concentration (0.71 mol fraction of methanol), were mixed on a silicon chip ($20*40mm^2$), covered with a glass lid. After mixing in a Y-shaped mixer, the bubble flow streams through a meandering channel to approach the equilibrium state. At the end of the meandering channel, the two phases were separated by gravity in a separation chamber ($8*8mm^2$). One concentration measurement was given in the paper with a vapour concentration of 0.815 mole fraction methanol (almost equilibrium at a temperature of 72 °C) and a liquid concentration of 0.65 mole fraction methanol. With an equilibrium concentration of about 0.54 mole fraction, the stage efficiency is about 65% for the liquid side. A better mixing in the meandering channel can enhance this value and contribute to a better separation. So, a good mixing process may enhance also a separation process due to the fast achievement of a thermodynamic equilibrium.

Within the frame of the MicroChemTec program of the German Ministry of Research and Education, the micro distillation in a single column was studied in a working group of Prof. Hampe, Technical University of Darmstadt. In two publications (Klemm et al. (2004) and MicroChemTec (2005)), some results were published on the reversible distillation in the micro device. The temperature profile over the length of the distillation slit determines the equilibrium concentration of the fluid mixture running down the heated wall. The heat conduction in the metal frame determines the temperature profile. In a prototype set-up, a volume flow of about 20 ml/h has been realized with 2.5 theoretical separation stages. Even small temperature changes lead to operational disturbances and uncontrolled concentration changes.

In an own design study, a distillation chip with various geometries was fabricated and tested by evaporation and condensation of a water-ethanol mixture. Due to the azeotropic nature of the mixture, the ethanol concentration was chosen to 25 weight-%. The chip consists of an anisotropic KOH-etched silicon wafer with a meandering separation channel (one inlet for the mixture and two outlets for the liquid and the condensed phase, respectively) and with a heating and cooling channel. Two Pyrex glass lids cover the silicon wafer, see Fig. 2 b). For the fluidic connections, holes were drilled into the bottom lid. The liquid transport in the separation channel happens in the triangular edges between KOH etched silicon and the glass lid by capillary transport. A syringe pump supplies the liquid mixture into the separation channel with a low flow rate of about 1 ml/h to 5 ml/h. From the heated side (hot water, 93°C) the mixture partly evaporates and condenses on the opposite side of the channel. In Fig. 2 c) and e), the droplet condensation on the cooled upper wall can clearly be seen. The condensation and evaporation takes place at the trays and results in a increasing concentration of ethanol to the colder, upper side. The liquid transfer to the next tray happens in the angular structure.

Unfortunately, the exact concentration measurement data are not yet available, because the experimental program has just started.

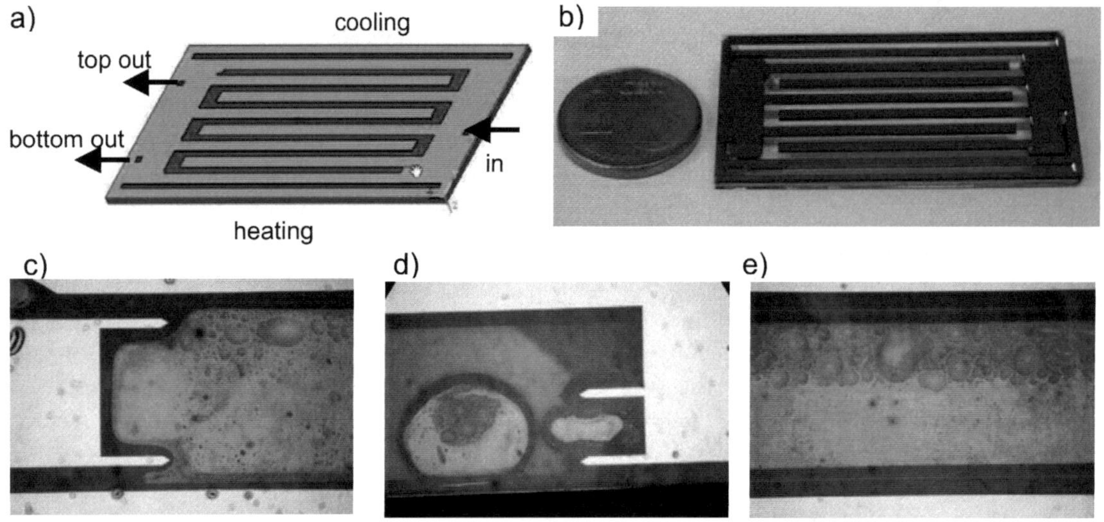

Fig. 2: a) Geometrical set-up of a distillation / rectification chip (20*40*1.6 mm3) made of a silicon wafer, anisotropic KOH-etched and covered with two glass lids, b) image of a rectification chip with meandering separatin channel; Bottom: Two phase flow and condensation of a water-ethanol mixture (25 weight-% ethanol) at the inlet and in the lower separation channel. c) One separation channel top and bottom outlet; d) Inlet meandering chip with bubble formation, e) first channel with droplet condensation at the glass wall.

The first experimental results show a low mass flow rate and a poor thermal insulation to the ambient. From the design study the following main issues have been identified:

- the fluidic transport can not rely on the gravity, other principles must be evaluated, the capillary transport delivers only a small amount of liquid.
- the contact between liquids of different concentration has to be avoided, which also limits the minimal size of the gaps for evaporation, condensation, and vapor flow.
- the fabrication of suitable geometries to guide the liquids and vapor currents is a crucial point. Besides the silicon techniques, alternative fabrication techniques like metal foil structuring should be investigated, see Klemm et al. (2004).
- an internal numbering up may help to increase the volume flow, which is a considerable drawback of the actual devices.

However, the potential of a high specific area provides a high transfer area and high exchange rate between the phases. A short distance between the phases and concentration differences allows a fast approach to the thermodynamic equilibrium and leads to smaller equipment, but the low throughput and complex fluid management has to be considered.

Thermal Separation – Thermodiffusion

The thermodiffusion describes the effect of different migration velocities of molecules with different size or mass in a thermal gradient field, which is exploited for their separation. Following the historical development, the thermodiffusion, assisted by natural convection, was firstly employed for the separation of isotopes in the Clusius-Dickel column, see London (1961). This process can be described with the help of the linear correlations from the thermodynamics of irreversible processes, see Langbein (1999). The Fourier law describes the coupling between the heat flux and temperature gradient. Fick's First law gives the linear relationship between the mass or species flow and the concentration gradient. The cross coupling between the heat flux and concentration gradient (Dufour effect L_{qD}) as well as the cross coupling between the species flux and a temperature gradient (Soret effect L_{Dq}) takes the following form.

$$\begin{aligned} J_q &= L_{qq}\,\Delta T + L_{qD}\,\Delta c \\ J_n &= L_{Dq}\,\Delta T + L_{DD}\,\Delta c \end{aligned} \tag{3}$$

The coefficients are formed in a way that the entropy generation $\dot{\sigma}$ is the product of the flux with the according driving forces.

$$\dot{\sigma} = \sum_i J_i \cdot X_i > 0 \tag{4}$$

The coefficient of the Soret effect L_{Dq} is 3 to 5 magnitudes smaller than the first order effects of heat conduction L_{qq} or diffusion L_{DD}. The separation of a fluid mixture with the help of a temperature gradient only needs a channel with adjacent splitting of the components. The mixture is flowing through a number of parallel micro channels with a superimposed temperature gradient. Smaller molecules tend to diffuse to the hotter side, larger diffuse to the colder side. At the end of the channel, the flow is split up and rearranged into the next channels. By cascading the parallel channels, the concentration of the smaller molecules is gradually enriched at the hot side.

In Fig. 3 an array of 5 * 3 channels is simulated with a given temperature profile at the walls and a resulting temperature profile in the channels (gas, 50% H_2, 50% CO_2). The simulations were performed with the commercial FVM-code CFD-ACE+ by ESI group. The thermodiffusion coefficients were calculated according the kinetic theory implemented in the code, see CFD-ACE (2002).

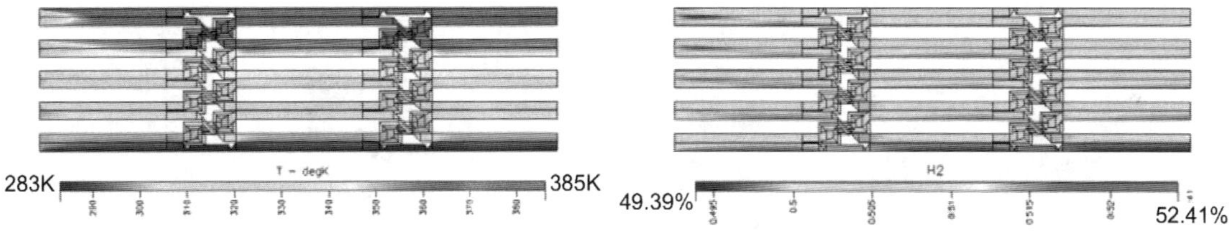

Fig. 3: Temperature profile (left) and concentration profile of H_2 (right) in a single channel with temperature gradient (100K), CO_2-H_2 system, 4000 Pa pressure difference, 120K temperature difference

The simulation gives the geometry optimization for an equal flow distribution in the channel network and an indication for the separation performance of the process. The separation effect depends on the temperature gradient as well as on the cascaded separation and collection of the species. The number of elements, which encounts almost 800,000 knots for the geometry of Fig. 3, limited the simulated geometry. With the help of the simulations and the available fabrication techniques, the structures were designed and fabricated in silicon to demonstrate the separation in liquid mixtures, see Fig. 4.

Four 40*20 mm^2 chips and four 20*20 mm^2 chips with different geometrical structures were arranged and fabricated on a 4-inch wafer. The mechanical strength of the RIE processed silicon wafer was weakened by the 400 μm deep channels which were additionally oriented in the crystal planes. The following fabrication steps like KOH etching, cleaning steps, and the bonding process have to be carried out very carefully. With the two glass lids, the entire chip was stiff enough to be integrated into the fluidic mount. Another difficult point was the drilling or grinding of the fluid connection with 1 μm diameter into the bottom glass lid. In Fig. 5 some results are shown where the cracks and split-ups in the glass are clearly visible. An optimization of the grinding speed and a very calm and vibration free tool holder lead to acceptable results for the grinded holes.

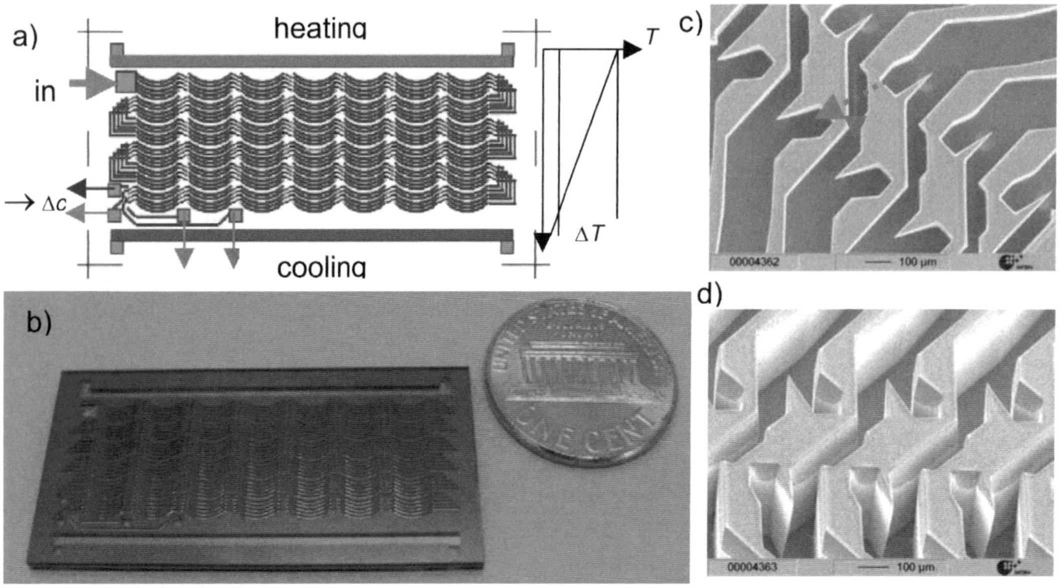

Fig. 4: a) Chip layout of a 40*20 mm^2 separation chip with 35 blocks of 5 bended channels each. The inlet and the four outlet connections are arranged on the left side, the heating and cooling channel on the top and bottom side, respectively. b) Fabricated silicon chip with glass lids. c) and d) SEM images of the RIE fabricated crossing structure, 400µm deep, the narrow gaps in the RIE crossing channel compensate the pressure loss in the KOH crossing (not displayed) on the bottom side. The arrows in image c) indicate the crossing direction between the channel blocks.

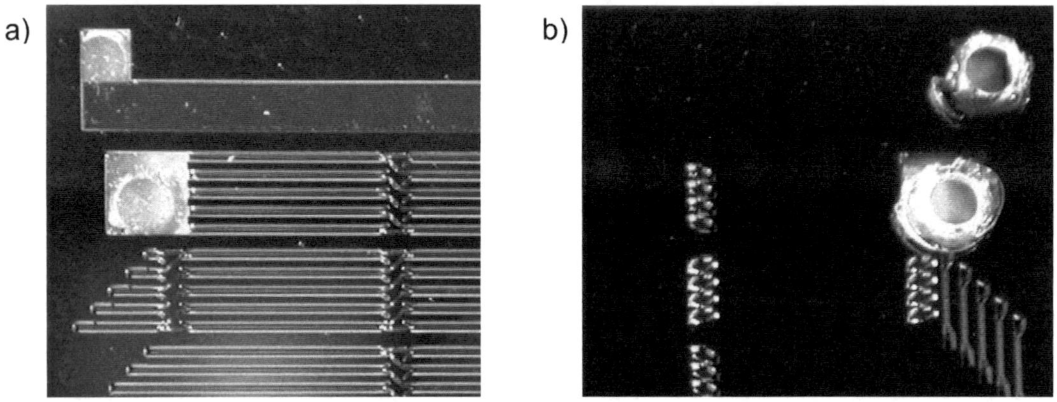

Fig. 5: a) Inlet heating channel and separation channels on the top left side, RIE fabrication; b) bottom view of the KOH etched crossing and transfer structures; the drilled holes in the bottom Pyrex lid can clearly be seen. This was one part of fabrication optimization as well as the KOH compensation structures

The next steps will be the design and implementation of the experimental set-up for the separation of liquid mixtures. The chip fabrication has to be optimized for an equal flow distribution in the channel network and a proper KOH crossing at the bottom. With the thermodiffusion a single-phase separation process was presented without unwanted interfacial forces, which may complicate or disturb the fluid flow and separation.

Summary and Outlook

The separation unit operations in micro structured equipment are the most neglected areas for process engineering, but are necessary for an entire plant in microstructures. Due to their small dimensions, the transport processes in microstructures are very fast, high gradients can be applied, elements can be cascaded very often, and various features can be integrated. In this paper three examples of mechanical and thermal separation units illustrate the benefits, but also the limits of miniaturization of unit operation and transport processes. With the micro cyclone a mechanical separation process was demonstrated, where a size reduction increases the separation factor, but also the pressure loss. The distillation and rectification in micro channels profits from the short diffusion length, but the liquid transport is complex and heat losses to the ambient have to be considered. A single-phase separation process like thermodiffusion exhibits no interfacial forces, which grow with smaller dimensions. The relatively weak effect is enhanced by parallel switching and cascading the separation channels.

With the benefits of miniaturization like a short transport distances and fast processes, many unit operations and separation processes offer a huge potential to be realized in micro structured devices comparable to micro reactors. High transfer rates can be achieved and the thermodynamic equilibrium (smoothing out temperature and concentration gradients) is approached very fast. Associated with the short distances, the unit operations in microstructures are also limited by a high pressure loss, a low mass flow rate in small devices, and a high fouling and clogging potential. Additionally, minor effects like interface forces and coupled processes have to be considered, which makes the design and operation of micro structured devices very complex. However, these minor effects can also be utilized for the individual purpose like fluid transport by capillary forces or single-phase separation by thermodiffusion. The role of unit operation should be extended to integrated processes in microstructures to provide a comprehensive toolbox for treating various tasks from process engineering.

For the future research, many aspects need to be addressed like liquid flow and transportation or two phase flow in capillary structures. The issues mentioned above should also be extended to other processes like extraction, adsorption, or electrokinetic processes. The integration of microstructures into macro equipment has to be carried out to use the benefits and opportunities of both fields and additional emergent effects. Hence, as guiding line for the next steps: Not only as small as useful for the designated application, but also as much micro as necessary for the actual process.

Acknowledgement

We gratefully acknowledge the enduring help of the clean room team, the mechanical workshop with Franz Richardt, as well as Jan Draheim and Johannes Kastner for their assistance during design, fabrication, and testing the thermodiffusion and rectification structures. Thank you very much.

Nomenclature

Name	Unit	Description
g	m/s^2	gravitational accelaration
J_n	mol/s	species flux
J_q	J/s	heat flux
J_n	mol/s	species flux
L_{DD}	mol/s	linear mass diffusion coefficient, proportional to the diffusivity
L_{Dq}	J/s	Soret coefficient
L_{qD}	mol/sK	Dufour coefficient
L_{qq}	J/sK	linear heat conduction coefficient, proportional to the heat conductivity

r	m	radius
w_t	m/s	tangential velocity
X	-	generalized force, unit depends on the force
z	-	separation factor
$\dot{\sigma}$	J/K	entropy production

Literature

Blass, E. (1997)
Entwicklung verfahrenstechnischer Prozesse – Methoden, Zielsuche, Lösungssuche, Lösungsauswahl, 2. Aufl., Springer-Verlag, Berlin.

Bockhardt, H.D., Güntzschel, P., Poetschukat, A. (1997)
Grundlagen der Verfahrenstechnik für Ingenieure, Deutscher Verlag für Grundstoffindustrie, Stuttgart.

CFD-ACE (2002)
CFDRC documentation, Chap 4 Chemistry module, CFDRC, ESI Group,

Cypes, S.H., Engstrom, J.R. (2003)
Construction, analysis and evaluation of a microfabricated stripping column, 7th International Conference on Microreaction Technology, IMRET7, Lausanne, Switzerland

Ehrfeld, W., Hessel, V., Löwe, H. (2000)
Microreactors: New technology for modern chemistry, Wiley-VCH, Weinheim.

Fang, Z. (1993)
Flow injection separation and preconcentration, Wiley-VCH, Weinheim.

Geankoplis, C. J. (2003)
Transport processes and separation process principles, 4th ed., Prentice Hall, Upper Saddle River.

Hessel, V., Hardt, S., Löwe, H. (2004)
Chemical Micro Process Engineering, Wiley-VCH, Weinheim.

Klemm, E., Rudek, M., Markovz, G., Schütte, R. (2004)
Mikroverfahrenstechnik, in: Winnacker–Küchler: Chemische Technik - Prozesse und Produkte, Vol. 2, Chap. 8, Wiley-VCH, Weinheim.

Kusakabe, K., Sotowa, K.I., Katsuragi, H. (2003)
Internal flow in a droplet formed in a microchannel contactor for solvent extraction, 7th International Conference on Microreaction Technology, IMRET7, Lausanne, Switzerland

Langbein, W. (1999)
Thermodynamik – Grundlagen und Anwendungen, Verlag Harri Deutsch, Thun, 2nd ed.

ltf-gmbh (2005)
Little Things Factory, Product catalogue,
URL: www.ltf-gmbh.de/deu/produkt/mikrofluidik/separieren/mikrozyklon.htm

London, H., Ed. (1961)
Separation of Isotopes, Georges Newnes Ltd., chap. 7, thermal diffusion

Melin, T., Rautenbach, R. (2003)
Membranverfahren – Grundlagen der Modul- und Anlagenauslegung, Springer-Verlag, Berlin

MicroChemTec (2004)
Abschlusspräsentation, 1st part, URL: www.microchemtec.de

Okubo, Y., Toma, M., Ueda, H., Maki, T., Mae, K. (2003)
Microchannel devices for the coalescence of dispersed droplets produced for use in rapid extraction processes, 7th International Conference on Microreaction Technology, IMRET7, Lausanne, Switzerland

Schönbucher, A. (2002)
Thermische Verfahrenstechnik, Springer-Verlag, Berlin.

Schütte, R. (1972)
Diffusionstrennverfahren, in Ullmanns Encyklopädie der technischen Chemie, Bartholomé, E. Ed. Vol. 2, Verfahrenstechnik I: (Grundoperationen).

Sowata, K.I., Kusakabe, K. (2003)
Design of microchannels for use in distillation devices, 7th International Conference on Microreaction Technology, IMRET7, Lausanne, Switzerland

TeGrotenhuis W.E., Cameron, R , Butcher, M.G., Martin, P.M., Wegeng, R.S. (1998)
Micro Channel Devices for Efficient Contacting of Liquids in Solvent Extraction, Separation Science and Technology. PNNL-SA-28743

Wegeng, R., Drost, M.K., Brenchley, D.L. (1999)
Process Intensification Through Miniaturization or Micro Thermal and Chemical Systems in the 21st Century, 3rd International Conference on Microreaction Technology, IMRET3, Frankfurt, Germany.

Formulation of Monodisperse Double Emulsions in Microfluidic Devices

Takasi Nisisako, Shingo Okushima, Toru Torii and Toshiro Higuchi
Department of Precision Engineering, Graduate School of Engineering, The University of Tokyo, 7-3-1 Hongo, Bunkyo-ku, Tokyo, 113-8656 Japan

Abstract

This paper describes the use of multiple-phase fluid flow in microfluidic systems to formulate highly monodisperse double emulsions. For producing water-in-oil-in-water (W/O/W) dispersion, aqueous droplets ruptured at the upstream hydrophobic junction are enclosed within organic droplets formed at the downstream hydrophilic junction. Droplets produced at each junction could have narrow size distribution with coefficients of variation in diameter less than 3 %. Internal drop size, external drop size and the number of enclosed droplets were flexibly controllable by variations in hydrodynamic conditions. Combination of the upstream hydrophilic junction and the downstream hydrophobic junction produced monodisperse oil-in-water-in-oil (O/WO) emulsion. W/O/W emulsions with two internal droplets of different compositions could be prepared using cross-shaped geometry as the upstream component. Monodisperse polymeric microcapsules were also engineered by curing intermediating phase in W/O/W droplets.

Introduction

A double emulsion is a multiple-phase dispersion in which droplets enclosing finer droplets are suspended in continuous liquid phase. Typical forms of such systems are the water-in-oil-in-water (W/O/W) emulsions and oil-in-water-in-oil (O/W/O) emulsions, and these have attracted considerable attention because of their potential applications in foods, cosmetics, pharmaceutics and many other fields. Two-step vigorous mixing [1] is mainly employed in practical formulation of double emulsions, but resultant products are generally polydisperse. Alternatively, narrower size distribution can be obtained by forcing single emulsions through membranes [2] or nozzles [3], but the number of internal drops and the volume fraction in each capsule are poorly controllable.

Here we present a new method for preparing monodisperse double emulsions by a two-step method of droplet formation in a series of T-junctions [4]. For a W/O/W emulsion, for example, to-be-enclosed aqueous drops are periodically formed upstream at the hydrophobic T-junction; then, in a continuing series, organic droplets that encase the aqueous droplets are formed downstream at the hydrophilic T-junction (Fig.1). Droplets of uniform sizes are reproducibly formed and their size is easily varied at the formation points by changing the flow conditions. By adjusting the relation between breakup frequencies at two junctions, the control in the number of enclosed drops can be achieved.

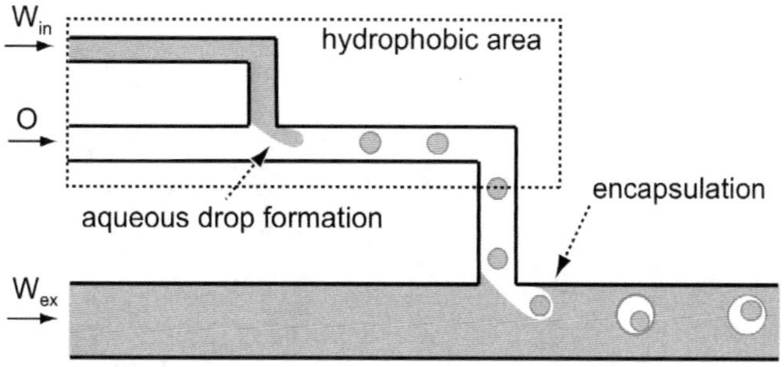

Figure 1. A schematic of the formation of double-emulsion droplets (W/O/W). W_{in}: internal aqueous phase; O: intermediating organic phase; W_{ex}: external aqueous phase.

Experimental

Microchannels were fabricated on glass (quartz glass or pyrex) plates by mechanical machining or lithographic techniques (isotropic wet etching or anisotropic dry etching). Typical channel sizes were 40-85 μm in width and 10-40 μm in depth for upstream junctions, and 130-225 μm in width and 65-100 μm in depth for downstream junctions. Two configurations are proposed for production of double emulsions; a one-chip type in which hydrophobic junction and hydrophilic junction are on the same plate, and a two-chip type in which each different property junctions on separate chips respectively (Fig.2). For aqueous droplet formation, channel surface was partially modified to be hydrophobic by siliconization.

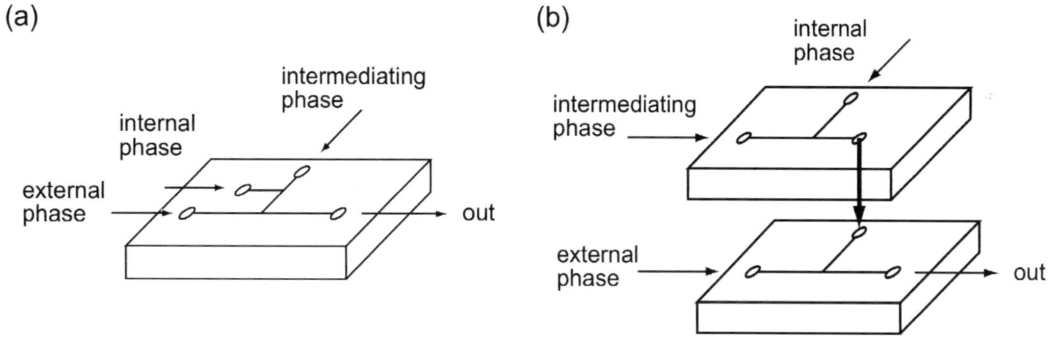

Figure 2. Schematic illustration of two configurations: (a) the one-chip module; (b) the two-chip module.

Deionized water was used as the aqueous phase. Corn oil, decane or an acrylic monomer was used as the organic phase. Surfactants were added to intermediating and external phases to stabilize the prepared multiple-phase dispersions. Lecithin or tetraglycerin condensed ricinoleic acid ester (CR-310, Sakamoto Yakuhin Kogyo Col, Ltd., Japan) was added to the organic phase as lipophilic surfactants. Sodium dodecyl sulfate (SDS), decaglycerol monostearate (MSW-7S, Sakamoto Yakuhin Kogyo Co., Ltd., Japan) or polyvinyl alcohol (PVA) was added to the aqueous phase as hydrophilic surfactants.

Results and Discussion

Figure 3(a) shows the formation of W/O/W droplets in a one-chip module. An array of monosized aqueous droplets (diameter = 52 um, coefficient of variation (CV) = 2.7%) was formed at the upstream hydrophobic junction. Each aqueous droplet was then reliably encased in an organic droplet (diameter = 83 um, CV=2.8%) at the downstream hydrophilic junction. By changing flow conditions slightly, ellipsoidal organic drops that encase two aqueous drops could be reproducibly produced (Fig.3(c)). O/W/O emulsions could be produced by reversing the order of hydrophobic and hydrophilic junctions. Of the two settings investigated, the one-chip module is better suited for precise control of encapsulation, whereas the two-chip approach has the feasibility in flexible device configurations for various double emulsions.

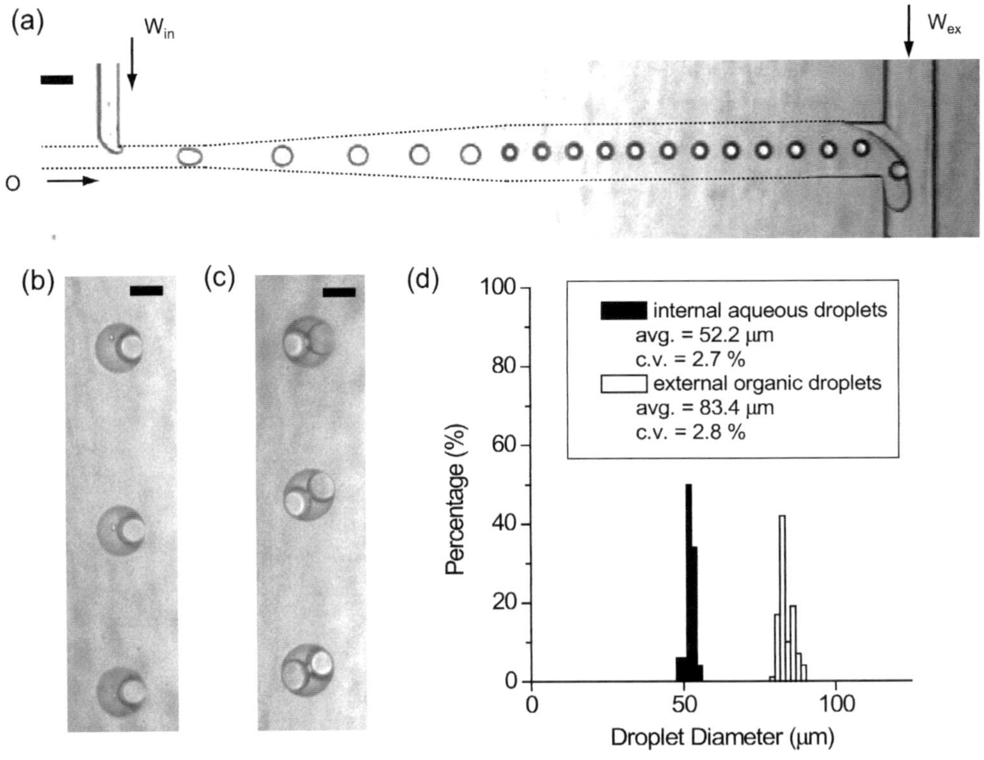

Figure 3 Preparation of W/O/W droplets in the one-chip module: (a) photomicrograph of droplet formation at two consecutive T-junctions (Breakup rate: ~22 drops/s); (b, c) Generated W/O/W droplets observed in the microfluidic device; (d) Size distributions of internal aqueous droplets and external droplets (n=100). Scale bars are (a) 100 μm, (b, c) 50 μm. W_{in}: deionized water, O: corn oil containing lecithin (0.5 wt%), W_{ex}: SDS aqueous solution (0.5 wt%)

The size of the droplets and the number of internal droplets could be varied by changing the flow conditions. The effect of the flow variation in the external aqueous phase stream is shown in fig. 4. As the flow rate of the external aqueous stream increased, the size of organic droplets formed at the downstream junction decreased. This reduction caused an increase of the breakup rate at the downstream junction, decreasing the number of encased droplets. Similarly, flow variations in the internal aqueous phase and the intermediating phase could change the form of double-emulsion droplets.

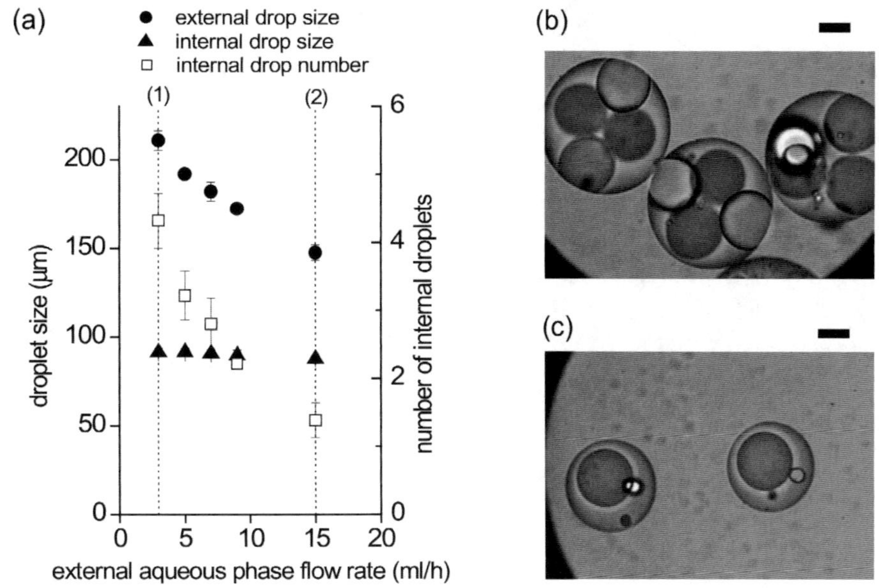

Figure 4. Effect of the external aqueous stream on the formation of W/O/W droplets: (a) Variation of the internal drop size, external drop size and the number of encased droplets; (b) W/O/W droplets formed in the condition (1) in the graph; (c) W/O/W droplets formed in the condition (2). Scale bars are 50 μm.

Double-emulsion droplets that encase internal droplets of different compositions were prepared using a cross-junction as the upstream component (Fig.5). Aqueous drops were alternately formed at the cross-junction [5, 6], and these arrayed droplets flew into the downstream junction to be jointly encased in organic droplets. This novel double emulsion may find applications in droplet-based biological screening [6, 7]. In addition, monodisperse polymeric microcapsules could be engineered by curing intermediating organic phase in W/O/W droplets (Fig.6).

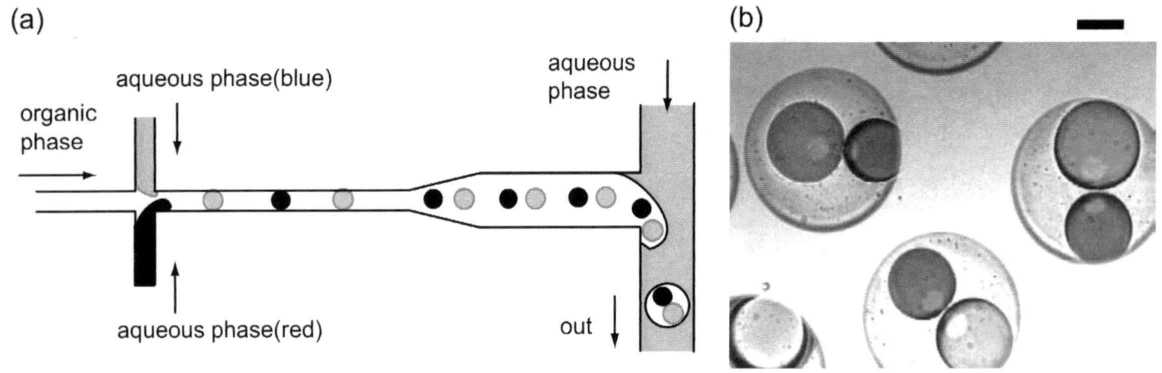

Figure 5. Formation of double-emulsion droplets enclosing two aqueous droplets from different sources. (a) Schematic illustration of the alternating formation of aqueous droplets and the subsequent encapsulation; (b) A photomicrograph of the prepared W/O/W droplets (diameter of the external drops: 175 μm). Scale bar is 50 μm.

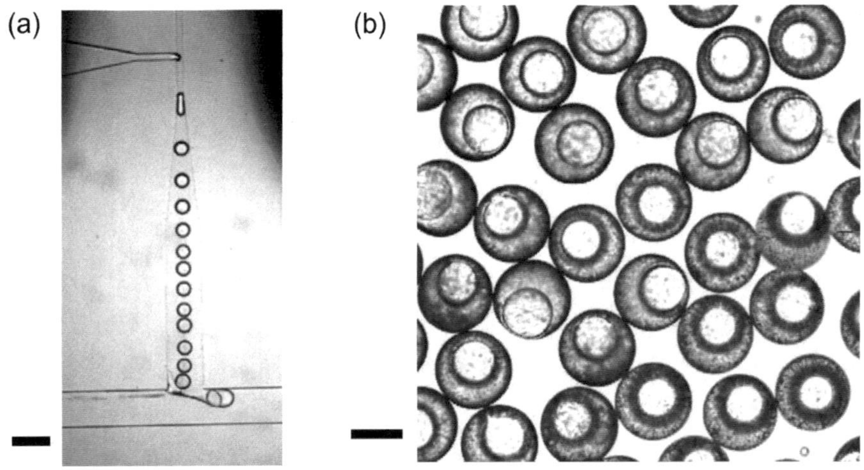

Figure 6. Production of monodisperse polymeric microcapsules: (a) formation of monomer droplets enclosing aqueous droplets; (b) Microcapsules polymerized by UV exposure. Scale bars are (a) 200 μm, (b) 100 μm.

Conclusion

We presented microfluidic systems for formulating monodisperse double emulsions. Both internal drops and external drops had narrow size distribution with CVs below 3%. Internal drop size, external drop size and the number of encased droplets were flexibly controlled by variation of hydrodynamic conditions. Both W/O/W and O/W/O emulsions could be prepared by interchanging the hydrophobic and hydrophilic components. Serial connection of a cross-shaped geometry and a T-shaped geometry produced organic droplets that encase two aqueous drops with different compositions. Monodisperse polymeric capsules could be engineered by photo-polymerization of W/O/W droplets.

We believe that the method we developed has a lot of promise to be functioned in various fields, such as analysis of confined chemical reactions, biological screening, drug delivery systems and three-dimensional colloidal assemblies.

Acknowledgement

This work was supported by a grant from the Research Association of Micro Chemical Process Technology (MCPT) of Japan.

References

[1] S. Matsumoto, Y. Kita and D. Yonezawa, *J. Colloid Interf. Sci.*, **57**, 353 (1976).

[2] T. Nakashima, M. Shimizu and M. Kukizaki, *Adv. Drug Deliver. Rev.* **45**, 47 (2000).

[3] S. Sugiura, M. Nakajima, K. Yamamoto, S. Iwamoto, T. Oda, M. Satake and M. Seki, *J. Colloid Interf. Sci.* **270**, 221 (2004).

[4] S. Okushima, T. Nisisako, T. Torii and T. Higuchi, *Langmuir* **20**, 9905 (2004).

[5] R. Dreyfus, P. Tabeling and H. Willaime, *Phys. Rev. Lett.* **90**, 144505 (2003).

[6] B. Zheng, J. D. Tice and R. F. Ismagilov, *Anal. Chem.* **76**, 4977 (2004).

[7] B. Zheng, J. D. Tice, L. S. Roach and R. F. Ismagilov, *Angew. Chem. Int. Ed.* **43**, 2508 (2004).

A Novel Production Microstructured Mixer – Characterization of Mixing Efficiency

P. Löb, V. Hessel, H. Löwe, B. Werner, Y. Men
Institut für Mikrotechnik Mainz GmbH, Carl-Zeiss-Str. 18-20, D-55129 Mainz, Germany
Tel.: +49 6131 990 377, E-mail: loeb@imm-mainz.de

The development of microstructured mixers plays a key role for carrying out advanced syntheses of fine chemicals as well as for the generation of dispersions, creams, foams, and emulsions [1]. So far, microstructured mixers were mainly limited to laboratory-scale or at best pilot plant-scale –typical maximum flow rates were from 2 – 100 l/h for watery fluid systems.

To fulfil industry's production demands, a new microstructured mixer was developed which is based on a multi-stack architecture of thin platelets [2, 3]. The platelets are disk-like with star-shaped feeding zone and a large inner circular opening which gives the mixing chamber (Figure 1). Owing to the star-shape of the platelets and the interdigital feeding scheme, the whole series of the mixer devices was named StarLaminator, abbreviated by the brand name StarLam. In this way an interdigital multi-nozzle array is created. Only two different types of platelets are needed to form the main feed tubing, the nozzle injection zone, and the central mixing chamber. Several hundreds of platelets form the final stack which is inserted in the recess of a housing (Figure 2). The StarLam mixer platelets were made via inexpensive fabrication techniques such as punching or laser ablation, rendering the costs of the device lower than its corresponding large-scale pumps.

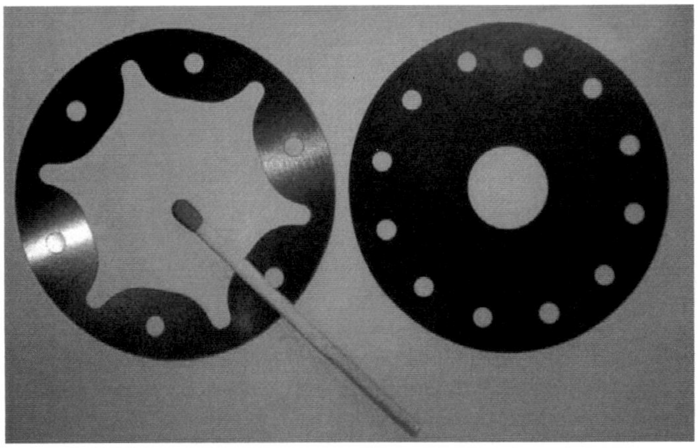

Figure 1. The two types of mixing foils for *StarLam3000* (*StarLam30*0 smaller but same design) which are assembled in an alternating sequence. The injection foil (left) is turned thereby by 30° in each step.

In this way, a simple numbering-up concept is given based on the increase of the number of the internal platelets. Experiments can be run with a few platelets at a laboratory-scale, while pilot processing can be done with the same platelets at higher overall repetition number. As a further possibility for scaling-out, the diameter of the platelets can be enlarged, including an increase of the inner mixing channel.

Two versions with differing platelet diameter actually were realised; the StarLam 300 of fist size and the StarLam 3000 of shoebox size (Figure 3). The StarLam 300 mixer achieves throughputs of watery flows of up to 1000 l/h at pressure drop of about 10 bar, while the

StarLam 3000 mixer can be operated at 3,500 l/h at a pressure drop of about 0.9 bar. In this case, the StarLam 300 contained 2 x 65 platelets with 100 µm thickness, the StarLam 3000 had 2 x 113 platelets with 250 µm thickness.

Figure 2. Inner part of *StarLam3000* with some exemplarily mixing foils.

Figure 3. Left: The new *StarLam*-mixers: *StarLam300* (left) and *StarLam3000* (right). Right: *StarLam3000* with flanges for integration into conventional macro plant environment.

The mixing was characterized by means of a competing reaction approach utilizing the Dushman reaction to monitor mixing efficiency in a photometric manner. It turned out when operating the StarLam 300 at the upper productivity limit, that the mixing efficiency is approaching that of the much smaller IMM's interdigital micro mixers with a capacity in the liter range and below (Figure 4). The same holds for the StarLam 3000, however, at much larger absolute flow rate. When referring the mixing efficiency to the pressure drop, all StarLam and other mixers (e.g. caterpillar device CPMM-R1200/8-pm) are on one declining line, which is indicative that the mixing occurs by turbulence.

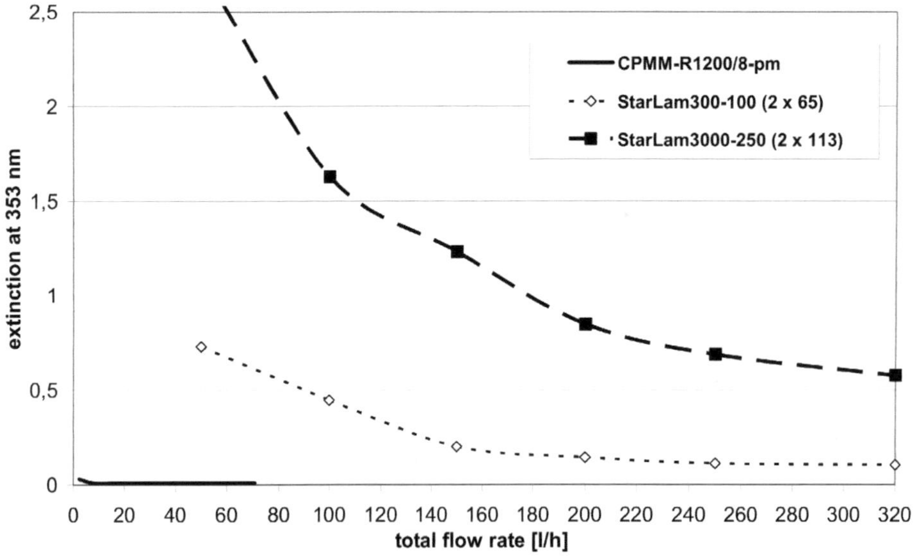

Figure 4. Dushman mixing quality as function of total flow rate for different mixers.

Having demonstrated that efficient, or at least reasonable, mixing is given for large-throughput microstructured mixers, detailed parametric studies were undergone with the devices of the StarLam series. The platelet thickness, i.e. dimensions of the nozzle injected stream, the number of platelets, the diameter of the platelets and other structural parameters were varied. In this way, it will be alluded to scaling-up issues, albeit this is more an empirical coarse report of increasing productivity, rather than giving a theoretical concept. Also, the fluid properties were changed, e.g. by contacting two viscous solutions or one viscous with a watery system. Reliability issues will be addressed as well such as achieving tightness for hundreds of platelets and the role of deviations from microfabrication on the production mixer performance. Pulse measurements finally will reveal the impact of the mixing device on the residence time distribution.

Both apparatus thus approach a performance of today's low-capacity (l/h) micro mixers. Therefore, continuity from the "real" micro mixers over the therein described high-throughput tools to conventionally manufactured static mixers with even higher flow rates is given. Accordingly, first industrial use of the StarLam production mixers is given.

References

[1] Hessel, V.; Löwe, H; Kolb, G., Müller, A., Chapter "Mixing of Miscible Fluids" in Chemical Micro Process Engineering – Processing and Plants, Wiley-VCH, Weinheim, 2005.

[2] Werner, B.; Hessel, V; Löb, P., Chemie-Ingenieur-Technik 76, 5 (2004) 567

[3] Werner, B.; Hessel, V; Löb, P., Chem. Eng. Technol. (2005) in press.

129d Mixing In Structured Microchemical Devices

Dionisios Vlachos, Soumitra Deshmukh

Introduction

Microdevices are being used in a wide variety of applications ranging from high throughput screening of materials and catalysts to microreactors for chemical and biological syntheses to portable power generation. Despite several advantages of microscale devices, including fast heat and mass transfer, operation of these devices is non-trivial. The small length scales in these devices imply laminar flow, where molecular diffusion is the primary mechanism for mixing of fluids. Diffusive mixing compounded with short residence times imply that mixing is challenging. Furthermore, large catalytic surface areas are difficult to achieve in compact devices.

Structured microreactors with small scale features, such as cylindrical posts, have recently been proposed to enhance the available surface area for catalytic reactions [1, 2] and potentially impart flow uniformity in static micromixers [3]. However, a systematic study of the effect of these microscopic features on mixing is lacking. In this work we characterize the flow in the presence of microscale structures along with their effect on overall chemical reactivity. Finally, guidelines for optimization of mixing of structured microdevices are proposed.

Materials and Methods

The occurrence of enhanced mixing at the microscale using posts is investigated using two-dimensional, time dependent computational fluid dynamics (CFD) simulations. Simulations are first performed to characterize the flow past a single post in a large 4 mm x 5 mm device as a function of Reynolds (Re) number. Prototype chemical reactions are then carried out in microdevices to investigate the coupling of reaction kinetics and flow dynamics. Mixing of un-premixed fluid steams is also investigated in arrays of structural elements (posts). The effect of different configurations of posts on the extent of mixing is also studied.

Results and Discussion

Oscillatory flow patterns are shown to occur in structured microdevices under certain conditions via a Hopf bifurcation instability. In particular, it is found that at Re number greater than a critical value, the boundary layer behind the post loses its symmetry, and an unsteady wake develops behind the post. The time period of the oscillations is very short (a fraction of the contact time).

It is shown that the oscillatory flow patterns in structured microstructured devices persist even in the presence of chemical reactions. A two-way coupling between fluid dynamics and chemistry occurs. It is demonstrated that the nonlinearity of the flow strongly affects chemical reactivity. On other hand, chemical reactions involving a change in number of moles affect the fluid density and thereby the velocity field and oscillation amplitude and dynamics. This effect is, however, small to moderate in magnitude and is observed for both volumetric as well as surface reactions. The reaction exothermicity or endothermicity also affects oscillatory instabilities and mixing though a change in temperature that in turn modifies the fluid density.

A systematic investigation of mixing of un-premixed streams in an array of posts is also studied. The effect of device configuration in terms of the number of posts as well as their local arrangement (segregation) is investigated. It is shown that the oscillatory wakes increase the interfacial area and the mixing by a significant factor. Guidelines for enhancing mixing in terms of the spatial arrangement and density of posts will be presented.

References

1. Ganley, J.C., Seebauer, E.G., and Masel, R.I. AIChE J. 50, 829 (2004).
2. Losey, M.W., Jackman, R.J., Firebaugh, S.L., Schmidt, M.A., and Jensen, K.F. J. MEMS 11, 709 (2002).
3. Norton, D.G., Wetzel, E.D., and Vlachos, D.G. Ind. Eng. Chem. Res. 43, 4833 (2004).

129e Microchannel Absorption for Portable Heat Pumps

Susie Stenkamp, Ward E TeGrotenhuis

Battelle has created a family of microchannel devices for gas-liquid processing, including phase separation, partial condensation, distillation, absorption, and desorption. For many of these unit operations, the primary resistance to mass transfer occurs due to diffusion in the liquid phase. Thus, intensification of mass transfer can be exerted through control of the liquid film thickness. In Battelle's devices, this thickness is controlled through the use of wicks. Diffusion occurs from an adjacent vapor channel. The wicks offer the additional advantage of orientation and gravitational independence since the fluid flow is controlled through capillary forces. The results of these attributes are devices that are highly suitable for portable applications. Two of these unit operations, absorption and desorption, are being developed for use in portable absorption cycle heat pumps for the U.S. Army. One application is for man portable cooling, while the second is for portable cooling of electronics on a vehicle or in a tent.

Currently Battelle is testing a 250 W breadboard heat pump that includes a microchannel absorber and ammonia/water as the refrigerant system. Due to its target use for combat in hot climates, the heat rejection temperature is 50C. The absorber is a multichannel device that is air cooled in cross-flow. The absorber and its principles of operation are presented, as well as results for the testing.

A Centrifuge-Based Microreactor

Stefan Haeberle, Hans-Peter Schlosser, Roland Zengerle, and Jens Ducrée

IMTEK – University of Freiburg, Laboratory for MEMS Applications, Georges-Koehler-Allee 103, D-79110 Freiburg, Germany, haeberle@imtek.de, Phone: +49 761 203 7476, Fax: +49 761 203 7539

Abstract

We introduce a novel modular platform for high-performance reactive micromixing. The platform operates in a continuous flow mode, driving the flow through microchannels rotating on a standard centrifuge at frequencies between 20 Hz and 120 Hz. Extraordinary throughputs of 1 ml per second and channel are achieved which can massively be enhanced to about 100 ml / s per disk unit by a generic parallelization scheme. The device also features superior mixing quality at high throughputs. Further benefits of the platform are the ease of fabrication due to the planar geometry of entirely static, low-aspect-ratio microstructures as well as the ease of replacing the mixing module due to the absence of pressure-tight micro-macro interfaces.

1. Introduction

The use of miniaturized channel networks for chemical process engineering has attracted a lot of scientific interest within the last decade [1]-[3]. Different mixing concepts have been presented, most of them utilizing high-aspect ratio microfluidic structures for multilamination of pressure driven laminar flows [4].

In this paper, we exploit the centrifugal "pumping" force to drive the liquid educts through the microchannels. Such centrifugal microfluidic systems have so far been chiefly employed for analytical applications that require the integration of several process steps in one microfluidic chip (so-called "disk") [5],[6]. However these functional elements aim at processing of small liquid volumes. We here present the extension of the platform towards continuous, high-discharge reactive micromixing. Special forces, only apparent on the non-inertial, rotating frame of reference yield to enhanced mixing conditions within simple microchannels as described in the following sections.

2. Centrifugal Micromixing

We make use of the Coriolis pseudo-force $\boldsymbol{F}_C$ to induce mixing between two flows through a rotating microchannel [7],[8]. As depicted in Fig. 1, we consider a centrifugal flow profile $\boldsymbol{u}(\boldsymbol{F}_\nu)$ through a radial microchannel rotating at an angular velocity vector ν. Due to the velocity-dependence of the transversal Coriolis force $\boldsymbol{F}_C \sim \boldsymbol{u} \times \nu$, the force field $\boldsymbol{F}_C$ peaks in the center of the parabolic flow profile and vanishes towards the no-slip boundary conditions ($\boldsymbol{u} = 0$) imposed by the walls.

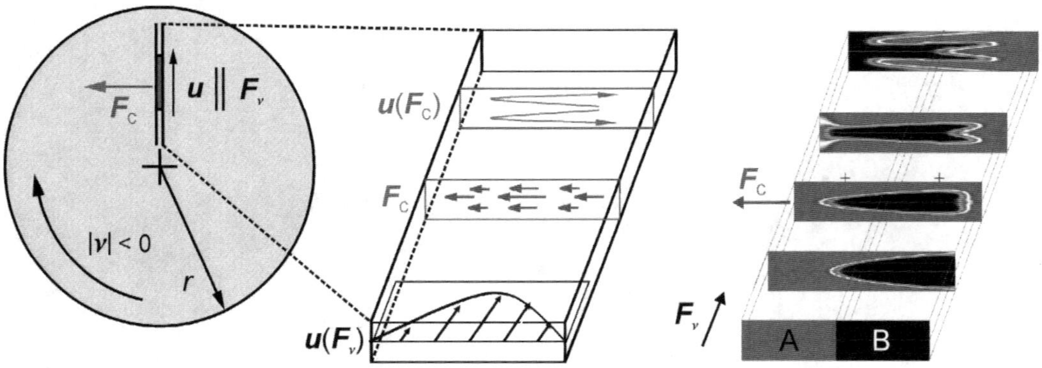

Fig. 1. Forces and flow patterns in a radial mixing channel. The simulations (CFDRC, ACE+ [9]) on the right show the Coriolis induced transversal stirring $\boldsymbol{u}(\boldsymbol{F}_C)$.

Above typical frequencies of about 16 Hz, the inhomogeneous force field $\boldsymbol{F}_C$ drives a transversal "stirring" $\boldsymbol{u}(\boldsymbol{F}_C)$ of the laminar flows. Since this effect scales with the flow velocity of the fluid, it is even more pronounced towards higher frequencies of rotations, i.e. higher throughput.
These considerations have been verified by CFD simulations (Fig. 1 right) [9]. Two liquids possessing the same physical properties but one marked with a (non-physical) scalar indicator are connected in the simulation model. The azimuthal motion of the liquid being pumped radial through the rotating channel becomes evident by the propagation of the A-B interface along the channel.

3. Setup and Flow Scheme

The Coriolis micromixer consists of a propulsion unit (e.g. standard lab centrifuge), a dispensing unit and a customized rotor (Fig. 2). The liquid educts are continuously delivered in a contact-free fashion into the rotor-mounted reservoirs. The frequency-dependent centrifugal force then drives the educts from separate inlet reservoirs into a radial mixing channel (length: 30 mm, width: 260 µm, depth: 195 µm). Finally, the reaction product is either centrifuged out of an orifice in the side surface of the rotor into a receiving vessel (**1**) or collected on a third reservoir on the rotor (**2**).
The modularity of the setup allows to integrate all moving parts as well as the actuation in a standard, reusable macro device while mixing is performed within an exchangeable module incorporating the (passive) microchannels. These channels are fabricated in a polymeric substrate ("disk"), using standard micromachining technologies. Thus, a huge variety of (polymer) materials is readily available for the mixing modules.

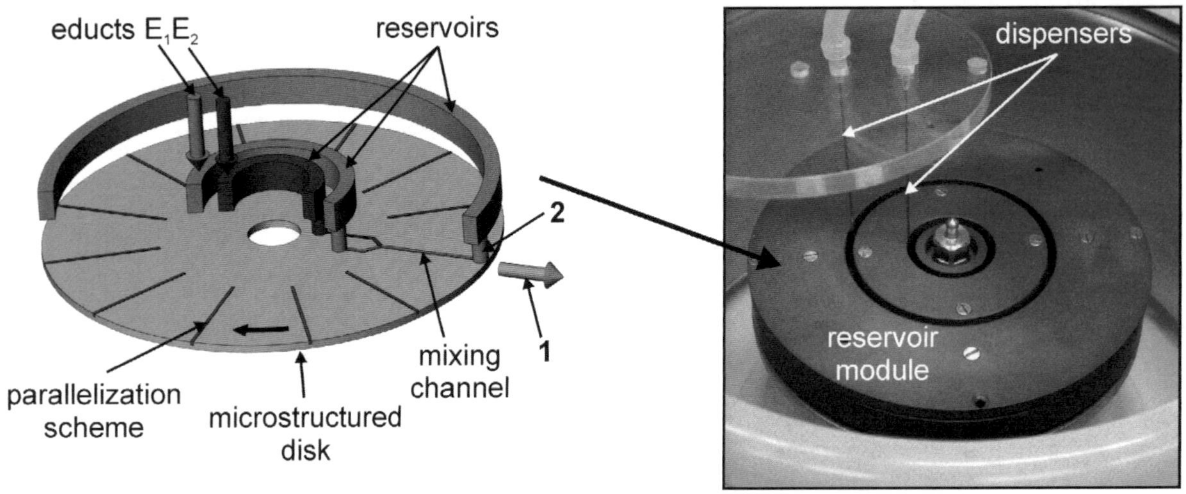

Fig. 2. **Left:** Flow scheme of the micromixer: two educts E_1 and E_2 are injected in a contact-free fashion into the rotating reservoirs on top of the rotor. Afterwards, they are merged at the junction-point of a Y-shaped mixing channel. At the end of this channel, the product is either centrifuged out of the module (continuous flow mode: **1**) or collected in a third reservoir (batch mode: **2**)

Right: The rotor of the micromixer is placed in a standard lab centrifuge. The microstructured disk featuring the microfluidic channels is mounted beneath the reservoir module.

4. Experimental Results

The Coriolis force induced "stirring" effect is visualized in Fig. 3. Water colored with ink and pure water respectively serve as sample liquids A and B. The Coriolis force $\boldsymbol{F}_C$ points to the right and pushes the black (ink-colored) liquid to the side-wall of the channel. The experimental results are in good agreement with the transversal flow-pattern predicted by the simulations (Fig. 3, right)

Diffusion is virtually set to zero in the simulation model, so, discarding the transversal stirring, the two liquids should not mix-up along the radial channel. The smearing of the A-B interface thus does not represent mixing but a numerical artefact of the simulation model ("numerical diffusion"). In order to quantify the performance of the system, flow rates as well as the quality of mixing have been experimentally determined. The overall flow rate is independent from the sense of rotation and scales with the square of the rotational frequency ν (Fig. 4, left).

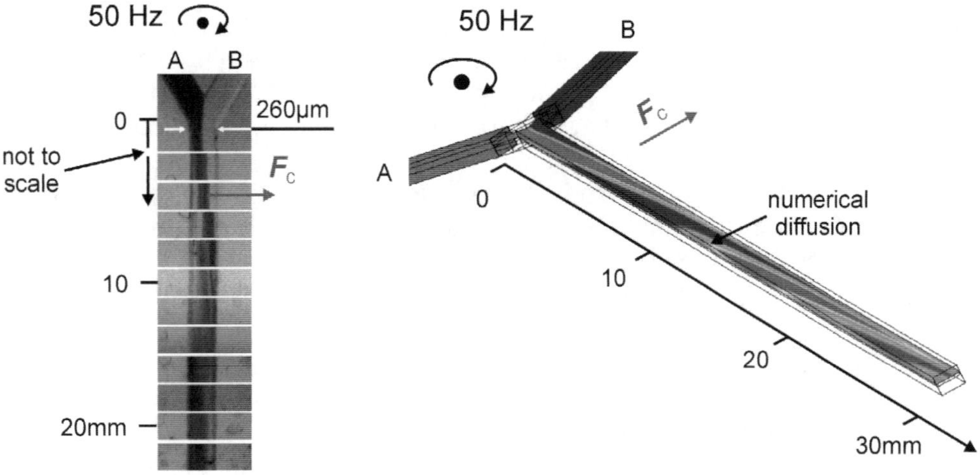

Fig. 3. Coriolis induced stirring in a simple Y-structure. The experimental picture series along the radial mixing channel shows the propagation of the interface between two liquids A and B (A: ink-colored water, B: pure water). It is in good agreement to the simulated flow patterning on the right (CFDRC, ACE+ [9]).

This is in excellent agreement with the theory of centrifugally driven flows. The measured volume flow through a single mixing structure of 0.3 ml / s at 50 Hz can be easily increased up to ~1 ml / s at 120 Hz. The turbulent flow regime (Reynolds number above 2300) is entered at approximately 80 Hz for the presented channel geometry.

The Villermaux reaction scheme [10],[11] has been used to determine the dimensionless micromixedness ratio α representing the quality of mixing. Thereby, the micromixer operated at frequencies from 10 Hz up to 120 Hz. The micromixedness ratio increases from $\alpha = 14$ at a volume throughput of 0.1 ml / s to $\alpha = 807$ at 0.97 ml / s (Fig. 4, right).

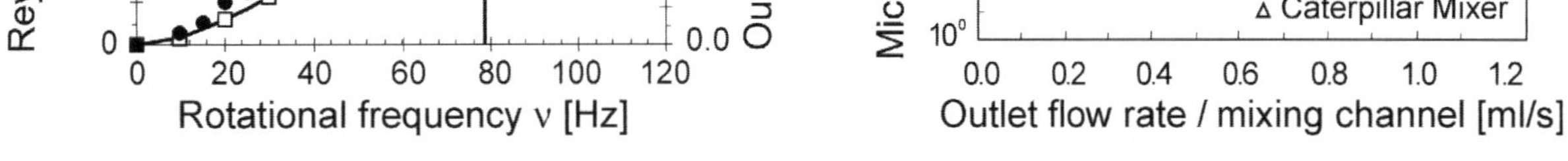

Fig. 4. **Left**: Flow rates per channel over frequency of the centrifugally driven micromixer. The corresponding Reynolds numbers show that the flow conditions enter the turbulent regime above frequencies of approx. 80 Hz.

Right: Experimental determined micromixedness ratio using the Villermaux reaction scheme [10],[11]. Higher micromixedness ratios indicate better mixing.

Besides the outlet flow rate through the mixing channel, also the ratio Φ_1/Φ_2 of the two inlet channel flow rates is an important process parameter. In the centrifugal micromixer, a base value of this ratio is seeded by the geometrical asymmetry of the inlet channels (channel dimensions and radial position). This base ratio is further adjusted dynamically by the sense and the frequency of rotation (Fig. 5). The simulated flow rate ratios between 1.38 and 1.22 at –5 Hz and –35 Hz, respectively, have been experimentally observed by a shift of the A-B interface within the radial mixing channel towards the right channel wall (again ink-colored and pure water served as phase A and B respectively).
The effect is caused by a different scaling of the centrifugal force $\boldsymbol{F}_\nu$, the Coriolis force $\boldsymbol{F}_C$ and the "clashing" force $\boldsymbol{F}_{clash}$ caused by the different kinetic energy (i.e. inertia) of the substreams Φ_1 and Φ_2 over the frequency of rotation (Fig. 5, right). In the clockwise sense of rotation, the Coriolis force $\boldsymbol{F}_C$ increasingly counteracts $\boldsymbol{F}_{clash}$ towards increasing frequencies, thereby "pushing" the A-B interface towards the right side of the channel. The geometry-dependent flow rate ratio can thus be modified "online" by controlling the frequency of rotation.
Inversing the sense of rotation results in a totally different behaviour. The Coriolis force now points into the same direction as $\boldsymbol{F}_{clash}$ supporting the dominance of Φ_1. This leads to an additional mixing effect ("thumb effect"), that occurs over a critical rotational frequency (e.g. ~ 30 Hz at the presented structure). It is caused by a hydrodynamic suppression of substream A, which is most pronounced in the center of the channel due to the parabolic flow profile. Phase A can only escape along the lower and upper channel wall where it experiences the lowest counter force. This way, a jet of the A-phase is vigorously injected into the B-phase, thus inducing an "instantaneous" advective inter-stream mixing right at the junction of the two inlet channels. This effect thus supports the downstream mixing caused by the "Coriolis-stirring".

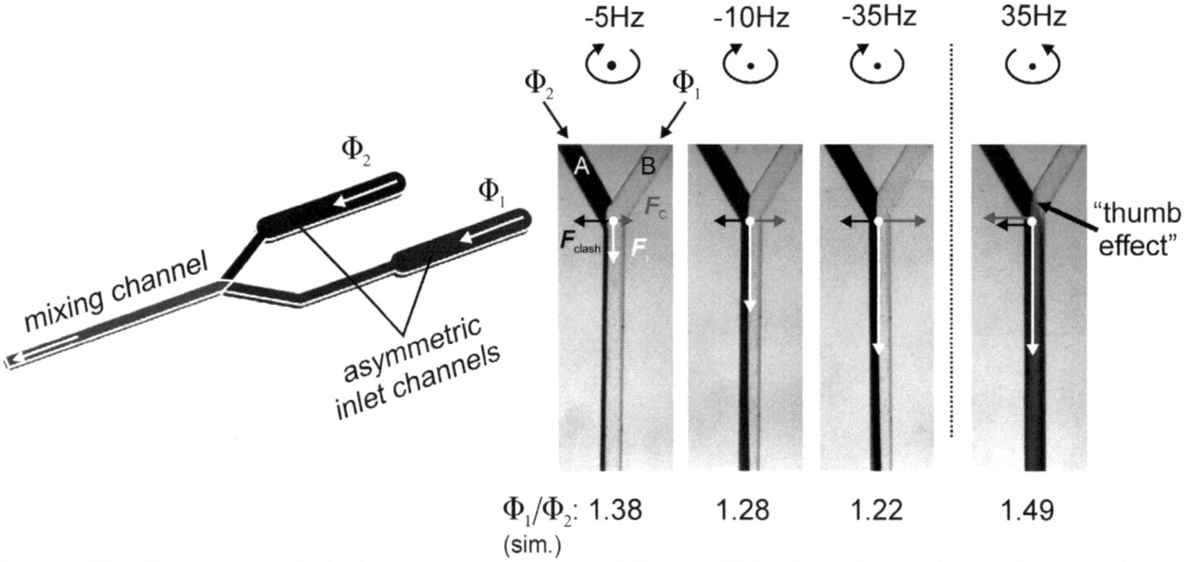

Fig. 5. The flow rate ratio between two concurrent flows within the mixing channel depends on the asymmetry of the inlet channels (static) as well as on the sense and the frequency of rotation (dynamic adjustment). The simulated change in flow rate ratio from 1.38 at -5 Hz to 1.22 at –35 Hz becomes visible by the shift of the A-B interface in the experimental pictures. Changing the sense of rotation leads to an additional mixing effect ("thumb effect").

5. Conclusion

In summary, our novel micromixing platform excels with its robust modular setup composed of a macroscopic actuator unit and a simple microfluidic mixing module featuring passive, low-aspect-ratio structures in a planar arrangement. Mechanism to mix non-uniform flow rates and online process control have been successfully implemented.

The low-cost mixing module can quickly be replaced and disposed to facilitate maintenance and minimize downtimes. Furthermore, the pulse-free centrifugal actuation principle allows high volume throughputs up to 10 liters per minute and mixing module in a generic parallelization scheme paired with an excellent mixing quality.

Acknowledgements

The authors are grateful to the financial support by the "Landesstiftung Baden-Württemberg" and enjoyed an excellent cooperation with HSG-IMIT and the group of Willi Bannwarth at the University of Freiburg.

References

[1] W. Ehrfeld, H. Löwe, V. Hessel, T. Richter, *Chemie Ingenieur Technik*, **69** (7), 931-934 (1997).

[2] V. Hessel, S. Hardt, H. Löwe, Chemical Micro Process Engineering – *Fundamentals, Modelling and Reactions*, Wiley-VCH (2004).

[3] H. Pennemann, P. Watts, S. J. Haswell, V. Hessel, and H. L. Löwe, *Org. Proc. Res. & Dev.*, **8** (3), 422-439 (2004).

[4] N-T. Nguyen, Z. Wu, *J. Micromech. Microeng.*, **15**, R1-R16 (2005).

[5] M. J. Madou and G. J. Kellogg, *Proceedings of SPIE – Systems & Technologies for Clin. Diagn. & Drug. Disc.*, **3259**, 80-93 (1998).

[6] A. Palm, S. R. Wallenborg, M. Gustafsson, A. Hedström, E. Togan-Tekin, and P. Andersson, *Proceedings of μTAS 2001*, eds. M. Ramsey, A. van den Berg, 216-218 (2001).

[7] J. Ducrée, T. Brenner, T. Glatzel, and R. Zengerle, *Proceedings of μTAS 2003*, eds. M. A. Northrup, K. F. Jensen, D. J. Harrison, 603-606 (2003).

[8] J. Ducrée, H-P. Schlosser, T. Glatzel, S. Haeberle, M. Grumann, T. Brenner, R. Zengerle, *Proceedings of μTAS 2004*, eds. T. Laurell, J. Nilsson, K. Jensen, D. J. Harrison, J. P. Kutter, 554-556 (2004).

[9] CFD Research Corporation, http:// www.cfdrc.com (accessed: March 2005)

[10] M.-C. Fournier, L. Falk, and J. Villermaux , *Chem. Eng. Sci.*, **51** (22), 5053-5064 (1996).

[11] S. Panic, S. Loebbecke, T. Tuercke, J. Antes, D. Boskovic, *Chem. Eng. Journal*, **101** (1-3), 409-419 (2004).

Experimental Benchmark of a Metallic Micro-Separator/Classifier Compared with Representative Hydrocyclone

Shinichi Ookawara, Nobuo Oozeki, Kohei Ogawa
Department of Chemical Engineering, Graduate School of Science & Engineering, Tokyo Institute of Technology, 2-12-1, O-okayama, Meguro-ku, Tokyo 152-8552, Japan

Abstract

Reduced grade efficiency curves of a micro-separator/classifier, whose principle was proposed at IMRET7 (Ookawara, et al., 2003, 2004a), are experimentally evaluated and compared with representative curves of classical hydrocyclones which are readily available in the literature. The main part of the device is a curved microchannel with a rectangular cross-section, which bifurcates at one end for separation/classification. For practical usage in industry, the metallic device is fabricated by means of photolithography and diffusion bonds, which enable one to achieve anti- high pressure/high temperature/chemicals, to mount many units operating in parallel, and to miniaturize the complete unit. The curved slit with a width of 200 microns and curvature radius of 20 mm is formed by photoetching both sides of an SUS304 plate with thickness 150 microns (presented by Dai Nippon Printing, Co., Ltd.). In most experiments, the plate is inserted between acrylic resin plates to create the channel with rectangular cross-section (200 × 150 microns). In addition, SUS304 plates with and without the slits are stacked in turns and diffusion-bonded. Since the numerical study predicts that the concentration profiles over the cross-section quickly converge in the entry part of the curved channel for the larger particles to be separated (Ookawara et al., 2005a), two arcs with 30 and 180 degrees are experimentally examined in this study. The examined suspensions consist of deionized water and acrylic monodisperse particles with density 1,190 kg/m^3, whose volume concentration is 0.05 vol%. The diameters are 1.8, 5, 10, 15, 20 and 30 microns, respectively. Slurry discharged from the outer branch of the microdevice mostly contains the larger particles to be separated corresponding to cyclone-separator underflow, while samples from the inner branch correspond to the overflow from the cyclone. Based on the similarity, a benchmark of the newly developed metallic micro-separator/classifier is performed by means of reduced grade efficiency curves usually used for the evaluation of cyclones. It is assumed that particle recovery reduced with liquid recovery to outer branch for each monodisperse particle could give the efficiency curve at a given flow rate condition of the feed. The examined feed flow rates are in the *Re* number range of 300-1500 (the *De* number range of 20-100). It is experimentally confirmed that both arcs with 30 and 180 degrees show superior performance comparable to that of a representative hydrocyclone. The energy consumption of the microdevice will be compared with that of hydrocyclone as well.

Introduction

The feasibility of a micro-separator/classifier for a solid-liquid system, whose main section is a curved rectangular microchannel, was experimentally and numerically confirmed in previous studies (Ookawara et al., 2003, 2004a, b, c). Those studies indicated that phase separation was caused by hydraulic interaction between entrainment by Dean vortices and the lift forces acting on the particles in the device. While the separation mechanism is being gradually understood, the performance has not been quantitatively evaluated from a practical standpoint. Of course, any micro device at least has advantages that it can operate with small amount of samples and the amount can be varied with constant performance by operating many devices in parallel; commonly denoted as the numbering-up methodology in the literature. For the novel micro- separator/classifier to be substituted for classical large devices and to gain popularity in industry it should have advantages of superior performance compared with other classical devices. For a separator, the particle size that can be separated and a required power for the particle size to be separated should be specified as the performance metrics. On the other hand, sharpness of the separation is an important factor for the classifier. It is also important, to decrease the fabrication costs, to achieve anti- high pressure/high temperature/chemicals, to mount many units operating in parallel, and to miniaturize the complete unit. Therefore, instead of acrylic microchannels mechanically formed in previous studies, a metallic micro-separator/classifier is fabricated by means of photolithography and diffusion-bonded to meet practical demands. Furthermore, in the present study those indices for the new metallic device are evaluated and compared with those of representative hydrocyclones that are available in the literature.

Experiments and analysis

As shown in Fig. 1 with a scale of 0.7, the curved slit with a width of 200 microns and curvature radius of 20 mm is formed by photoetching both sides of an SUS304 plate with a thickness of 150 microns (presented by Dai Nippon Printing, Co., Ltd.). Small holes connected to slit ends form through-holes for inlet and outlets of the microchannel. Large holes are also formed to bolt acrylic plates on both sides of an SUS plate (or diffusion-bonded SUS plates). Every hole of the SUS plate(s) and acrylic plates is positioned coaxially. Slurry is fed into the inlet and discharged from the outlets of either the outer or the inner branches. Since the previous numerical study predicts that concentration profiles over the cross-section converge quickly in the entry of the curved channel for the larger particles to be separated (Ookawara et al., 2005a), two arcs with (a) 30 and (b) 180 degrees are also examined experimentally in this study. An SUS plate with a slit is inserted between acrylic plates to create a single channel with rectangular cross-section (200 × 150 microns). A seal plate is additionally inserted between an SUS plate and acrylic plate in the experiments. These devices are named arc30-single and arc180-single, respectively. In addition, SUS304 plates with and without the slits are stacked in turn and diffusion-bonded to examine the numbering-up device with three independent channels working in parallel, but only for the arc with 180 degrees (arc180-triple).

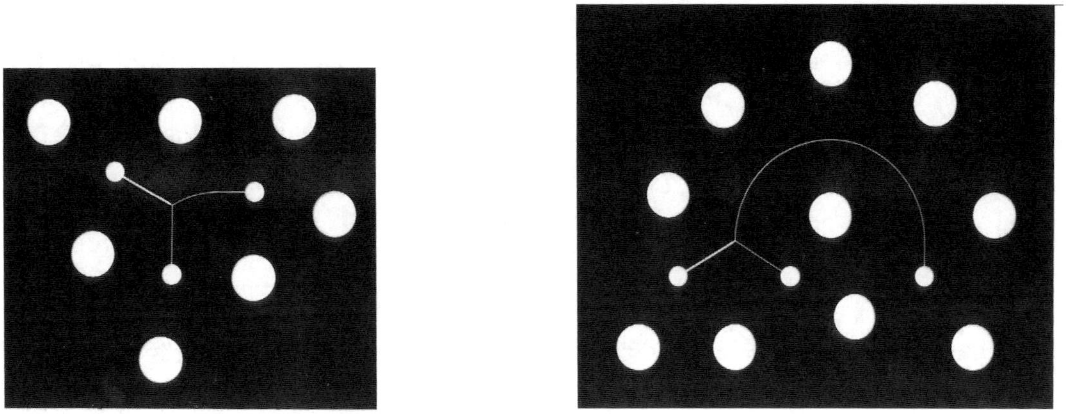

(a) Arc slit with 30 degrees for arc30-single (b) Arc slit with 180 degrees for arc180-single&triple

Fig. 1 SUS304 plates with micro slits.

The examined suspensions consist of deionized water and acrylic monodisperse particles with density 1,190 kg/m^3, whose volume concentration is 0.05 vol%. The diameters are 1.8, 5, 10, 15, 20 and 30 microns, respectively. The device is operated at Dean numbers in the range of 20 to 100 (Reynolds number range of 300-1500). In this study, Dean and Reynolds numbers are respectively defined as;

$$De = Re\sqrt{\frac{D_h}{2R}} \quad (1)$$

$$Re = \frac{UD_h}{\nu} \tag{2}$$

where U, R, ν and D_h are the mean velocity, curvature radius, kinematic viscosity and hydraulic diameter, respectively.

Slurry discharged from the outer branch of the microdevice mostly contains the larger particles to be separated corresponding to cyclone-separator underflow, while samples from the inner branch correspond to the overflow of cyclone. Based on the similarity, a benchmark of the newly developed metallic micro-separator/classifier is performed by means of reduced grade efficiency curves usually used for the evaluation of cyclones. It is assumed that particle recovery reduced with liquid recovery to outer branch for each monodisperse particle could give the efficiency curve at a given flow rate condition of the feed. By weighing samples obtained from the inner and outer branches in a given time W_{in} and W_{out}, liquid recovery to the outer branch can be calculated as follows since the slurry is very dilute.

$$R_f = \frac{W_{out}}{W_{in} + W_{out}} \tag{3}$$

The absorbance of the samples I_{in} and I_{out} are measured to determine the particle recovery to outer branch R as well.

$$R = \frac{I_{out} W_{out}}{I_{in} W_{in} + I_{out} W_{out}} \tag{4}$$

The reduced grade efficiency R_c is calculated for each monodisperse particle as follows.

$$R_c = \frac{R - R_f}{1 - R_f} \tag{5}$$

It is often the case that the reduced grade efficiency of a particular cyclone is fitted by a generalized expression and the parameter value for the classification sharpness is compared with that of a representative cyclone, despite the particle size and density, throughput and the flow rate ratio of under to over flow. For instance, Coelho and Medronho (2001) evaluated the sharpness of classification by means of the following expression.

$$R_c = \frac{\exp(\alpha d / d'_{50}) - 1}{\exp(\alpha d / d'_{50}) + \exp(\alpha) - 2} \tag{6}$$

where d and d'_{50} are the particle diameter and the reduced cut size, respectively. When the value of α is large, the classification is sharp. Coelho and Medronho cited the work of Rietema (1961) as a classical work and made the comparison in terms of α. The present work also determines the α of the micro device and compares it with that of the Rietema-type cyclone given by Coelho and Medronho. On the other hand, the work of Kelsall (1953) is cited in this study for the comparison of the required energy to separate particles with a particular size in terms of pressure loss. Since the work of Kelsall was also cited in the above works (Rietema,1961; Coelho and Medronho, 2001), his apparatus can also be regarded as one of representative cyclones. It should be noted that Kelsall examined cyclones by using a slurry of acrylic particles in water, whose diameters and density (specific gravity is 1.18) are similar with those of the present study.

Results and discussion

Fig. 2 shows a typical example of the reduced grade efficiencies at De = 45, which are created by curve-fitting the relation between De and R_c of each particle diameter. The solid lines indicate the curve fitting based on Eq. (6). The α values of arc180-single&triple and arc30-single are listed in Table 1. It is noted that the selected *De* condition for the comparison gives stable results for all examined device. One of the devices shows extremely good performance, but another device does not in the higher *De* range. It can be clearly seen that classification of arc180-single and –triple is much sharper than that of arc30-single. Further, it is found that even arc30-single can give better classification compared with the Rietema-type cyclone.

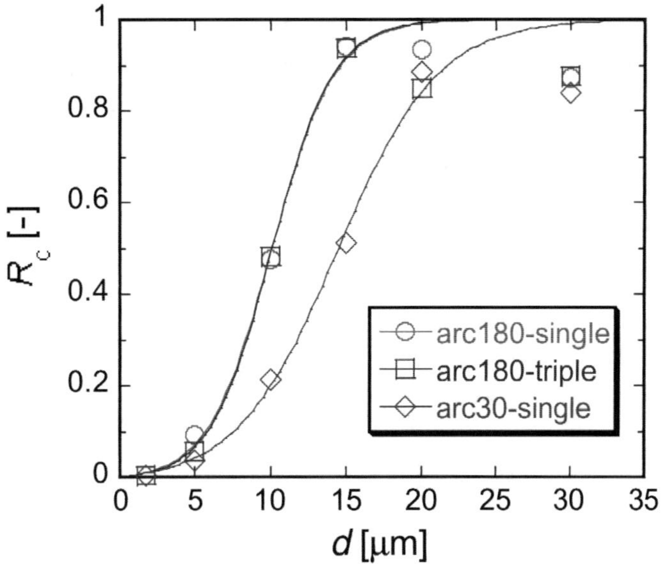

Fig. 2 Reduced grade efficiency at $De = 45$.

Table 1 Comparison of α values

Device	α
arc180-single	4.89
arc180-triple	5.05
arc30-single	4.40
Rietema-type	4.23

Table 2 Comparisons of d'_{80} and necessary power.

Device	d'_{80} [microns]	Pressure loss [kPa] in curved section	Feed pressure [kPa]
arc180-single	13	198	-
arc180-triple	13	198	-
arc30-single	19	33	-
Kelsall-type	21	-	276

In this study, a diameter of particle separated to 80 % in reduced grade efficiency, d'_{80}, and pressure loss to achieve the separation are compared with those of the Kelsall-type cyclone. The run that gives a smallest d'_{80} among a series of experiments by Kelsall is selected for the comparison. For the micro-separator/classifier, the increase of friction factor due to the appearance of secondary flow patterns in the curved section can be estimated as about 25 % compared with a straight channel with same dimensions based on the work of Cheng (1976). In this study, pressure losses in the curved section are estimated as listed in Table 2 with the feed pressure specified by Kelsall. It is remarkable that the pressure loss of arc30-single is approximately 12 % of the feed pressure for the Kelsall-type cyclone to obtain similar d'_{80}. Even arc180-single&triple, whose d'_{80} is much smaller, need a power that is smaller than that of the Kelsall-type. In addition, the feed pressure of the present micro device will not be so different from the pressure loss in the curved section because it is possible to further reduce the length of straight feed section. Due to the notable energy-saving effect, the novel micro-separator/classifier is expected to be substituted for the hydrocyclone broadly in industry.

Conclusions

It is shown in this study that the novel micro-separator/classifier consisting of a curved microchannel has many superior features such as sharp classification, small cut size and considerable energy-saving effects. The dependence of these performance indices on the operational condition and geometrical design will be examined in future work.

Acknowledgements

The authors are most grateful to Professor Lars Nyström and Dr. David Street for their valuable advice on the present study.

References

Cheng, K. C., Lin, R. C., Ou, J. W., 1976. Fully developed laminar flow in curved rectangular channels. Transactions of the ASME. J. Fluids Eng. March, 41-48.

Coelho, M. A. Z., Medronho, R. A., 2001. A model for performance prediction of hydorcyclones. Chem. Eng. J. 84, 7-14.

Kelsall, D. F., 1953. A further study of the hydraulic cyclone. Chem. Eng. Sci. 2, 254-272.

Ookawara, S., Higashi, R., Street, D., Ogawa, K., 2003. Feasibility study on concentration of slurry and classification of contained articles by microchannel. Book of Abstracts of 7th International Conference on Microreaction Technology, 231-233.

Ookawara, S., Higashi, R., Street, D., Ogawa, K., 2004a. Feasibility study on concentration of slurry and classification of contained articles by microchannel. Chem. Eng. J. 101, 171-178. (Special issue for IMRET7)

Ookawara, S., Higashi, R., Street, D., Ogawa, K., 2004b. The Influence of channel depth on the performance of a micro-separator/classifier. Kagaku Kougaku Ronbunshu 30, 135-141.

Ookawara, S., Street, D., Ogawa, K., 2004c. A practical application of the Euler-granular model to a micro-separator/classifier. Proceedings of 5th International Conference on Multiphase Flow, CD-ROM, #206.

Ookawara, S., Street, D., Ogawa, K., 2005a. A Numerical Modeling of Particle Concentration Profiles in a Curved Microchannel. Int. J. Multiphase Flow, submitted.

Rietema, K., 1961. Performance and design of hydrocyclones-I, II, III, IV. Chem. Eng. Sci. 15, 298-325.

Pinched Flow Fractionation: A New Method for Continuous Particle Classification Using Microfabricated Structures

*Masumi Yamada[1, 2], Yuushi Sai[1], Junya Takagi[1], Hirosuke Maenaka[1], Masahiro Yasuda[1] and Minoru Seki[1]**

[1]Department of Chemical Engineering, Graduate School of Engineering, Osaka Prefecture University, 1-1 Gakuen-cho, Sakai, Osaka 599-8531, Japan

[2]Department of Chemistry and Biotechnology, School of Engineering, The University of Tokyo, 7-3-1 Hongo, Bunkyo-ku, Tokyo 113-8656, Japan

**Tel: +81-72-254-9296, Fax: +81-72-254-9911, E-mail: seki@chemeng.osakafu-u.ac.jp*

A concept of continuous classification of particles using microfabricated channels, named pinched flow fractionation (PFF) has been proposed and demonstrated as one of the micro unit operations. Accurate size separation of particles is one of the most important technologies in the fields of chemical process, analytical measurement, and chemical or biological research. There have been proposed many methods concerning particle separation in micro-scale, using field flow fractionation, hydrodynamic chromatography, capillary hydrodynamic fractionation, or split-flow thin fractionation. However, these methods have their merits and demerits. Our proposed method utilizes only the laminar flow profile in specific microchannel geometry, not necessitating any kinds of outer field.

Principle of the pinched flow fractionation is shown in Figure 1. In this method, particles suspended in liquid were continuously introduced into a microchannel having a pinched segment, and were aligned to one sidewall in the pinched segment by another liquid flow without particles. The particles were then separated perpendicularly to the flow direction according to their sizes by the spreading flow profile inside the microchannel.

Microdevices were fabricated using usual rapid prototyping and replica molding methods, and the material was PDMS. The various shapes and sizes of micro-channels were used for particle separation to clarify the effects on the separation efficiency of the proposed method. Also, various kinds of particles, including polymer beads with micro/nano meter in diameter, emulsions, and biological cells, were tested and successfully separated with this method. In addition, separated particles were collected independently by making branches at the end of the pinched segment.

Since this method utilizes only the laminar flow profile inside a microchannel, complicated outer field control could be eliminated, which is usually required for other kinds of particle separation methods such as field flow fractionation. Therefore, this method is applicable for continuous preparation of monodispersed particles in microscale systems, since the separation can be rapidly performed in a simple structure.

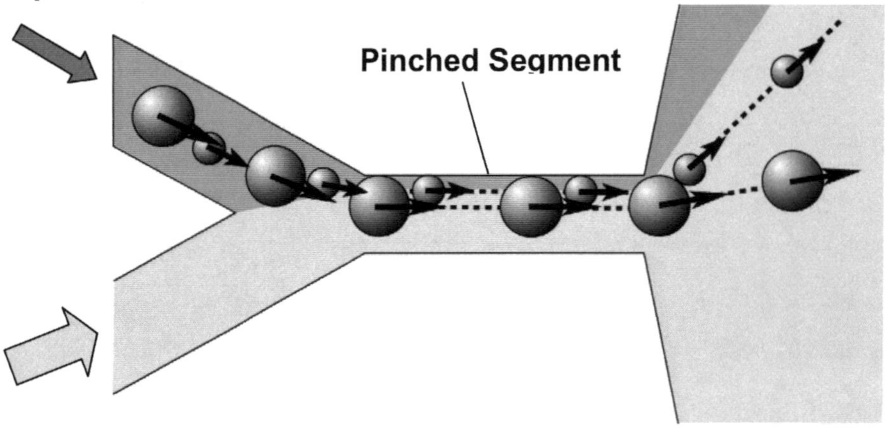

Figure 1. Principle of pinched flow fractionation: In the pinched segment, particles are aligned to one sidewall regardless of their sizes by controlling the flow rates from two inlets. Next, particles are separated according to their sizes by the spreading flow profile at the boundary of the pinched and the broadened segments. The liquid containing particles is dark-colored.

ACKNOWLEDGEMENT

This research was supported in part by Grant-in-Aids for JSPS Fellows and Priority Areas (A) (No. 13025216) from the Ministry of Education, Science, Sports and Culture of Japan, by the Nanotechnology Project of the Ministry of Agriculture, Forestry and Fisheries of Japan, and by the Research Association of Micro Chemical Process Technology, Japan.

REFERENCES

Yamada, M., M. Nakashima, and M. Seki (2004) Pinched flow fractionation: continuous size separation of particles utilizing a laminar flow profile in a pinched microchannel. *Anal. Chem.*, 76: 5465-5471.

130a Hydrogen Production from Ammonia Decomposition: Hierarchical, Multiscale Microkinetic Modeling and Microreactor Simulation

Ashish Mhadeshwar, Dionisios Vlachos, Soumitra Deshmukh

Introduction

Ammonia synthesis is one of the most extensively studied catalytic processes due to its use in the manufacture of agricultural fertilizers. Ammonia decomposition chemistry, on the other hand, has only recently become of interest due to the desire to produce hydrogen for fuel cells. In contrast to reforming processes, ammonia decomposition has the advantage of eliminating the downstream processing of CO removal. Ruthenium (Ru) based catalysts have been studied due to their higher activity at lower H2/N2 ratios and lower tendency to poisoning by NH3 compared to reduced Iron [1,2]. Ru appears to be the best single metal catalyst [3,4].

In this contribution, first we study the mechanism of NH3 decomposition on Ru. Quantum mechanical density functional theory (DFT) calculations on Ru(0001) are presented to estimate the N-N and N-H interactions. Subsequently, a microkinetic model is developed and parameter optimization is carried out. The important reactions are elucidated, and comparison of computational results with experimental data from a microreactor is presented.

Model development

DFT calculations are carried out on (2x2) repeated surface cells of 3 layers thick Ru(0001) slabs, with all the 3 layers being relaxed. A plane wave basis set with an energy cutoff of 350 eV is employed with 18-k Chadicohen point mesh. The PW91 exchange-correlation functional is applied and the Vanderbilt ultrasoft pseudopotential [5] is used to describe the core electrons. The DACAPO [6] code has been employed.

Ammonia chemistry is modeled with a network of six elementary-like, reversible reactions. Parameter estimation follows the approach outlined in [7,8]. In brief, the orders of magnitude of the pre-exponential factors are first assigned using transition state theory (TST). The heats of chemisorption of N and H on a bare surface are obtained from experimental data reported and are slightly modified to better describe TPD data. The N-N interactions are taken from the DFT calculations. Heats of chemisorption of NHx species are calculated using the semi-empirical unity bond index-quadratic exponential potential (UBI-QEP) theory [9,10]. UBI-QEP is then used to calculate the activation energies using the heats of chemisorption as input. The temperature dependence of heats of chemisorption is incorporated using statistical mechanics. The initial estimates of the pre-exponentials in each step of the mechanism are the only parameters optimized to capture the NH3 decomposition data of [11]. The parameters of the mechanism are thermodynamically consistent at the free energy level at all temperatures and have been validated against different sets of experimental data.

Analysis of the dominant reactions paths is carried out using reaction path analysis (RPA) and sensitivity analysis (SA). Chemistry reduction is also performed. The reduced model is incorporated in computational fluid dynamics (CFD) simulations, and the results are compared to experimental data from a microreactor.

Results and Discussion

Based on the results of the microkinetic model with UBI-QEP predicted activation energies, the rate-determining step (RDS) in NH3 decomposition is the N-H bond cleavage in NH2*, whereas the associative desorption of nitrogen contributes to the overall rate to a much less extent. Under typical decomposition conditions, H* followed by N* are the most abundant reactive intermediates (MARI). Overall, our analysis shows that N-N interactions considerably alter the identity of the MARI and the RDS in NH3 decomposition chemistry, giving rise to different conclusions about the kinetically significant steps from most previously published work. While NH3 decomposition data in fixed bed reactors can adequately be described using mechanisms that lack adsorbate-adsorbate interactions, the latter are essential to capture the N2-TPD data, the submonolayer saturation N*-coverage, and the physics of the important NH3/Ru chemistry.

Armed with the developed microkinetic model, the ammonia decomposition for hydrogen production in a newly fabricated, aluminum framework post microreactor is modeled. A PFR model is used as a low level hierarchical tool for developing a reduced rate expression using a computer-aided methodology. This reduced one-step expression is used in CFD simulations of a microreactor, and good agreement with experimental data is observed. It is shown that the posts play an important role in enhancing the transverse mass transfer and providing high catalyst surface area. Interesting flow patterns and back diffusion occur at lower flow rates. Finally, parametric studies are performed and a trade off between pressure drop and conversion is observed under certain conditions.

References

[1] W. Tsai, W. H. Weinberg, J. Phys. Chem. 91 (1987) 5302.

[2] S. Dahl, P. A. Taylor, E. Tornquist, I. Chorkendorff, J. Catal. 178 (1998) 679.

[3] C. J. H. Jacobsen, S. Dahl, B. S. Clausen, S. Bahn, A. Logadottir, J. K. Norskov, J. Am. Chem. Soc. 123 (2001) 8404.

[4] J. C. Ganley, E. G. Seebauer, R. I. Masel, (2004) in preparation.

[5] D. Vanderbilt, Phys. Rev. B 41 (1990) 7892.

[6] B. Hammer, O. H. Nielsen, J. J. Mortensen, L. Bengtsson, L. B. Hansen, A. C. E. Madsen, Y. Morikawa, T. Bligaard, A. Christensen, DACAPO version 2.7 (CAMP, Technical University, Denmark)

[7] Y. K. Park, P. Aghalayam, D. G. Vlachos, J. Phys. Chem. A 103 (1999) 8101.

[8] P. Aghalayam, Y. K. Park, D. G. Vlachos, AIChE J. 46 (2000) 2017.

[9] E. Shustorovich, H. Sellers, Surf. Sci. Rep. 31 (1998) 1.

[10] E. Shustorovich, Adv. Catal. 37 (1990) 101.

[11] T. V. Choudhary, C. Sivadinarayana, D. W. Goodman, Cat. Lett. 72 (2001) 197.

130b Effects of Heat Transfer and Flame Stretch in Microcombustors

Craig M Miesse, Rich Masel, Shaurya Prakash

Sub-millimeter microcombustors have been proposed for a variety of purposes, including heat sources for microengines and small fuel converters for portable power production, micro scale detectors, heaters for chip scale gas chromatographs for environmental testing, and chemical and biological weapons defense.

Currently, flame dynamics in sub-millimeter channels are not well defined. Under conventional conditions, a fuel/air flame cannot propagate in a cavity narrower than a millimeter or two. Though our recent work shows that gas-phase combustion can occur in sub-millimeter channels, the interaction between heat and mass transfer at this length scale, and the resulting flame characteristics, flame structure, and stability criteria are still unclear. Further, a major challenge of micro-combustor device integration is to keep an optimum balance between sustaining combustion and maximizing the heat output.

Previously we have reported that the homogeneous combustion that occurs in sub-millimeter burners can exhibit flame structures that are entirely different than those seen at a larger scale. Instead of observing a continuous flame, a laminar diffusion flame splits into a series of periodically spaced, robust, nearly spherical flame cells. Other phenomena, such as periodic pulsation and patterned splitting of individual flame balls, are observed with fuel rich mixtures. Varying the contacting pattern of the fuel and oxidant has also been shown to play a pivotal role in flame structure, burner temperature profile, and burner efficiency.

In recent work, the structure and dynamics of flames in micro scale burners are examined in further detail. The effects of varying the burner fuel and adding components to both the fuel and oxidant to alter their kinetic and transport properties have been characterized. Novel manipulation of the flow patterns have been explored. Finally, the effects of changing the thermal environment via manipulation of burner heat losses and temperature distribution have been elucidated.

In this talk we present experimental results that show the effects of flame stretch rate and that demonstrate the interaction of heat and mass transfer in determining structure and stability of laminar fuel/oxidant diffusion flames in sub-millimeter scale microcombustors. Our intent is to describe the fundamental determining factors for flame structure and stability for homogeneous phase microcombustion and to provide design guidelines for burner integration to MPG devices as heat sources.

Quantitative Prediction of Separation Efficiency of a Micro-Separator/Classifier by an Euler-Granular Model

Shinichi Ookawara[1], David Street[2], Kohei Ogawa[1]

[1]Department of Chemical Engineering, Graduate School of Science & Engineering, Tokyo Institute of Technology, 2-12-1, O-okayama, Meguro-ku, Tokyo 152-8552, Japan

[2]Fluent Asia Pacific, Nittochi Nishishinjuku Bldg. 18F, 6-10-1, Nishishinjuku, Shinjuku-ku, Tokyo 160-0023, Japan

Abstract

The feasibility of a newly developed micro-separator/classifier was experimentally shown at IMRET7 (Ookawara et al., 2003, 2004a). In this study the separation efficiency of the microdevice using an Euler-granular model is quantitatively validated based on experimental benchmark tests (Ookawara et al., 2005b). The computational domain for the simulation has the same geometry as that of the experimental device, which is a semicircular microchannel with width, depth and curvature radius of 200 microns, 150 microns and 20 mm, respectively. The end of the semicircular sections separates to form two collection branches. These collection branches, which are present in the experimental apparatus, are not modeled computationally and instead the semicircular section is connected to straight channels at both ends. Six monodisperse particles with density 1,190 kg/m^3 and diameters 1.8, 5, 10, 15, 20 and 30 microns respectively are examined. The uniform volume fraction is specified as 0.0005 at the inlet of the straight channel where a uniform velocity profile of fluid is also specified for the Reynolds numbers (*Re*) given used in the experiments. The Euler-granular model, which treats both particles and fluid as two interpenetrating continuous phases, implemented in the commercial CFD code FLUENT 6.1 is applied for the coupled particle and fluid flow modeling. Cross sectional planes used for can be formed within the curved microchannel at particular angular positions. The mathematical model enables one to evaluate the volume fraction, velocity profiles of both phases and therefore the mass flow rates of each phase passing through the plane. The plane under consideration is also divided into an inner section and an outer section. The outer section corresponds to the region of the fluid that would exit through the outer exit branch of the experimental device if the exit bifurcation were placed at this position. The inner section corresponds to the region of the fluid that would exit through the inner branch. Thus the mass flow rate ratio of the outer region to the total region of the particulate phase is calculated as the particle recovery and it is reduced with the recovery of the liquid phase. This value is known as the reduced grade efficiency. Plots for several particle sizes constitute a reduced grade efficiency curve as is commonly utilized for the evaluation of a cyclone separator. It will be shown that proper coefficient values of a lift force model exist, which give accurate numerical predictions. Since there seems to be no lift force model validated in the very high shear rate range expected in the micro device so far, the coefficients are based on the comparison between the efficiencies obtained experimentally and numerically. The model is expected to be useful for the further optimization of the device.

Introduction

The micro-separator/classifier is a novel micro device whose main component is a curved microchannel with a rectangular cross section. The semicircular channel with curvature radius of 20 mm is connected to a straight inlet channel at the upstream end and bifurcates to form two exit branches at the downstream end in order to separate/classify and collect particles contained in the slurry. The channel width and depth typically fall into the range of 100 to 400 microns. The feasibility as a micro device for particle separation/classification was experimentally confirmed by using polydisperse acrylic particles whose average diameters are 7, 10 and 20 microns in previous studies (Ookawara et al., 2004a, b).

In the curved microchannel, even moderate laminar flow conditions result in high streamwise velocity and consequently high centrifugal force acting on the particles in the slurry. However, the centrifugal force cannot be the main factor that concentrates particles near the outer channel wall (Ookawara et al., 2004c, 2005a). In the curved channel, the laminar secondary flow pattern known as Dean vortices are formed over the cross sectional plane. Since the circulating velocity of Dean vortices is much larger than that of the particle slip velocity caused by the centrifugal force, the particles may become entrained and dispersed over the channel cross section by the secondary flow. It was also shown that lift forces that occur near the walls of the microchannels must be taken into consideration in order to model phase separation in the channel. In those numerical studies, the lift force model proposed by Drew and Lahey (1993), which assumes inviscid liquid as the continuous phase, was adopted. Since the experimental conditions of our research always fall into moderate laminar flow range, however, the sign of the formula was altered so that the direction of the acting force agreed with that of Saffman's lift force model, which assumes very viscous liquid as the continuous phase (Saffman, 1965). It was shown that the Drew and Lahey model with the opposite sign successfully predicted the phase separation in the cross sectional plane of the curved microchannel. The previous studies were very useful to understand the phase separation mechanism in the device and how the geometrical and operational parameters influence the separation efficiency. To the authors' knowledge, however, no lift force model has been quantitatively validated in the extremely high shear rate ranges that can be seen in the curved microchannel. In order to optimize the geometrical configuration and operating conditions of the device the validation of the lift force model proposed in previous studies, is carried out especially in terms of the magnitude to predict accurate separation efficiency.

In the present study, therefore, the separation efficiencies are evaluated at the downstream end of the curved section using an Euler-granular model, which treats both fluid and particles as continuous phases. The particulate phase is specified as spherical monodisperse particle. The coefficient of the lift force model was varied in the range of 0 (without lift force) to 0.5 (the values adopted in previous studies). The results are compared with newly obtained experimental results (Ookawara et al, 2005b). It will be shown that there exists an appropriate coefficient value to give the separation efficiency that agrees with experimental results.

Numerical methods

In this study, a commercial CFD code FLUENT 6.1, which is based on finite volume method, is employed to perform the Euler-granular multiphase calculation. The Euler-granular model treats the liquid phase and particles as continuous phases. This multiphase model uses a two fluids approach and takes into consideration the granular nature of the solid secondary phase.

Fig. 1 schematically shows the numerical geometry. The width and depth of the rectangular channels are 200 microns and 150 microns, respectively. The curvature of the main semicircular channel is 20 mm. These sizes are identical to the dimensions of the microchannel examined in experiments (Ookawara et al., 2005b). Although the actual shape of the cross section is not exactly rectangular, due to the photo-etching procedure use for the device fabrication, it is assumed that any slight deformations of the shape do not affect the predictions of separation efficiency. The bifurcation at the downstream end of semicircular section is not modeled mathematically and both the ends are connected to straight sub-channels with a length of 5 mm. This geometrical modeling makes it possible to examine the effect of the flow rate ratio leaving from the outer and inner branches as discussed later in this study. The meshes are generated at 10 microns/mesh cell over the cross sectional plane and at about 50 microns/mesh cell in the mainstream direction. The geometry contains 437,100 computational fluid cells. The density ρ_f and viscosity μ_f of water, defined as the primary phase or continuous phase are 995.65 kg/m^3 and 0.7973 mPa.s, respectively. The density ρ_p of the secondary phase, or dispersed phase, is specified as 1,190 kg/m^3. Monodisperse spherical particles of diameters d_p are 1.8, 5, 10, 15, 20 and 30 microns are examined in each numerical simulation. The properties of the liquid and particle phases are also the same as the experimental conditions.

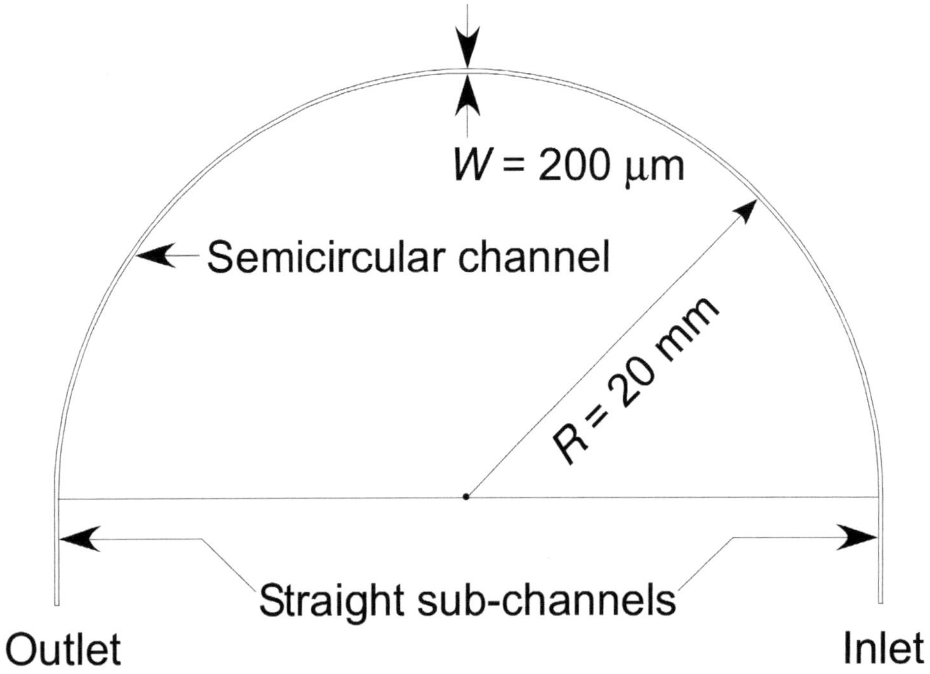

Fig. 1 A schematic diagram of the numerical geometry used for simulation.

Steady flow condition and incompressibility of both phases can reduce the governing equations to be solved as follows.

Conservation of mass:

Fluid phase: $\nabla \cdot (\alpha_f \mathbf{u}) = 0$ (1)

Particulate phase: $\nabla \cdot (\alpha_p \mathbf{v}) = 0$ (2)

where α_f and α_p denote the volume fractions of fluid and particulate phases, and **u** and **v** are velocities of fluid and particle, respectively.

Conservation of momentum:

Fluid phase:

$$-\nabla \cdot (\alpha_f \mathbf{uu}) - \frac{\alpha_f}{\rho_f}\nabla p + \frac{1}{\rho_f}\nabla \cdot \underline{\underline{\tau_f}} + \alpha_f \mathbf{g} + \frac{K}{\rho_f}(\mathbf{v}-\mathbf{u}) + \alpha_f \mathbf{F}_{\mathrm{Lift,f}} = 0 \quad (3)$$

Particulate phase:

$$-\nabla \cdot (\alpha_p \mathbf{vv}) - \frac{\alpha_p}{\rho_p}\nabla p - \frac{1}{\rho_p}\nabla p_p + \frac{1}{\rho_p}\nabla \cdot \underline{\underline{\tau_p}} + \alpha_p \mathbf{g} + \frac{K}{\rho_p}(\mathbf{u}-\mathbf{v}) + \alpha_p \mathbf{F}_{\mathrm{Lift,p}} = 0 \quad (4)$$

where $\underline{\underline{\tau_f}}$ and $\underline{\underline{\tau_p}}$ are the stress-strain tensors of fluid and particle.

$$\underline{\underline{\tau_f}} = \alpha_f \mu_f (\nabla \mathbf{u} + \nabla \mathbf{u}^T) \quad (5)$$

$$\underline{\underline{\tau_p}} = \alpha_p \mu_p (\nabla \mathbf{v} + \nabla \mathbf{v}^T) \quad (6)$$

Here μ_f and μ_p represent shear viscosities of fluid and particulate phases. The gravity is specified as 9.8 m/s^2 in the perpendicular direction to the plane of the semicircle. The lift force acting on the particle in fluid $\mathbf{F}_{\mathrm{Lift,p}}$ is computed as follows.

$$F_{\mathrm{Lift,p}} = k_{\mathrm{Lift}} \rho_f \alpha_p (\mathbf{u}-\mathbf{v}) \times (\nabla \times \mathbf{u}) \qquad (0 \le k_{\mathrm{Lift}} \le 0.5) \tag{7}$$

The equation is derived assuming an inviscid continuous phase, but originally with opposite sign as k_{Lift} = -0.5 (Drew and Lahey, 1993). Since the flow condition in our study always falls into moderate laminar flow regime, the sign was altered, but with constant magnitude, viz., k_{Lift} = 0.5 so that the acting direction becomes the same as that of Saffman's lift force (1965), which is derived by assuming a very viscous liquid as the continuous phase. It was confirmed that the Drew-Lahey lift force model with k_{Lift} = 0.5 predicted particle concentration profiles at the end of semicircular section that qualitatively agree with experimental results (Ookawara et al., 2004c, 2005a). Further, to predict separation efficiency of the device with quantitative accuracy, k_{Lift} is systemically varied as 0 (without lift force), 0.125, 0.25 and 0.5 in this study. The fluid always experiences the force of the same magnitude, but in the opposite direction as

$$\mathbf{F}_{\mathrm{Lift,f}} = -\mathbf{F}_{\mathrm{Lift,p}} \tag{8}$$

Generally, the exchange coefficient K in equations (3) and (4) for the momentum exchange between fluid and solid phases is expressed as follows.

$$K = \frac{\alpha_p \rho_p f}{\tau} \tag{9}$$

Here the particle relaxation time τ is defined as

$$\tau = \frac{\rho_p d_p^{\ 2}}{18\mu_f} \tag{10}$$

For the drag function f, the Syamlal-O'brien model (Syamlal and O'brien, 1989) is adopted as follows.

$$f = \frac{C_D Re_p \alpha_f}{24 v_{r,p}^{\ 2}} \tag{11}$$

where the drag function has the following form (Dalla Valle, 1948).

$$C_D = \left(0.63 + \frac{4.8}{\sqrt{Re_p / v_{r,p}}}\right)^2 \tag{12}$$

This model is based on measurements of the terminal velocities of particles in fluidized or settling beds, with correlations that are a function of the volume fraction and relative Reynolds number defined as

$$Re_p = \frac{\rho_f |\mathbf{v}-\mathbf{u}| d_p}{\mu_f} \tag{13}$$

Consequently, the exchange coefficient is expressed as

$$K = \frac{3\alpha_p \alpha_f \rho_f}{4 v_{r,p}^{\ 2} d_p} C_D |\mathbf{v}-\mathbf{u}| \tag{14}$$

where $v_{r,p}$ is the terminal velocity correlation for the solid phase (Garside and Al-Dibouni, 1977).

$$v_{r,p} = 0.5\left(A - 0.06 Re_p + \sqrt{(0.06 Re_p)^2 + 0.12 Re_p (2B - A) + A^2}\right) \tag{15}$$

with

$$A = \alpha_f^{4.14} \tag{16}$$

and

$$B = 0.8\alpha_f^{1.28} \qquad (\alpha_f \le 0.85) \tag{17}$$

and

$$B = \alpha_f^{2.65} \qquad (\alpha_f > 0.85) \tag{18}$$

This model is appropriate when the solid shear stresses are defined according to Syamlal et al. (1993) as described below.
Solid pressure p_p is calculated as follows:

$$p_p = \alpha_p \rho_p \Theta_p + 2\rho_p (1 + e_{pp}) \alpha_p^2 g_{0,pp} \Theta_p \tag{19}$$

where e_{pp} is the coefficient of restitution for particle collision specified as 0.9 in this study, $g_{0,pp}$ is the radial distribution function, and Θ_p is the granular temperature. The granular temperature is proportional to the kinetic energy of the random fluctuating component of the particle velocity. The radial distribution function $g_{0,pp}$ is a correction factor that modifies the probability of collisions between grains when the solid granular phase becomes dense. In this study, the following model is adopted (Ogawa, et al., 1980).

$$g_{0,pp} = \left[1 - \left(\frac{\alpha_p}{\alpha_{p,max}}\right)^{\frac{1}{3}}\right]^{-1} \tag{20}$$

where $\alpha_{p,max}$ is an allowed maximum value of the solid volume fraction, which is specified as 0.63 in this study.
Solid shear viscosity is composed of collisional, kinetic and frictional parts.

$$\mu_p = \mu_{p,col} + \mu_{p,kin} + \mu_{p,fr} \tag{21}$$

The collisional part of the shear viscosity is modeled as (Gidaspow et al., 1992; Syamlal et al., 1993).

$$\mu_{p,col} = \frac{4}{5} \alpha_p \rho_p d_p g_{0,pp} (1 + e_{pp}) \left(\frac{\Theta_p}{\pi}\right)^{\frac{1}{2}} \tag{22}$$

The kinetic viscosity is specified as follows (Syamlal et al., 1993).

$$\mu_{p,kin} = \frac{\alpha_p d_p \rho_p \sqrt{\Theta_p \pi}}{6(3 - e_{pp})} \left[1 + \frac{2}{5}(1 + e_{pp})(3e_{pp} - 1)\alpha_p g_{0,pp}\right] \tag{23}$$

The frictional viscosity is specified as zero in this study. Further discussion on the granular temperature for solid phase is described elsewhere (Fluent Inc., 2003).

At the inlet to the straight section of the microchannel a uniform velocity U is assumed in the liquid phase. The velocity is chosen so that Dean numbers (De) are 20, 40, 60 and 80, which cover most of the range examined in the experiments (Ookawara et al., 2005b). In this study, De and Re are defined as;

$$De = Re\sqrt{\frac{D_h}{2R}} \tag{24}$$

$$Re = \frac{UD_h}{\nu} \tag{25}$$

where R, ν and D_h are the curvature radius, kinematic viscosity and hydraulic diameter, respectively.

The volume fraction α_p of 0.0005 and the same velocity, U, are specified as the inlet boundary conditions of the particulate phase. These boundary conditions are identical with the experimental conditions. The granular temperature is specified as zero at the inlet since there is no reason to specify any particular value. The granular temperature profile is expected to become fully developed within the straight inlet channel before reaching the semicircular separator section. The no-slip condition is applied on the channel walls.

For the pressure-velocity coupling and discretization schemes for convection terms, the Phase-Coupled SIMPLE algorithm (Vasquez and Ivanov, 2000) with a second-order upwind scheme is adopted. For the volume fraction, a second-order upwind scheme is employed as well.

Results and discussion

Fig. 2 illustrates the predicted profiles of the particulate phase (a) volume fraction and (b) streamwise velocity over the transverse plane at the end of semicircular section (De = 40, particle diameter d_p = 10 microns). The left edge of each contour map corresponds to the outer wall of the semicircular channel. The Euler-granular calculation also gives the velocity and volume fraction profiles of the continuous phase in the same manner. These profiles immediately enable one to evaluate the mass flow rate of the particulate and liquid phases passing through any bounded plane. The cross sectional plane is divided into inner and outer planes as shown in Fig. 2 so that the slurry flow rate ratio of outer to inner planes agree with the slurry flow rate ratio of outer to inner branches of the device based on the experimental results (Ookawara et al., 2005b).

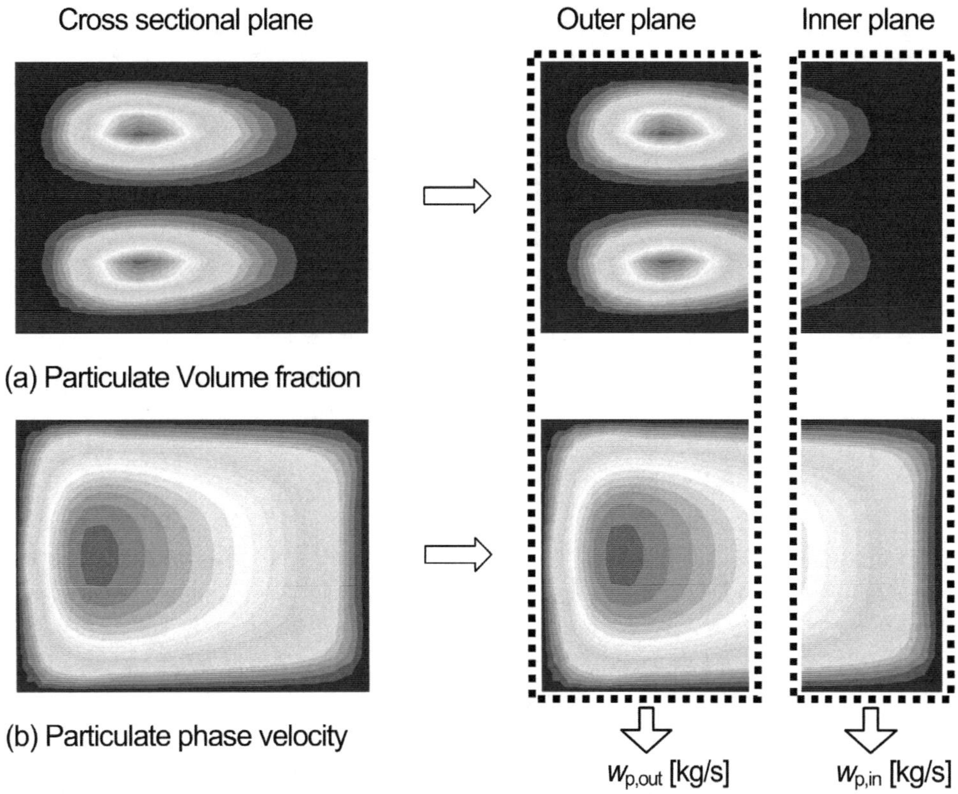

Fig. 2 Particulate (a) volume fraction and (b) streamwise velocity at the end of semicircular section. (De = 40, d_p = 10 microns)

In the bulk region where the particles are concentrated, the streamwise velocity reaches several m/s and the velocity of the secondary flow, viz., Dean vortices, is too small to circulate particles within a small distance such as the space of the bifurcation (Ookawara et al., 2004a, c). Consequently, due to the large inertia that the particulate phases exhibit as they pass through those inner and outer planes these are assumed to correspond to the particles flowing into inner and outer branches of the experimental device. On the other hand, the flow adjacent to the upper and lower walls may not be split vertically as shown in Fig. 2 due to the small inertia. However, the concentration of larger particles to be separated is extremely low because of the large lift force in the near wall region (Ookawara et al., 2004c, 2005a). It is assumed that, therefore, the particle separation efficiency is accurately evaluated by vertically splitting the slurry flow at the end of the semicircular section of the numerical domain. In this study, the ratio is fixed at 2.8 which is an average value given by the experiments using water.

The particulate mass flow rates passing through inner and outer planes, $w_{p,in}$ [kg/s] and $w_{p,out}$ [kg/s] gives particle recovery to outer flow for the particular particle size, viz., grade efficiency, R.

$$R = \frac{w_{\mathrm{p,out}}}{w_{\mathrm{p,in}} + w_{\mathrm{p,out}}} \tag{26}$$

The fluid recovery R_f is also given by means of liquid mass flow rates passing through the inner and outer planes, $w_{f,in}$ [kg/s] and $w_{f,out}$ [kg/s].

$$R_{\mathrm{f}} = \frac{w_{\mathrm{f,out}}}{w_{\mathrm{f,in}} + w_{\mathrm{f,out}}} \tag{27}$$

The reduced grade efficiency R_c, which is generally utilized as the performance index of hydrocyclones, is calculated for each examined particle size, viz., 1.8, 5, 10, 15, 20 and 30 microns as follows.

$$R_{\mathrm{c}} = \frac{R - R_{\mathrm{f}}}{1 - \mathrm{R_f}} \tag{28}$$

In experiments, the slurry recovery R_s, which is defined as follows, is substituted for the fluid recovery R_f to evaluate R_c for the convenience.

$$R_{\mathrm{s}} = \frac{w_{\mathrm{f,out}} + w_{\mathrm{p,out}}}{w_{\mathrm{f,in}} + w_{\mathrm{f,out}} + w_{\mathrm{p,in}} + w_{\mathrm{p,out}}} \tag{29}$$

Since the difference between R_f and R_S is very small for the dilute slurry whose feed volume fraction is 0.0005, the difference in R_c due to the substitution is negligible and in the order of 10^{-4}. Therefore, the reduced grade efficiencies obtained by experiments and numerical simulations can be compared directly.

Highly monodisperse particles with diameters of 1.8, 5, 10, 15, 20 and 30 microns were examined in the experiments (Ookawara et al., 2005b) as well as in the numerical simulations. It was found that the reduced grade efficiency of 1.8 micron particle was always nearly zero, while it reached almost the maximum value and practically remained constant for particles above 15 microns, despite the *De*. In this study, therefore, the comparisons of reduced grade efficiencies are shown for the particle diameters of 5, 10 and 15 microns in Figs. 3 (a), (b) and (c), respectively. The experimental results are shown in the *De* range of 30 to 50 because stable and consistent results are obtained by both the single channel device (Experiments-1) and the numbering-up device (Experiments-2) where three independent channels that work in parallel (See Ookawara et al., 2005b). The numerical and experimental conditions are indicated by using the same plot shapes and colors through Fig. 3(a) to (c) although the legend is only given in Fig. 3(a).

It is shown that the predicted R_c increases with the magnitude of the lift force despite the particle size and *De*. For the 5 micron particle shown in Fig. 3 (a), R_c increases linearly with *De* when the lift force is applied. For 10 micron particles, on the other hand, R_c seems to reach the maximum value at *De* = 60 and then remains nearly constant afterwards, when k_{Lift} is constant as shown in (b). For 15 micron particles in (c), however, it seems that R_c is nearly constant in the *De* range of 20 to 60, with the k_{Lift} of 0.25 and 0.5, although the slight asymptotic tendency, which is similar with that of 10 micro particles, can be seen. Although the experimental results in these figures are slightly scattered, it is expected that they would be in good agreement with the line constructed for plots predicted with some constant k_{Lift} value. For instance, a value between 0.25 and 0.5 will accurately predict the R_c of 5 and 15 micron particles, while another value between 0.125 and 0.25 will quantitatively predict R_c of 10 micron particle in the present *De* range. It can be said that, therefore, the proper coefficient value of Drew and Lahey lift force model, which could give the accurate numerical prediction of R_c, exists for the micro-separator/classifier in this study. It can be easily understood that this kind of discussion based on numerical simulations is very useful to optimize operational conditions for the device. In the present case, for instance, sharper separation can be achieved in the lower *De* range since the R_c of 15 micron particle is constantly high, while 5 micron particles are not separated. In the higher *De* range, the particle, whose size is smaller than 5 microns, will be separated and consequently the reduced grade efficiency curve will become broader. Associated with experimental results a further decrease of *De* causes unstable separation. A value of around 30 is the optimum *De* condition if 15 microns is the size to be separated. This is a useful strategy since better separation will be achieved without higher pressure drop.

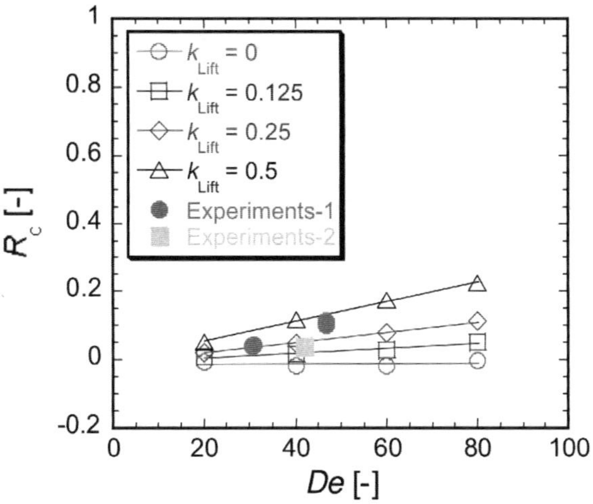

(a) 5 micron paritcle

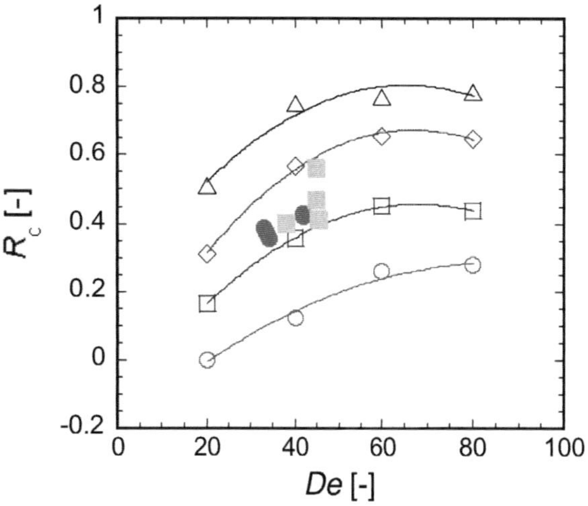

(b) 10 micron particle

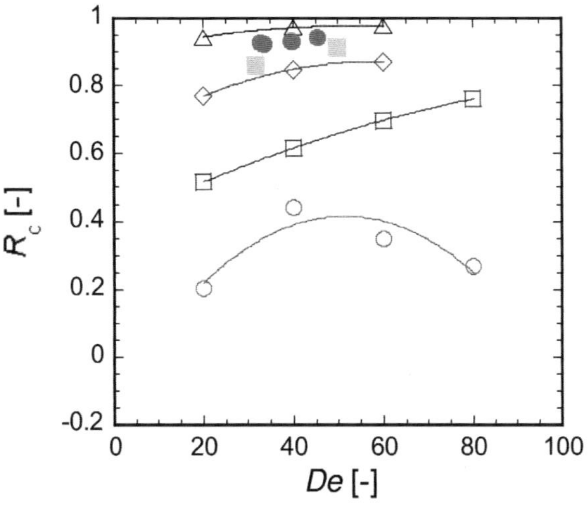

(c) 15 micron particle

Fig. 3 Comparison of reduced grade efficiencies obtained by experiments and numerical simulations for the particle diameters of (a) 5 microns, (b) 10 microns and (c) 15 microns.

Conclusions

The Euler-granular multiphase model, which treats both liquid and particles as continuous phases, is employed to predict the reduced grade efficiency of the micro-separator/classifier. Although there is no lift force model that is validated in the very high shear rate in the curved microchannel of this device, to authors' knowledge, it is shown in this study that there exists a proper coefficient value of a lift force model that is able to quantitatively predict the efficiency that agrees with experimental results. Based on the results obtained both numerically and experimentally, a strategy of how to use the device is discussed. By validating the lift force model, it is expected that the model will be useful to intensify the device performance and to adjust the operational conditions and geometrical configuration for the particles to be separated.

References

Dalla Valle, J.M., 1948. Micromeritics. Pitman, London.

Drew, D.A., Lahey, R.T.Jr., 1993. Analytical modeling of multiphase flow. Particulate two-phase flow, Butterworth-Heinemann, Boston, 509-566.

Fluent Inc., 2003. Fluent 6.1 User's Guide.

Garside, J., Al-Dibouni, M.R., 1977. Velocity-voidage relationships for fluidization and sedimentation. I & EC Process Des. Dev. 16, 206-214.

Gidaspow, D., Bezburuah, R., Ding, J., 1992. Hydrodynamics of circulating fluidized beds, kinetic theory approach. Fluidization VII, Proceedings of the 7th Engineering Foundation Conference on Fluidization, 75-82.

Ogawa, S., Umemura, A., Oshima, N., 1980. On the equation of fully fluidized granular materials. J. Appl. Math. Phys. 31, 483.

Ookawara, S., Higashi, R., Street, D., Ogawa, K., 2003. Feasibility study on concentration of slurry and classification of contained articles by microchannel. Book of Abstracts of 7th International Conference on Microreaction Technology, 231-233.

Ookawara, S., Higashi, R., Street, D., Ogawa, K., 2004a. Feasibility study on concentration of slurry and classification of contained articles by microchannel. Chem. Eng. J. 101, 171-178.

Ookawara, S., Higashi, R., Street, D., Ogawa, K., 2004b. The Influence of channel depth on the performance of a micro-separator/classifier. Kagaku Kougaku Ronbunshu 30, 135-141.

Ookawara, S., Street, D., Ogawa, K., 2004c. A practical application of the Euler-granular model to a micro-separator/classifier. Proceedings of 5th International Conference on Multiphase Flow, CD-ROM, #206.

Ookawara, S., Street, D., Ogawa, K., 2005a. A Numerical Modeling of Particle Concentration Profiles in a Curved Microchannel. Int. J. Multiphase Flow, submitted.

Ookawara, S., Oozeki, N., Ogawa, K., 2005b. Experimental Benchmark of a Metallic Micro-Separator / Classifier Compared with Representative Hydrocycolne. 2005 Spring Meeting Conference Proceedings CD-ROM, New York, AIChE, 129g.

Saffman, P.G., 1965. The lift on a small sphere in a slow shear flow. J. Fluid Mech. 22, 385-400.

Syamlal, M., O'Brien, T. J., 1989. Computer Simulation of Bubbles in a Fluidized Bed. AIChE Symp. Series 85, 22-31.

Syamlal, M., Rogers,W., O'Brien, T. J., 1993. MFIX Documentation: Volume 1, Theory Guide. National Technical Information Service, Springfield, VA, DOE/METC-9411004, NTIS/DE9400087.

Vasquez, S. A., Ivanov, V. A., 2000. A phase coupled method for solving multiphase problems on unstructured meshes. Proceedings of ASME FEDSM'00: ASME 2000 Fluids Engineering Division Summer Meeting, Boston.

130d Experimental Investigation of Micromixing in T-shaped Micromixers using μ-LIF and μ-PIV

Marko Hoffmann, Michael Schlüter, Norbert Räbiger

Micromixers are central components of microfluidic systems for chemical and biochemical applications. The fluid flow in tubes and channels on a microscale has emerged as an important area of research. This has been motivated by their use in such diverse applications as micromixers for controlled mixing of reactants in pharmacy or microreactors for unique process regimes and for optimizing conventional processes.

Common passive micromixers are working in a laminar flow regime and perform by splitting the fluid into small flow filaments to mix then by recombination and diffusion. This multilamination typically requires complicated three dimensional channel geometries. In our research we use a T-shaped micromixer with a rectangular cross section (dimensions for example: channel depth 200 μm, channel width 400 μm). This kind of mixer uses the flow-induced creation of secondary vortices inside the micro channels to enhance the mixing performance. Up until now investigation in micromixing have consisted largely of bulk measurements, such as volume flow rates, overall mixing qualities and pressure losses. Because of a lack of fundamental knowledge, microreactors are mostly laid out according to "trial-and-error".

For modern technical applications e.g. reactions with high selectivity, a precise analysis of local mass transfer and hydrodynamics for different flow regimes, i.e. different velocities in the mixing channel and different viscosities, is necessary by means of microscope-based measurements of the flow and concentration fields in non-reacting and reacting liquid flows.

In our research micro-resolution Particle Image Velocimetry (PIV) as well as micro-resolution-Laser Induced Fluorescence (LIF) is used to study the flow field and mixing performance along the channel length in dependency of the inlet conditions as well as the physical and chemical properties.

In order to investigate the local mixing performance in different depths of the reactor the concentration field is visualized by Laser Scanning Confocal Microscopy (LSCM). This enables image slices a few microns thick (minimum slice thickness of 2 μm) to be taken within a sample. By taking image slices at different depths and assembling these slices together, a three-dimensional rendering of the mixing channel is possible.

The results obtained from these experiments show that in T-shaped micromixers through simple hydrodynamic mixing very fine structures (approx. 3μm) can be achieved even under laminar flow conditions. Consequently large contact areas between the reactants can be achieved. These experimental results allow a profound understanding of the mass transfer processes in micro-mixing by using a T-shaped micromixers as an example and a desired improvement in the design rules for static micromixers. This necessitates new models for mixing performance taking into account the phenomenology of local micro-mixing which were performed in our research group ("Department of Chemistry and Chemical Engineering", University of Paderborn, Germany).

Acknowledgement:

This work is supported by the German Research Foundation (Deutsche Forschungsgemeinschaft - DFG), grant RA 574/17.

Towards a Modular Split&Recombine Micro Mixer

C. Cupelli, P. Koltay, Jens Ducrée, Roland Zengerle, and M. Santer
IMTEK – University of Freiburg, Laboratory for MEMS Applications,
Georges-Koehler-Allee 103, D-79110 Freiburg, Germany,
cupelli@imtek.de, Phone: +49 761 203 7327, Fax: +49 761 203 7539

Abstract

In this paper we discuss a novel approach towards a system-level description for micro reactors. In particular, a compact model for laminar micro mixers for fluids with different viscosities is presented. We show how the well known principle of laminar Split-and-Recombine mixing can be modeled by three elementary geometrical building blocks each of which is treated as separate compact model. This way, the analytical and numerical treatment can basically be reduced to multi-laminar flow in straight channels with square cross sections to a good approximation. The presented model reliably predicts the flow field, the fluidic resistances and the hydrodynamic spreading for multi-laminar flows. The model is validated in comparison with experiments and computational fluid dynamics (CFD) simulations for viscosity ratios as large as 1:20. This way it is possible to map relevant characteristics of a mixer onto a compact model description in order to provide fast network simulations instead of time consuming CFD-simulations.

Introduction

Microfluidic systems have an increasing relevance in life sciences, where the well known benefits such as minimized sample consumption and the potential use in integrated designs have lead to many concepts of "lab-on-a-chip" (LOAC) devices. Micro mixers are essential components for many of these systems [5], e.g., in chemical process engineering. The high surface to volume ratio, short diffusion lengths and an enhanced heat transport improve the quality of chemical reactions in micro dimensions, thus reducing the effort for time consuming successive steps like purification of the chemical products from undesired by-products [9].
Despite all potential and practical advantages, a generalized description of LOAC systems at an adequate level of abstraction is one of the remaining challenges, in particular for micro mixers. In contrast to microelectronics, where large circuits may be represented, designed and tested by lumped element networks, the techniques aiding the design for many fluidic components are mostly based on time consuming, iterative CFD-simulations, often leading to only monolithic applications. Additionally, in arbitrarily specified geometries CFD is often error-prone towards, e.g., numerical diffusion [4] or discretization artifacts. With this work we demonstrate how inefficient use of CFD might be avoided through enhancing the modular character of the whole micro-fluidic environment.

Modular Concept

The concept we pursue may be divided into three stages: first, defining a set of fluidic components with possibly restricted, but well-defined operations that are amenable to an analytic description. Second, optimizing and characterizing these operations by a series of "smart" CFD-simulations, establishing strict modularity and connectivity and third, the integration of these informations in a compact model, yielding reliable estimates about efficiency, but also hints for the layout to the designer.
In the following, we will discuss how the so-called Split-and-Recombine (S&R) principle for micro mixers may be reorganized in this way in order to arrive at a lumped element

representation, suitable for integration in a pressure driven fluidic environment. The design may then be reconfigured solely via the compact model.

Fluidic Setup

Many different Split-and-Recombine mixers, are partly based on complex geometries ([2,4] and references therein). We chose to decompose a whole S&R-step into three different geometric structures, attributed to defined operations on entering fluid streams, see Fig. 1a : i) combining two streams, ii) twisting the flow profile by 90° and iii) symmetric split.

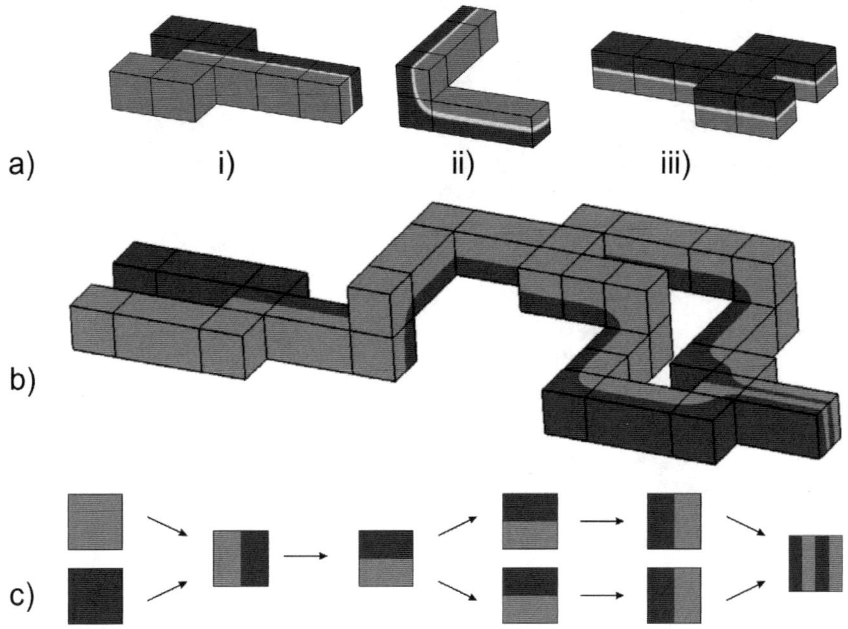

Fig. 1. a): Modular components of the S&R mixer: i) combine, ii) twist and iii) split structure.
b):CFD-simulation of the S&R mixer.
c): Schematic working principle of the mixer.

Fig. 1b illustrates how the three components are used to create a complete S&R-step, i.e., a doubling of the number of fluid lamellae. Most of the CFD computing time is spent in identifying and optimizing the elementary components, and establishing their modularity. An important point is that in order to allow for an analytic description of the mixing progress, the fluid interfaces should be kept parallel to the channel walls throughout the whole geometry. This is achieved by requiring that each component separately complies with this requirement. Furthermore, the particular assembly of components in Fig. 1b leads to a symmetric distribution of fluidic resistances. Thus, the assembly of fluid lamellae *leaving* the S&R-step may be treated as if the four fluid streams had been introduced into a single channel symmetrically. This is where the strategy outlined above leads to a rather simple analytic description of the mixing process, as described in the next section.

Analytical Approach and Compact Model

Since a complete analytical solution of the problem of multi-laminar flow in rectangular geometries is not possible [3], we firstly consider a corresponding 2D-approximation of a multi-laminar flow in a slit. This is useful because the flow profile in a square channel is parabolic near the channel center, where the largest part of the total flux is transmitted. Furthermore, diffusion (and thus a possible blur of the fluid interfaces) is neglected. This is also justified since the averaged lamella thickness is halved after each S&R-step, and the diffusion time decreases quadratically with decreasing lamella thickness; thus one expects significant mixing to set in suddenly, essentially within the time spent in a single output channel after, say, the n-th S&R-step. The governing equation for the fluid motion is the Navier-Stokes equation, which for the case of two-dimensional, fully developed slit flow of parallel fluid lamellae at low Re numbers, can be reduced to the Poisson equation [6]. For a slit flow with multiple fluid lamellae and possibly different viscosities, the solution for the flow field can be calculated by applying the Poisson equation for the particular domain, i.e.

$$\frac{\partial p}{\partial y} = \eta_i \cdot \frac{\partial^2 u_{y_i}}{\partial x^2} \tag{1}$$

where we assume that the domains are separated by a fluid interface at position x_i. The different domains are coupled at x_i via equilibrium of shear forces [3],

$$\eta_i \cdot \frac{\partial u_{y_i}}{\partial x} = \eta_{i+1} \cdot \frac{\partial u_{y_{i+1}}}{\partial x} . \tag{2}$$

and the continuity of the flow field,

$$u_{y_i}\Big|_{interface} = u_{y_{i+1}}\Big|_{interface} . \tag{3}$$

The solution of the eqn. (1-3) allows to determine all characteristic quantities entering the compact model. In addition to external parameters such as the diffusion coefficients D_i and the channel length $l_{ch,}$, these are the fluidic resistance, the diffusion lengths l_{Di} as estimated from the distribution of lamellae thicknesses, and the mean residence time $l_{ch}/<u_i>$ through the average flow velocity $<u_i>$ for each fluid lamellae. The point where notable progress of mixing sets in may then be estimated by evaluating the Fourier Number Fo,

$$Fo = \frac{D_i \cdot l_{ch}}{\langle u_i \rangle \cdot l_{Di}^{\ 2}} > 1 , \tag{4}$$

with a value greater than 1 indicating considerable diffusion in passing a channel segment of length l_{ch}. Eqn. (1-3) may be solved formally assuming a piecewise polynomial solution for u with stick boundary conditions, resulting in a system of implicit non-linear algebraic equations for the x_i, the pressure drop Δp and the flow rate ϕ ,

$$\Delta p \cdot R^{-1}(\eta_1, \eta_2, d, x_{i,interface}) = \phi , \tag{5}$$

that can be mapped to a network model.

Numerical Implementation and Validation

Equations (5) constituting the core of the compact model, were implemented in the hardware description language VHDL-AMS with all computations performed with the commercial software SMASH 5.2.0 from Dolphin Integration. The analytical model was validated by CFD-simulation with the commercial software CFDRC-ACE+ for square channels applying the Volume of Fluid (VOF) method. In order to extend the analytical model to a full 3D-model the fluidic resistance and the diffusion lengths for square channels were investigated and compared with the analytical model.

The system (5) has many solutions; in order to stick to the physical one, parameters such as the η_i are varied continuously with time, starting from a known solution for the x_i at a defined operation point of the network model. Changes in η_i , for instance, due to temperature changes, may be emulated by ramping the corresponding parameter to the new value. This process is illustrated in Fig. 2, where a change in viscosity ratio β from 1 to 4 is accomplished, with the x_i changing from a symmetric configuration to an asymmetric one due to hydrodynamic spreading. The time scale at which the transitions occur may be chosen either according to a specific model or in order not to conflict with other time scales in the system. This leaves some space for the level of abstraction the model is operated at.

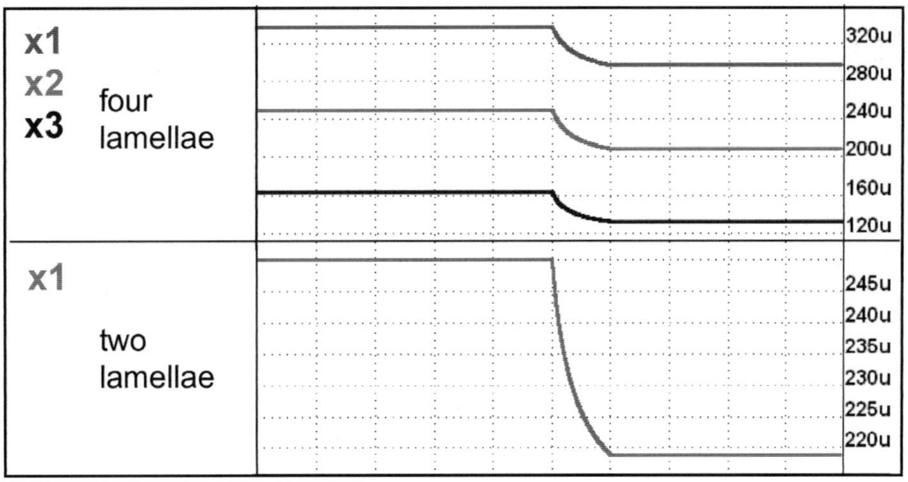

Fig. 2. Network simulation of the position of fluid interfaces x_i for a complete S&R-step with β changing from 1 to 4 (1u corresponds to 1μm,with the right channel wall located at 0 u and the left one at 500 u).

The convergence of the network model was verified for a variety of parameters, and up to viscosity ratio as large as 1:20. The analytical model for the slit flow was compared to a full three-dimensional CFD-simulation for two as well as four fluid lamellae.

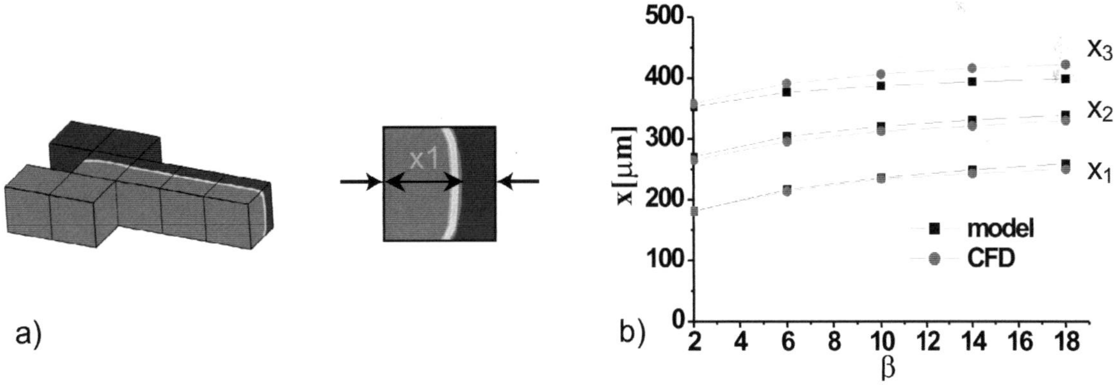

Fig. 3. **a):** CFD-simulation for a viscosity ratio β = 10. The position of the fluid interface was evaluated at the middle of the channel cross section.
b): Positions x_i of fluid interfaces of the network model vs. CFD-simulation.

Fig. 3a shows that in contrast to a slit, the fluid interface between two adjacent lamellae are characteristically bended [5]. This effect influences the hydrodynamic spreading in confined geometries and consequently the accuracy of the model. The accuracy of the model for the fluidic resistance and the diffusion lengths was validated up to a viscosity ratio 1:20. Fig. 3b sketches the positions of the fluid interfaces for an increasing ratio of viscosity. The model reliably predicts the asymmetry of the lamination pattern for four fluid lamellae as β increases. In order to account for the three-dimensional geometry, the fluidic resistance of the compact model was adjusted by a geometric factor which for uni-laminar flow is [11]

$$\gamma = \frac{\Delta p_{square}}{\Delta p_{slit}} \approx 2.4 \; . \tag{6}$$

The pressure drop Δp in the model scaled by γ was compared to a full 3-D simulation, and only a weak dependency of the geometric factor γ on the viscosity ratio β was found.

Experiments

In order to validate the model, experiments were performed with water and glycerol mixtures with different mass fractions of water and glycerol. A rapid prototyping technique based on SU-8/PDMS [7] was used to manufacture the microfluidic structures with a lateral size of 200μm and 500μm, respectively. The channel structures were fixed with a PMMA bracket in order to access the fluidic channels by external connection. In order to investigate the multi-lamination for different conditions, a tree-shaped channel structure in one plane was realized, see Fig. 4a, with flows being controlled by a syringe pump. For optical measurements a microscope (Zeis Axio Phot) was used. For visualization of the fluid lamellae, the glycerol-water mixture with higher mass fraction of glycerol, i.e. the fluid of higher viscosity was died with ink.

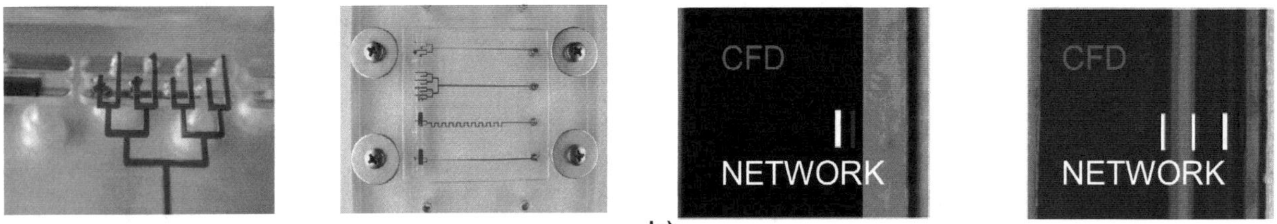

a) b)

Fig. 4. a): Tree-structures of microfluidic channels for generating regular flow profiles including up to eight fluid lamellae.
b): Lamination pattern of multi-laminar flow for a viscosity ratio β=10. The flow direction is in plane.

The agreement between numerical results and experiments is best between CFD-simulation and experiments, see Fig. 4b. As predicted by the model and CFD-simulation, the lamination pattern shows an increasing asymmetry for an ascending viscosity ratio β. Deviations between model and experiments are mainly caused by the curvature of fluid interfaces not taken into account by the compact model.

Summary and Conclusion

In summary, our results suggest that it is possible to implement compact models and reliably predict the performance even of complex microfluidic systems such as mixers, by keeping their design modular and reconfigurable. The procedure was demonstrated for a specific design of a laminar S&R micro-mixer. Through the modular approach, the analysis of the flow patterns could be reduced to regular, multi-laminar flow in straight channel segments, which in turn was approximated by two-dimensional slit flow. This way, a model of nonlinear algebraic equations was developed that estimates the thickness of fluid lamellae, the fluidic resistance and the flow field even for strongly varying viscosities at least up to a ratio of 1:20. Good agreement between compact model, CFD-simulation and experiments was achieved. Using this information, effective diffusion times can be estimated, allowing to trace the progress of mixing. Reconfiguration of the mixer layout in order to optimize the design is possible within the framework of the compact model parameters.

References

[1] S. Hardt , V. Hessel and C. Hofmann, Lab-on-a-Chip, 4, pp. 65-69, 2004.

[2] N. Schwesinger, T. Frank and H. Wurmus, Journal of Microengeneering, 6, pp. 99-102, 1996.

[3] P. J. Styles and D. F. Fletcher, Lab-on-a-Chip, 4, pp. 121-124, 2004.

[4] S. Hardt, H. Pennemann and F. Schönfeld, accepted for publication in the Journal of Microfluidics and Nanofluidics.

[5] P. Yager, E. Oosterbroek, A. van den Berg, Lab-on-a-Chip, Elsevier, 2004.

[6] E. Truckenbrodt, Strömungsmechanik, 1, Springer, 1980.

[7] J. Cooper et al., Electrophoresis, 21, pp. 27-40, 2000.

[8] E. L. Cussler, Cambridge University Press, 1997.

[9] J. Ducrée et al., Micro Total Analysis System Conference Proceedings, 1, 2004.

[10] F. Schönfeld et al., Nanotech Conference Proceedings, 1, pp. 378 – 381, 2004.

[11] H. M. Schaedel, Friedr. Vieweg & Sohn, 1979.

High-speed videography of microchannel array evaporators

Torsten Henning, Juergen J. Brandner, and Klaus Schubert
Forschungszentrum Karlsruhe, Institute for Micro Process Engineering IMVT
Hermann-von-Helmholtz-Platz 1, DE-76344 Eggenstein-Leopoldshafen, Germany

Prepared for Presentation at the 2005 Spring National Meeting, 2005-04-10/14, Atlanta/GA
8th International Conference on Microreaction Technology (Session TK008)
Unpublished

In an array of 68 mechanically machined microchannels, each of them 200 μm wide and 100 μm deep, the generation of vapor from water flowing at a mass flux of 61 kg m^{-2} s^{-1} was observed by high-speed videography. Image analysis showed that vapor generation at the boiling front occurred via the nucleation of bubbles at a few sites, the subsequent downstream expansion of these bubbles into vapor slugs, and the periodic loosening and downstream movement of the entire vapor slugs. A thin water film remained on the walls in the vapor filled portions of the channels. The locations of the nucleation sites and the frequencies with which bubbles were generated at them showed no correlation. The low frequency (sub-audio) pulsation generated in microchannel array evaporators is related to the practically random characteristics of individual microchannels.

1 Introduction

The evaporation of liquids and the generation of dry and overheated steam in an evaporator device may, dependent on the flow rate of the liquid and its enthalpy of evaporation, require a high heat flux to be applied. Constraints on the allowable overall device size, or requirements such as precise temperature control or short ramping times, or the demand for features of inherent safety, make microstructure devices [4], especially electrically powered devices with multi-microchannel arrays [1], an obvious choice for such tasks.

Since no analytical or numerical tools can to date calculate or simulate the processes of evaporation and condensation, of the liquid-vapor two phase flow, and of their interactions, the optimization of microchannel array evaporators is an empirical task that relies on suitable characterization methods. High-speed videography has been shown earlier [3, 2] to be a valuable experimental method, since it provides good spatial resolution, while at the same time being able to cover many parallel channels of a device.

An important requirement from the application point of view is that the evaporation should be complete at all times, that is, that no liquid (in the following, this working fluid is always water) should reach the device outlet. Another desirable property of evaporators is that they introduce as little pressure fluctuations (pulsation) as possible. The combination of strong pulsation with incomplete evaporation is most undesirable, and the lower the frequency with which such events occur, the worse the performance of the device is to be rated.

The temporal resolution of high-speed videography with a commercial off-the-shelf camera and continuous (non-stroboscopic) illumination is a suitable tool for the investigation of such low-frequency phenomena occurring in microchannel array evaporators [2]. It is the objective of the investigation reported in the following to find out where and how low frequency pulses are generated in a microchannel array water evaporator. This question is connected to the problem of how important the individual characteristics of the microchannels (shape and surface variations) are for the device as a whole.

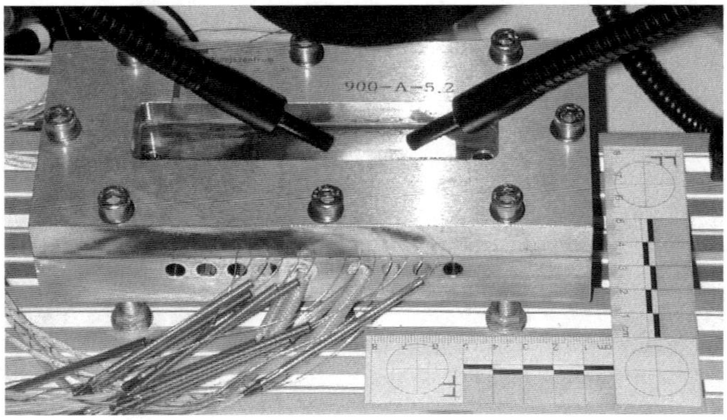

Figure 1: Photograph of the test device.

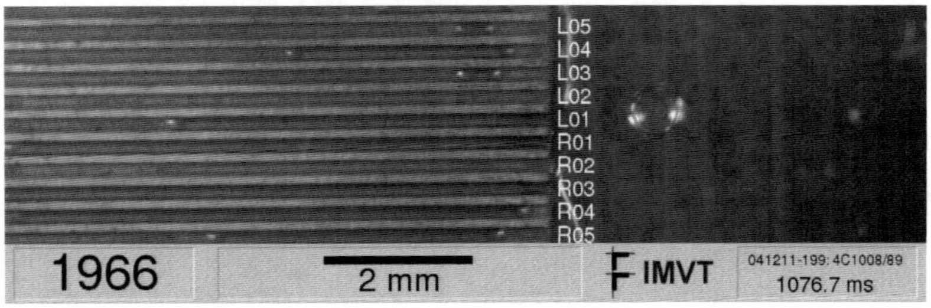

Figure 2: Photograph of the portion of the microchannel array seen by the camera. Direction of flow is from left to right, the channels are numbered to the left (L) and right (R) starting with 01 closest to the lateral center of the array.

2 Device and experimental procedures

The test device was described earlier in detail [3, 2]. It consisted of an electricaly heated base plate, onto which exchangeable microstructure foils could be mounted, and of a glass window pressed onto the microstructure foil's top side by a sturdy frame via strongly compressed polymer seals. Figure 1 shows a photograph of the test device, mounted under the objective of a stereo microscope (which has been retracted for the photograph), and illuminated by two 250 W halogen lamps via gooseneck light guides pointing in upstream and downstream directions, respectively.

A variety of microstructure foils were manufactured and tested in this device. In the following, we restrict ourselves mostly to results from a foil with straight channels, shown in fig. 2. The array consisted of 68 channels, each of them 200 μm wide and 100 μm deep. The array of microchannels and dividing ridges had a total width of 20.3 mm and a length of 40.0 mm. The sample was made of brass by mechanical micromachining.

3 Results and discussion: straight channels

During the experiment, a constant electrical heating power was applied to the device, and data for the microstructure foil temperature $\vartheta_{\rm foil}$, the outlet temperature $\vartheta_{\rm out}$ and the pressure drop across the device Δp were taken after a sufficiently long waiting period, and averaged over 30 seconds for each set of parameters. Figure 3 shows the data for pressure drop and outlet temperature, at a mass flow rate of 0.3 kg h^{-1} (corresponding to a mass flux of 61 kg m^{-2} s^{-1}). These data have been plotted against the averaged microstructure foil temperature $\bar{\vartheta}_{\rm foil}$, which has been calculated as the arithmetic mean of the readings of five thermocouples, spaced at intervals of 5 mm between the longitudinal center of the microchannel array and its downstream end.

It is obvious from the right panel in fig. 3 that overheated steam could be produced in this device. Even

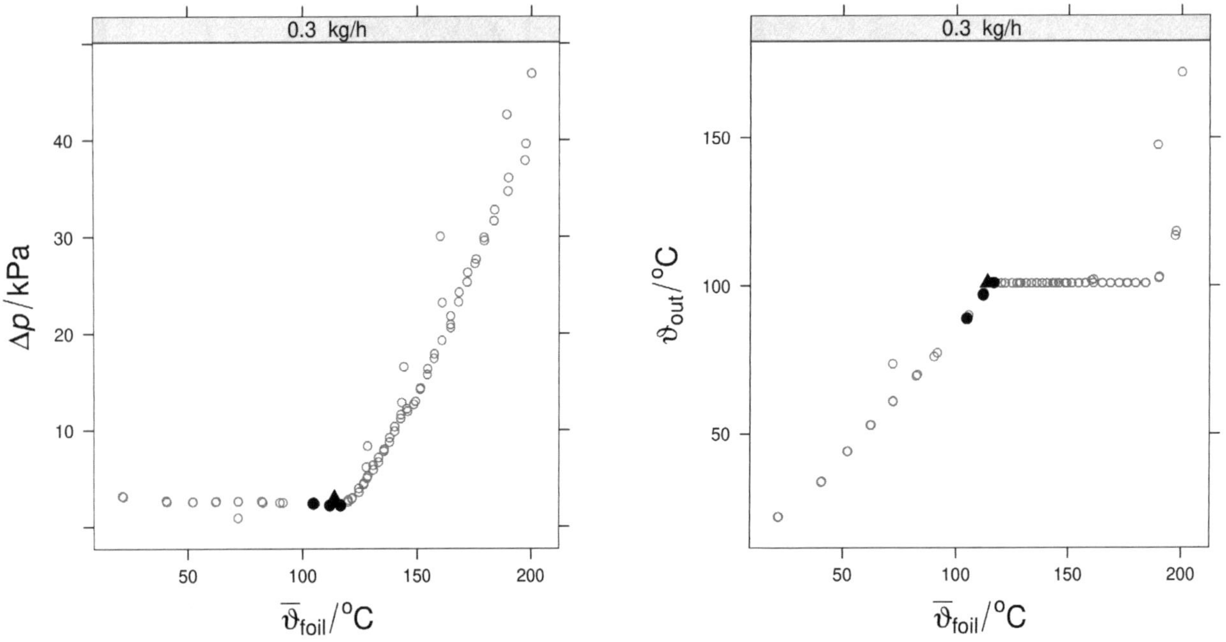

Figure 3: Pressure difference across the device Δp (left) and outlet temperature ϑ_{out} (right) as a function of the mean microstructure foil in the heated region $\bar{\vartheta}_{\text{foil}}$. The filled circles indicate the three parameter sets at which the images analyzed in fig. 4 were taken, the filled triangle symbols show the parameters corresponding to the data presented in fig. 5.

at the highest powers and temperatures, however, liquid water still left the microchannel array, especially near its lateral center, and complete evaporation did not occur until the remaining water contacted the hot walls of the outlet plenum.

In the straight microchannel array, a boiling front with approximately parabolic shape could be observed, extending further downstream in the center of the array than at its edges. With increasing power, the location of this boiling front moved further upstream. When the boiling front had reached the inlet plenum and boiling in the plenum occurred first at the outer edges, the boiling front still reached into the array around the center, until finally, with the power increased even more, boiling occurred in front of all of the microchannels. Since we are here primarily interested in the two phase behavior occurring at the boiling front, the field of view for the experiments discussed in the following was placed in the downstream section of the microchannel array (cf. fig. 2), and the measurements discussed in the following were taken at relatively low power input, when the boiling front was passing through this downstream section of the microchannel array (cf. fig. 3).

To study the generation and movement of bubbles, slugs or plugs, the images taken by the high-speed camera were analyzed for the presence and location of well-defined liquid-vapor interfaces. For the liquid-vapor interface trace plots discussed in the following, the spatial derivative of the pixel intensity along the center lines of the channels was calculated and translated into a grayscale map, with black corresponding to a high value of the derivative, and white corresponding to a flat intensity profile. Figure 4 shows the result of such a calculation for the microchannel no. L04 (cf. fig. 2) under three different conditions. At the lowest of the three foil temperatures, just below 105 °C, this particular channel was mostly filled with water, and during the first second of observation, only two bubbles streaked through this channel.

At a foil temperature of approximately 112 °C (center panel in fig. 4), a bubble nucleation site became active in this channel, located about 5 mm from the channel exit. Bubbles were generated at a certain frequency, and grew until they filled the entire channel cross section. During this growth, they travelled a

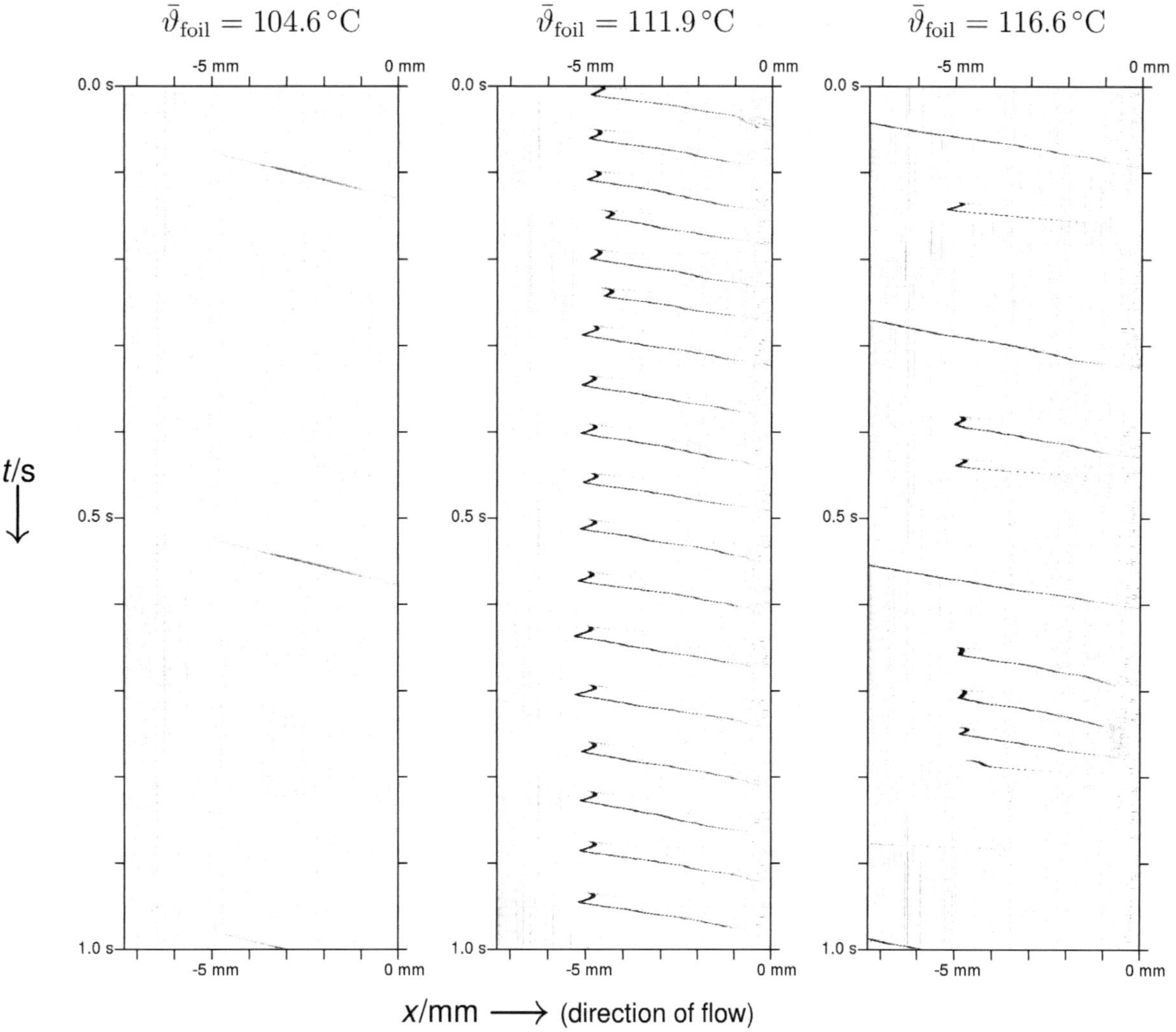

Figure 4: Traces of sharp liquid-vapor interfaces in channel no. L04 for three different average microstructure foil temperatures, plotted over the first second of each of the respective high-speed video sequences.

short distance against the general direction of flow. Once a bubble filled the entire channel, its downstream end moved very quickly (within a few milliseconds) until vapor filled the channel between the nucleation site and the channel exit. The downstream ends of these bubbles, expanding into vapor slugs, travelled so fast that they appeared blurred in the images, and therefore are not visible in the plots in fig. 4. After a vapor slug extended to the channel exit, its upstream front started moving in the direction of flow. Once the channel was filled with water again, a new cycle began with the nucleation of another bubble at the same location as before.

The frequency with which bubbles were generated at this particular nucleation site increased as the temperature of the microstructure foil was raised by another 5 °C (right panel in fig. 4). The regular pattern was, however, destroyed by the transit of bubbles generated further upstream.

Figure 5 shows a similar plot of the liquid-vapor interface locations against time for three adjacent microchannels at the same nominal foil temperature. It is obvious from these plots that the nucleation sites were located randomly in the individual microchannels. In channel no. L05, the only active nucleation site was located slightly less than 2 mm from the channel exit. Bubbles generated at this particular site expanded into vapor slugs relatively slowly, so that the downstream ends did not become blurred and thus appear as light gray traces in the left panel of fig. 5, originating from the same points as the dark traces that mark the positions of the upstream ends.

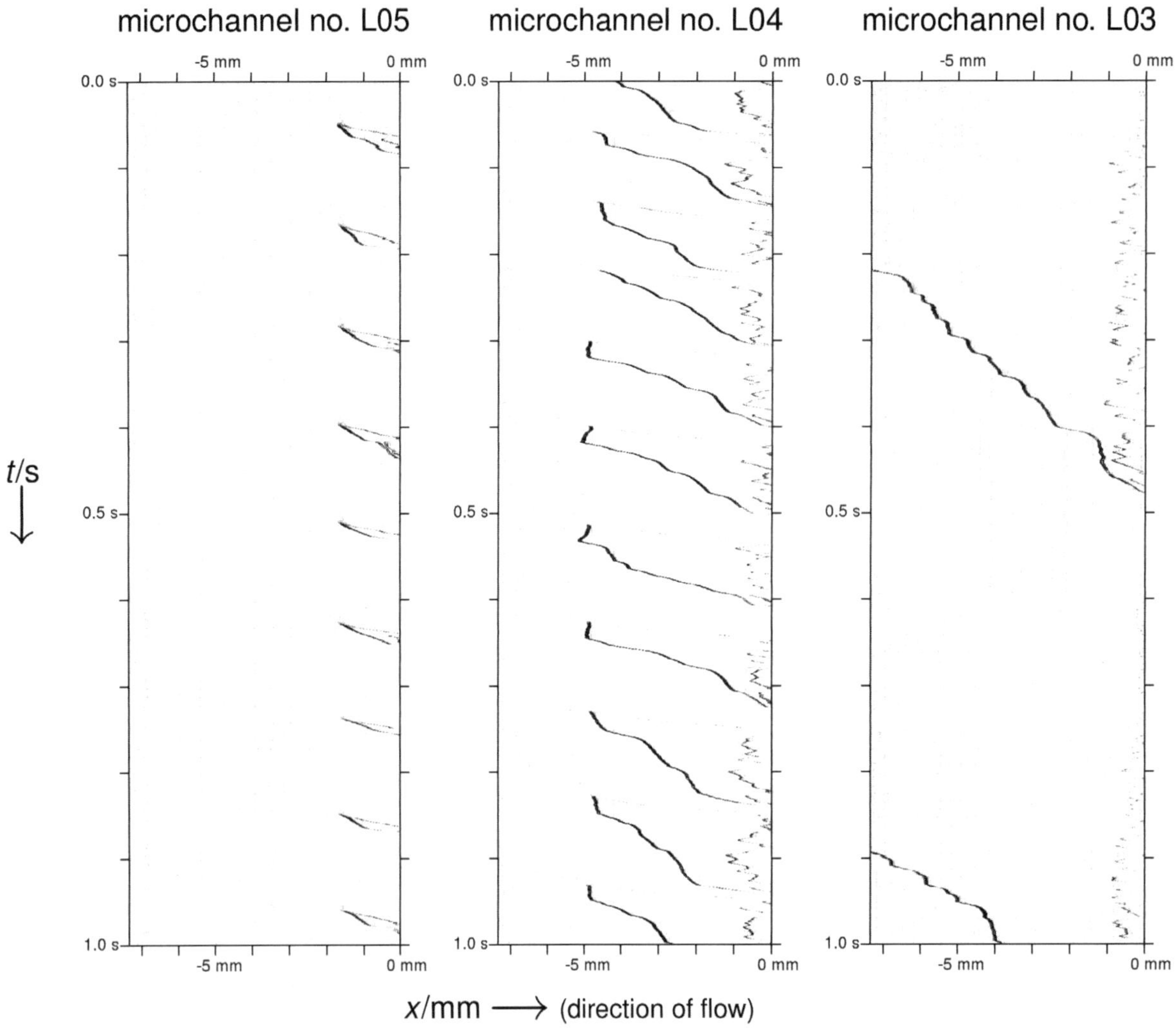

Figure 5: Traces of sharp liquid-vapor interfaces in three adjacent microchannels, plotted over the first second of each of the high-speed video sequence. The average microstructure foil temperature was 113.9 °C.

In microchannel no. L04, the same nucleation site as discussed in fig. 4 was active. The motion of the upstream ends was, however, less uniform as in the earlier measurements. Another difference to the measurements discussed earlier is that here the vapor slugs did not extend to the outlet plenum at all times. Instead, another vapor-liquid interface appeared at the channel exits, oscillating at a high frequency up to about 1 mm into the microchannel from its exit. In microchannel no. L03, the bubble nucleation site lay upstream from the field of vision. The upstream liquid-vapor slug interface moved slower in this channel than in the other two discussed above, and these vapor slugs exhibit most clearly the nonuniform "stick-slip" movement. The appearance of the oscillating vapor-liquid interface at the exit is very prominent from the right panel of fig. 5. It should be noticed that there seems to be no apparent correlation between the location of a bubble nucleation site, measured from the channel exit, and the frequency with which bubbles were generated at this site.

4 Other channel geometries

Apart from the straight channels, two other channel geometries were tried. First, the straight channels were augmented by three staggered rows of impactor blade structures situated in the outlet plenum, immediately in front of the microchannels. Second, a microstructure foil was made in which the channels were not straight, but followed a sinusoidal path.

Analysis of the high-speed videos showed potential for improvement in the design of the impactor blade structures. At rather incomplete evaporation, liquid water accumulated on the downstream (leeward) sides of the impactor blades, coalescing into large drops. Pulses originating from the microchannel array led to turbulent movement of the two phases behind the blades, loosening and dragging along the drops. Hence, pulsation, as determined by the time-dependent amount of liquid water in the outlet plenum, was amplified rather than dampened under certain conditions. As complete evaporation was approached with further increasing power input and foil temperature, jets of water originating at the microchannel exits moved through the impactor blade array in an angled but straight path, demonstrating the need for a more elaborate staggering pattern.

With the microchannels following a sinusoidal path, the overall frequency distribution of the pulsation was slightly narrower than for straight channels [2]. Further experiments are needed to find out whether this might be correlated to a slightly higher pressure drop observed in the sinusoidal channel array.

5 Conclusions

In mechanically micromachined multichannel array evaporators, vapor generation starts by the formation of bubbles at individual nucleation sites whose locations are apparently random. These bubbles expand into vapor slugs, which then move downstream, often at non-constant speed (accelerating or in a stick-slip pattern). The downstream ends of the vapor slugs can extend to the channel exits, or oscillate in their vicinity. There is no apparent correlation between the location of a bubble nucleation site and the frequency with which bubbles are generated at this particular site. Although the vapor slugs fill almost the entire channel cross section, a thin water film remains on the walls of the microchannels, and the water flowing in this film can coalesce into large drops in the outlet plenum, which may be detrimental for evaporator performance.

Acknowledgements

The authors thank the Micromachining Group at IMVT for the fabrication of the exchangeable microstructure foils. Data, images and video sequences were processed with Open Source Software.

References

[1] Henning T, Brandner JJ, Schubert K (2004). Characterisation of electrically powered micro heat exchangers. Chem Eng J 101:339–345. doi:10.1016/j.cej.2003.10.020.

[2] Henning T, Brandner JJ, Schubert K (2005). Comparison of microchannel array water evaporators by high-speed videography. Submitted for Proc. 3rd Int. Conf. on Microchannels and Minichannels (ICMM2005), Toronto/ON, Canada, June 13–15, 2005.

[3] Henning T, Brandner JJ, Schubert K (2005). High-speed imaging of flow in microchannel array water evaporators. Microfluidics and Nanofluidics doi:10.1007/s10404-004-0020-9. In press.

[4] Schubert K, Brandner J, Fichtner M, Linder G, Schygulla U, Wenka A (2001). Microstructure devices for applications in thermal and chemical process engineering. Microscale Thermophys Eng 5:17–39. doi:10.1080/108939501300005358.

130g DNS Based Evaluation of Residence Time Distribution for Bubble Train Flow in a Square Mini-Channel

Alexandru Onea, Braduts Ghidersa, Martin Woerner

Bubble train flow is a flow pattern often observed for co-current gas-liquid flows in narrow straight channels. It consists of a regular sequence of bubbles of identical shape filling almost the entire channel cross-section (Taylor bubbles). The individual bubbles move with the same axial velocity. Therefore, bubble train flow is fully described by a single flow unit cell. Bubble train flow is of considerable technical relevance and occurs e.g. in micro-bubble columns and monolithic reactors. When chemical reactions are performed in these devices, the knowledge of the residence time distribution of the liquid phase is of great importance to provide hydro-dynamically optimal conditions for the kinetics of the respective chemical process.

At the last IMRET conference we presented the first 3D direct numerical simulations of the bubble train flow of air and silicon oil in a square vertical channel of 2 mm width and obtained good agreement with experimental data from literature. Since then, a detailed numerical study on the influence of the length of the flow unit cell has been performed. In the present contribution we use this DNS data base to determine residence time distributions for the liquid phase. For that purpose mass-less particles are introduced in the centre of each mesh cell entirely filled with liquid. Taking into account that the velocity field in the frame of reference moving with the bubble is steady, the particle trajectory is computed in the fixed frame of reference. For each particle we determine the time the particle needs to travel an axial distance equal to the length of the flow unit cell. By normalizing the resulting histogram we obtain the residence time distribution (RTD). We find that independent of the length of the flow unit cell the RTD shows a characteristic shape. Because no fluid particle moves faster than the bubble, the RTD is zero for time values t smaller than T , defined as the ratio between the unit cell length and the bubble velocity. The RTD obeys its maximum for time values slightly larger than T and thus shows a sharp peak. For larger values of t the RTD exhibits an exponential decay. We thus conclude that for bubble train flow the RTD can well be represented by a functional relationship of type c exp(-ct) where c 0 is a fitting parameter.

130h Compact, Efficient Microreactors for Steam Reforming of Methanol

Abhaya Datye, Ayman Karim, Jaime Bravo, Travis Conant

Compact fuel cell systems operating with hydrocarbon fuels possess high volumetric and gravimetric energy density and could outperform batteries in low-power, portable electronics [1]. The challenge is to develop efficient micro reformers for converting the chemical fuel into H2 which can be fed to a proton exchange membrane fuel cell to produce electricity. Among the fuels considered, methanol is an attractive candidate since it is sulfur free, and most importantly, can be reformed at low temperatures (200-250 C). For methanol steam reforming (MSR), in order to achieve high conversions, the reformer has to operate with low gas velocity. The vast majority of studies on MSR have been done using packed-bed reactors of inner diameters greater than 4mm and reactor to particle diameter ratio of less than 25. The low gas velocity dramatically lowers the effective thermal conductivity of the catalyst bed [2]. Furthermore, the low ratio of reactor to catalyst particle diameter, and the low gas velocity tends to introduce a large heat transfer resistance between the wall and the catalyst particles where the void fraction is largest [3] and in turn lead to temperature gradients in the bed. For an endothermic reaction such as MSR, temperature gradients in the packed-bed lead to lowered catalyst activity and could also lead to falsified kinetics.

The importance of heat transfer from the wall, particularly at low gas velocities is not well appreciated and is often neglected. There are no reported experimental measurements of the impact of heat transfer on the MSR reaction, or recommendations for reactor dimensions that would overcome transport limitations. In this work, we have performed an experimental study of the rate of MSR under conditions where we attempt to minimize the effect of transport limitations. The work was complemented by reactor modeling using a 2-D pseudo homogeneous model that allows us to calculate temperature profiles and to assess their impact on catalyst reactivity. The agreement between the experimental data and the reactor model confirms our initial hypothesis that MSR can be subject to severe heat transfer limitations in packed-bed reactors. We have therefore wall-coated the MSR catalyst into quartz capillary tubes ranging from 1mm diameter down to 230microns [4] to overcome the heat transfer limitations for packed-bed reactors. The presentation will compare the performance of wall-coated and packed-bed microreactors.

This work has been funded by the U. S. Army Research Laboratory under the Collaborative Technology Alliance Program, Cooperative Agreement DAAD19-01-2-0010

References

[1] Hu, J., Wang, Y., VanderWiel, D, Chin, C., Palo, D., Rozmiarek, R., Dagle, R., Cao, J., Holladay, J., Baker, E. Chem. Eng. J. 93, 5560 (2003)

[2] Yagi, S. and Kuni, D AIChE J. 3, 373 (1957)

[3] Mears, D.E. J. Catal. 20, 127 (1971).

[4] Bravo, J., Karim, A., Conant, T., Lopez, G.P., Datye, A. Chem. Eng. J. 101, 113 (2004)

130i Modeling a Structured Reactor Based on a Plate Heat Exchanger Design

Jean Pierre Leclerc, Mahvand Vafaei

Open Plate Reactor (OPR) is a new chemical reactor which is based on a structure of a plate heat exchanger. It is then a flexible construction that consists of reaction modules and heat transfer modules assembled in a frame. Reaction modules can be connected in parallel and/or in series depending on reaction. The flexible structure of OPR allows to inject the reactants at different points of this reactor. Heating or cooling is achieved through heat transfer modules, designed from high-performance heat exchangers and located on both sides of the reaction modules. The direction of flowing of utility fluid in heat transfer modules can be easily modified and adapted to the reaction. This makes possible to adapt the configuration of the reactor in function of the process of the reaction and controlling the evolution of temperature.

Modelling the behaviour of this tructured reactor is the aim of this work. The model provides a classical global approach to the process engineering that can be interesting for chemical industries.

The measurements of residence time distribution are used for characterisation of the reactant flow. Then the flow is modelled on the basis of series of stirred tank reactors for low viscosity fluids. When the viscosity increases, an exchange with the stagnant zone is taken into account. The proposed model has been validated for three homogeneous liquid phase reactions.

These reactions are :

Alkaline hydrolysis of ethyl acetate

Alkaline hydrolysis of ethylene glycol diacetate

Oxidation of sodium thiosulphate by hydrogen peroxide

The first reaction is a single reaction, it is only sensitive to global hydrodynamic of the reactor. The second one is a series of competitive consecutive reactions, it allows studying selectivity problems. The third one is a very exothermic reaction and permits to study the thermal capacity of the reactor.

After validation of the model, predictive simulations are done. These simulations study the influence of the conversion for the reaction 1, the selectivity for the reaction 2 and the effect of stagnant regions in the case of eventual local overheating

131a A Catalytic Micro-Structured Reactor Assembly for Generating Hydrogen from Complex Hydrocarbons

Jeff S Pickles, Jeffrey Harrison, Patricia Irving

A micro-channel fuel processor for catalytic hydrogen production based on steam reforming and membrane separation technology has been developed for multiple fuel types. After successful demonstration of our first prototype system in 2002, the system has been further optimized by redesign of the major components, emphasizing appropriate micro channel sizing, balance between heat transfer and catalyst volume, compactness, improved thermal and mass transfer, and system efficiency. The reactor, heat exchangers, membrane separation assembly, and manifold are built on a microchannel design, engineered to provide a very high surface to volume ratio for improving heat and mass transfer while maintaining the compactness of the system.

For the catalytic reactor, an innovative approach using catalytic channels that alternate with cross-flowing heat exchange channels provides the conditions for very high space velocity to meet our requirements for high efficiency, small volume, quick start up and rapid response to changes in hydrogen demand. Finite element and CFD modeling was used to determine the dimensions and number of heat exchange and catalytic microchannels needed for optimum chemical processing. A novel fabrication approach was utilized to cost-effectively produce complex micro-structured prototypes for testing. This approach included the use of iconel shim stock as the fundamental material of construction. The shims were chemically etched to achieve the micro-structure geometry required, then stacked, diffusion bonded and cut using wire EDM. This approach allowed us to include several chemical process components into one integrated reactor. The integrated components included:

- Burner that provided hot exhaust for the heat exchange channels
- Endothermic catalytic reactor
- Exothermic reactor heat exchanger
- Steam generator
- Steam superheater
- Air preheater
- Steam condenser
- Membrane feed heater

Several configurations were built using this process and tested to verify and adjust modeling parameters. Results from these tests will be presented along with the process we used to arrive at the optimum arrangement and grouping of the system subcomponents into a completely integrated portable chemical plant.

Other components of the reforming system also incorporate microchannel design elements including a miniature diesel and steam mixer/injector of novel design, and a membrane module for hydrogen purification. The entire system is manifolded together using a unique approach that eliminates fittings and piping between components. A history of our design iterations showing our progressive development to a fully integrated multi-component micro-structured hydrogen production plant for portable and distributed applications will be described.

Assessment of the Ecological Potential of Micro-reaction Technology

Dana Kralisch , Günter Kreisel, Annegret Stark
Institute for Technical and Environmental Chemistry, Friedrich-Schiller-University Jena, Germany

Abstract

By means of life cycle assessment (LCA), the question whether the transfer of a chemical synthesis from a macro-scale discontinuous batch process to a continuous micro-scale set-up would bring about ecological advantages, was assessed. For the comparison, the reaction was carried out both in a discontinuous double-walled reactor and in the continuous Cytos® Lab system of CPC Systems GmbH, Germany. The two-step synthesis of *m*-anisaldehyde from *m*-bromoanisole was taken as a model reaction. This synthesis, due to its highly exothermic behavior, can only be carried out under significant safety precautions and with high cooling energy demand in conventional batch reaction equipment. The application of micro-structured devices allows for a continuous isothermal reaction set-up.

The results of the LCA, expressed as environmental impact potentials, clearly indicate the significant ecological advantage associated with the continuous synthesis in the micro-reactor for the selected model reaction. This outcome can be clearly related to the savings in energy consumption, the reduction of solvents, and the increase of reaction yield in the micro-reactor. Compared to these saving potentials, the fabrication of reactors, thermostats, distillation units etc. only plays a minor role. Upon frequent exchange of micro-reactors, toxicity potentials, resulting from the stainless steel production, become increasingly important. These toxicity potentials surmount in certain cases those of the alternative discontinuous process.

Introduction

For several years now, micro-reaction technology represents a serious alternative to conventional macroscopic methods [1, 2]. Increased surface-to-volume ratios lead to lower cooling- and heating power requirements, which favorably contribute to ecologically significant reductions of power consumption. Excellent control of heat transfer may additionally result in higher yields and selectivities of chemical conversions [3, 4]. Therefore, a great reduction of the environmental burden is expected from the implementation of micro-reaction technology in industry.

The advantages of micro-reaction technology mentioned are, however, also accompanied by disadvantages. The manufacture of micro-reactors is more material- and energy consuming than that of conventional equipment. This holds especially for the production of micro-structures from stainless steel. In addition, micro-reactors might have shorter life times than conventional stirred tank reactors due to fouling (clogging) phenomena in the micro-channels.

The pros and cons of micro-reaction technology were investigated using life cycle assessment (LCA) methodology [5]. The ecological advantages associated with the transfer of a chemical synthesis from a macro-scale discontinuous batch process to a continuous micro-scale set-up were investigated, and the potential environmental effects of both alternative methods compared.

The system boundary spans from the earlier life cycle stages of the chemicals used, over the energy supply, the realization of the model reaction, work-up, and transports, to the disposal of wastes. Additionally we were interested in the dependency of the choice of reactor (stainless steel micro-reactor vs. discontinuous borosilicate glass double-walled reactor) on the additional expenditure of materials and energy necessary. This aspect was investigated on the example of different case scenarios. In order to provide a holistic view of the influence of the peripheral paraphernalia, the reaction vessels, control elements, thermostats, distillation equipment *etc.* were included into the balance (Figure 1).

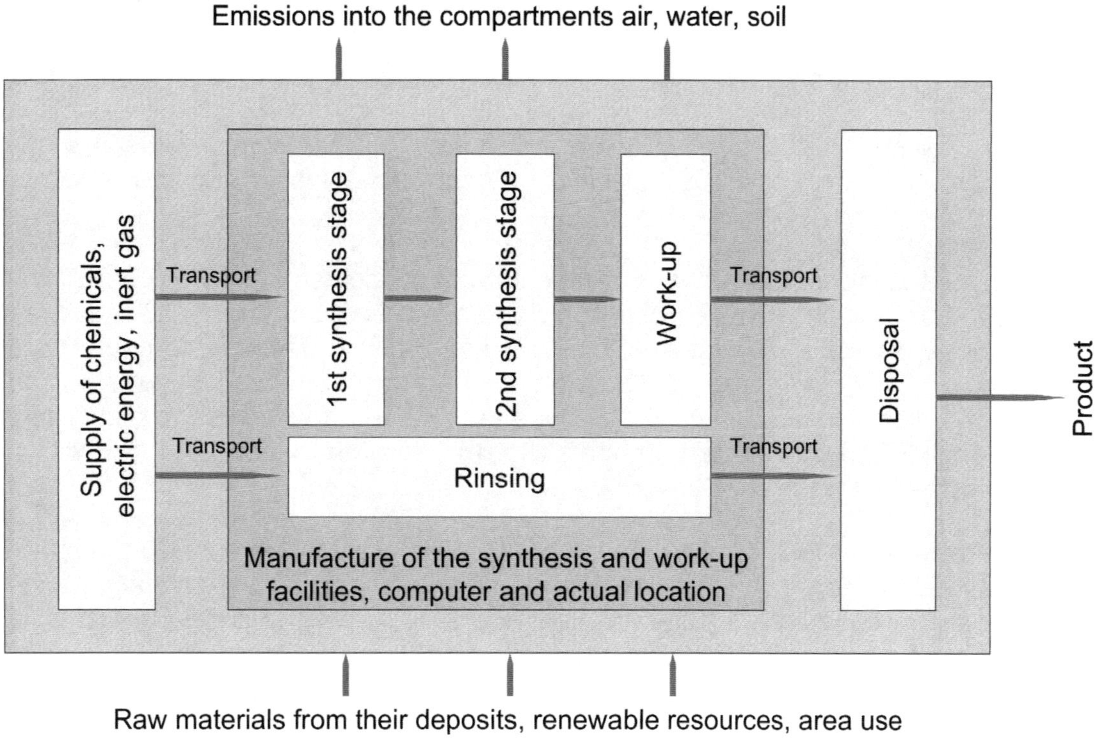

Figure 1: System bounday

Model reaction and its realization

A very exothermic lithium-organic reaction (Figure 2) was chosen as basis for the comparison of the alternative methods. In a macro-scale discontinuous batch reactor, this reaction can only be carried out under extreme safety precautions and high energetic effort.

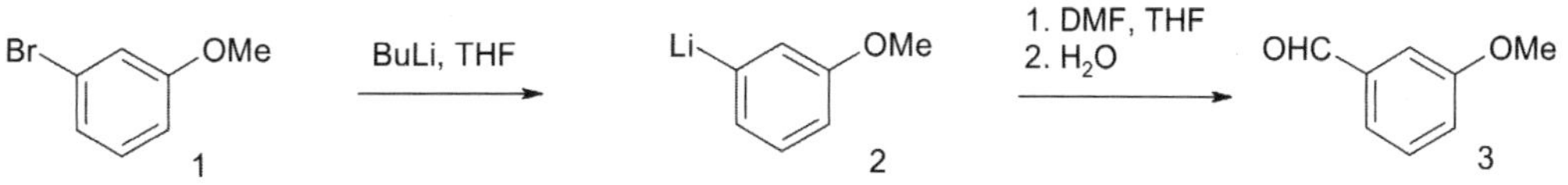

Figure 2: Model reaction

At stage I of the synthesis, *m*-bromoanisole (1) and *n*-butyllithium are converted *via* brom/lithium exchange to yield *m*-lithium anisole (2) and *n*-bromobutane. At stage II, the reaction mixture is treated with the electrophilic dimethylformamide, after which the reaction is quenched using 3M hydrochloric acid. At both synthetic steges, tetrahydrofurane is used as solvent.

The discontinuous experiment was conducted in a 10 L double-walled reactor (manu-facturer: QVF). A temperature of 323.15 K (- 58°F) was maintained using a thermostat. Both, the two reaction steps and the first work-up step were carried out in the above-mentioned apparatus. The organic phase was distilled *in vacuo* after phase separation. In order to obtain pure *m*-anisaldehyde, the residue was submitted to two further distillation procedures after removal of the low-boiling solvents. This procedure resulted in a yield of 60.3 % of the theory (0.4 kg within 11.5 h), with a purity of 96 % (GC). Cleaning and rinsing between batch runs required 4.75 h.

It should be noted that the reaction can hardly be controlled in reactors of larger capacity, *e.g.* 50 L, due to the extremely exothermic nature of this reaction. Additionally, a strong dependency of the batch size on the reaction outcome has been observed [4], in so far as the yield of *m*-anisaldehyde is reduced in larger reactors.

In continuous mode, the two-step synthesis was conducted at 373.15 K (32°F) in two Cytos®-Lab-System modules connected in series. Besides the micro-structured reactor, these include also control elements and pumps. A personal computer was used to control the overall set-up. The work-up of the reaction mixture was carried out as in the discontinuous reaction by phase separation and successive distillation. For this reaction mode, the yield amounted to 87.7 % of the theory (1.4 kg), with a purity of 96 % (GC), after a run-time of 24 h.

When transferred to a continuous micro-scale set-up, the increased surface-to-volume ratio allows for an improved control of the heat supply and –removal, and the reaction can be conducted isothermally. A decreased energy demand for cooling processes and an increase in yield resulted, which can be attributed to a short residence time of the instable lithium-containing intermediate (2) (a few seconds) before being submitted to stage II of the reaction. The short residence time of the intermediate (relative to its rate of decomposition) leads to a higher yield when compared to the batch process. In this latter instance, several hours are necessary for the dosage of the *n*-butyllithium, depending on the batch size [4].

Scenarios of service life of micro-structured reactors

As opposed to laboratory glass-ware from borosilicate glass, there is hardly any knowledge about the service life of micro-structured stainless steel appliances, which is dependent on its chemical resistance and breakage safety. Moreover, its applicability depends on the type of synthesis performed. In case of "fouling" phenomena within the micro-channels, an immediate exchange of the Cytos® becomes necessary. In order to elucidate the band-width of exchange cycles possible with respect to the resulting potentials of environmental burden, four scenarios were envisaged. The "worst case scenario" prescribes as minimum prerequisite, a life-time of one week. In order to allow for comparison of the two synthetic methods, the presumed "best case" (scenario 3) would be equal to the service life of a double-walled glass reactor, which is in average thirty years, according to the manufacturer QVF. Scenario 1 and 2 represent life expectancies of three month and three years, respectively, thus ranging between the worst and best case scenario. A comprehensive assessment of the influence of the supply of Cytos® micro-structured reactors on the overall result is thus feasible.

Selected results of the life cycle impact assessment (LCIA)

During a life cycle impact assessment (LCIA), results generated in a life cycle inventory (LCI) are condensed and assigned to potential environmental effects, with the aim of generating a manageable number of impact categories and thus increasing the comparability of the data. Using an equivalence factor, each value is related to a reference compound. For this contribution, this LCIA includes the impact categories: cumulative energy demand (CED), raw material demand and area use, global warming potential (GWP), ozone depletion potential (ODP), photochemical ozone creation potential (POCP), acidification potential (AP) und nutrification potential (NP) according to UBA [6]. Additionally, an assessment of the human toxicity potential (HTP), fresh aquatic eco-toxicity potential (FAETP), marine aquatic eco-toxicity potential (MAETP) and of the terrestrial eco-toxicity potential (TETP) according to Huijbregts et al. [7] was taken into account.

On the examples of the impact categories CED, GWP, AP and HTP, some of the results of the LCIA obtained are presented in the following. All results refer to a production of 10 kg of *m*-anisaldehyde as functional unit (FU). The CED is calculated from the sum of the total energy input for the whole process chain. For this purpose, the energy input is either described or experimentally determined *via* the heating value, potential efficiency, material requirements *etc.* [8]. Figure 3 (top left) shows a comparison of the CED of the process scenarios „batch" and „continuous", and highlights that the transfer of the model reaction from batch to continuous mode in a micro-structured reactor leads to a reduction of CED of between 51 to 55 %. The life-span of the micro-scale set-up plays a subordinate role.

Anthropogenic green-house gases in the atmosphere influence the heat radiation of the earth, and thus lead to a warming of the troposphere. The potential of the greenhouse gases are given in carbon dioxide equivalents, which serves as a reference. The effects of the respective gases are dependent on their longevity, and the conversion factor to carbon dioxide equivalents was determined for a time scale of 100 years. The greenhouse gases carbon dioxide, methane, trichlorofluoro-methane, dichlorodifluoromethane, and perfluoro-methane were included into the calculations. Figure 3 (top right) shows that the transfer of the model reaction to a continuous mode using a micro-reactor results, when compared to the batch-mode, in a reduction of the GWP by 52 to 57 %, again independent on the life-span of the reactor.

The acidification of the compartments water and soil due to the imission of acidogenic compounds may lead to damage to forests, and to the release of toxic aluminum ions. After their generation during the combustion of fossil fuels, these compounds are partly oxidized in the environment and react with water to yield acid. The estimation includes the acidification potential of sulfur dioxide, nitrogen oxides, ammonia, hydrogen chloride and hydrogen fluoride. A reduction of the acidification potentials by 42 to 45 % was determined, if the synthesis of *m*-anisaldehyde was conducted in the continuous micro-reactor mode rather than in batch-mode.

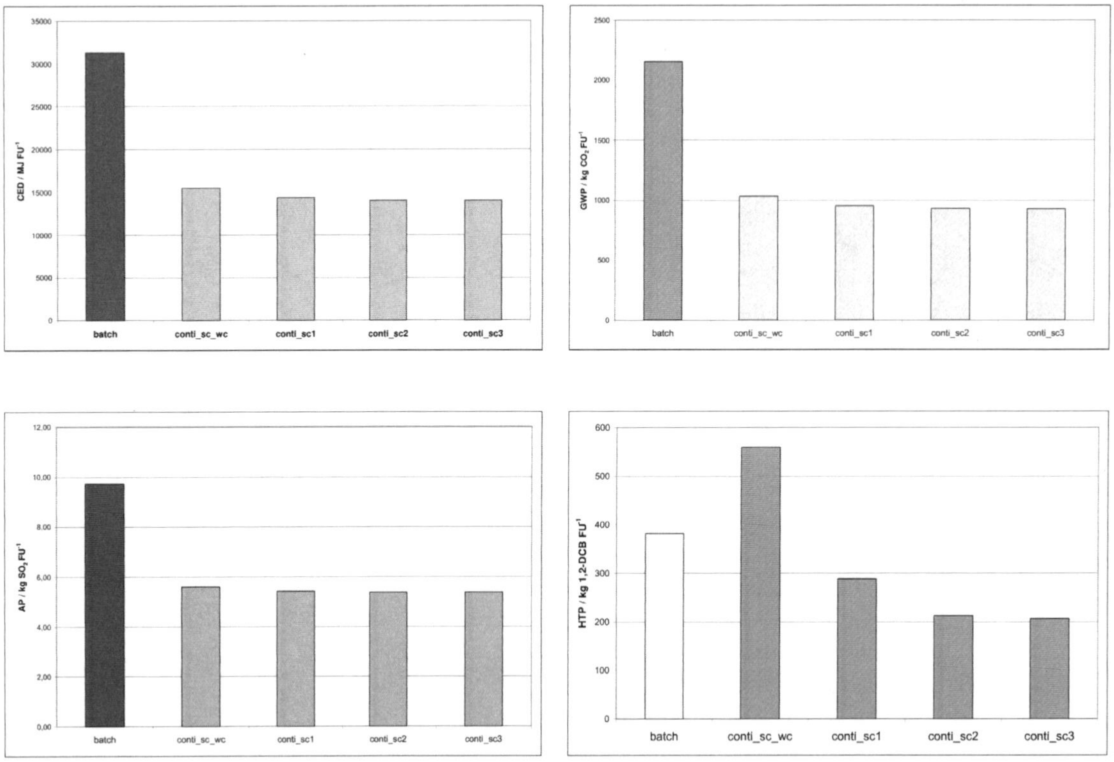

Figure 3: Results obtained in the impact categories: CED (top left), GWP (top right), AP (down left) and HTP (down right).

HTP includes human intake of toxic compounds by inhalation of contaminated air, and from drinking water and food. Threshold values such as the MAC-, MIC- or NOEL-values are used as basis for the determination of the HTP [7]. As far as data was available, the initial emission compartments air, fresh- and sea water, and agriculturally used area were taken into account. The equivalence factors of this impact category were referred to the substance 1,4-dichlorobenzene. The results are based on the toxicity potentials of the heavy metals antimony, arsenic, barium, beryllium, lead, cadmium, chromium III, chromium VI, cobalt, copper, nickel, mercury, selenium, thallium and vanadium. Additionally, both inorganic compounds, such as ammonia, hydrogen chloride, sulfur dioxide and nitrogen oxides, and organic compounds, such as benzo(a)pyrene, benzene, hexachlorobenzene and PAH (polycyclic aromatic hydrocarbons) were considered.

Figure 3 (down right) indicates that the HTP is increased by 47 % for the „worst case"-scenario of the continuous process as opposed to the batch-mode. This effect results mainly from the emission of chromium and nickel during stainless steel production. However, realistically, longer life-spans of the Cytos® Lab System are expected, thus leading to a significant reduction of the HTP of up to 46 %. A similar result was also obtained for both, FAETP and the TETP, while the MAETP was decreased in the continuous production mode in each scenario considered.

Analysis of the sensitivity of selected results

The presented results of the impact assessment are valid for the model synthesis and the presumptions made in the course of the balance. These results may be used as indication for similar problems, but should not be generalized. In order to validate the effect of single factors on the overall result, both the CED and yield are examined more closely in the following.

Breakdown of the CED

Figure 4 shows the break-down of the CED into single modules of the process chain, illustrating the great influence of both, the supply of chemicals and the energy demand during synthesis and work-up. The results of the „worst case"-scenario were used for the comparison. It shows again that the supply of the production unit only plays a minor role from an energy point of view, while the supply of the peripheral equipment exhibits almost no effect.

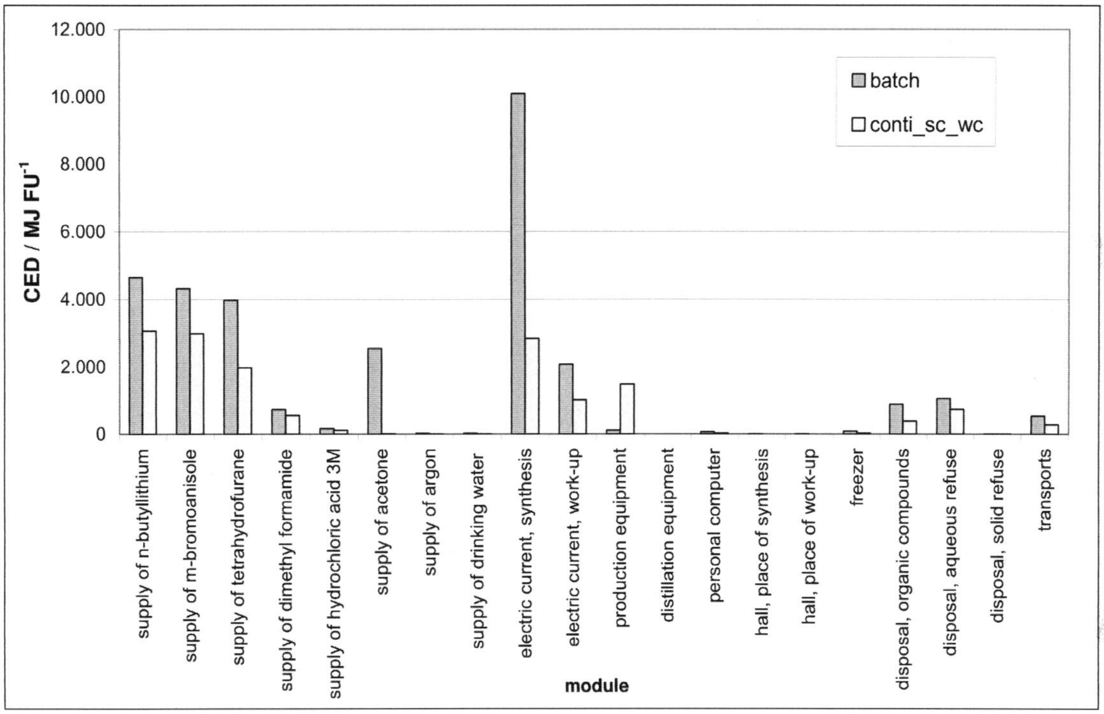

Figure 4: Break-down of the CEDs of the scenarios „batch" and „worst case continuous" mode in single modules.

Influence of the yield on the results of the LCIA

The transfer of the model reaction from batch-mode to continuous micro-scale set-up resulted in a yield increase of 27 %. This effect influences considerably the results of the LCIA. A break-down of the two most dominant aspects, *i.e.* the chemical demand and the energy demand during the synthesis, allows for the quantification of this influence. The main target was to scrutinize whether a decrease of the energy demand obtained in continuous mode would lead to a reduction of the overall environmental burden, even if this was juxtaposed by an increased expenditure for the supply of the micro-structured devices. For this purpose, a comparable yield of 88 % was presumed for both, the batch-mode and continuous-mode.

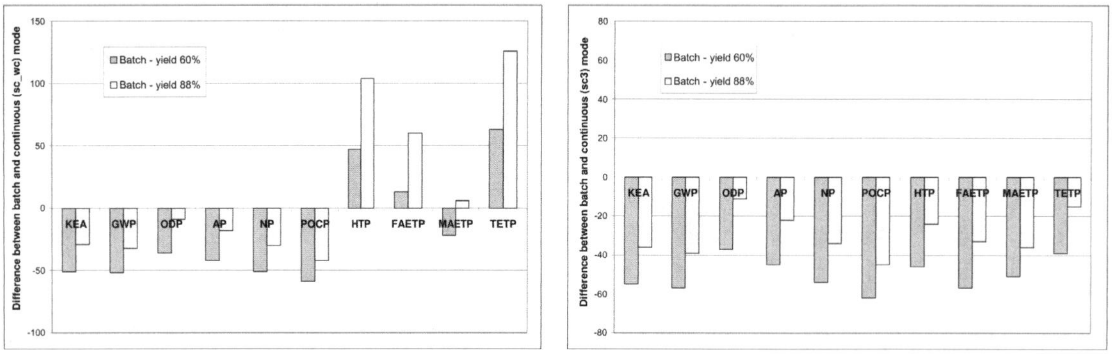

Figure 5: Effect of the difference between batch- and continuous mode on the impact categories. Yellow bars: overall batch yield 60 %, Light yellow bars: (presumed) overall batch yield 88 %. Left: worst case scenario, right: best case scenario).

Figure 5 shows the alterations of each single impact category for presumed batch-mode yields of 60 and 88 %, resulting from the transfer to the continuous Cytos® Lab System, for the „worst case"-scenario, in which a weekly change of micro-structured device is necessary. It could be shown, that the increase in reaction temperature results in a reduction of energy demand for most impact category, while the negative effects of the impact categories HTP, FAETP and TETP becomes even more pronounced. This can be counteracted by a longer life time of micro-reactors. Thus, Figure 5 (right) illustrates for the best case scenario (30 years), a reduction of the resulting environmental burden in all impact categories, which is brought about by the continuous reaction mode in the micro-reactor system.

As an example, the break-even point for the HTP in the case of 60 % yield, *i.e.* where the HTP of the continuous system is equal to that of the batch mode, lies at about two months operational time. In case of the (presumed) yield of 88 % in the batch mode, the operational time at equal HTP is almost six months.

Conclusions and future prospects

The results of the LCA, expressed as environmental impact potentials, clearly indicate the significant ecological advantage associated with the continuous synthesis in the micro-reactor. This outcome can be related to the savings in energy consumption, the reduction of solvents, and the increase of reaction yield achieved in the micro-scale set-up. Compared to these saving potentials, the fabrication of reactors, thermostats, distillation units *etc.* only plays a minor role. Upon frequent exchange of micro-reactors, toxicity potentials, resulting from the stainless steel production, become increasingly important. These toxicity potentials surmount in certain cases those of the alternative discontinuous process. Therefore, for reasons of both environmental impact potentials and economic aspects, special attention should be paid to the longevity of the micro-structured devices from stainless steel.

An analysis of the various impact categories showed that even at equivalent conversions obtained in the batch- or continuous mode, the latter mode is more ecologically advantageous. Thus, less energy is required, which leads to a lower environmental burden in most of the impact categories. However, the effect of higher toxicity- and eco-toxicity potentials (mainly due to the stainless steel production) appears more blatantly.

The comparison of a continuous reaction mode in a micro-structured reactor and a discontinuous reaction mode in a double-walled reactor presented here is based on data collected on mini-plant scale. In our future work, this comparison will be conducted for industrial production scale.

The authors thank the Deutsche Bundesstiftung Umwelt (DBU), Osnabrück, Germany, who is sponsoring this work (ref. no. 19 348), and the partners (CPC Systems GmbH, Bayer AG) involved in this cooperation for fruitful team-work as well as for providing the data for this life cycle assessment.

[1] W. Ehrfeld, V. Hessel, H. Löwe, Microreactors - New Technology for Modern Chemistry, 1st ed. Wiley-VCH, Weinheim **2000**.

[2] S. J. Haswell, R. J. Middleton, B. O'Sullivan, V. Skelton, P. Watts, P. Styring; The application of micro reactors to synthetic chemistry; Chem.Commun. **2001**, 391-398.

[3] K. Jähnisch, V. Hessel, H. Löwe und M. Baerns; Chemie in Mikrostrukturreaktoren; Angew. Chem. **2004**, 116, 410-451

[4] T. Schwalbe, V. Autze, M. Hohmann, W. Stirner; Novel Innovation Systems for a Cellular Approach to Continuous Process Chemistry from Discovery to Market; Org. Proc. Res. & Dev. **2004**, 8, 440-454

[5] D. Kralisch, G. Kreisel. Bewertung der ökologischen Potenziale der Mikroverfahrenstechnik, Chem. Ing. Tech., **2005,** *in press*

[6] Umweltbundesamt, Methodik der produktbezogenen Ökobilanzen – Wirkbilanz und Bewertung, UBA-Texte 23/95, Berlin, **1995**

[7] M.A.J. Huijbregts, U. Thissen, J.B. Guinée, T. Jager, D. Kalf, D. van de Meent, A.M.J. Ragas, A. Wegener Sleeswijk, L. Reijnders, Priority assessment of toxic substances in life cycle assessment. Part I: Calculation of toxicity potentials for 181 substances with the nested multi-media fate, exposure and effects model USES-LCA, Chemosphere **2000**, 41, 541

[8] U. R. Fritsche, W. Jenseit, Ch. Hochfeld, Methodikfragen bei der Berechnung des Kumulierten Energieaufwands (KEA), Arbeitspapier im Rahmen des UBA-F&E Vorhabens Nr. 104 01 123, Darmstadt, **1999**

131c Concept of Microreaction Plants for Aggressive Chemicals

Andreas Freitag, Ralf Scholz, Thomas R Dietrich

The concept of a microchemical plant will be explained. The described plants have been designed and built to carry out different chemical reactions. The microfluidic reaction devices are made from glass, because of its ideal properties, e.g. the chemical stability to use reactive and aggressive chemicals. For the same reason, all modules to run a glass microreactor, e.g. pumps, valves, tubings, temperature-, pressure-, flow-sensors, analytical modules, safety installations, etc. have to have the same chemical stability.

In a conventional chemical production the apparatus is fixed and build out of standard equipment. The chemistry is forced to work in this reactor. In contrast, microreactors give the opportunity to build a fluidic system which is optimized for the chemical reaction ! Therefore, a flexible material and structurization process is necessary. For the current work the photoetchable glass FOTURAN has been used. The material itself is photosensitive and allows to form channels, holes, or chambers with a high aspect ratio of approximately 20:1. The glass is exposed by UV-light through a mask which contains all necessary structures. In a subsequent heat treatment the glass crystallizes in the exposed areas. Afterwards, these crystalline areas can be etched out of the remaining glass by hydrofluoric acid. With this process microreactors, mixers (for liquid-liquid as well as for gaseous-liquid reactions), heat exchangers, dwell devices, analytical modules etc. can be built. Even very complicated three-dimensional structures can be manufactured by etching different geometries into different layers and bond up to 20 of these layers together to form a microreactor.

These microfluidic modules out of glass can be connected to supplying modules via commercially available tubes and fittings, e.g. from HPLC. As long as temperatures are not above 400 K, all tubings and valves can be made out of teflon.

There are no perfect pumps for microfluidic systems on the market, yet. Most pumps show a relatively high pulsation while others are made out of materials which do not resist aggressive chemicals. An optimum seems to be a rotary piston pump with a ceramic head, combined with a pulsation damper. Flows between 0,01 ml/min and 100 ml/min and pressures up to 6 bar can be handled. With other pumps, e.g. syringe pumps, even higher pressures can be applied if necessary.

microreaction systems have to be controlled by an automatic control software, like the SIMATIC S7 from Siemens. This control unit needs the input of the relevant process parameters like temperature, pressure, or flow. These sensors, too, should show the same chemical resistivity as the glass reactor itself. The sensors should be within or as close as possible to the interesting area. mikroglas has developed together with Buerkert a so-called "sensorblock", which enables the measurement of different parameters within the microfluidic system. In addition, the sensorblocks have various safety features like a pressure relief valve or a non-return valve. Other chemical sensors, like pH sensors can be integrated easily.

The whole microreaction system is designed in a flexible modular way, so that it is very easy to replace reactors as well as supplying units like pumps, sensors, or filters. The goal with this "multi-purpose-concept" is to be able to change the setup within one or two days to switch from one chemical production to another. This enables the user to carry out different small volume productions within one system. It could be shown that with such a microreaction system of a size of a table top apparatus, it is possible to handle up to 2 tons of chemicals per day.

Using different examples (e.g. nitrations, sulfonizations, ozonizations, metal organic reactions, photo reactions) it will be shown how flexible such a system can be. In combination with peripheric instrumentation, like auto-sampler, IR-spectroscopy, the number of applications can be increased.

As an outlook, a new started project, funded by the German Federal Ministry of Education and Research, will be presented. The goal of this project is to show how microreactors and plants can be used for cost-effective and safe development and production in the chemical industry.

Numbering-up of Microreactors for Radical Polymerization

Takeshi Iwasaki [a], *Jun-ichi Yoshida* [b]

[a] *The Research Association of Micro Chemical Process Technology, Kyoto University, Kyoto 615-8530, Japan*

[b] *Department of Synthetic Chemistry and Biological Chemistry, Graduate School of Engineering, Kyoto University, Kyoto 615-8510, Japan*

Free radical polymerization is an important process for the synthesis of macromolecules since free-radicals are compatible with a wide variety of functional groups, which are not compatible with ionic and metal-catalyzed polymerization. It is important to note that the precise temperature control is essential for carrying out free radical polymerization in highly controlled manner, because free radical polymerization reactions are usually highly exothermic. Therefore, polymerization reactions in conventional macro-scale bath reactors often suffer from inefficient heat removal and the lack of homogeneity of the reactor temperature, which eventually give rise to low level of molecular weight distribution control. Thus, a major concern with free radical polymerization is the controllability of the reaction temperature from both academic and industrial view points.

Among inherent advantages of microreactors, efficient heat transfer seems to be one of the most important features for free radical polymerization. Generally, the heat generation rate in polymerization increases in proportion to the reactor volume, while the heat removal capacity decreases with the increase of the reactor volume. Because the heat of polymerization is removed through the wall of the reactor, a wall-surface area / reactor-volume ratio, which decreases with the increase of the size of the reactor, plays a crucial role. Thus, in polymerizations in macro-scale reactors, heat removal capacity often becomes a limiting factor. Therefore, the advantage of the polymerization in microreactors is obvious. As the reactor size decreases, heat transfer becomes faster and the temperature control becomes more efficient, leading to better control of molecular weight distribution.

It is important to note that the temperature control problem described above may arise as a serious problem during the scale-up from a bench process to an industrial production using conventional macro-scale reactors. Micro chemical plants, in which the increase of the production scale can be achieved by the numbering up of microreactors without changing the size of each of them, are expected to solve this problem. The advantages of microreactors of easy modulation and the possibility of combining reactors in parallel promise a quick means for making chemical plants for large scale production.

We first examined the radical polymerization in a single micro tube reactor which inner diameter was 0.5mm. For the polymerization of butyl acrylate (BA), the polydispersity index (PDI) of the polymer obtained using the microreactor was much smaller than that obtained with a macro-scale batch reactor. The result can be explained in terms of much higher heat removal efficiency of the microreactor compared with the macro-scale batch reactor. For the polymerization of benzyl methacrylate (BMA) and methyl methacrylate (MMA), the effect of microreactor on PDI was smaller than the case of BA. For the polymerization of vinyl benzoate (VBz) and styrene (St), no appreciable effect of microreactor on PDI was observed. These experimental results indicate that the microreactor is quite effective to the molecular weight distribution control for highly exothermic free radical polymerizations (BA, BMA, and MMA), but that it is not so effective for less exothermic polymerizations (VBz and St).

In order to expand production scale, we started the research on numbering-up reactor devices for radical polymerization. We have developed a shell and tube reactor which involves 94 peaces of micro tubes (inner diameter = 0.5 mm, length = 60 cm). We examined the polymerization of MMA using this reactor, and obtained same polymerization yield, molecular weight, and PDI compared to a single micro tube reactor.

Reference

T. Iwasaki, J.-I. Yoshida, Macromolecules, **2005**, *38*, 1159-1163

Flexible Plant Concepts in Microchemical Engineering

D. Kirschneck, W. Linhart, [1]M. Kober, [1]R. Marr
Microinnova Dr. Kirschneck KEG, Feuerbachgasse 30c, 8010 Graz, Austria;
tel. +43316 8734970, e-mail: office@microinnova.com, www.microinnova.com.
[1]Technische Universität Graz, Institut für Thermische Verfahrenstechnik und Umwelttechnik, Inffeldgasse 25, 8010 Graz, Austria;
tel. +43316 8737470, e-mail: marr@tvtut.tu-graz.ac.at;

Microreaction technology and microchemical engineering are fast growing disciplines in chemistry and chemical engineering. A large number of microstructured devices have been developed and production techniques for producing them [1]. A lot of work has been done investigating different reactions [2]. But only a few concepts for plants with microstructured devices are available up to now. Reaction processes or separation processes can be intensified by using this technology. Even new processes become possible by using these plant concepts based on microstructured devices [3]. Traditional chemistry usually does not care too much about mass transport and mass transfer. Microreaction technology and microchemical engineering provide a new way for process intensification by optimisation of transport processes. 20 % of all chemicals traded in the European Union are produced in quantities of less than ten tons per year. They usually produced in inefficient stirred vessels, where the chemistry will be designed for the existing equipment and not the other way round. New plant concepts based on microstructured devices offers a new way to produce this kind of chemical by means of a new concept.

A new concept "Chemical Production Anywhere" for on-site-production of chemicals was developed. The aim was to construct a small portable unit to produce hazardous or explosive chemicals, where ever they are needed. This avoids dangerous transporting or storing of these chemicals. Based on these considerations a complete ready-to-use plant is fixed into a suitcase. The suitcase was equipped with two storing bottles filled with educts substances, pumps, reactor, heat exchangers and typical sensors like pressure, temperature and throughput. All this is operated by a build-in state-of-the-art process control unit. The user interface was designed. Since the integrated web engine has its own IP-address experiments or production runs can be controlled or monitored by a normal web browser. Only power supply is needed for operation. An even this can be provided by a mobile generator. Data of model applications and validation of all components of the system are presented

The second new concept "Lab Experiment Toolbox" was designed to speed up process development and reduce the time-to-market of chemicals or compounding products. A modular toolbox has been developed to lab scale experiments. Small tube connections are used to connect various unit operation devices to each other. More than 30 different devices are available to carry out various unit operations like mixing, heat exchanging, precipitation or reactors. A new experimental setup takes dependent on the number of devices, which have to be integrated several minutes up to one hour. The integration of the sensors into system is different to the other concept before. The process control unit has got a front panel, where common sensors can be added on plug-and-play basis. All experimental data are stored in spreadsheets, which can be transferred to common spreadsheet programs for further data processing. Online analytics like UV-spectroscopy, IR- spectroscopy or micro gas chromatography can be integrated easily into the system. .Due to small residence times a lot of experiments can be carried out in a very short time. Multistep synthesis can be carried even automated reporting for standard experiments are possible.

This work shows how plant concepts based on microstructured devices can be beneficial for chemistry and chemical engineering. Especially fine chemicals and pharmaceuticals can be produced with higher yields and smaller, more efficient plants. Development can be speed up and plant safety can be increased because of small reaction volumes.

[1] Madou, M.J.; Fundamentals of Micro Fabrication: The Science of Miniaturisation, 2nd Ed., CPC Press LLC, Boca Raton/Florida, 2002

[2] IMRET7, Proceedings, Lausanne 2003

[3] Ehrfeld, W.; Hessel, V.; Löwe, H.;Microreactors-New Technology for Modern Chemistry, Wiley-VCH, Weinheim, 2000

DEMiS®: Results from the development and operation of a pilot-scale micro reactor on the basis of laboratory measurements

F. Becker, J. Albrecht, G. Markowz, R. Schütte, Degussa AG, Hanau
St. Schirrmeister, K.J. Caspary, Uhde GmbH, Dortmund
H. Döring, Th. Schwarz, E. Dietzsch, E. Klemm, TU Chemnitz, Chemnitz
Th. Kruppa, F. Schüth, MPI Mülheim, Mülheim

Introduction

Many investigations in the past decade evaluated prospects and limitations of micro reactors for the production of chemicals. Various advantages of micro reaction technology have been identified and demonstrated on the laboratory scale. Those potentials include higher space time yield (i.e. smaller reactor), higher selectivities (i.e. higher feedstock efficiency), full conversions (i.e. no recycle stream), or safety benefits (i.e. safe operation with explosive mixtures). Especially for heterogeneously catalyzed gas phase reactions only very few attempts have been made so far to transfer promising laboratory results into industrial production using micro reactor technology. The project DEMiS® (Demonstration Project for the Evaluation of Micro reactor technology in industrial Systems) is an important contribution to reduce the barrier between laboratory and industrial production. Of course, it is mandatory that significant economic, ecological and/or safety benefits can be identified.

Focus of the DEMiS® project are heterogeneously catalyzed gas phase reactions for the production of chemicals and intermediates with at least 10.000 mt per year. The epoxidation of propylene with hydrogen peroxide vapour was selected as model reaction.

Development strategy

The development strategy of the DEMiS® project was a equal-down approach, i.e. the planned design of the industrial micro reactor dictated the design of the laboratory reactor and not vice versa (see Figure 1). The reactor development was driven by process and reaction engineering aspects within the constraints of economical construction and manufacturing. A further requirement was flexibility with regard to wide range of applications and production capacities. Eventually the laboratory design is equal to the industrial design with respect to the governing micro effects which implies similarity with respect to key geometries, fluid dynamics, reaction engineering and heat management. In the DEMiS® reactor design this was realized by using slit-like reaction channels, Thus, the micro effects are induced only in one dimension, the other two dimensions can be macro-sized. Slit-like reaction chambers are realized for example by spacer elements or by grooves. Capacity is increased by stacking plates with integrated slits.

Laboratory reactor, experiments and transfer into production

As a representation of the industrial design (equal-down) the laboratory reactor comprised just one reaction chamber. Whereas the channel height is similar to the industrial reactor the channel length may be much smaller, provided that certain similarity parameters are kept constant. If mass transfer resistance lateral to the flow direction is negligible the modified residence time is the crucial parameter for the transfer of lab results to the technical scale.

The laboratory reactor platelets were first used to develop methods for the wall preparation of the catalyst. General requirements for the coating method were the ability to realize a reproducible layer thickness of several 100 µm and the transferability of the method into the industrial scale. Next, the coating parameters and process conditions were optimized in lab scale experiments. The optimum parameters were then used for the detail design of the pilot reactor ("equal-up", see Figure 1).

Results from pilot plant operation

The pilot plant periphery was equipped with conventional components in order to test the performance of the reactor in a typical industrial environment. The pilot reactor was run in different campaigns over a period of several 100 hours. With respect to operational characteristics the process and the reactor showed very good process stability and controllability. Laboratory results were reproduced satisfactorily. Differences between laboratory and pilot scale experiments could be traced to the different decomposition of vaporous hydrogen peroxide on the lab scale and under pilot conditions.

The presentation will report on the reactor design principle, the development of the catalyst preparation method, the optimization in the lab and the results from pilot plant operation.

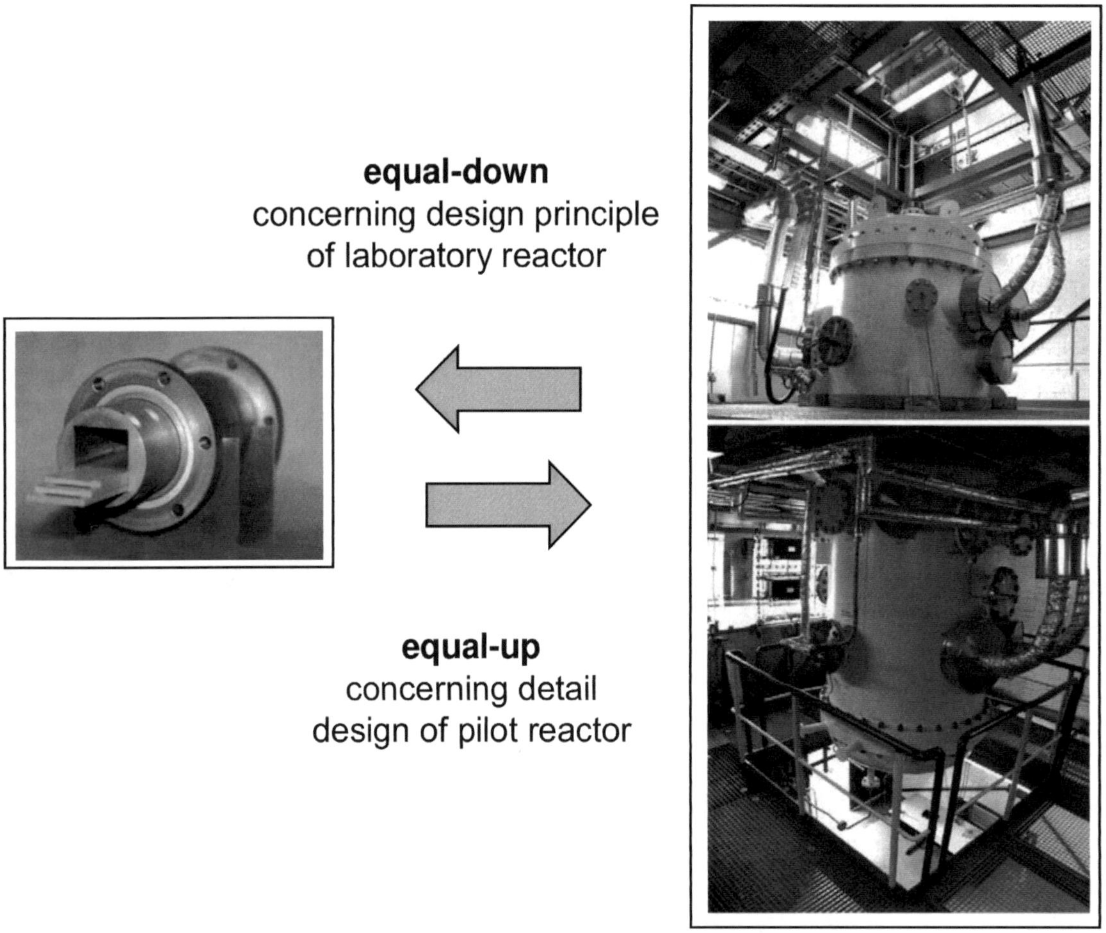

Figure 1: Development strategy in the DEMiS® project

Design of Fluid Segments in Microreactors: Influences of Design Factors on Mixing by Diffusion and Product Compositions

Nobuaki Aoki, Shinji Hasebe, and Kazuhiro Mae

Department of Chemical Engineering, Graduate School of Engineering, Kyoto University
Kyoto-daigaku Katsura, Nishikyo-ku, Kyoto 615-8510, Japan
aoki@cheme.kyoto-u.ac.jp

Prepared for the Presentation at the 2005 AIChE Spring National Meeting (IMRET8 – 8th International Conference on Microreaction Technology), April 10–14, Atlanta, GA.
Topic: TK013 [132] IMRET8 Poster Paper, Paper No. 132a.

Unpublished.

Abstract

The mixing operation that reactant fluids are split into many fluid segments to shorten diffusion length and thus mixing time is often used in microreactors. In this paper, we introduce two dimensionless numbers that represent effects of geometric design factors for the fluid segments, that is, arrangements and shapes of the fluid segment cross section at the reactor inlet, on the product yield for the reactors using this mixing operation. We also devise methods to determine the parameters. The dimensionless numbers are the ratio of reaction rate to diffusion rate ϕ and the aspect ratio of the mean diffusion length to the two-dimensional directions in the fluid segment w. To examine the validity of these parameters, we compare the yield of the desired product for a reactor having an arrangement and a shape of fluid segment cross sections at the reactor inlet with that for the reactor having rectangular cross section arranged in two-dimensional directions of the same dimensionless numbers using CFD (Computational Fluid Dynamics) simulations. The results show that the effect of arrangements for square cross section can be expressed by w, and that the order of ϕ allows us to determine whether a reactor is operated under the reaction-controlled condition even when the cross-sectional shapes are changed. Moreover, the effect of w on the product yield for cross-sectional shapes applied in this study, namely rectangle and right-angled triangle, is different when ϕ is fixed.

Keywords: microreactor, arrangements and shapes of fluid segments, mixing by molecular diffusion, yield and selectivity of multiple reactions

Introduction

Mixing performance in microreactors is the key issue for efficient production, namely high yield and selectivity of desired products. Mixing in microreactors is mainly driven by molecular diffusion, since their small dimensions lead to a low Reynolds number and thus to laminar flow in each reactor channel. The mixing time by molecular diffusion is proportional to the square of diffusion length. Hence, to shorten diffusion length and thus mixing time is necessary for fast mixing in microreactors, and the mixing operation that reactant fluids are split into many micro fluid segments is often used for this purpose. The interdigital micromixer (Ehrfeld et al., 1999), the static V-micro-jetmixer (Ehlers et al., 2000), the SuperFocus mixer (Hardt et al., 2003; Löb, 2004), and the K-M mixer (Nagasawa et al., in press) are examples of micromixers using this mixing operation.

CFD (Computational Fluid Dynamic) simulations have been used to model the flow with reaction, diffusion, and convection in reactors. CFD is especially useful for modeling the mixing and reaction in microreactors since laminar flow allows us to precisely predict the phenomena in microreactors. CFD simulations are also effective to evaluate mixing performance and determine design factors for this mixing operation. Ehlers et al. have studied the progress of the mixing and the product composition for the parallel competitive reactions in the jetmixer using CFD simulations. Hardt et al. have examined the effects of the mixer chamber geometries on the mixing performance by experiments and CFD simulations.

However, a design method for the mixing operation has not been established. This method is needed to appropriately design the operation to produce desired products at high yield and selectivity. We have been focusing on this mixing operation through addressing the design method for fluid segments. We have examined the qualitative effects of the geometric design factors on the mixing performance and the product yield for a multiple reaction system (Aoki et al., 2004), and the results have indicated that these factors greatly affect the product yield. The geometric design factors studied in the present research are arrangements and shapes of the fluid segment cross section at the reactor inlet. When the shape of the inlet channels can be changed arbitrarily, the arrangement and the shape can be regarded as the geometric design factors of the microreactors. Designing the fluid segments also leads to control the mixing rate by molecular diffusion on the plain perpendicular to the axial direction of reactors. Therefore, identifying the quantitative effect of the design factors on the mixing rate by molecular diffusion and relative rate to reactions is needed for the design of fluid segments.

To this end, we introduce two dimensionless numbers that represent effects of the geometric design factors for fluid segments on the product yield for reactors using this mixing operation. We also devise methods to determine the parameters for an arrangement and a shape of the fluid segment cross section at the reactor inlet. To investigate the validity of these parameters, we compare the yield of the desired product of a multiple reaction system for a reactor having an arrangement and a shape of the fluid segment cross sections at the reactor inlet with that for the reactor having rectangular cross section arranged in two-dimensional directions of the same dimensionless number using CFD simulations.

Definition of Dimensionless Numbers

Definition

Figure 1 shows the schematic diagram of a reactor where reactant fluids are fed with the form of fluid segments. First, we mention assumptions for the reactor. The reactor inlet plane corresponds to the *xy* plain, and the axial direction corresponds to the *z* axis. Reactant fluids for reactants A and B are split into many fluid segments before the inlet plane, and then fed into the reactor. The reactants mutually mix only by molecular diffusion, and then react in the reactor. The reaction takes place from the interfaces between each reactant fluid. Influences of channels that split reactant fluids into fluid segments are omitted in this study. Thus, the reactant fluids are fed into the reactor through the cross sections of the fluid segments in the reactor inlet plane. In this example, the

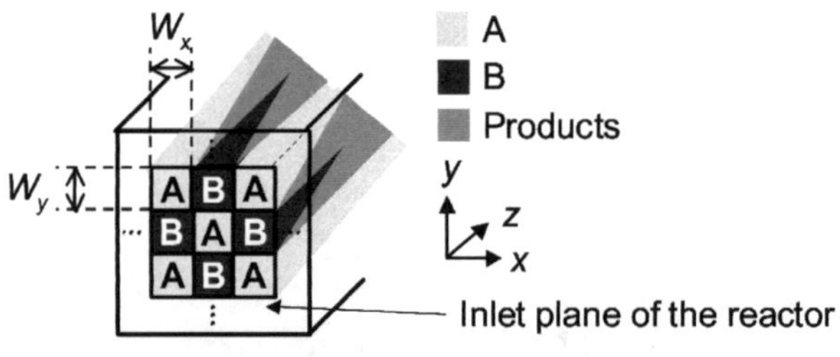

Figure 1. Reactor where reactant fluids are fed with the form of fluid segments.

cross section whose characteristic sizes, which relate to diffusion length and will be defined in the next section, are W_x and W_y are arranged in two-dimensional directions. To identify the effect of the geometric design factors of fluid segments on the mixing rate by molecular diffusion, we also assume flat velocity distribution in the reactor.

The reaction formulas and the rate equations of the multiple reactions proceeding in the reactor are as follows:

$$\left.\begin{aligned} \mathrm{A}+\mathrm{B}\rightarrow\mathrm{R},\quad r_1 = k_1 C_\mathrm{A} C_\mathrm{B} \\ \mathrm{B}+\mathrm{R}\rightarrow\mathrm{S},\quad r_2 = k_2 C_\mathrm{B} C_\mathrm{R} \end{aligned}\right\} \tag{1}$$

where A and B are the reactants and the key components (component species; R is the desired product; S is the by-product; r_i and k_i are the reaction rate [$\mathrm{kmol{\cdot}m^{-3}{\cdot}s^{-1}}$] and the rate constant of the ith step [$\mathrm{m^3{\cdot}kmol^{-1}{\cdot}s^{-1}}$], respectively; and C_j is the molar concentration of component j [$\mathrm{kmol{\cdot}m^{-3}}$]. When the axial dispersion is negligible, the mass conservation equations in a dimensionless form for the key components can be written as

$$-\frac{\partial c_\mathrm{A}}{\partial \tau}+\left(w\frac{\partial^2 c_\mathrm{A}}{\partial X^2}+w^{-1}\frac{\partial^2 c_\mathrm{A}}{\partial Y^2}\right)-\frac{k_1 C_\mathrm{B0} W_x^{\ 2}}{D_\mathrm{A}} w c_\mathrm{A} c_\mathrm{B}=0\,, \tag{2}$$

$$-\frac{C_\mathrm{B0}}{C_\mathrm{A0}}\frac{\partial c_\mathrm{B}}{\partial \tau}+\frac{D_\mathrm{B}}{D_\mathrm{A}}\frac{C_\mathrm{B0}}{C_\mathrm{A0}}\left(w\frac{\partial^2 c_\mathrm{B}}{\partial X^2}+w^{-1}\frac{\partial^2 c_\mathrm{B}}{\partial Y^2}\right)-\frac{k_1 C_\mathrm{B0} W_x^{\ 2}}{D_\mathrm{A}} w\left(c_\mathrm{A} c_\mathrm{B}+\frac{k_2}{k_1} c_\mathrm{B} c_\mathrm{R}\right)=0 \tag{3}$$

where X, Y, and Z are the dimensionless coordinates; $C_{\mathrm{j}0}$ is the initial concentration of component j; c_j and D_j are the dimensionless concentrations and the diffusion coefficient [$\mathrm{m^2{\cdot}s^{-1}}$] of component j, respectively; w is the aspect ratio of the characteristic sizes, that is, w is equal to W_y/W_x. From these equations, dimensionless concentration profiles of the components in the reactor depend on the following dimensionless numbers: $C_\mathrm{B0}/C_\mathrm{A0}$, $D_\mathrm{B}/D_\mathrm{A}$, k_2/k_1, $\phi = k_1 C_\mathrm{B0} W_x^{\ 2}/D_\mathrm{A}$, and w. The dimensionless number ϕ represents the ratio of the reaction rate to the mixing rate by molecular diffusion and is called the Damköhler number. When the first three dimensionless numbers are fixed, the profiles can be determined by ϕ and w. To determine these two parameters, we need methods to determine the characteristic sizes for the fluid segment.

Methods to determine the characteristic sizes

We then define the method to determine the characteristic sizes for fluid segments so that the dimensionless numbers represent the phenomena of the diffusion and reaction in the reactors. To correlate the characteristic size with the diffusion rate, namely the diffusion length, we first consider the mean diffusion length to each two-dimensional direction. The shortest distance to the x direction between any point in the segment cross section of a reactant and the interface between reactants as the diffusion length to the x direction. A cross section, whose area is A, has some interfaces and is divided into domains where any point takes the common interface as the end point of the diffusion length. The diffusion length at a point in each domain $S_{x,j}$, whose cross-sectional area is $A_{x,j}$, is denoted by $l_{x,j}$. The mean diffusion length to this direction for each divided area, $L_{x,j}$, is then obtained by integrating $l_{x,j}$ over $S_{x,j}$ and then divided by $A_{x,j}$. Finally, the mean diffusion length for the whole cross section L_x is obtained by area-weighted arithmetic average of $L_{x,j}$. These procedures are expressed as follows:

$$L_{x,j}=\frac{\iint_{S_{x,j}} l_{x,j}\,\mathrm{d}A}{A_{x,j}},\quad L_x=\frac{\sum_{j=1}^{n} A_{x,j} L_{x,j}}{A} \tag{4}$$

These procedures are repeated for the y direction, and then we have the mean diffusion lengths L_x and L_y. From the definition by Eq. (4), $L_{i,j}$ is the first moment of area divided by the area of each domain and thus equal to the distance to the i direction between the center of mass of the domain and the interface between the reactants for the domain (Street et al., 1996).

Then, L_i is correlated with W_i. We first calculate L_i for the reactor whose fluid segments has the rectangular cross section and are arranged in two-dimensional directions as the reference. The rectangle has the width W and the height H. Here, we assume that W is larger than H. Figure 2 shows the coordinate and the symmetric properties of the rectangle, denoted by the broken lines, used in the calculations. Using Eq. (4) and Figure 2, we can obtain the diffusion length for this reactor as follows:

$$L_x = \frac{\int_0^{H/2}\int_0^{W/2} x\,dx\,dy}{A/4} = \frac{W}{4}\,(A = WH),\quad L_y = \frac{\int_0^{W/2}\int_0^{H/2} y\,dy\,dx}{A/4} = \frac{H}{4} \tag{5}$$

To equalize the characteristic size with the size of the rectangle, we thus define the characteristic size as $W_x = 4L_x$, $W_y = 4L_y$, that is $W_x = W$, and $W_y = H$ in this instance. Then, we determine the value of w. We can choose the x and y directions arbitrarily. Thus, w is defined as the ratio of the characteristic size for the direction that has the larger W_i to that for the direction that has the larger W_i. In other words, w is always 1 or more and means the aspect ratio of the mean diffusion lengths to the two-dimensional directions. In this example, $w = W/H$. Therefore, we can conclude that $\phi = k_1C_{B0}(4L_y)^2/D_A = k_1C_{B0}H^2/D_A$, and $w = W/H$.

When the geometric design factors are changed, more methods to determine the dimensionless numbers are needed. If each cross section has different L_i, we first obtain L_i for each cross section, and then L_i for the all cross sections by area-weighted arithmetic average. Using the methods described above, W_i for a reactor of a cross-sectional arrangement and a cross-sectional shape can be obtained.

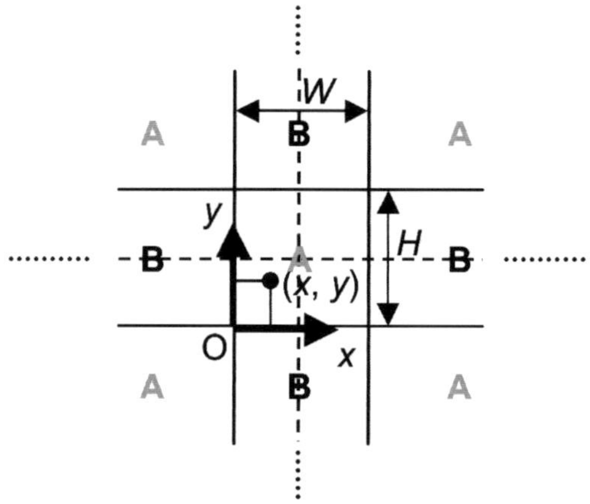

Figure 2. Coordinates and symmetric properties for the calculation of L_i.

Arrangement

Figure 3 shows the reactors having different arrangements of the fluid segments. Each reactor is hereafter called by the name shown in the figures. Fluid segments are aligned periodically in a line for the 1 array, in two lines for the 2 arrays, and in three lines for the 3 arrays, respectively. For the 2D, fluid segments are arranged regularly in two-dimensional directions. For the reactors shown in Figure 3(a)−(c), the cross-sectional shape of fluid segments at the reactor inlets is the square, whose side length is denoted by H.

The dimensionless numbers for these reactors are then determined. The characteristic sizes for the x direction W_x are equal to H for these three reactors as shown in the previous section. On the other hand, W_y is different for each reactor. First, for the 1 array, the reactants cannot diffuse to the y direction since no concentration gradient to this direction exists. Thus, W_y, and w correspond to infinity, and $\phi = k_1C_{B0}H^2/D_A$, $w = \infty$. Moreover, when the flat velocity distribution to the y direction is assumed, this reactor can be modeled in the two-dimensional domain as shown in Figure 3(a). Second, for the 2 arrays, the reactants can diffuse to the upward or downward direction. Thus, the diffusion length to the y direction becomes twice as compared to that of the 2D whose cross-sectional shape is identical since the symmetric property shown by the line parallel to the x direction in Figure 2 disappears in the 2 arrays. Therefore, $L_y = H/2$, and $\phi = k_1C_{B0}H^2/D_A$, $w = 2$. Third, for the 3 arrays, the reactants can diffuse to the upward or downward direction for the segments aligned in the first and third arrays. For these segments, $L_y = H/2$. The diffusion length for the segments in the second array is the same as that for the reactor of two-dimensional arrangement described in the previous section, that is, $L_y = H/4$. Since L_y for each array is not identical, L_y for all segments is determined using the procedure explained in the last part of the preceding section. Hence, $L_y = (H/2)\times(2/3)+ (H/4)\times(1/3) = 5H/12$, and $W_y = 4L_y = 5H/3$; we conclude that $\phi = k_1C_{B0}H^2/D_A$, and that $w = 5/3$.

Extending to the n-line arrangement ($n \geq 2$), L_y for the first and nth array is $H/4$ and for the other $n - 2$ arrays is $H/2$. Therefore,

$$L_y = \frac{H}{2}\frac{2}{n} + \frac{H}{4}\frac{n-2}{n} = \frac{n+2}{4n}H,\ W_y = \frac{n+2}{n}H \quad (n \geq 2) \tag{6}$$

When n is infinity, $L_y = H/2$, and $W_y = H$. They are equal to those for the 2D of the square cross section whose side length is H.

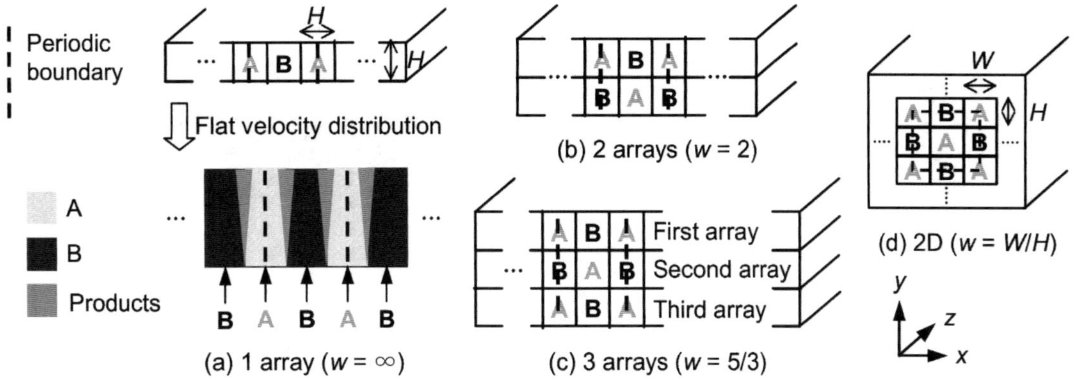

Figure 3. Arrangement of fluid segments.

Shape

To illustrate the calculation of the dimensionless numbers for the reactor of the cross-sectional shape different from the rectangle, we apply the right-angled triangle shown in Figure 4 in this instance. The cross sections are periodically arranged in two-dimensional directions. In Figure 4(a), H denotes the height and W denotes the width of the right-angled triangle. We assume that W is larger than H. In the study on the cross-sectional shapes, we call the reactors by the kind of the shape, that is, the Rectangle or the Right-Angled Triangle.

Using Figure 4(b), we first calculate L_x. The right-angled triangle $E_1E_2E_3$ is divided into the two domains by the bisector line E_1F. For the left-hand side domain of the right-angled triangle E_1E_3F, the shortest distance between the point P and the interface between the reactants is the segment PQ_1; for the right-hand side domain of the shape E_1FE_2, the shortest distance is the line segment PQ_2. From Eq. (4), $L_{i,j}$ for each domain is the distance between the center of mass for E_1E_3F or E_1FE_2 and the interface between the reactants. Both domains have the same distances equal to $W/6$, and also have the same area. We thus have L_x for the whole cross section, that is, $L_x = (W/6)\times(1/2)+ (W/6)\times(1/2) = W/6$. Similarly, $L_y = H/6$. Therefore, $\phi = k_1C_{B0}(4L_y)^2/D_A = k_1C_{B0}(2H/3)^2/D_A = 4k_1C_{B0}H^2/9D_A$, and $w = W/H$.

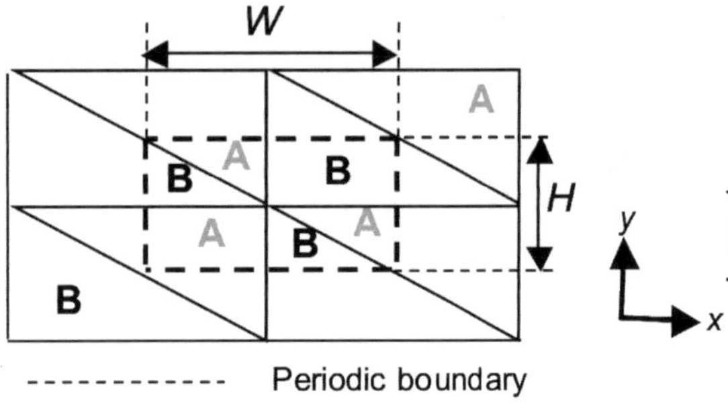

(a) Simulation domain.

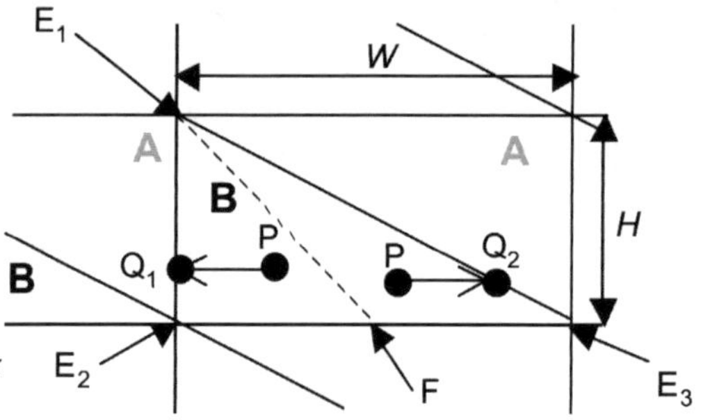

(b) Diffusion length to the x direction.

Figure 4. Right-angled triangle as the cross-sectional shape of fluid segments.

Methods of CFD simulations

First, the settings for all CFD simulations of this study are explained. We use a commercial CFD code FLUENT 6.1.18. This code solves the flow field in the reactors by the finite-volume method (Fluent inc., 2003). The laminar flow and the finite-rate models are applied in the modeling of the reactors. We simulated the mixing by diffusion and reaction in the reactor for various arrangements and shapes of fluid segments that will be mentioned in the subsequent sections. The reaction formulas and the rate equations of the multiple reactions proceeding in the reactors are as follows:

$$\left.\begin{aligned} &\mathrm{A}+\mathrm{B}\rightarrow\mathrm{R}, \quad r_1 = k_1 C_\mathrm{A} C_\mathrm{B} \\ &\mathrm{B}+\mathrm{R}\rightarrow\mathrm{S}, \quad r_2 = k_2 C_\mathrm{B} C_\mathrm{R} \end{aligned}\right\} \tag{7}$$

where A and B are the reactants; R is the desired product; S is the by-product; r_i and k_i are the reaction rate [$\mathrm{kmol \cdot m^{-3} \cdot s^{-1}}$] and the rate constant of the ith step [$\mathrm{m^3 \cdot kmol^{-1} \cdot s^{-1}}$], respectively, and C_j is the molar concentration of component j. The reactants were assumed to react isothermally, and the rate constants were fixed at $k_1 = 1\ \mathrm{m^3 \cdot kmol^{-1} \cdot s^{-1}}$ and $k_2/\mathrm{k}_1 = 0.1$. The initial concentrations of the reactants were as follows: $C_{\mathrm{A}0} = 10\ \mathrm{kmol^{-1} \cdot m^3}$; $C_{\mathrm{B}0} = 20\ \mathrm{kmol^{-1} \cdot m^3}$. The diffusion coefficient of every component D_j was assumed to be $10^{-9}\ \mathrm{m^2 \cdot s^{-1}}$, which is the typical order of the diffusion coefficient for liquid phase reactions. The inlet velocity of the reactant fluids was fixed at $0.001\ \mathrm{m \cdot s^{-1}}$, and the flow in the reactors has flat velocity distribution. This velocity distribution is obtained by setting the share stress on the reactor wall equal to 0. The reactor length L was set at 1 cm for $\phi = 10^1$, 10^2; L was set at 10 cm for $\phi = 10^3$, 10^4, that is, the mean residence time of the reactants in the reactors τ was 10 s for $\phi = 10^1$, 10^2; τ was 100 s for $\phi = 10^3$, 10^4. The vessel dispersion number, D/uL, is equal to 10^{-4} or less, and thus the influence of axial dispersion is negligible (Levenspiel, 1998).

Using the settings mentioned above and the geometric design factors explained in the subsequent sections, we conducted the CFD simulations to obtain concentration profiles of the components in the reactors and then calculated relations between the yield of the desired product R Y_R and the conversion of the reactant A x_A in the reactors as a reactor performance measure. These relations were obtained from the mass-weighted average concentration of each component on the plane perpendicular to the axial direction.

Arrangement

We adjusted the cross sectional size for the 2D so that the 2D and each reactor shown in Figure 3(a)–(c) have the same dimensionless numbers. We then compared the yield of R in both reactors to examine the validity of these dimensionless numbers on the segment arrangement.

The side length of the square cross section of each segment for the reactors shown in Figure 3(a)–(c) was 316 μm. The dimensionless number ϕ was fixed at 10^3 for all reactors. Table 1 shows the sizes of the adjusted rectangle for the 2D. The sizes *H* and *W* correspond to the sizes denoted by the same letters in Figure 3. The definite value of *w* for 1 array is infinity. The corresponding side length of the longer side of the rectangle in 2D is also infinity. However, such size is inapplicable. Thus, we set at *w* = 16 for 2D in the comparison of the yield of R between 1 array and 2D.

The periodic boundary conditions shown in Figure 3 were applied. The simulation domain for 1 array was the area between the periodic boundaries. The domains for 2 arrays and 3 arrays were the volume surrounded by the periodic boundaries and the walls of the reactor. Thus, effects of the reactor walls of the vertical sides were neglected in the simulations for these reactors. The domain for 2D was the volume surrounded by the periodic boundaries. The effects of the reactor walls of the vertical and horizontal sides are neglected in the simulations for 1 array and 2D.

Table 1. Adjusted Cross-sectional Sizes of the Rectangle for 2D

Arrangement	*H* [μm]	*W* [μm]	*w* = *W*/*H*
1 array	316	5056	16
2 arrays	316	632	2
3 arrays	316	527	5/3

Shape

We investigated the validity of the dimensionless numbers on the second geometric design factor, the cross-sectional shape of fluid segments at the reactor inlet. In other words, we compared the yield of R in the reactors that take the same dimensionless numbers, but take the different cross-sectional shape. The rectangle shown in Figure 3(d) and the right-angled triangle shown in Figure 4 were employed as the cross-sectional shape. The fluid segments for the reactors of both shapes were periodically arranged in two-dimensional directions. First, the yield of R in the reactors for both shape were calculated with changing ϕ and fixed *w*. The dimensionless number ϕ was changed from 10^1 to 10^4, and *w* was fixed at 1. Then, effects of *w* were examined. The applied values of *w* were 1, 2, 4, or 8. In this study, ϕ was fixed at 10^3. Table 2 shows the sizes of each shape for each set of the dimensionless numbers. The sizes *H* and *W* in this table correspond to the same sizes in Figure 3(d) and Figure 4.

We applied the periodic boundary conditions shown in Figure 3(d) and 4. The simulation domains for the reactors of both shapes were the volumes surrounded by the periodic boundaries. Effects of the reactor walls of the vertical and horizontal sides are neglected in the simulations.

Table 2. Sizes of the Rectangle and the Right-Angled Triangle for Each ϕ and *w*

Rectangle				Right-angled triangle			
ϕ	*w*	*H* [μm]	*W* [μm]	ϕ	*w*	*H* [μm]	*W* [μm]
10^1	1	32	32	10^1	1	47	47
10^2	1	100	100	10^2	1	150	150
10^3	1	316	316	10^3	1	474	474
10^3	2	316	632	10^3	2	474	948
10^3	4	316	1264	10^3	4	474	1896
10^3	8	316	2528	10^3	8	474	3792
10^4	1	316	316	10^4	1	1500	1500

Results and discussion

Arrangement

Figure 5 shows the relation between Y_R and x_A in the reactors shown in Figure 3. The results show that the yield of R for each reactor shown in Figure 3(a)–(c) agrees well with that for the 2D that has the corresponding dimensionless numbers. When the 2D takes the value of w = 16, the 2D give the comparable yields of R in the reactors to that in the 1 array. In the 2D whose w is 16, the mixing by the molecular diffusion to the direction giving the smaller mean diffusion length, namely the y direction in Figure 3, proceeds much faster than that to the x direction. The height of the rectangular cross section and the diffusion length to the y direction are uniform at any x coordinate. Thus, the diffusion to the y direction at any x coordinate except for the both ends of the rectangle completes at the same time, and the mixing in the whole reactor volume also completes. At the both ends, the diffusion to the x direction also affects the mixing rate, and the mixing completes earlier than the other reactor volume. Thus, the further increase in w for the 2D with fixed ϕ has no effect on the mixing in the whole reactor volume and thus the yield of R in the reactor, and we can infer Y_R for the 2D whose w is infinity is also comparable to that for the 1 array.

In conclusion, the dimensionless number w can represent the effect of the segment arrangement for square cross sections. This also indicate that if relations among the dimensionless numbers and yield of desired products are determined in advance, we can estimate the product yields for designing reactors using these relations. When we choose the arrangement of the fluid segment, we first calculate the dimensionless numbers for each reactor of different arrangement and then have the product yield using the relations. For instance, Figure 5 enable us to figure out the difference of the maximum of Y_R in the 1 array and the 3 arrays simulated in this study at 10%.

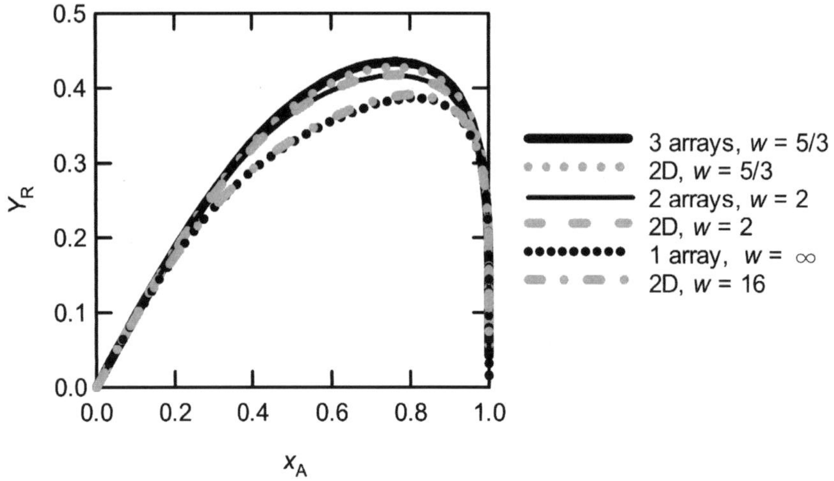

Figure 5. Effects of the arrangements on the yield of R in the reactors. $\phi = 10^3$.

Shape

Figure 6 shows the effect of ϕ for the Rectangle and the Right-Angled Triangle on Y_R when w is fixed at 1. The results show that the yield of R for the Rectangle and the Right-Angled Triangle differs when these reactors take the same dimensionless numbers. The Right-Angled Triangle gives higher Y_R than that for the Rectangle especially when $\phi = 10^3$ and 10^4. Compared the maximum yield of R, the Right-angled Triangle gives higher value of 4% than that for the Rectangle when $\phi = 10^3$, and 8% when $\phi = 10^4$. We speculate that this difference is caused by the definition of the diffusion length for fluid segments. For the Right-angled Triangle, L_x is a mean value of the diffusion length, and the diffusion length to the *x* direction varies by the *y* coordinate. The width in the domain over the center of mass of the right-angled triangle $E_1E_2E_3$ in Figure 4(b), namely the diffusion length to the *x* direction, is shorter than that of the rectangle for the same dimensionless numbers. We can say similarly for L_y and the height of the right-angled triangle. The area over the center of mass or the right-hand side of that point of the right-angled triangle is more than the half of the area of the right-angle triangle and has the shorter diffusion length to at least one direction than that of the rectangle. Consequently, we can infer that the Right-Angled Triangle gives the faster mixing by molecular diffusion and the higher Y_R than those in the Rectangle especially when ϕ is large.

However, the relations between Y_R and ϕ for both reactors approximately correspond, and thus ϕ serves as the index for the estimate of the product yield. Moreover, we can identify whether reactions proceed under reaction-controlled condition from the value of ϕ. When $\phi = 10^1$, the yield of R for the reactors of both cross-sectional shapes agree with that for the reaction controlled case (Levenspiel, 1998). Small values of ϕ mean that the relative rates of reaction to diffusion are slow. Therefore, $\phi = 10^1$ is the threshold for the reaction controlled condition when *w* is fixed at 1.

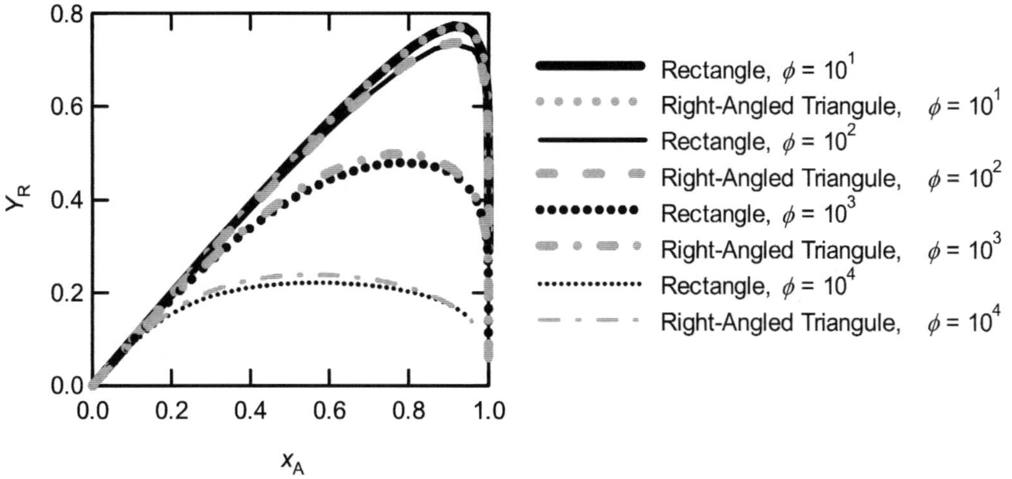

Figure 6. Influences of ϕ on Y_R for the Rectangle and the Right-Angled Triangle. $w = 1$.

Figure 7 shows the yield of R in the reactors for each *w* with fixed ϕ. The maximum yield of R for the Right-angled Triangle is higher than that for the Rectangle when $w = 1$ and 2. On the other hand, when *w* is larger than 2, the Rectangle have the higher maximum yield of R. The Right-Angled Triangle gives the lower Y_R than that for the Rectangle especially at high conversion of A. The difference in Y_R increases with increasing *w*.

This difference is caused by the difference in the transition of the concentration profiles of the reactants along with the diffusion for each cross-sectional shape. Figure 8 shows the contour plots of C_A/C_{A0} in the reactors. As shown in Figure 8(a) and mentioned in the preceding paragraph, when *w* is large, the mixing in the Rectangle is mainly proceeded by the diffusion to the direction giving the shorter diffusion length, that is, the *y* direction, and the influence of the change in *w* on the mixing rate and the product yield is small for large values of *w*. On the other hand, the influence of the change in *w* is not small in large values of *w* for the Right-Angled Triangle. In the right-angled triangle cross section, the diffusion length to the *y* direction is changed by the *x* coordinate. The reactant A or B in the area circled in Figure 8(b) has shorter diffusion length and completes the mixing and the reactions earlier than that in the other area. Beside the circled area, much counterpart reactant remain unreacted, and the reactants are detached to the *x* direction, which gives the longer diffusion length, as shown in Figure 8(b-3). The distance between the remained reactants to the x direction increases with increase in w. Thus, the mixing between the remained reactants at high value of w proceeds slowly and results in lower Y_R at high conversion of A. We can conclude that the effect of w on product yield for the Right-Angled Triangle is grater than that for the Rectangle, and w for the Right-Angled Triangle should be close to 1 for fast mixing. Besides the decrease in the product yield, since fabricating channels to produce fluid segments whose cross-sectional shape is the right-angled triangle of a high value of w is difficult, such fluid segments are impracticable.

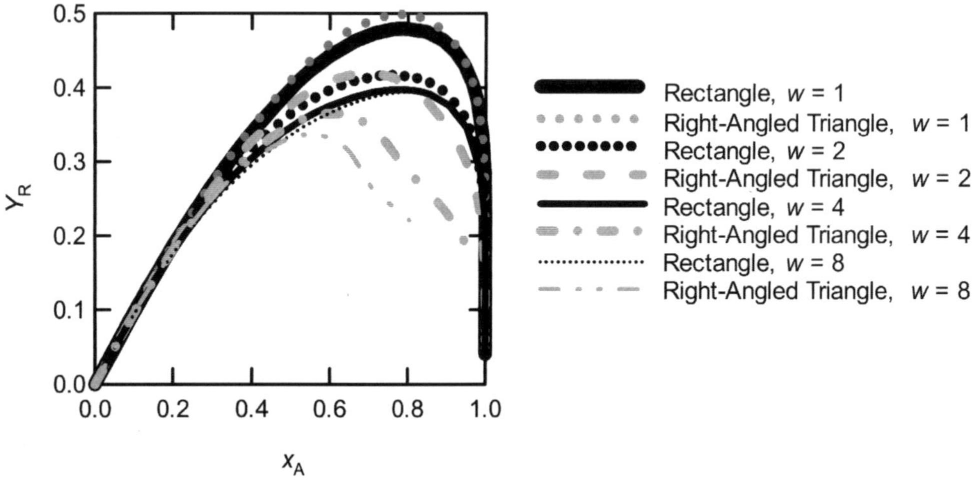

Figure 7. Influences of w on Y_R for the Rectangle and the Right-Angled Triangle. $\phi = 10^3$.

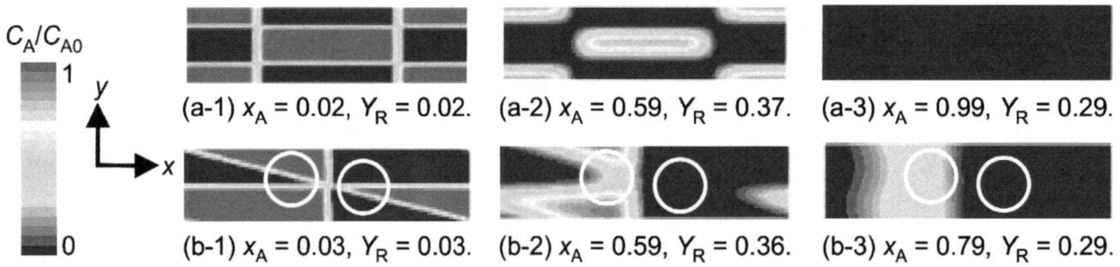

Figure 8. Mixing by diffusion in the reactor of each cross-sectional shape: (a) Rectangular; (b) the Right-Angled Triangle. The figures show the contour plots of C_A/C_{A0} at (a-1), (b-1) z = 0.0001 m, (a-2), (b-2) z = 0.004 m, and (a-3), (b-3) z = 0.02 m. $\phi = 10^3$, and w = 4.

Conclusions

We have addressed quantitative effects of the geometric design factors, namely arrangements and shapes of the cross section of fluid segments at the reactor inlet, on product yield for the mixing operation in microreactors that reactant fluids are split into many micro fluid segments and then fed into the reactors. We have introduced two dimensionless numbers so that we can estimate effects of the geometric design factors on the product yield for reactors using this mixing operation. The dimensionless numbers are the ratio of reaction rate to diffusion rate ϕ and the aspect ratio of the mean diffusion length to the two-dimensional directions w. We have also devised methods to determine the parameters for a arrangement and a shape of the fluid segment cross section at the reactor inlet. To investigate the validity of these parameters, we have compared the yield of the desired product of a multiple reaction system for a reactor having an arrangement and a shape of fluid segment cross sections at the reactor inlet with that for the reactor having rectangular cross section arranged in two-dimensional directions of the same dimensionless number using CFD simulations. The results show that the effect of arrangement for square cross section can be expressed by w, and that the order of ϕ allows us to determine whether the mixing operation is reaction controlled even when the cross-sectional shape is changed. Moreover, the effects of w for cross-sectional shapes applied in this study are different. This dimensionless number has the greater effects on the product yield for the reactor whose cross-sectional shape is the right-angled triangle than that of the rectangle.

The results also indicate that we can predict the product yield for the reactor of a cross-sectional arrangement and shape of fluid segments using the chart showing relations among product yields and the dimensionless numbers for base shapes of the cross-section for the fluid segments. Beside product yield, effects of the geometric design factors on the operability are also essential issues. For example, the isothermal condition has been assumed in this study. The facility of the isothermal condition differs when the geometric design factors are changed. For the design of fluid segments, the relation between the fabrication cost and the geometric design factors is also needed. Thus, this mixing operation would be designed from trade-off between product values, fabrication and operation costs.

Acknowledgments

We have conducted this research within the Project of Micro-Chemical Technology for Production, Analysis and Measurement Systems financially supported by the New Energy and industrial Development Organization (NEDO). We would also appreciate the Micro Chemical Plant Technology Union (MCPT) for their support.

References

Aoki, N., S. Hasebe, and K. Mae; "Design of Feed Segments: Influences of Sizes and Shapes on Product Composition of Multiple Reactions," *2004 AIChE Annual Meeting CD-ROM*, Omnipress, Madison, USA, Paper 534b (2004).

Ehlers, St., K. Elgeti, T. Menzel, and G. Wießmeier, "Mixing in the Offstream of a Microchannel System," *Chem. Eng. Proc.*, **39**, 291 (2000).

Ehrfeld, W., K. Golbig, V. Hessel, H. Löwe, and T. Richter, "Characterization of Mixing in Micromixers by a Test Reaction: Single Mixing Units and Mixer Arrays," *Ind. Eng. Chem. Res.*, **38**, 1075 (1999).

Fluent 6.1 Use's Guide, Fluent inc., Lebanon, NH, Chapter 24 (2003).

Hardt, S., and F. Schömnfeld, "Laminar Mixing in Different Interdigital Micromixers: II. Numerical Simulations," *AIChE J.*, **49**, 578 (2003).

Levenspiel, O., *Chemical Reaction Engineering*, 3rd ed., John Wiley & Sons, New York, USA, pp. 190, 314–315 (1998).

Löb, P., K. S. Drese, V. Hessel, S. Hardt, C. Hofmann, H. Löwe, R. Schenk, F. Schönfeld, and B. Werner; "Steering of Liquid Mixing Speed in Interdigital Micro Mixers—From Very Fast to Deliberately Slow Mixing," *Chem. Eng. Technol.*, **27**, 340 (2004).

Nagasawa, H., N. Aoki, and K. Mae, "Design of a New Micromixer for Instant Mixing Based on Collision of Micro Segment," *Chem. Eng. Technol.*, in press.

Street, R. L., G. Z. Watters, and J. K. Vennard, *Elementary Fluid Mechanics*, 7th ed., John Wiley & Sons, New York, USA, pp. 50, 697 (1996).

Application of a Microplant for the Sulfonation of Toluene with Gaseous Sulfur trioxide

Klaus Jähnisch[1], *Norbert Steinfeldt*[1], *Uwe Dingerdissen*[1], *Andreas Müller*[2], *Benjamin Horn*[2], *Vania Cominos*[2], *Volker Hessel*[2], *Alexis Bazzanella*[3], *Volker Großer*[4], *Volker Hillmann*[4], *Günter Rinke*[5], *Manfred Kraut*[5]

[1] *Institut für Angewandte Chemie Berlin-Adlershof e.V., 12489 Berlin, Germany, E-mail: jaehnisch@aca-berlin.de*
[2] *Institut für Mikrotechnik Mainz GmbH, 55129 Mainz, Germany*
[3] *DECHEMA e.V., 60486 Frankfurt, Germany*
[4] *Fraunhofer Institut für Zuverlässigkeit und Mikrointegration, 13355 Berlin, Germany*
[5] *Forschungszentrum Karlsruhe GmbH, Institute for Micro Process Engineering, Postfach 3640, 76021 Karlsruhe, Germany*

The use of microstructured components for process engineering has gained increasing importance in chemical, pharmaceutical and life science applications during the last years. Several obstacles in the introduction of this technology resulted from frequent incompatibility of the hardware components. Therefore, the German Federal Ministry for Education and Research (BMBF) funded a strategic research project, led by an industry platform, with the goal to develop and use a modular hardware and software concept for compatible modules for microprocess engineering technology. This strategy allows for a flexible construction of microprocess engineering equipments. Within the framework of this project a "backbone kit" for microprocess engineering technology was developed and used to construct complete microplants. The goal of the present study is to demonstrate the performance of such a microplant for multi-step processes.

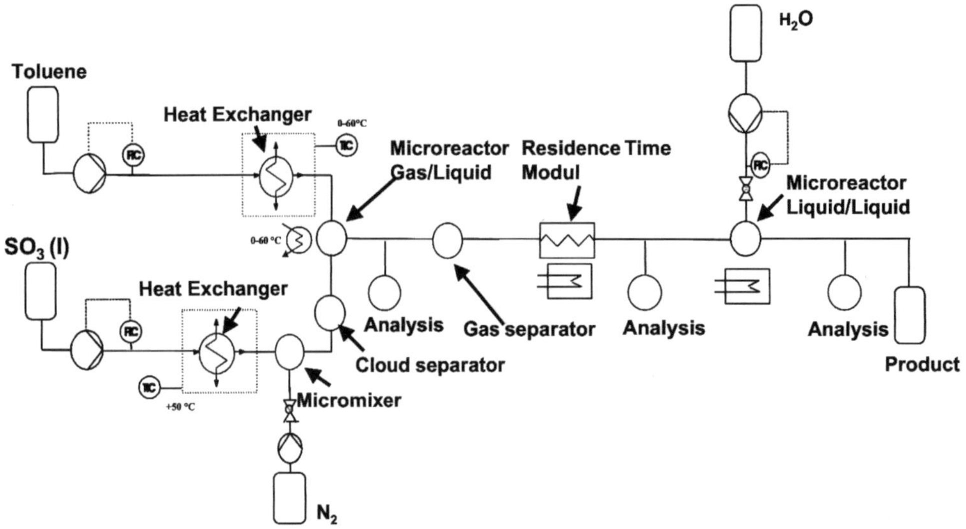

Figure 1. Process flowsheet for sulfonation of toluene with gaseous SO_3 in a microplant

The highly exothermic sulfonation of toluene with gaseous sulfur trioxide has been investigated in such a microplant. Figure 1 shows the process flowsheet of the microplant used.

Toluene is heated up to reaction temperature using a microstructured heat exchanger. Liquid sulfur trioxide is heated up to 50° C to evaporate it and nitrogen is further added to dilute the SO_3. The stream flows through to falling-film microreactor where it reacts with the liquid toluene (reaction 1 – 4). As shown in Scheme 1 sulfonic acid (reaction 1) is produced in the first reaction step. By consecutive reactions a mixed anhydride (unstable and undetectable) (reaction 2), di-tolylsulfone (reaction 3) and sulfonic acid anhydride (reaction 4) are also formed. In the residence time module the reaction of mixed anhydride with toluene to sulfonic acid (reaction 5) is completed. The last reaction step, the hydration of sulfonic acid anhydride to sulfonic acid (reaction 6), is carried out in a microstructured reactor specifically constructed for liquid phase reactions.

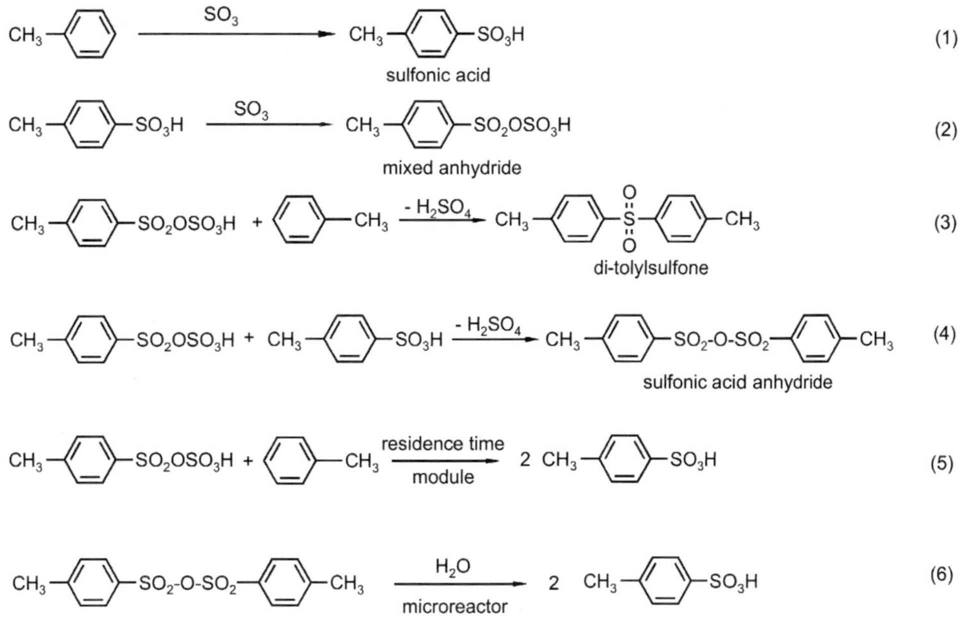

Scheme 1: Reaction steps in the sulfonation of toluene with gaseous SO_3

The dependence of product selectivity on toluene conversion at different temperatures after the gas/liquid reaction in the falling-film microreactor, after the residence time module and after the hydrolysis step has been determined. Figure 2 shows the results at 26 °C after the reaction in the falling-film microreactor. At a toluene conversion of 9 % the selectivity of sulfonic acid is 68 % and drops to 58 % at a conversion of 14 %. The selectivity of sulfonic acid anhydride decreases slightly with increasing conversion and the selectivity of di-tolylsulfone is independent of the toluene conversion.

Figure 3 shows the selectivity of the products as a function of conversion measured at the outlet of the residence time module. The conversion of mixed anhydride with toluene has been completed and therefore the selectivity of sulfonic acid is about 20 % higher than at the outlet of the falling-film microreactor. The selectivities of di-tolylsulfone and sulfonic acid anhydride are approximately the same at both positions in the process.

By addition of water in the last step of the process the sulfonic acid anhydride is hydrolyzed completely (see figure 4).

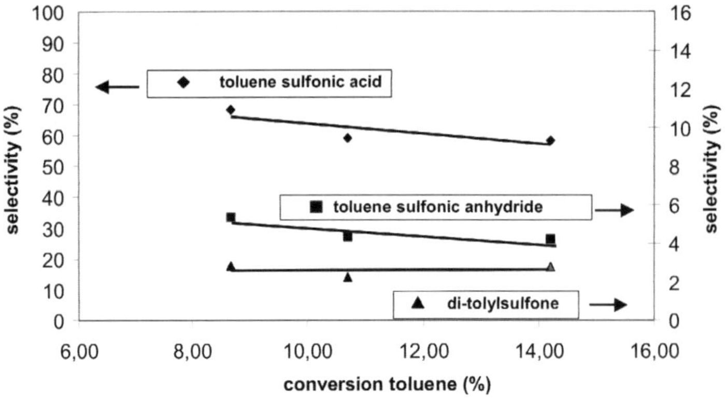

Figure 2. Sulfonation of toluene at 26 °C - after the gas/liquid reaction

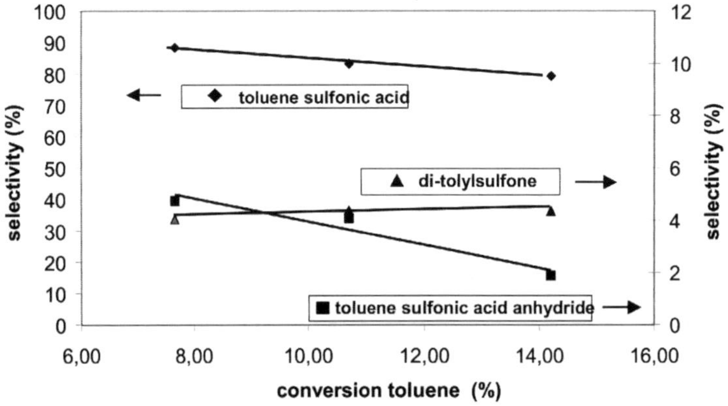

Figure 3. Sulfonation of toluene at 26 °C - after the residence time module

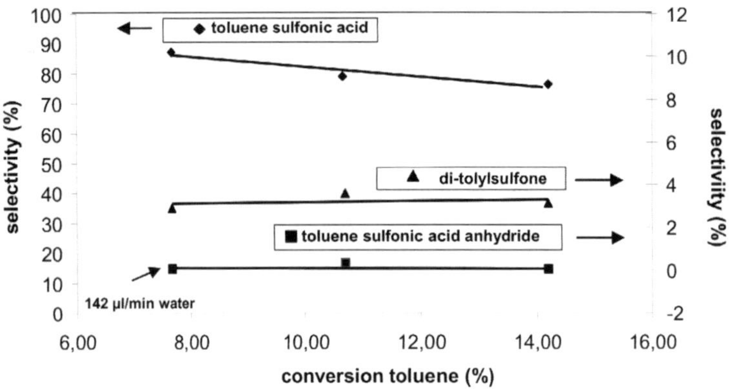

Figure 4. Sulfonation of toluene at 26 °C - after the hydrolysis of the anhydride

In the studied range of toluene conversions a high selectivity to the desired product sulfonic acid has been obtained and the only observed side product, di-tolylsulfone, has been formed in small amount. The results show that the backbone kit is suitable for construction of microplants and for the investigation of multi-step chemical processes.

Literature

[1] A. Müller, V. Cominos, V. Hessel, B. Horn, A. Ziogas, K. Jähnisch, V. Hillmann, V. Großer, K.A. Jam, A. Bazzanella, G. Rinke, M. Kraut, Chem Eng. Journal, 2005, in press

Development of the Integrated Hydrogen Production System Using Micro-structured Devices

Y.C. Byun[2], Y. Kwon[2], J. Choe[*2] and K.H. Song[1]*

1) Dept. of Chemical & Biological Engineering,
Korea University, Seoul 136-713, Korea
2) Corporate R&D, LG Chem / Research Park
Daejeon 305-380, Korea
**Corresponding author. E-mail address: choej@lgchem.com*

Abstract

An integrated fuel processor, which produces hydrogen for fuel cell from aqueous methanol solution, has been developed by employing wall-coated microchannel catalytic reactors for steam reforming and preferential oxidation reaction. The developed fuel processor can produce the amount of hydrogen equivalent to a few hundred watts of electricity, when supplied to a fuel cell. We fabricated methanol fuel processor consists of a catalytic combustor, heat exchangers, evaporators, a reformer, and a PrOx(preferential oxidation) reactor. $CuO/ZnO/Al_2O_3$ catalyst was used for reforming and Pt/Al_2O_3 catalyst was used for both preferential oxidation and catalytic combustion. Besides the reformer and the PrOx reactor, the whole system also required three micro-structured evaporators. The main evaporator for the reformer feed vaporization was integrated with the catalytic combustor, and the heat recuperated from combustor vent gas is used to vaporize the methanol feed for two additional evaporators. The integrated system was operated consistently with methanol conversion of over 99.0 mol% at 300 °C, and the carbon monoxide concentration under 20 ppm in the final product flow.

Introduction

Polymer electrolyte membrane fuel cell (PEMFC) has been offered as an alternative to conventional batteries. Since fuel cells can operate on hydrogen, fuel-processing systems are necessary to use of hydrocarbon fuels. However, the size and efficiency of fuel processor is one of the technological hurdles to overcome. Recently microreaction technology is gaining increasing importance in the chemical industry since the size reduction in chemical devices allows the significant enhancement of heat and mass transfer rate. Several research groups have tried to apply microreaction technology to develop the compact fuel processor for fuel cells. Holladay et al. developed an integrated micro fuel processor composed of two vaporizer/preheaters, a reformer, a catalytic combustor, and a heat exchanger, which produced 200 mW_t of hydrogen [1]. TeGrotenhuis et al. demonstrated an integrated gasoline fuel processor including an inconel 625 steam reforming microreactor with combustor and recuperative heat exchanger [2]. Park et al. developed a microchannel methanol steam reformer with vaporizer unit [3]. The objective of this study is to develop a micro methanol fuel processor without external heating.

Experimental

The developed methanol micro fuel processor consists of a catalytic combustor, heat exchangers, an evaporator, a reformer, and a PrOx reactor. Two methanol evaporators were operated by exchanging heat with exit gas from combustors. Microchannels on the reaction plate of methanol steam reformer were patterned by a wet etching process on stainless steel sheet. The dimension of stainless steel sheet was 60 mm × 154 mm × 300 μm. Microchannels were 450 μm wide, 150 μm in deep, and 100 mm long. The $CuO/ZnO/Al_2O_3$ catalyst was used for the steam reforming of methanol. The catalyst slurry was prepared by adding NH_4OH to the mixed solution of $Cu(NO_3)_2{\bullet}3H_2O$, $Zn(NO_3)_2{\bullet}6H_2O$, and $Al(NO_3)_3{\bullet}9H_2O$ and coated on the microchannels. $Pt/\alpha\text{-}Al_2O_3$ catalyst for combustion and preferential oxidation reaction was prepared by washcoating of $\alpha\text{-}Al_2O_3$ on stainless steel sheet [4] and loading 3.3 % (wt/wt) Pt.

The channels coated with reforming catalyst and those with the combustion catalyst were stacked alternatively as shown in Fig. 1 in order to transfer heat from combustor into the reformer.

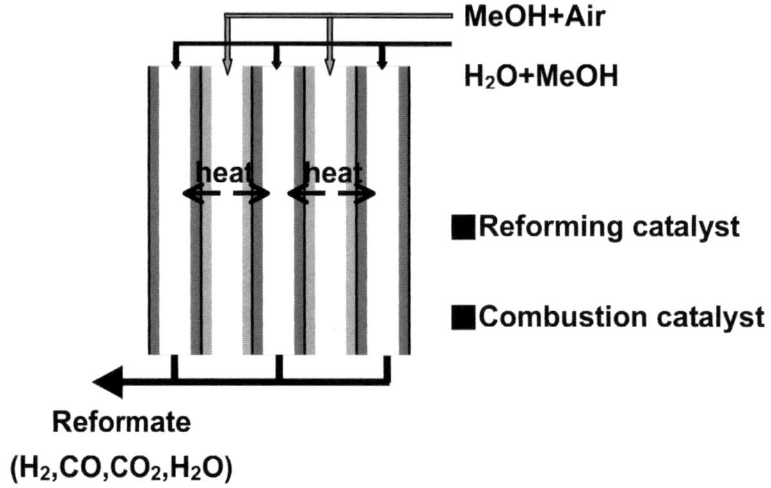

Fig. 1. Methanol steam reformer integrated with combustor

The preferential oxidation reactor was connected to the exit of the methanol steam reforming section. Air was supplied to remove CO gas in the reformate. $Pt/\alpha\text{-}Al_2O_3$ catalyst was used for PrOx reaction and the reaction temperature was attained by exchanging heat from reformate.

The products of methanol steam reformer were analyzed by GC (HP 5890 series) equipped with chromosorb 102 (80~100mesh) column and TCD. The products of PrOx reaction were analyzed by micro-GC (Varian 4700) equipped with molecular sieve 5A column and TCD.

Results and Discussion

The performance of the integrated methanol reformer is shown in Fig. 2. The required heat for feed evaporation and steam reforming was supplied from the combustion reaction. After 90 min, the conversion of methanol was stabilized at 99.5%, and the reformate stream was composed of 74.8% H_2, 24.4% CO_2 and 0.84% CO on a dry basis. Overall efficiency of our integrated reformer can be defined as follows [1].

$$\text{Efficiency of reformer} = \frac{\Delta H_c \text{ hydrogen}}{\Delta H_c \text{ reformer} + \text{combustor methanol feed}} \quad (1)$$

$$CH_3OH(l) + 1.5O_2 = CO_2 + 2H_2O(l) \quad \Delta H = -726 \text{ kJ/mol} \quad (2)$$

$$H_2 + 0.5O_2 = H_2O(g) \quad \Delta H = -242 \text{ kJ/mol} \quad (3)$$

Equation (2) is the combustion reaction of methanol at 25 °C, 1 atm, and equation (3) is the combustion reaction of hydrogen at 80 °C, which is the operating temperature of PEMFC. The calculated efficiency of our integrated system without PrOx reactor was 59.5% based on equation (1).

Assuming 85 % utilization of hydrogen and 60 % efficiency of fuel cell, the estimated electric power that the developed reformer can produce is 257 W_e and the expected overall efficiency is 30.3 %.

When the PrOx reactor was connected to the exit of the reformer, the overall efficiency of a fuel processor can be decreased due to the hydrogen loss in the PrOx reactor. However, the efficiency of a fuel processor increased by reusing of PEMFC anode effluent hydrogen (15%) at combustion reaction. Increasing the capacity of reformer can also reduce heat loss in combustor. Palo et al. reported that the efficiency of the fuel processor increased from 20% to 58% when the capacity of the integrated methanol reformer increased from 14W to 80W [5].

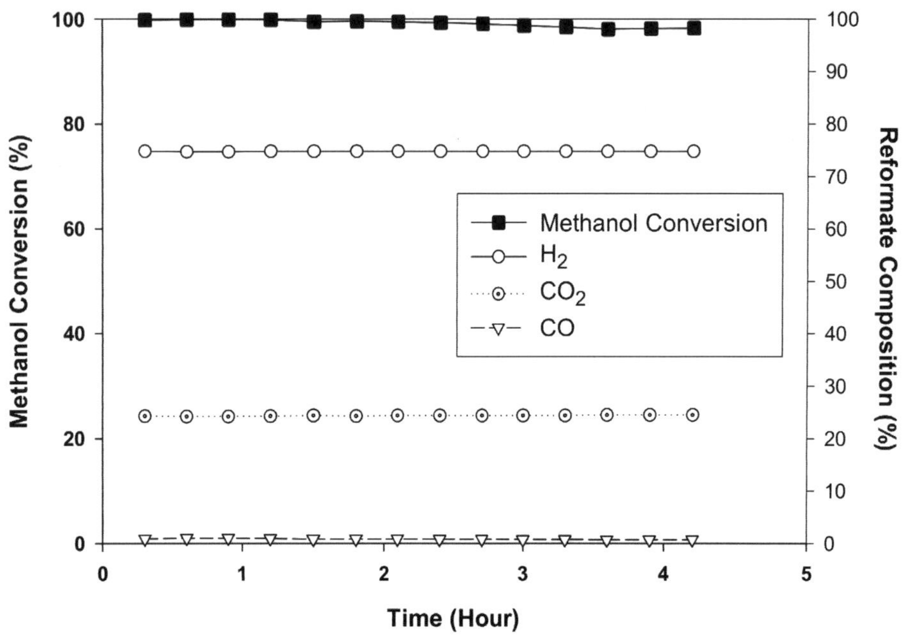

Fig. 2. Performance of the integrated methanol reformer
(50wt% $MeOH/H_2O$ mixture 3ml/min, combusted Methanol 1.1ml/min)

The performance of the integrated fuel processor was investigated. When PrOx reactor was operated at 160 °C, and O_2/CO molar ratio of 3 with air flow rate of 250 ml/min, the concentration of CO in the reformate was 40 ppm. When the air flow rate was increased to 370 ml/min, the concentration of CO in the reformate decreased to 30 ppm. It showed that only 0.07% of the oxygen in the additionally supplied air (120ml/min) was used for oxidation of CO and rest of the oxygen (99.93%) reacted with the hydrogen. To increase overall efficiency, we tested preferential oxidation reaction with two PrOx reactors in series. When the first PrOx reactor was operated at O_2/CO molar ratio of 2 and the second operated at O_2/CO molar ratio of 1, we observed that the concentration of CO in the reformate decreased to 15 ppm. As a result, the efficiency of a fuel processor including the PrOx reactor was 56.7 % and lower heating value of the produced hydrogen was 480 W_t.

Conclusions

The methanol fuel processor was developed without any external heat supply. The estimated electric power of the integrated fuel processor is 257 W_e. In the case of the fuel processor including PrOx reactor, the estimated electric power is 245 W_e and the overall efficiency of our integrated reformer system is 56.7%.

References

[1] J.D. Holladay, E.O. Jones, M. Phelps, J. Hu, J. Power Sources, **2002**, 108, 21
[2] W.E. TeGrotenhuis, K.P. Brooks, J.M. Davis, C.M. Fischer. Proceedings of the 7th International Conference on Microreaction Engineering (IMRET 7). **2003**. 19-21
[3] G. Park, D.J. Seo, S. Park, Y. Yoon, C. Kim, W. Yoon, Chemical Engineering Journal, **2004**, 101, 87~92
[4] A. Rouge, B. Spoetzl, K. Gebauer, R. Schenk, A. Renken, Chemical Engineering Science, **2001**, 56, 1419~1427
[5] D.R.Palo, J.D.Hollday, R.T.Rozmiarek, C.E.Guzman-Leong, Y. Wang, J. Hue, Y. Chin, R.A.Dagle,E.G.Baker, J. of Power Sources, **2002**, 108, 28~34

Potential of microreactor in production of phthalic anhydride from o-xylene

E. Klemm, *G. Mathivanan*

Department of Chemical Technology, Chemnitz University of Technology, Chemnitz, Germany

Introduction

Phthalic anhydride (PA), a chemical of considerable importance, is mainly produced by V_2O_5/TiO_2 catalyzed gas-phase oxidation of o-xylene in Multi-Tubular Reactors (MTR). All PA production processes are forced to work in low o-xylene concentration (1 mol %), due to very high exothermic ($\approx$1142 kcal mol^{-1}) nature of the reaction. Even a slight change in inlet conditions can lead to thermal run-away problems.

It is well known that microreactors are more efficient in removal of excess heat of reaction due to their high specific heat transfer coefficient $\left(\alpha/d_h\right)$ values [1]. So, it allows to operate the process at higher o-xylene concentration and/or at higher inlet temperatures. In this work the potential of microreactor on o-xylene oxidation to PA production process is investigated theoretically.

Reaction and Kinetics

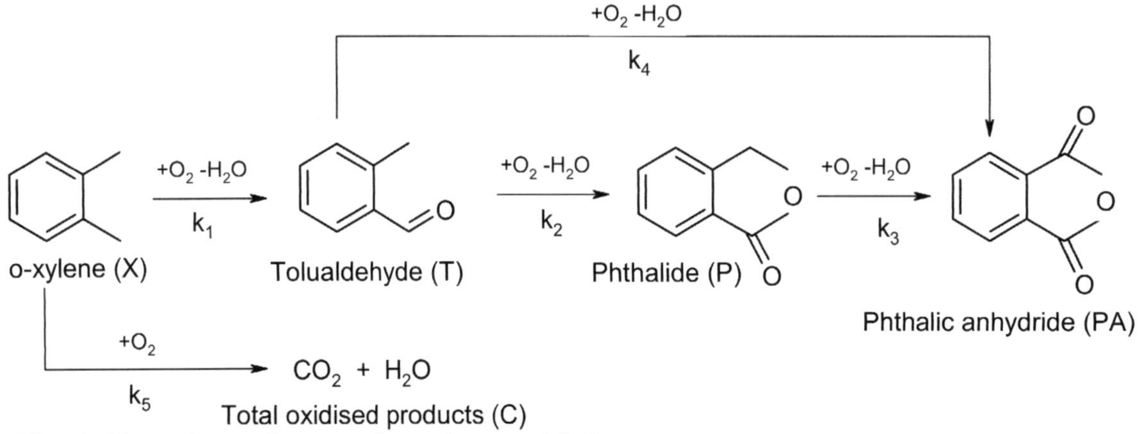

Fig. 1. Reaction scheme of o-xylene oxidation

For the kinetic model for PA synthesis both oxidized and reduced sites have been considered and the rate determining step is aromatic-compound oxidation reactions [2].

$$R_X = (k_1 + k_5) \cdot \alpha \cdot p_X \cdot p_{O_2}$$
$$R_T = \left(k_1 \cdot p_X \cdot p_{O_2} - p_T\left(k_2\sqrt{p_{O_2}} + k_3 \cdot p_{O_2}\right)\right) \cdot \alpha$$
$$R_P = (k_2 \cdot p_T - k_4 \cdot p_P) \cdot \sqrt{p_{O_2}} \cdot \alpha$$
$$R_{PA} = \left(k_3 \cdot p_T \cdot p_{O_2} + k_4 \cdot p_P \cdot \sqrt{p_{O_2}}\right) \cdot \alpha$$
$$R_C = k_5 \cdot p_X \cdot p_{O_2} \cdot \alpha$$
$$k_i = k_i^0 \cdot e^{\frac{-E_A}{R \cdot T}}$$

$$\alpha = \frac{1}{\left[1 + K_O \cdot \sqrt{p_{O_2}} + K_A \cdot (p_X + p_T) \cdot \sqrt{p_{O_2}} + K_B \cdot (p_T + p_P + p_{PA})\right]^2}$$

The values for rate constants (k_j) and activation energies (E_{Aj}) have been taken from J.N. Papageorgiou et al. [2].

Reactor model

Assumptions

The following assumptions have been made to develop the reactor model:

1. Steady-state plug flow reactor with negligible radial profiles (one-dimensional model).
2. Pseudo homogeneous phase.
3. Constant velocity profile.

Mass balance

$$u_f \frac{dp_i}{dz} = \frac{M \cdot P \cdot \rho_b}{\rho_g(T)} \cdot R_i(p_i, T)$$

Energy balance

$$u_f \cdot \frac{dT}{dz} = \sum_{j=1}^{m} \frac{(-\Delta H_r)_j \cdot \rho_b}{c_p \cdot \rho_g(T)} \cdot r_j(p_i, T) - 4 \cdot \frac{\alpha}{d_h} \cdot \frac{1}{c_p \cdot \rho_g(T)}(T - T_w)$$

Only the term $\left(\alpha/d_h\right)$ is different when comparing microreactor with MTR. For a microreactor with a hydraulic diameter of $100\,\mu m$, the ratio of surface area to volume $\left(4/d_h\right)$ is 40000 m^{-1} whereas the ratio of surface area to volume in a tubular reactor with 0.0254 m ($\approx$1 inch) diameter is only 157.48 m^{-1}. So, the ratio of surface area to volume in microreactor is 254 times larger than in 1 inch tubular reactor. Thus a major contribution to the enhancement of heat removal in microreactor comes from its higher ratio of surface area to volume. The other contribution to heat removal enhancement in microreactor comes from heat transfer coefficient. Heat transfer coefficient in tubular reactor with catalyst bed is normally in the range of 100-150 W m^{-2} K^{-1}. For microreactors α can be calculated using following Nusselt number correlation [3]:

$$Nu = 0{,}453 \cdot Re^{1/2} \cdot Pr^{1/3}$$

For fully developed laminar flow, when we substitute the value of Nusselt number, the above equation becomes as follows,

$$Nu = 3.657 = \frac{\alpha \cdot 100^{-6} m}{\lambda},$$

$$\alpha = \frac{3.657 \cdot 0.044}{100^{-6}} \quad [W\ m^{-2}\ K^{-1}]$$

So the value of heat transfer coefficient in microreactor is approximately 1609 W m^{-2} K^{-1}. This is approximately 10 times higher than the heat transfer coefficient of packed bed tubular reactor. Thus altogether the specific heat transfer coefficient $\left(\alpha / d_h\right)$ of microreactor is about 2500 times larger than for the conventional MTR. The above ordinary differential equations have been solved using PRESTO© software.

Validation of the model

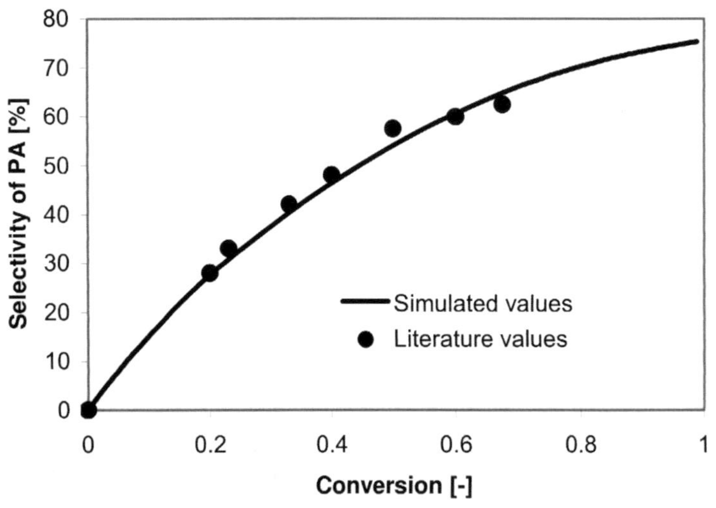

Fig. 2. Comparison of simulated and experimental results

Inlet o-xylene concentration (C_{Xo})= 1 mol%, Inlet o_2 concentration= 20 mol%

The model results have been compared with the experimental results of *J.N. Papageorgiou, M.C. Abello and G.F. Froment* [2] and both the results are shown in the following figure.

Simulation results

1. Temperature profile

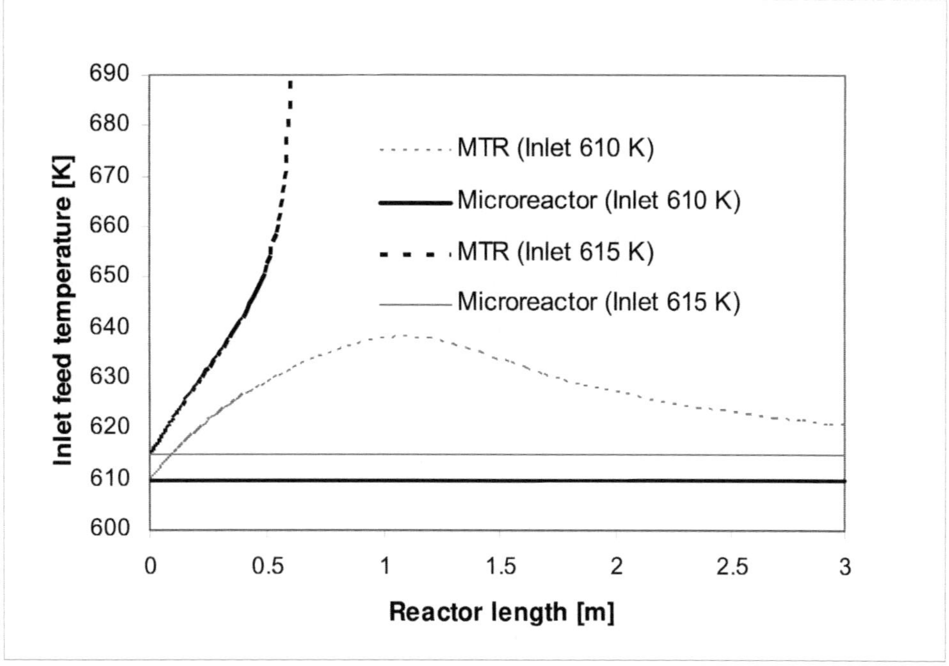

Fig. 3. Temperature profiles of microreactor and MTR

Inlet o-xylene concentration (C_{Xo})= 1 mol%, Inlet o_2 concentration= 20 mol%

Fig. 3 shows temperature profiles of microreactor and MTR. Two kinds of curves are shown for MTR. One curve is for inlet temperature 610 K and the other is for 615 K. Thermal runaway in MTR occurs when the inlet temperature is raised only about 5 K. It is clear from Fig. 3 that the calculated temperature profile of microreactor is isothermal even if temperature is increased. So, it gives more freedom to raise the inlet temperature and/or inlet feed concentration. In our work inlet feed temperature is raised and results are analyzed.

2. Selectivity of PA

Up to 610 K inlet temperature selectivity of PA in MTR is slightly lower compared to microreactor. In both cases selectivities of PA in MTR decrease with the increase in temperature, since higher temperatures encourage total oxidised product formation. In MTR selectivity of PA decreases drastically when the inlet temperature exceeds 610 K, because runaway occurs. Only microreactor can be operated at these elevated temperatures without runaway.

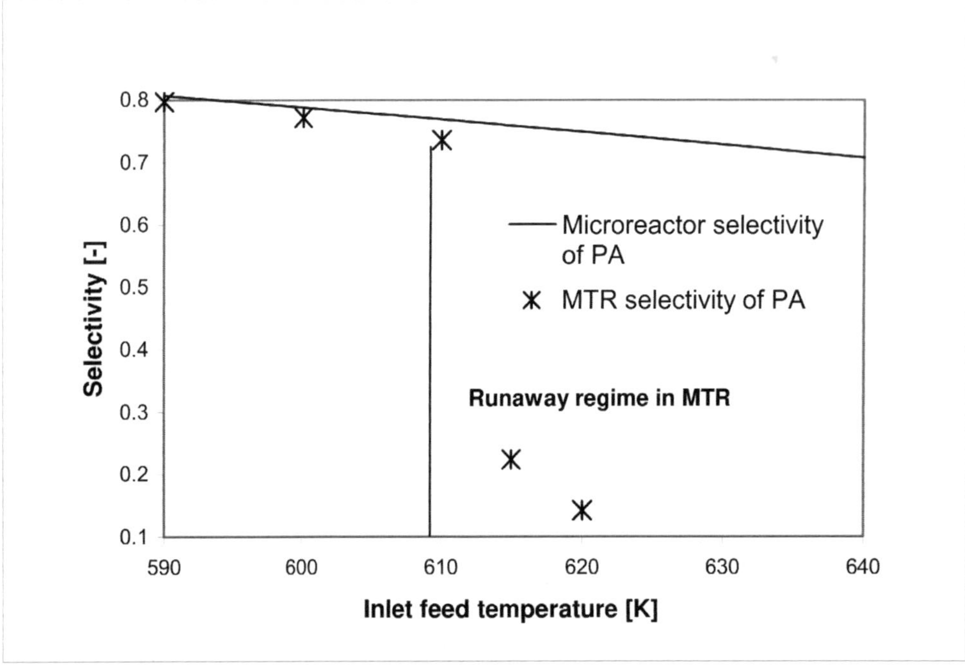

Fig. 4. Influence of inlet feed temperature on selectivity of PA

Inlet o-xylene concentration (C_{Xo})= 1 mol%, Inlet o_2 concentration= 20 mol%, conversion of o-xylene= 99%

3. Space-time-yield (STY) of PA

STY of PA at a fixed o-xylene conversion of 99 % increases with the increase in temperature for microreactor as well as for MTR. But, in MTR thermal runaway occurs when the inlet feed temperature exceeds 610 K, whereas microreactor can be operated under isothermal conditions even at higher inlet temperatures. Thus the STY of PA can be increased significantly by operating at higher temperatures. For example, at 625 K microreactor offers 3 times larger STY of PA than MTR and at the same time selectivity of PA remains same to that of MTR.

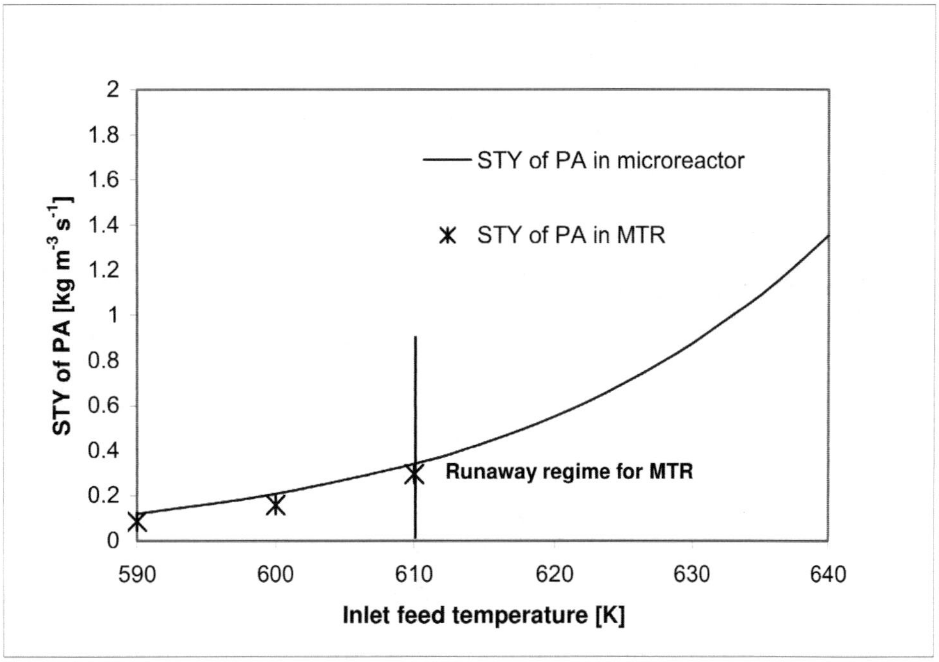

Fig. 5. Influence of inlet feed temperature on STY of PA

Inlet o-xylene concentration (C_{Xo})= 1 mol%, Inlet o_2 concentration= 20 mol%, conversion of o-xylene= 99%

Conclusions

In microreactors, oxidation of o-xylene to PA can be operated at higher inlet temperatures without the fear of thermal runaway. The problem of hotspot can be avoided by using microreactor since microreactor operates under complete isothermal condition. Thus the selectivity and STY of PA can be increased.

Abbreviations

C	Concentration [mol m^{-3}]
c_p	Specific heat [kJ kg^{-1} K^{-1}]
d_h	Hydraulic diameter [m]
E_a	Activation energy [J mol^{-1}]
k_1, k_3, k_5	Rate constant [mol kg^{-1} h^{-1} atm^{-2}]
k_2, k_4	Rate constant [mol kg^{-1} h^{-1} $atm^{-1,5}$]
K_O	Adsorption coefficient for dissociative chemisorption of oxygen [$atm^{-0,5}$]
K_A	Adsorption coefficient for adsorption of aromatic compounds on oxidised site [$atm^{-1,5}$]

K_B	Adsorption coefficient for adsorption of aromatic compounds on reduced site [atm^{-1}]
M	Mean molecular weight [g $moles^{-1}$]
Nu	Nusselt number [-]
Pr	Prantdl number [-]
P	Total pressure [atm]
p	Partial pressure [atm]
p_x	Partial pressure of xylene [atm]
p_T	Partial pressure of tolualdehyde [atm]
p_P	Partial pressure of phthalide [atm]
p_{O_2}	Partial pressure of oxygen [atm]
p_{PA}	Partial pressure of phthalic anhydride [atm]
Ri	Rate of depletion and rate of formation of a component [mol kg^{-1} h^{-1}]
R	Universal gas constant [J mol^{-1} K^{-1}]
T	Temperature [K]
T_w	Wall temperature [K]
U_f	Fluid velocity [m s^{-1}]
z	Axial coordinate [m]
α	Heat transfer coefficient [W m^{-2} K^{-1}]
λ	Thermal conductivity [W m^{-1} K^{-1}]
ρ_b	Bulk density [kg m^{-3}]
ρ_g	Fluid density [kg m^{-3}]

References

[1] Elias Klemm, Markus Rudek, Georg Markowz, Rüdiger Schütte
Mikroverfahrenstechnik, in: Winnacker - Küchler: Chemische Technik, Wiley-VCH, Weinheim, (2003) 759-815

[2] J.N. Papageorgiou, M.C. Abello, G.F. Froment
Applied Catalysis A: General 120 (1994) 17-43

Design and Development of a Low-Cost, High-Temperature Silicon Carbide Micro-Channel Recuperator

Merrill A. Wilson of Ceramatec, Inc., Salt Lake City, Utah
Kurt Recknagle of Pacific Northwest National Laboratory, Richland, Washington
Kriston Brooks of Pacific Northwest National Laboratory, Richland, Washington

ABSTRACT

Typically, ceramic micro-channel devices are used for high temperature heat exchangers, catalytic reactors, electronics cooling, and processing of corrosive streams where the thermo-mechanical benefits of ceramic materials are desired. These benefits include: high temperature mechanical and corrosion properties and tailorable material properties such as thermal expansion, electrical conductivity and thermal conductivity. In addition, by utilizing Laminated Object Manufacturing (LOM) methods, inexpensive ceramic materials can be layered, featured and laminated in the green state and co-sintered to form monolithic structures amenable to mass production.

In cooperation with the DOE and Pacific Northwest National Labs, silicon carbide (SiC) based micro-channel recuperator concepts are being developed and tested. The performance benefits of a high temperature, micro-channel heat exchanger are realized from the improved thermal efficiency of the high temperature cycles and the improved effectiveness of micro-channel heat and mass transfer. In designing these structures, the heat and mass transfer within the micro-channels are being analyzed with heat transfer models, computational fluid dynamics models and validated with experimental results. As an example, a typical micro-turbine cycle was modified and modeled to incorporate this ceramic recuperator and it was found that the overall thermal efficiency of the micro-turbine could be improved from about 27% to over 40%.

Process improvements require technical advantages and cost advantages. These LOM methodologies have been based on well-proven industry standard processes where labor, throughput and capital estimates have been tested. Following these cost models and validation at the prototype scale, cost estimates were obtained. For the micro-turbine example, cost estimates indicate that the high-temperature SiC recuperator would cost about $200 per kWe.

The development of these heat exchangers is multi-faceted and this paper focuses on the design optimization of a layered micro-channel heat exchanger, its performance testing, and fabrication development through LOM methodologies.

INTRODUCTION

Microturbines have been introduced into commercial operation over the last 10 years. However, the acceptance of this distributed power source has been somewhat limited due to their economics. The Department of Energy goal for microturbines is to reach 40% thermal efficiency[1]. The current systems (generation 1) operate at about 30%[2]. Bridging this gap will require much higher operating temperatures and require a new material set (ceramics) for the hot components[3], turbine, combustor and recuperator. Under a Small Business Innovative Research grant from the DOE, the advancement of high temperature, high efficiency ceramic heat exchangers were studied.

The design process is an iterative cycle where concepts are derived and validated through component fabrication and empirical testing. This paper will discuss these 3 facets of design development and summarize with the expected performance metrics of this high temperature silicon carbide heat exchanger.

DESIGN DEVELOPMENT

Process Modeling

Process modeling software was used to define the recuperator operating conditions necessary to obtain 40% thermal efficiency in the microturbine cycle. Based on these process models, it was found that the recuperator effectiveness had to exceed 90%, the feed side pressure drop must be below 2 psid, and the exhaust side pressure drop should remain below 0.5 psid. Additionally, these models were exercised to "rough-out" the design by parametrically varying the channel geometry and flow rates. Given these flow conditions, the resultant pressure drops and heat exchange was estimated from empirical correlations. These analyses led to a counter-flow micro-channel design wherein high heat transfer coefficients could be attained without excessive pressure drops. Figure 1 summarizes the effects of the channel geometry on cycle efficiency, pressure drop and recuperator effectiveness.

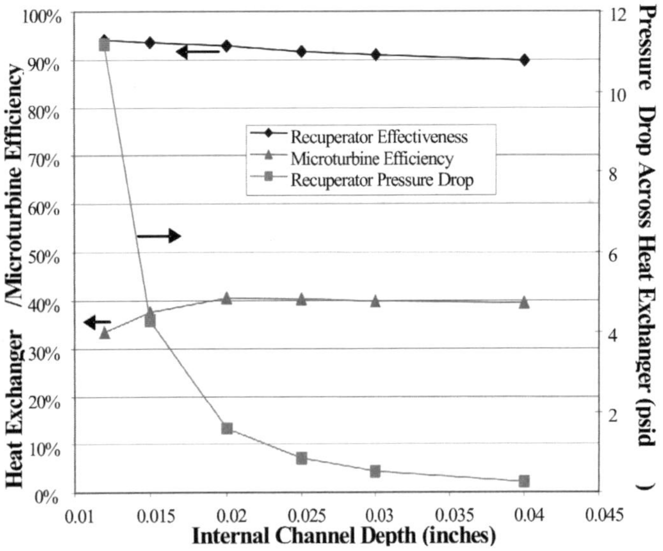

Figure 1. Cycle and Heat Exchanger Performance as a Function of Micro-Channel Geometry

The results from the process model developed into a "shell and plate" concept for the microturbine recuperator. In this design, the low pressure drop exhaust flows on the "shell-side" of the recuperator; the high pressure feed flows through the internally manifolded plates, much like a plate-and-fin heat exchanger. The exhaust flow enters downstream of the turbine at a temperature of 955C and exits at 293C; the feed gas enters the recuperator downstream of the compressor at 198 C and leaves at 895C. Figure 2 illustrates this design with flows streams indicated.

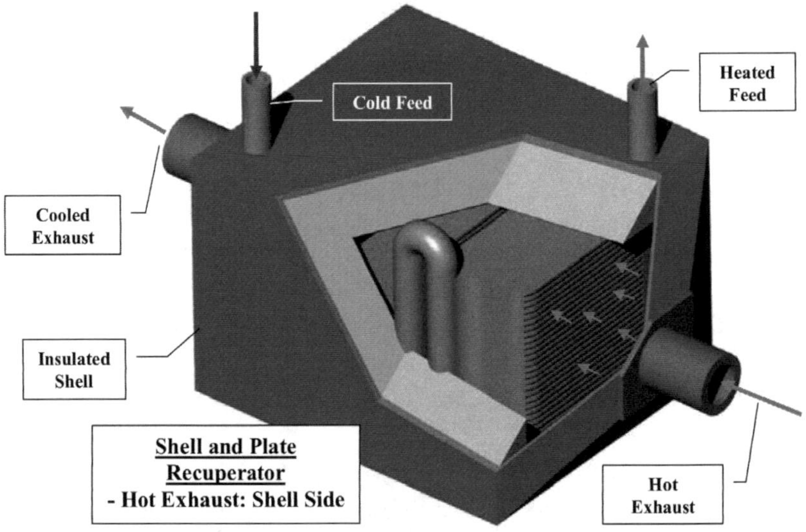

Figure 2. Shell and Plate Recuperator

Micro-channel Design Refinement using Finite Element Models

As with most micro-channel devices, the internal and external flow conditions are laminar and simplified flow and heat transfer correlations were used to refine the design of the "shell-and-plate" recuperator. Shih et al[4] has correlated the friction factor and Reynolds number for various aspect ratio rectangular ducts throughout the laminar flow regime. Likewise, the average Nusselt number ($Nu=hd_h/k$) in a rectangular duct varies with aspect ratio of the channels; square channels (1:1) have a Nusselt number of 3.61 and high aspect channels (approximately 20:1) have a Nusselt number of 8.23[5]. Based on these correlations, the heat transfer coefficient can be estimated for a micro-channel, the flow rate can be maximized to the permissible pressure drop and the heat transfer area calculated for the required duty.

Using the correlations described above, micro-channel designs were investigated using finite element models. These models have the capability of tracking the energy balance between the gas stream(s) and the solid. Thus with this coupled fluid/solid model, the parameters describing the micro-channels (flow rate, length, width and depth) and the parameters defining the heat transfer in the solid (thermal conductivity, channel pitch and membrane thickness) could be investigated and optimized. For example, the high thermal conductivity of silicon carbide reduces the conduction losses between counter-current gas streams, allowing for thicker membranes to resist differential pressure loads. However as an isotropic material, the silicon carbide also conducts heat axially (hot end to cold end) and reduces the total heat transfer due. During the optimization process, it was found that when the fluid flow was low or the solid cross-section was high, axial heat conduction through the solid became significant such that the heat exchanger tended towards isothermal conditions. This problem was mitigated by instigating a two-stage recuperator where an axial "discontinuity" in the solid existed. These analyses predicted that a two-stage recuperator could meet the >90% effectiveness goal. Figure 3 shows the resultant temperature profiles of the gas streams and the micro-channel walls of the two-stage recuperator.

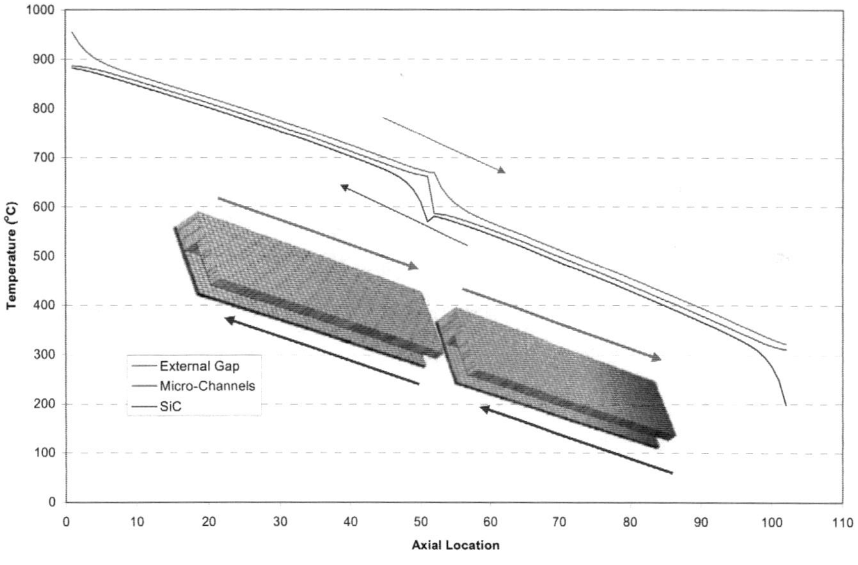

Figure 3. Axial Thermal Profile in 2-Stage Micro-Channel Recuperator

It was determined that a conjugate flow and mass transfer analysis would be required in order to integrate these micro-channels into an entire internally manifolded heat exchanger plate. These models incorporated the momentum losses associated with internal manifolds and entrance/exit effects associated with micro-channel flow. These models were also used to calculate the local heat transfer characteristics (flux, Nusselt number) throughout the component.

Preliminary analyses concluded that the predicted performance fell short of the desired 90% effectiveness goal. After interrogating the results, it was found that three primary mechanisms accounted for the reduced performance. 1) The dense cross-section for the entire plate was excessive and allowed for axial thermal dispersion. This axial conduction tended to equilibrate the hot end and cold end, reducing the log mean temperature differential or heat exchanger driving force. 2) With relatively long micro-channels, the heat flux was high near the entrance of the channels but quickly degraded along the length of the channels. 3) Lastly, the internal manifolds were undersized creating an excessive pressure drop with its associated lateral temperature gradient and flow mal-distribution within the channels. Parametric analyses indicated that these factors could be minimized by increasing the flow rate (thermal mass) through the heat exchanger and/or by reducing the axial conductivity within the solid.

These parametric analyses were used to optimize the operating conditions of the "shell-and-plate" heat exchanger. These lead to an increased hydraulic diameter within the internal channels (manifolds and heat transfer channels) and increased flow rates per component. These flow rates required wider internal manifolds and more effectively used the length of the micro-channels. These results indicated that the pressure drop goals were maintained, flow distribution improved and the effective heat transfer area was increased. The optimized design was predicted to have a five fold increase in flow over the original design concept. Figure 4 illustrates temperature profiles from the original concept to the improved design as predicted from these analytical tools.

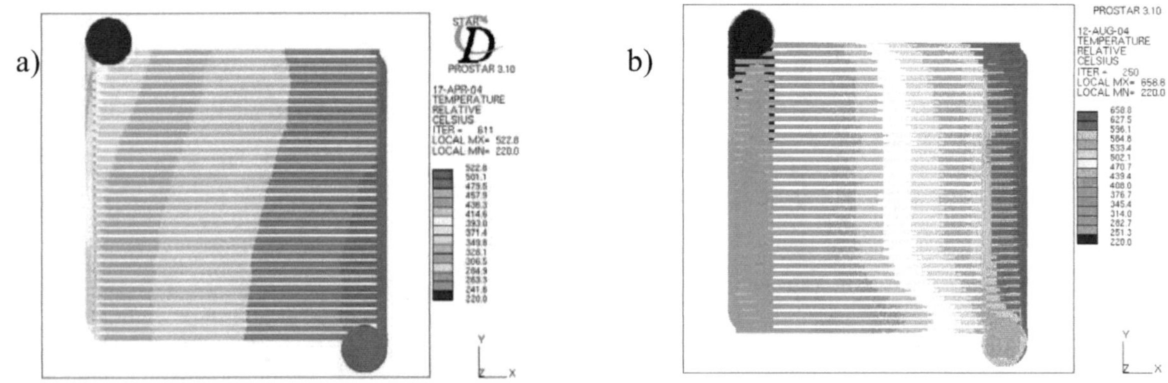

Figure 4. Temperature Profiles as Predicted from Star-CD® a) Original Concept, b) Optimized Design

DESIGN VALIDATION

Using the LOM method symmetric flow/thermal test coupons were fabricated. These coupons had a row of approximately 12 flow channels near the upper surface and an additional row of parallel channels on the under half of the coupon (see Figure 5). In this manner, the flow and thermal characteristics could be measured in a co-flow or in a counter-flow configuration. These rows of micro-channels were separated by a thin, dense layer of silicon carbide that acted as the primary heat transfer surface. In order to assess the performance sensitivity

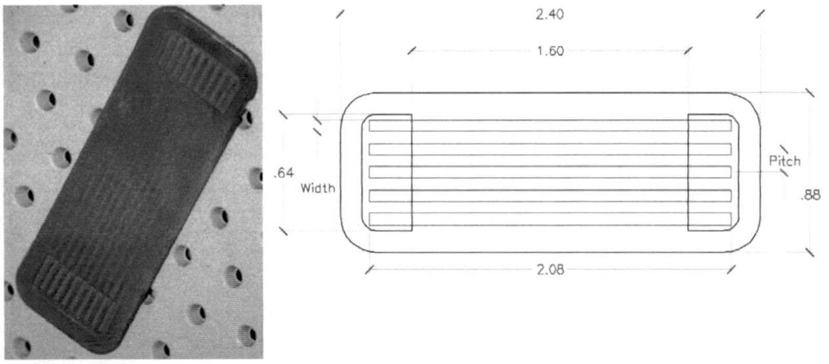

Figure 5. Flow/Thermal Coupons for Micro-Channel Validation

to the micro-channel geometry, various channel designs were fabricated and tested; these included micro-channel depth (200μ-400μ), width (1025μ-2050μ) and pitch.

The testing of these coupons included first 1) a leak check to verify flow integrity, 2) a flow/pressure drop test and 3) a flow/heating test is the test rig shown in Figure 6. This test rig was designed to provide one heated gas stream (air) and an additional cool gas stream (air). The inlet and exit temperatures and pressures were monitored in order to determine the pressure drop and heat transfer between the fluids. The entire test was encapsulated with insulating material to minimize the heat lost to the surroundings.

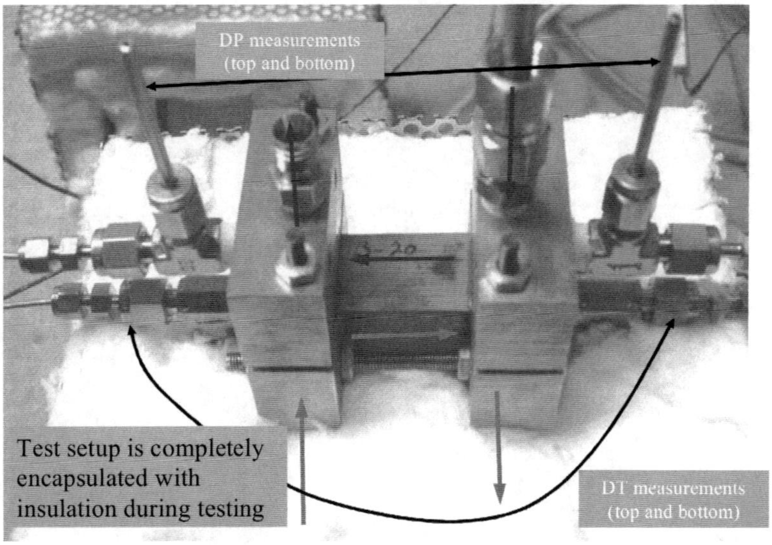

Figure 6. Test Rig for Micro-Channel Validation

These various test conditions were also modeled with finite elements to assess the validity of the modeling effort and to interpolate temperatures and heat transfer characteristics between the entrance and exit conditions. It was found that even though there were parasitic heat losses from the surfaces, over a wide range of flow rates and temperatures the data and models were in good agreement. In Figure 7 is an example of these results where the analytical results are shown as the solid lines and the empirical data are shown as markers with error bars. For the case illustrated, the hot gas stream (lower channels) was feed at flow rates between 2 and 4 liters per minute. This inlet temperature (30C to 120C) was increased with an increasing flow rate. The cold feed stream was heated using similar flow rates in a counter-flowing configuration. As seen in this figure, the parasitic heat losses accounted for about 50% of the energy transferred from the hot stream.

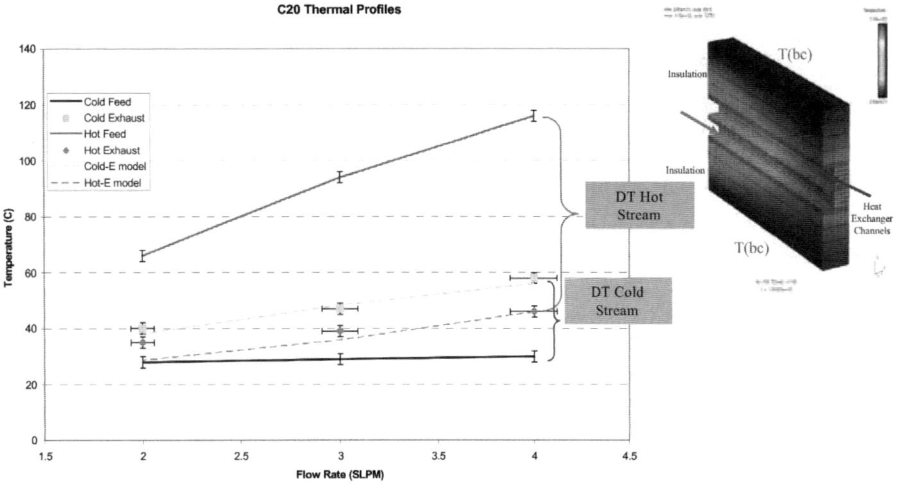

Figure 7. Comparison of Analytical and Empirical Results.

COMPONENT FABRICATION USING LAMINATED OBJECT MANUFACTURING (LOM)

These optimized micro-channels were designed into a plate that could be assembled into a module as part of the "shell-and-plate" heat exchanger. These plates are the primary building blocks of the heat exchanger and include internal features for gas manifolding and large ports as gas headers (see Figure 8). By stacking and bonding these plates with spacers, narrow gaps are formed and define the gas channels for the shell-side flow. Thus, a stack of plates becomes a modular unit (or module) that is assembled into the heat exchanger shell.

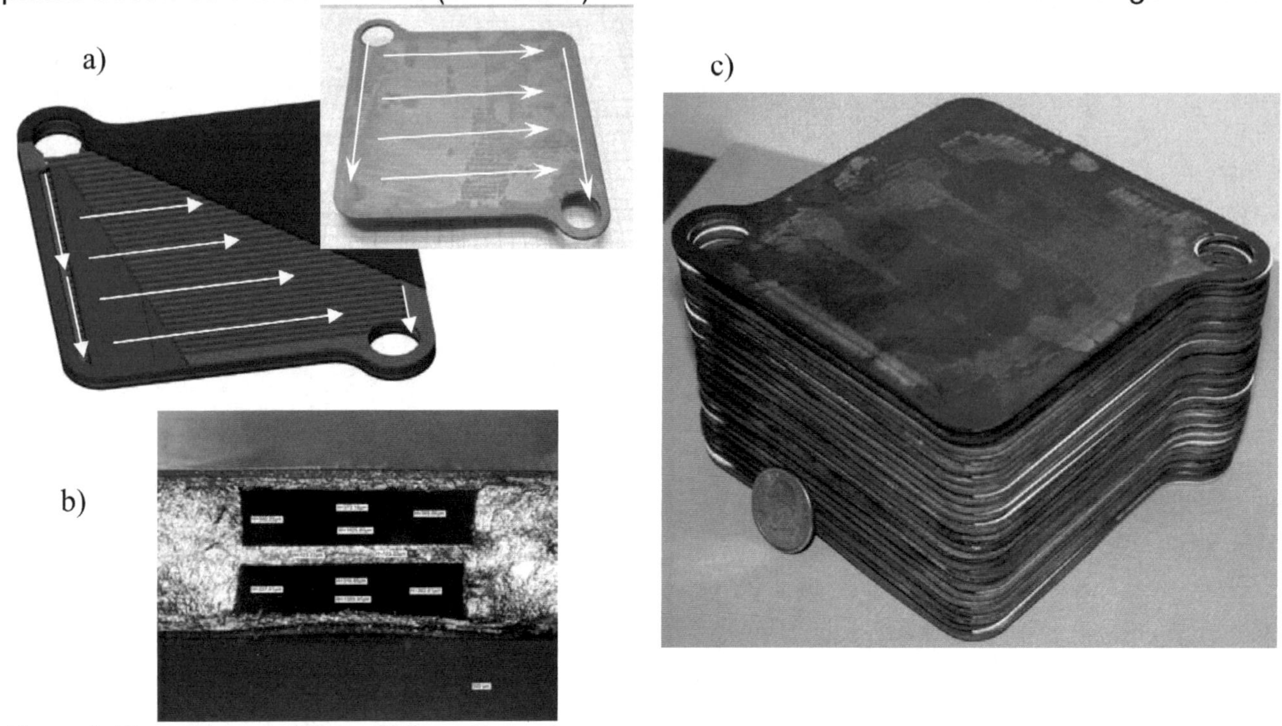

Figure 8. Planar Array or Plate of Micro-Channels.
a) Sketch and Picture of Full-Size Plate, b) Internal Features, c) Modular Stack

The primary delay in migrating to microchannel devices has been the ability to mass produce the microchannel components in the preferred materials at a commercially viable cost. In order to attain dimensional tolerances, a Laminated Object Manufacturing (LOM) method has been developed. This LOM process begins with ceramic powders (oxides and/or non-oxides) within a specified process window for surface area and particle size. These powders are blended with a mix of organic binders, plasticizers and solvents to a consistency similar to paint. This blend (ceramic slip) is then cast onto a mylar film that dries and cures into a flexible, ceramic-filled plastic tape. This process is called tape casting. Once cured, these tapes can be featured by numerous machining technologies; CNC laser machining is used for fine features and punching for larger features. The featured, plastic layers are laminated and bonded together using pressure, temperature and solvents. Once these

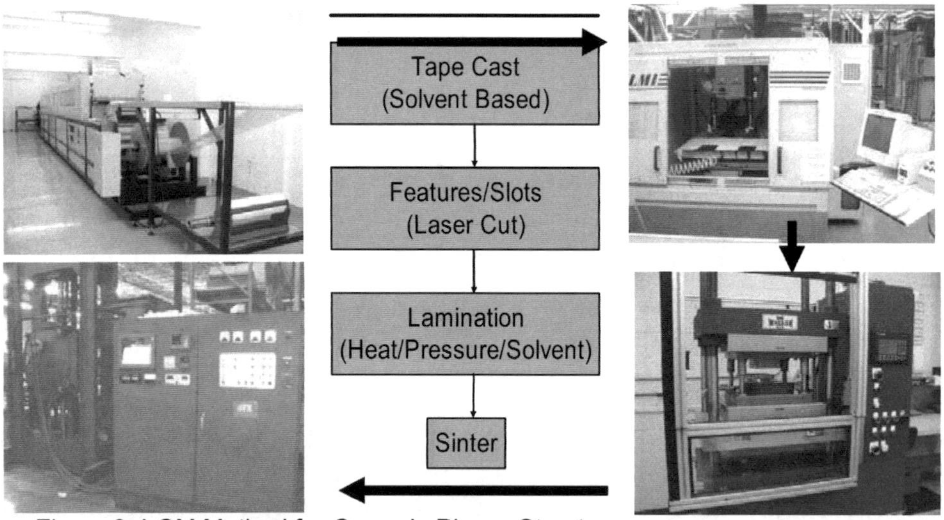

Figure 9. LOM Method for Ceramic Planar Structures

layers are laminated, they are heated to first remove the organics and subsequently fired to densify the remaining ceramic powders. The densification step is called sintering. Typically the sintering process results in a shrinkage of about 20% in all dimensions. This simplified process is shown in Figure 9.

PERFORMANCE METRICS

Heat Exchanger Performance

Once the design of this high temperature ceramic heat exchanger was optimized, several metrics were computed. These are listed in Table 1.

Table 1. Performance Metrics of High Temperature, Ceramic Heat Exchangers

Performance Metric	Value
Thermal Duty	107 kW (heat)
Hydraulic Diameter - Feed	975μ
Hydraulic Diameter - Exhaust	2100μ
Pressure Drop – Feed	2.0 psid
Pressure Drop – Exhaust	0.5 psid
Effectiveness	90%
Temperature Span (Inlet to Inlet)	198C to 955C
Volume	0.5 ft^3
# of Stages	2
# of Plates	300

Heat Exchanger Economics

This LOM methodology was developed such that all unit processes were commercially implemented, production-scale processes. By following this strategy, estimates for the capital equipment, through-puts, yields and labor costs can be extrapolated. Tape casters are widely used in the electronics industry for making electronic packages and capacitors. CNC Laser cutters are used extensively in metal and plastics machining. Lamination presses are used in plastics forming and lamination. High temperature sintering furnaces are the foundation of ceramics processing. By accessing these industries and equipment manufacturers, the manufacturing costs could then be estimated and scaled according to production capacity.

The economic analyses we performed by formulating a bottoms-up manufacturing plan where direct, indirect, capitalization and facilitization were considered. This manufacturing plan included cost models that were based on commercially available information as stated above. These cost models were scaled to assess the

Table 2. Economic Feasibility for Ceramic

Item	Description	Amount
Direct Labor	Labor (80 man years)	$2,093,077
Indirect Labor	Management / QC (68 man years)	$4,127,548
Benefits	All Labor (30% of labor)	$1,866,188
Overhead	(20% of labor)	$1,617,363
Sub-total	Labor related variable costs	$9,704,176
Direct Materials	Material used in product (SiC powder)	$5,216,013
Consumables	Other Materials (mylar, solvents etc)	$3,662,557
Other Materials	HX packaging ($500 each)	$5,000,000
Sub-total	Materials related fixed costs	$13,878,570
Capital Depreciation	10 year straight-line	$10,879,826
Sub-total	Annual expenses	$34,462,572
Management	G&A (20% of expenses)	$6,892,514
Sub-total	Cost of production	41,355,086
Profit	15% of production costs	$6,203,263
Total	Revenue	**$47,558,349**
Number of Units	10,000 per year	
Price per Unit	Total/Number of Units	**$4,755.83**
Price per kWe	26 kW electric micro-turbine cycle	**$183**
Price per kWh	107 kW heat recuperation	**$44**

commercial feasibility of a ceramic heat exchanger for the micro-turbine power cycle. These economic estimations can also be extended to other similar high temperature applications.

The cost model for which this economic analysis was done is based on a recuperator as per the design of this report. This system assumed a 26kWe micro-turbine cycle (107kW of heat recuperation) at a quantity of 10,000 units per year. Included in this analyses were the capital expenses for production, facilities and custom fitting of a facility for the capital under a 10 year straight-line depreciation. Direct and indirect labor with their associated benefits and overheads were also included. The material costs consisted of direct materials and consumables. The cost of doing business (G&A) was added into the cost analysis. Table 2 summarizes these cost analyses.

SUMMARY

A high-temperature, high efficiency ceramic heat recuperator was designed as a "shell and plate" heat exchanger. The scope of this design work also included performance modeling, empirical testing and component fabrication. Additional engineering analyses included the manufacturing economics based on current commercial ceramic processing practices and equipment.

Using these high temperature materials in the gas to gas recuperator, the microturbine thermal efficiency can be increased to 40% (assuming other hot components, turbine and combustor are capable). This high temperature recuperator shows promise in that the energy in the hot turbine gas exhaust can be captured to preheat the compressed gas feed.

It was found that by using micro-channel structures, the heat transfer is significantly enhanced and minimizes the overall volume and cost of the recuperator. The estimated volume being about 0.5 ft^3 and the capital cost being below $45/kW heat.

Albeit the next steps in developing this technology require prototype fabrication and performance validation testing, this technology shows great promise for high temperature cycles and may even find applications as moderate temperature recuperator. Through prototype testing, this technology can be validated and process applicability understood.

ACKNOWLEDGMENTS

This work was done under a DOE funded SBIR (#DE-FG03-01ER83211).

REFERENCES

[1] US Departmant of Energy, Energy Efficiency and Renewable Energy Fact Sheets, Microturbine Systems – DOE Program Overview, Adavanced Generation: Microturbine Systems, pp. 2. http://www.eere.energy.gov/de/pdfs/microturbines.pdf

[2] Energy and Environmental Analysis, Inc., "Advanced Microturbine System: Market Assessment", pp 14. Report submitted to Oak Ridge National Laboratory, May 2003. http://www.eere.energy.gov/de/pdfs/micro_market.pdf

[3] Maziasz, Philip, et al., "Advanced Alloys for High Temperature Recuperators", DER Peer Review, March 13, 2002, pp. 2. Fairfax, VA. http://www.eere.energy.gov/de/pdfs/micro_indgas_conf_02/phil_maziasz.pdf

[4] Shih, F. S., 1967, "Laminar Flow in Axisymmetric Conduits by a Rational Approach," *Can. J. Chem. Eng.* 45, 285-294.

[5] Holman, J.P., "Heat Transfer", Fifth Edition. McGraw-Hill Book Company, New York, New York, 1981, pp. 223.

132c Control of Oxidation of Aromatics with Hydrogen Peroxide under Severe Conditions using Microreactor System

Kazuhiro Mae, Kunio Yube

Hydrogen peroxide is probably the best oxidant after dioxygen with respect to environmental and economic considerations. However, in most oxidations with hydrogen peroxide, it is known that the reactions are carried out under very mild conditions; lower temperature, lower concentration of oxidant, and nearly atmospheric pressure. Higher temperature and higher concentration of oxidant are preferable from the point of view of the reaction acceleration, while such severe conditions also accelerate unfavorable side reactions: consecutive over-oxidations, parallel side-chain oxidations, decomposition of hydrogen peroxide, and so in. Therefore, in traditional semi-batch processes, oxidants are added slowly in course of removing the heat of exothermic reactions. Hydrogen peroxide is generally added dropwise in order to maintain high selectivity of desired products and high utility of oxidant. Both temperature and concentration of oxidant in the flask have to be kept low throughout the reaction operation, so the oxidation cost long reaction time.

We were trying to construct efficient oxidation processes using microreactor systems, which have the potentials of cost benefits and timesaving in manufacturing and process development of functional chemicals and pharmaceutical ingredients. As a model reaction, we performed oxidation of 2-methylnaphthalene to 2-methyl-1,4-naphthoquinone (vitamin K3) with hydrogen peroxide solution in acetic acid as solvent. For this homogeneous reaction, we adopted simple micro-tube reactors. Commercial static mixer was connected before a reactor tube of several kinds of sizes. 1 mole of substrate dissolved in 4 liters of glacial acetic acid and 3 moles of hydrogen peroxide in 60% aqueous solution were fed by non-metallic pumps. The range of residence time could change precisely from 1 second to 10 minutes.

Using these microreactors having highly efficient heat transfer, it is possible to mix substrate and oxidant at predetermined higher concentrations immediately. Reaction time could be greatly shortened compared with the case of semi-batch process. In the presence of palladium acetate and sulfuric acid, the reaction almost finished in 10 minutes at 70 degree C. It was confirmed that reaction temperature in the micro-tube was kept in error by less than 1 degree C. However, the selectivity of 2-methyl-1,4-naphthoquinone was lower than the case of flask.

To improve the selectivity, we challenged to control active species of oxidation. In this oxidation system, reaction mechanism is still ambiguous, but it is said to involve two different types of active species, i.e. hydroxyl radical species and hydroxyl cation species. The former are highly reactive species and cause undesired by-products: carboxylic acid by the side-chain oxidation, polymerized quinone compounds, and so on. It is proposed that the latter induce the favorable first step of the reaction, aromatic hydroxylation. The hydroxyl cation species are generated easily from peracetic acid, which is formed from the equilibrium reaction of hydrogen peroxide and acetic acid.

Practically, feeding peracetic acid as "actual oxidant" at given concentration into microreactor gave good selectivity of 2-methyl-1,4-naphthoquinone. As the mole ratio of peracetic acid to substrate increased, conversion of 2-methylnaphthalene and yield of 2-methyl-1,4-naphthoquinone showed monotonically increasing, and both reached plateau at the stoichiometric ratio of oxidant to substrate, just 3 times. Under extremely hard conditions, even more than 100 degree C, the efficiency of the oxidant was maintained high, not to mention safety.

In conclusion, it was clarified that microreactor system gives us a new route to control oxidation of aromatics under severe conditions.

Synthesis of Terminally Modified Polymer with a Micromixer

Toshiharu Kuboyama [a], Jun-ichi Yoshida [b]

[a] The Research Association of Micro Chemical Process Technology, Kyoto University, Kyoto 615-8530, Japan

[b] Department of Synthetic Chemistry and Biological Chemistry, Graduate School of Engineering, Kyoto University, Kyoto 615-8510, Japan

Polycondensation reactions are important processes for the synthesis of highly thermal resistant polymers such as polyamide and polyimide. Because these reactions are quite exothermic, it is generally difficult to control molecular weights and molecular weight distributions precisely in a macro-scale batch reactor. Although it was reported that the premixing with a micromixer is quite effective for the regulation of molecular weight distribution in radical polymerization and anionic living polymerization, there is no report on the polycondensation reactions in a microreactor. Thus, we examined the two-step reaction, polycondensation and terminal modification reactions in a microreactor, in which the reaction temperature can be precisely controlled. The rate of the polymerization and the molecular weight distribution were compared with those for batch type polymerization.

The reaction in the microreactor was found to be faster than that in the batch system. Higher mixing efficiency of monomers seems to be responsible for faster reaction. It is also important to note that the molecular weight distribution of terminally modified polymers obtained in the microreactor system is slightly narrower than that obtained in the batch reactor. Precise control of reaction temperature in the microreactor seems to be responsible for narrower molecular weight distribution.

Design of Functionality of Three-Dimensionally Bent Channels Based on Deformation and Rotation of Fluid Interface

Kazuo Ohkawa[1,2], Takashi Nakamoto[2], Yoshiro Inoue[2] and Yushi Hirata[2]

[1] *Chemical Development Laboratories, Fujisawa Pharmaceutical Co., Ltd., Osaka, Japan*

[2] *Division of Chemical Engineering, Graduate School of Engineering Science, Osaka University, Japan*

E-Mail: Kazuo_ohkawa@po.fujisawa.co.jp, hirata@cheng.es.osaka-u.ac.jp

Introduction

In designing a microdevice such as microreactor or micromixer it is important to grasp functionalities of microchannels since such devices are composed of microchannels. Although design data for channel configuration or flow condition that provides functionalities such as mixing or interflow-bifurcation for microchannels are being accumulated [1,2], fundamental investigation on flow analysis of microchannels is still insufficient for designing such functionalities. We have conducted experimental and CFD studies to elucidate the deformation and rotation process of a virtual interface of fluids with the same physical properties while they are passing through a three-dimensionally bent channel, and we have clarified characteristics and functionalities of the flow through the bent channel by paying attention to the deformation of the fluid interface with Reynolds number and its rotation in the cross-section with respect to the crossed axis angle of the bent channel. In our poster, it will be demonstrated that an element of plate static mixer can be constructed by combining three-dimensionally bent channels, in which fluids are divided, rotated and merged to make static mixing progress systematically in the element.

Experimental and CFD method

The configuration of a three-dimensionally bent channel is shown in Fig. 1. Two rectangular channels are placed on parallel planes at a vertical spacing, and they are vertically

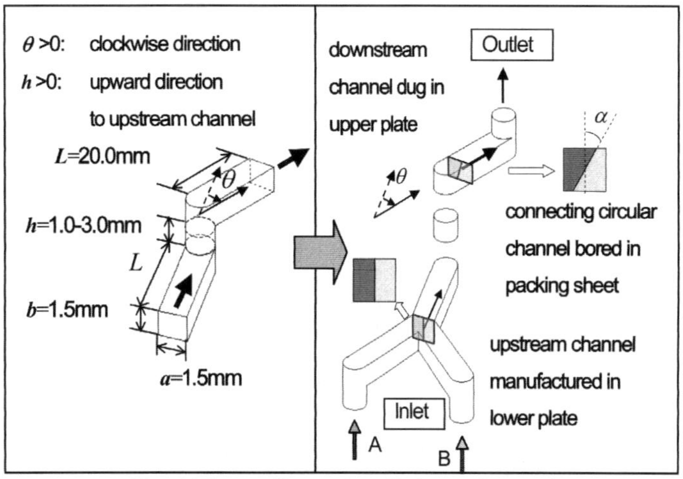

Fig. 1 Three-dimensionally bent channel

interconnected with a circular channel of h in length. The axis of the downstream rectangular channel is crossed an angle θ from that of the upper channel when it is projected on the plane of the upper channel. The upstream channel provides a Y-shaped branch for introducing two fluids into it. Thus a three-dimensionally bent channel was constructed by sandwiching open rectangular channels dug on plates of acrylic resin with a packing sheet with a hole, as shown in the figure. A straight fluid interface parallel to the channel sidewall was formed in the central region of the upstream channel by introducing two fluids with the same properties through the Y-shaped branches. The fluid interface in the cross-section of the downstream channel was visualized by means of the laser induced fluorescence method using fluorescein sodium solution, as shown in Fig. 2. Its deformation and rotation angle α with respect to the coordinates set in the downstream cross-section were investigated in the range of Reynolds number from 0.503 to 118. Flow analysis was conducted by using CFD software RFLOW and the fluid interface was also obtained from the trajectories of the fluid particles by using CFD velocity data.

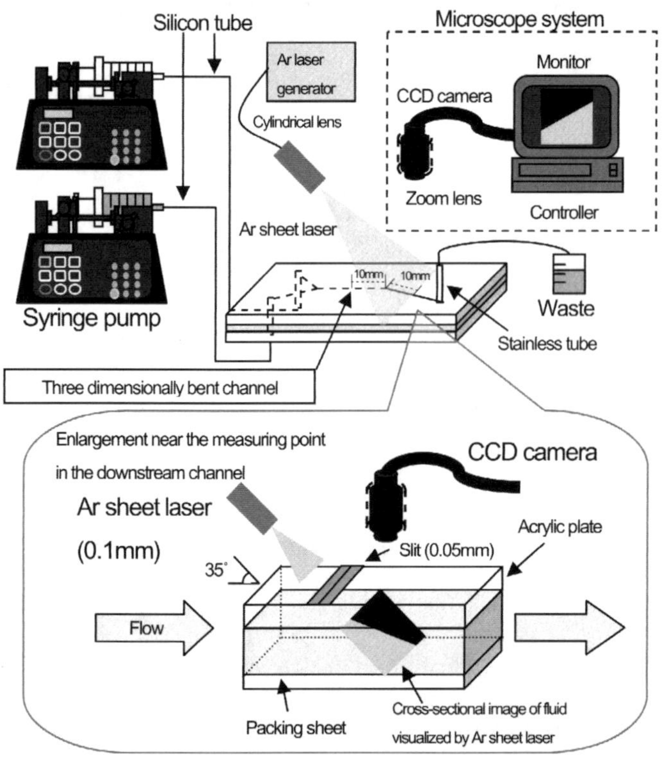

Fig. 2 Experimental apparatus and method for visualizing the fluid interface in downstream channel

Results and Discussion

The fluid interfaces in the downstream channel at θ = 60 degrees and h = 3.0 mm are shown in Fig. 3. The particle trajectories obtained by CFD are in good agreement with the image of fluid interface visualized by experiment at each Reynolds number. The interface was curved and deformed with increasing Reynolds number, and part of the interface was folded up at Re = 56.4. In case of low Reynolds number less than 2, the fluid interface channel rotates in the downstream cross-section from the initial perpendicular line in the upstream cross-section while keeping its deformation small. Figure 4 shows the visualized images of the fluid interface in the downstream cross-section for various crossed angle θ between the upstream and downstream channel axes at Re = 0.503. The rotation angle α of the fluid interface is almost equal to the crossed axis angle θ. Rotation of the interface is clockwise if $h\theta$ >0 and counterclockwise if $h\theta$<0, where h>0 when the downstream channel is set at an upper position from the upstream channel. From these results, rotation of the fluid interface at a given angle in the desired direction can be easily performed by combining three-dimensionally bent channels. This is a new finding on the functionalities of microchannels. Using the rotation of the fluid interface described above, a splitting and recombining type micromixer can be constructed by connecting a Y-shaped bifurcation channel with two bent channels so as to rotate the fluid interfaces by 180 degrees and merge them to form fluid layers. We will demonstrate a plate static micromixer using this function to rotate the interface.

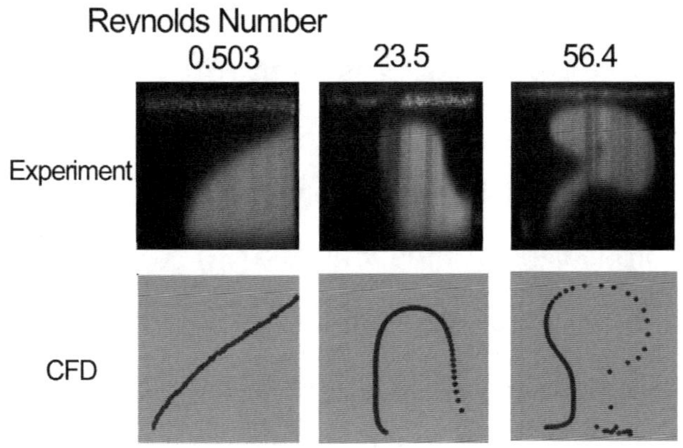

Fig. 3 Rotation of fluid interface in the downstream channel at h = 3.0 mm, θ= 60°

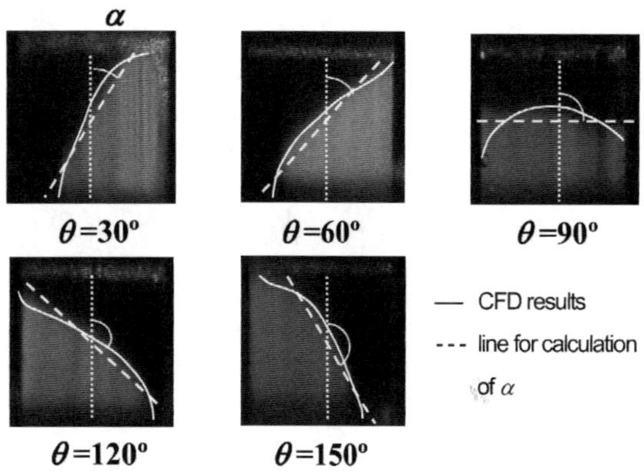

Fig. 4 Rotation of fluid interface in the downstream channel at Reynolds number of 0.503

Reference

1. Yi, M. and H. H. Bau, "The Kinetics of Bend-Induced Mixing in Micro-Conduits", *Int. J. Heat and Fluid Flow*, **24**, pp.645-656, 2003
2. Inoue, Y., T. Yanase, N. Yoshio and Y. Hirata, "Flow Separation in a Bifurcating Minichannel Equipped with a Flow Controlling Channel", Proceedings of the 2nd International Workshop on Micro Chemical Plants, Hyogo, Japan, Feb. 3rd-4th, pp. 30, 2004

Halogen-Lithium Exchange Reactions of Aryl- and Heteroaryl Bromides in Microreactor Systems

Shigeki Hikage [a], *Jun-ichi Yoshida* [b]

[a] *The Research Association of Micro Chemical Process Technology, Kyoto University, Kyoto 615-8530, Japan*

[b] *Department of Synthetic Chemistry and Biological Chemistry, Graduate School of Engineering, Kyoto University, Kyoto 615-8510, Japan*

In this presentation we wish to report halogen-lithium exchange reactions of aryl bromides and heteroaryl bromides using microflow systems. The organolithium compounds have been used widely in laboratory synthesis, but their reactions should be carried out at low temperatures such as -40 °C and -100 °C, because organolithium compounds are highly reactive and unstable. Although laboratory-scale reactions can be carried out easily at such a very low temperature, industrial-scale reactions are difficult to perform at such a low temperature.

Efficient heat transfer and precise control of the residence time in microflow systems are expected to diminish the decomposition of organolithium intermediates.

Thus, we examined the halogen-lithium exchange reactions of aryl bromides and heteroaryl bromides with butyllithium followed by the trapping with an electrophile such as ketones and chlorosilanes using a micro flow systems, and found that the reactions took place with high selectivity even at 0 - 20 °C to give desired compounds in good yields.

132g Production of Bicolored Polymeric Microspheres Using Sheath Flow Regimes in Microfluidic Devices

Takanori Takahashi, Takasi Nisisako, Toru Torii, Toshiro Higuchi, Yoichi Takizawa

Bicolored polymeric microspheres for particle-based displays were engineered in a microfluidic device using sheath flow geometry. Preparation of bichromal beads was accomplished by rapid formation of bicolored microdrops at a sheath-flow junction in a glass microdevice and subsequent polymerization. The size of the spheres could be varied in the range 50-190 um by changing the parameters of the aqueous/organic streams in the microchannels, giving highly monodisperse spheres with less than 5 % variation. Futhermore, in order to increase the productivity, multiple channels having similar dimensions were integrated on a quartz chip by lithographic technique. We confirmed that bicolored drops of uniform size could be reliably obtained at each of the multiple junctions. The resulting spheres could be rotated to display images between a set of planar ITO electrodes by applying suitable voltages. The relation between the symmetry in flow regimes and the color distribution in droplets was also studied.

Micromotive - Development and Fabrication of Miniaturised Components for Gas Generation in Fuel Cell Systems

[1]P. Pfeifer, [1]O. Görke, [1]K. Schubert, [2]D. Martin, [2]S. Herz, [3]U. Horn, [3]Th. Gräbener

[1]Institute for Micro Process Engineering (IMVT), Forschungszentrum Karlsruhe GmbH, Hermann-von-Helmholtz Platz 1, 76344 Eggenstein-Leopoldshafen, Germany, www.fzk.de/imvt-en

[2]Ätztechnik Herz GmbH & Co., Industriegebiet Kilbigswasen, 78736 Kilbigswasen, Germany, www.aetztechnik-herz.de

[3]Gräbener Pressensysteme GmBH & Co. KG, Wetzlaer Str. 1, 57250 Netphen Werthenbach, Germany, www.schulergroup.de

Prepared for Presentation at 2005 Spring National Meeting / IMRET 8 / April 10-14 / Poster Session

Unpublished

Introduction

Within the joint project "MicroMotive", supported by the German Federal Ministry of Education and Research (BMBF), the project partners (see above) are engaged in fundamental systems work and are dealing with cost effective mass production of microchannel systems for automotive applications like steam reforming of gasoline for fuel cell systems e.g. auxiliary power units (APU). Such small fuel cell systems (up to 10 kW electrical power) have to compete with other systems for power generation in terms of efficiency, system size and costs. Microchannel systems are just the right tools to meet the technical requirements. Considerably enhanced heat and mass transfer in and between microchannels will definitely enhance heat recuperation and efficiency and by this also reduce the system size.

Results

The partners contribute to the project with their respective capabilities. The Institute for Micro Process Engineering (IMVT) at the Forschungszentrum Karlsruhe is one of initiators in development of microchannel systems since the beginning of the 90'ies (mainly by micromachining) and is able to fabricate microchannel system prototypes also for high fluid throughputs (range up to tons per hour) [1]. Microstructured metal foils are usually joined by diffusion bonding. Catalyst incorporation into the microchannels was also demonstrated by IMVT previously [2].

During the project IMVT paid most attention to the mass fabrication suitability of the microdevices like different system heat exchangers and microstructured gasoline reformers as well as fabrication methods.

In a first series of prototypes it was demonstrated that it is possible to meet the requirements of heat transfer quantities up to 3 kW, small system size and weight as well as small pressure drops down to 20 mbar even with hot burner off-gas. Although the prediction of device properties for gas-phase heat transfer at elevated temperatures were good, some problems like pressure fluctuations in water vaporisation and stability of the devices up to 950°C were obvious in this first series.

In a second series of prototypes the latter problems were solved and it was possible to standardise system design to increase further mass production suitability. Temperature resistance was established by examination of joining parameters and their influence on the material properties. Vaporisation was complete without significant pressure variation. The design standardisation is obvious in Figure 1 - a comparison of the well known standard cross flow heat exchangers from IMVT compared to the new design for *Micromotive*.

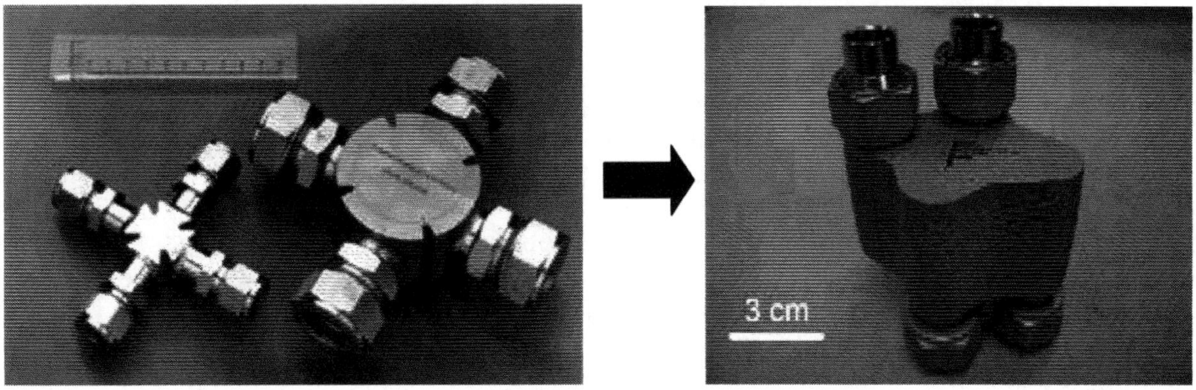

Figure 1: Comparison of standard cross flow heat exchangers from IMVT (left) and the new design for *Micromotive* (right)

Ätztechnik Herz, one of the leading companies in Europe in etching technology, is working together with the IMVT for a few years now in order to enhance microetching technology for low cost microfabrication.

One of the major tasks of Herz was to decrease tolerances of the microstructures in combining laser cutting and etching technologies during prototype fabrication for the different substrate materials used and to demonstrate a continuous etching procedure shown in Figure 2 (schematically).

Although both series of micro heat exchangers for system tests were built with foils made by microetching another option of microfabrication, namely microembossing was evaluated in the project.

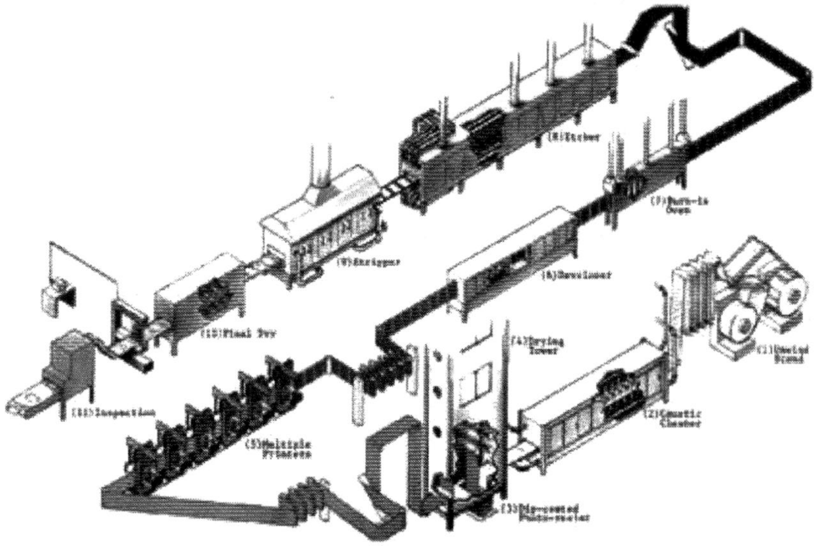

Figure 2: Schematic of the continuous etching procedure

Microembossing of metals is a possibility of cheap microfabrication at large numbers of microstructured foils. This task is pursued by the SchulerGroup (Gräbener Pressensysteme) within the project. Gräbener is one of the world leading companies in fabrication of knuckle-joint presses. These types of presses have a low acceleration in the stationary point and are judged to be the best tool for embossing of structures in the micron-range.

Starting from very basic experiments of real embossing in the micron-range a fabrication method of microstructured foils was established. The basic experiments were necessary due to extreme material displacement compared to metal deep drawing. Together with IMVT it was recently possible to fabricate first leak tight prototypes by diffusion bonding with microembossed foils. In Figure 3a one of these heat exchangers which are non-optimised with respect to the weight is shown.

DaimlerChrysler is joining the project for guidelines and in order to characterise the microsystems for the fuel cell application. In collaboration of DaimlerChrysler and IMVT, alternatives of joining the microstructured foils by soldering or laser welding as well as catalyst incorporation for mass production were under development so far. In the DaimlerChrysler group (Dornier GmbH) it was possible to manufacture heat exchangers by soldering microembossed foils with solder paste (Figure 3b).

Compared to the paste soldering, the soldering experiments of etched microfoils with high temperature soldering foils were conducted more successfully at IMVT in terms of leak tightness. This can be explained by irregularities in paste application and paste grain size. Problems by irregularities of foil thickness for combining soldering foils and microembossed foils are expected. However a solution for this specific combination is currently under development.

Figure 4 illustrates the tests of the heat exchangers and the microreformers at the DaimlerChrysler group. The tests were performed at inlet temperatures up to 850°C and 10 bar successfully. The leak tightness of the devices was proven afterwards and pressure drop as well as heat transfer specifications were reached according to specifications.

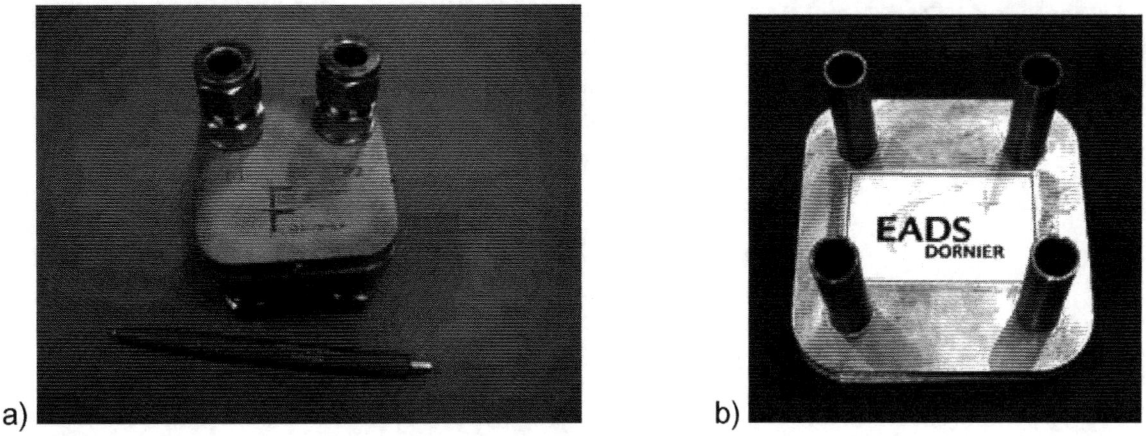

Figure 3: Micro heat exchangers with microembossed foils a) by diffusion bonding and b) by soldering.

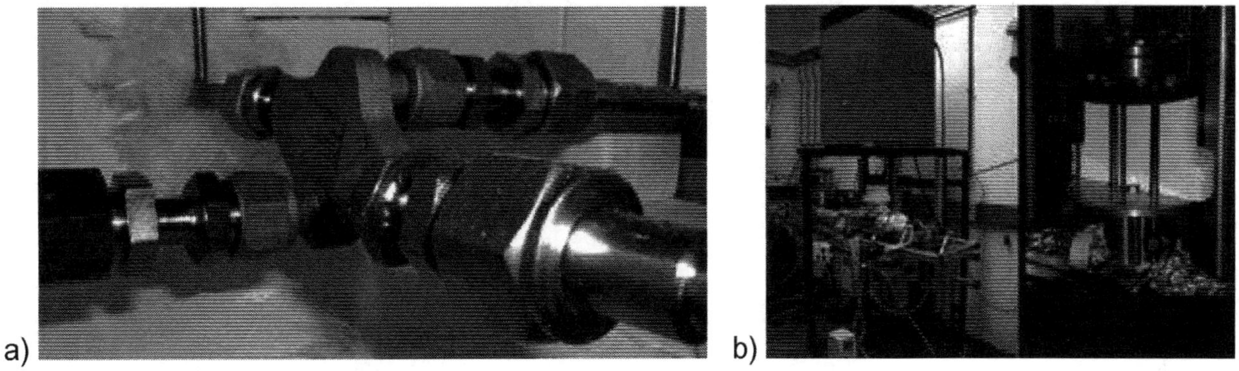

Figure 4: a) Micro heat exchanger and b) microstructured reformer in test at the facilities at the DaimlerChrysler group.

Conclusion

The improvement of the fabrication methods and the microstructure design for mass fabrication within the project *Micromotive* enables microtechnology feasibility for low cost and dynamic operation of gas generation systems for hydrogen generation in mobile fuel cell systems.

Acknowledgement

The authors thank the German Ministry of Education and Research for financial support (AZ 2218).

Literature

[1] K. Schubert, J. Brandner, M. Fichtner, G. Linder, U. Schygulla, A. Wenka; Microscale Thermophysical Engineering; 5:17-39, 2001.

[2] K. Haas-Santo, O. Görke, P. Pfeifer, K. Schubert; Chimia - Chimia Report, 56(2002), 605-610.

132i Rapid and Continuous Separation of Satellite Droplets Using A Branching Microchannel Configuration

Takasi Nisisako, Toru Torii, Toshiro Higuchi

In recent years, microfluidic devices with T-junctions have shown a lot of promise in making emulsion droplets of highly uniform size. However, interfacial instability usually causes the formation of satellite drops when main droplets are formed. These unwanted by-products constitute a hindrance in many applications. We report here a methodology for hydrodynamic fractionation of droplets by a simple microfluidic geometry. By exploiting the positional relation immediately after breakup at a T-junction, main droplets and satellite droplets can be separated downstream in a rapid and continuous manner. Droplet rupturing at a T-junction and subsequent sorting downstream were successfully performed in both oil-in-water (O/W) and water-in-oil (W/O) systems. We studied the flow conditions needed for successful separation.

Microchannels were fabricated in poly(methylmetacrylate) (PMMA) or quartz glass by conventional machining techniques and lithographic techniques. The less hydrophilic properties of the PMMA surface were exploited in generating a W/O dispersion. The hydrophilic glass devices were used for the preparation of O/W dispersions. The configuration includes a T-junction (100 um x 100 um) for breakup of drops and a downstream droplet separator where the channel broadens gradually and divides into two branch channels.

At low Reynolds number (Re~10⁻¹) condition in a PMMA module, satellite drops of multiple sizes (D = 1, 3, 5 um), which were generated rapidly (~1.5kHz) along with main droplets (D = 70 um), were fractionated in size and entered into the branch channel. This follows that monosized droplets of several specified sizes can be prepared at the same time by using this technique. Next, we investigated the relation between flow conditions and fractionation performance at the sorting area by a glass module. Several flow regimes were observed at the sorting area, and a specific flow condition could be characterized in which sorting takes place. Correct choice of geometry was found to be vital for efficient separation.

Using a simple microfluidic configuration, the rapid and continuous separation of satellite drops could be demonstrated. Flow regimes for successful purification were studied experimentally in both W/O and O/W dispersion systems. This technique is capable of producing genuinely monodisperse dispersions without any fractionation processes outside channels. Satellite drops themselves may find applications because they also have very narrow size distribution.

Fabrication of a PCBMEMS Electrochemical Sensor for Detecting Explosive Residues in the Field

X. J. Fu, H. A. Broadbent, G. T. Steimle, and D. Fries
University of South Florida
College of Marine Sciences
Center for Ocean Technology
St. Petersburg, Florida 33701
xfu@seas.marine.usf.edu

ABSTRACT

Direct patterning of channels in different photoreactive materials is a common technique. In this paper, we present the use of electronic masks as a new opportunity to quickly create a wide variety of electrical and photonic circuits, mechanical acoustic waveguides, RF waveguides and gas or liquid flow channels. The electronic mask technique permits an easier route to fabricate multilayer channel systems for sensors, actuators, reactors and combination of all three. When such a patterning technique is combined with post pattern transfer processes, such as chemical etching of the dielectric and post electroforming of metallic layers or thin films, a variety of system can be made in particularly the electrochemical system. In this paper, we are applying the process sequence to fabricate a solid polymer electrolyte based electrochemical sensor for detecting explosive residues. This fabrication system is based on a PCBMEMS (printed circuit board MEMS) process flow using liquid crystal polymer (LCP) and Polyimide(PI) materials. The PCBMEMS approach permits a system with integrated fluidic and electronic function made in an economical format [1,2].

Electrochemical sensors are being developed for both vapour phase detection of nitro group explosives, hydrazine and TNT in sea water in the field. These sensors are based upon a three-electrode cell and an amperometric technique [3] for continuous detection. Appropriate selection of solid polymer electrolyte (SPE) membranes improves the sensor's selectivity, and enhances the capability for detecting vapour phase explosive residues. Sensing at 10-100 ppm levels are achieved for these representative functional groups. Desirable features of the SPE sensors include a rapid response and relaxation (about 1-2 minutes) upon exposure to vapours for the appropriate membrane on the sensor. Optimization of the sensor configuration, including a microfluidic inter digitated sensor array is discussed, and prospects for using the electrochemical sensor for real-time monitoring are discussed.

1. Introduction

Microfabrication is a process used to construct physical objects with dimensions in the micrometer to millimeter range. Microfabrication devices, also known as microelectromechanical systems (MEMS), micromachines, lab-on-a-chip, microsystems, and micro-total analysis systems (micro TAS), have existed for over 30 years, with several applications attaining commericial and/or scientific success. When applied in the right instances, microfabrication can either significantly enhance a device in relation to its conventional counterpart or enable entirely new devices (J.Voldman, M.L.Gray and M.A.Schmidt, 1999). In this paper, a micron-sized resolution maskless photoimaging /patterning tool that permits the creation of small, arbitrary features, for constructing devices, structures and packages, in photoimageable material Liquid Crystal Polymer (LCP), along with additive plating and subtractive etching is employed. The fabrication technology can provide features down to 10 microns simultaneously over a field of 2cm height and 2 cm width. The instrument relies on the use of microoptics and spatial light modulation to create the required 2D pattern aerial image for photoimprinting, the electroplating and etching process steps used are conventional.

Electrochemical sensors, holding the advantages such as sensitive and selective towards electroactive species, fast and accurate, compact, portable, inexpensive, and amenable to miniaturization, have been receiving attention for on-site environmental analysis. The classical electrochemical sensors with liquid electrolytes are usually rather bulky, therefore, the electrochemical sensor based on the solid polymer electrolyte (SPE) are the ones that fit in this category. In this type of sensor, an indicator electrode is either created directly on the SPE surface, or is in intimate contact with it [3-5]. The SPE provides an electrolytic connection of the indicator electrode with the other electrodes in an electrochemical cell, and serve for separation of the sample space to which the indicator electrode is exposed from the reference and counter electrode compartment. The advantages results from the SPE arrangement have made it possible for applying the microfabrication technology in producing a planar shaped three-electrode electrochemical sensor.

In this paper, a SPE based MEMS electrochemical planar sensor for detecting gas phase explosive residues is described. The sensor is composed of carbon fiber electrode, Ag reference electrode and Platinum counter electrode. The electrochemical reactions that determine the sensor signal, i.e., the current obtained in an amperometric circuit or the potential in a potentiometric circuit, occur at the three-phase boundary where the sample, the indicator electrode and the SPE meet and thus the properties of this boundary exert the decisive influences on the sensor behavior. The length of this boundary is an important parameter that determines, together with the indicator electrode catalytic activity, the measuring sensitivity of the sensor and primarily depends on the indicator electrode perimeter. The inter digitated microsensor is designed based on this guideline. The fabrication of the sensor is using PCBMEMS based on liquid crystal polymer (LCP) with the various sensing elements made on top or within the LCP laminate.

1.1 Maskless Lithography

We have employed a unique microfabrication prototyping technology (Model SF-100, Intelligent Micro Patterning LLC, St Petersburg FL) that allows us to create micromechanical structures rapidly and with ease, a schematic is shown in figure 1.The technology is a maskless lithographic phototool that permits direct writing of whole planar light onto a photoreactive compound such as photoresist. This approach yields an advantage in design and also in manufacturing since the technique writes directly using parallel planar light, which is more efficient than rastering a single laser spot over the same field of view. The lithographic technique eliminates masks from the microfabrication process and can additively create microstructures of non-linear geometries.

We have also coupled the direct write approach with pattern transfer processes to impart patterns into a variety of metals, polymers, and ceramics necessary for MEMS fieldable systems. A key point to be made is that the creation of microstructures in non-standard (i.e non-Silicon) MEMS materials is of importance for fieldable MEMS devices since environmental microsystems are most likely to be made from a variety of materials and silicon is not considered a harsh environment compatible material.

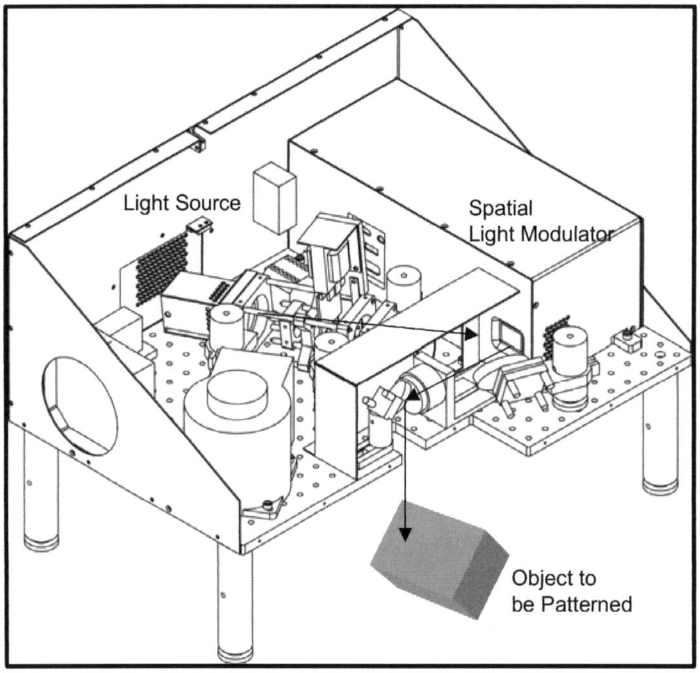

Figure 1. Maskless Lithographic Exposure Workstation- major subsystems

1.2 Introduction to LCP materials

The liquid crystal polymer is a thermoplastic polymer material with unique structural and physical properties. It contains rigid and flexible monomers that link to each other. When flowing in the liquid crystal state, rigid segments of the molecules align next to one another in the direction of shear flow. Once this orientation is formed, their direction and structure persist, even when LCP is cooled below the melting temperature [6,7]. This is different from most thermoplastic polymers (e.g., Kapton), whose molecules are randomly oriented in the solid state [8].

As a result of the unique structure, LCP offers a combination of electrical, thermal, mechanical and chemical properties unmatched by other engineering polymers. It was originally used as a high-performance thermoplastic material for high-density printed circuit board (PCB) fabrication [9,10] and semiconductor packaging [7]. Results from high-frequency test show that LCP has a uniform relative dielectric constant of 3 in the range 0.5 to 40 GHz and an extremely low loss factor of ~0.004[7]. LCP has very low moisture absorption (~0.02%) and low moisture permeability. For other gases, including oxygen, carbon dioxide, nitrogen, argon, hydrogen and helium, LCP also exhibits above-average barrier performance. Further, the permeation of gases through LCP is not affected by humidity, even in an environment with elevated temperature (e.g., 150°C). The LCP film also shows excellent chemical resistance. Research shows that LCP is virtually unaffected by most acids, bases and solvents for a considerably long time and over a broad temperature range [10].

In terms of mechanical properties, the thermal expansion coefficient of the LCP material can be controlled during the fabrication process to be both small and predictable [17].

Commercial LCP material is supplied in thin film with predefined thickness ranging from 25 μm to 3mm. One or both sides of the LCP film may have 18 μm thick copper cladding. This copper layer is laminated in a vacuum press at a temperature around the melting point of LCP [7]. LCP can be subsequently laminated into multilayer layer constructed for a wide range of designs and purposes.

2. Experiment

2.1Chemical Reagents

Chemicals used to produce the SPE for detecting nitro group can be found in reference [11]. In this paper, PVC cement is used in substitute of the high-molecular polyvinyl chloride (PVC, Sigma), and 2-nitrophenyloctyl ether (NPOE, Sigma) has been found out that is not good for sensing nitro group when using PVC cement to form a SPE. Nitromethane (Sigma-Aldrich, St. Louis, MO), Hydrazine (Sigma-Aldrich, St. Louis, MO) and 2,4,6-trinitrotoluene (TNT) (Sigma-Aldrich, St. Louis, MO) were used as the detection target. Multiple interlaced carbon fibers have been chosen as working electrode, and the detailed selection and pretreatment procedures were used in accordance to [12]. Thin film layer of silver, served as a psudo reference electrode, and thin film layer of platinum served as counter electrode are electroplated on the copper surface which is a grown layer on LCP.

2.2Electrochemical sensor fabrication and process flow

Starting with a liquid crystal polymer (LCP) substrate, a thin layer of copper or platinum or gold was grown on it to a thickness of about 18μm (Figure I). In Figure II, photoresist was coated on the LCP wafer. Then the maskless lithographic technique was applied for creating the interdigitated microelectrode and eliminates the use of masks (Figure III). Followed by using ferric chloride to perform the chemical etch (Figure IV), the photoresist was removed and the designed electrochemical sensor is fully developed on LCP substrate.

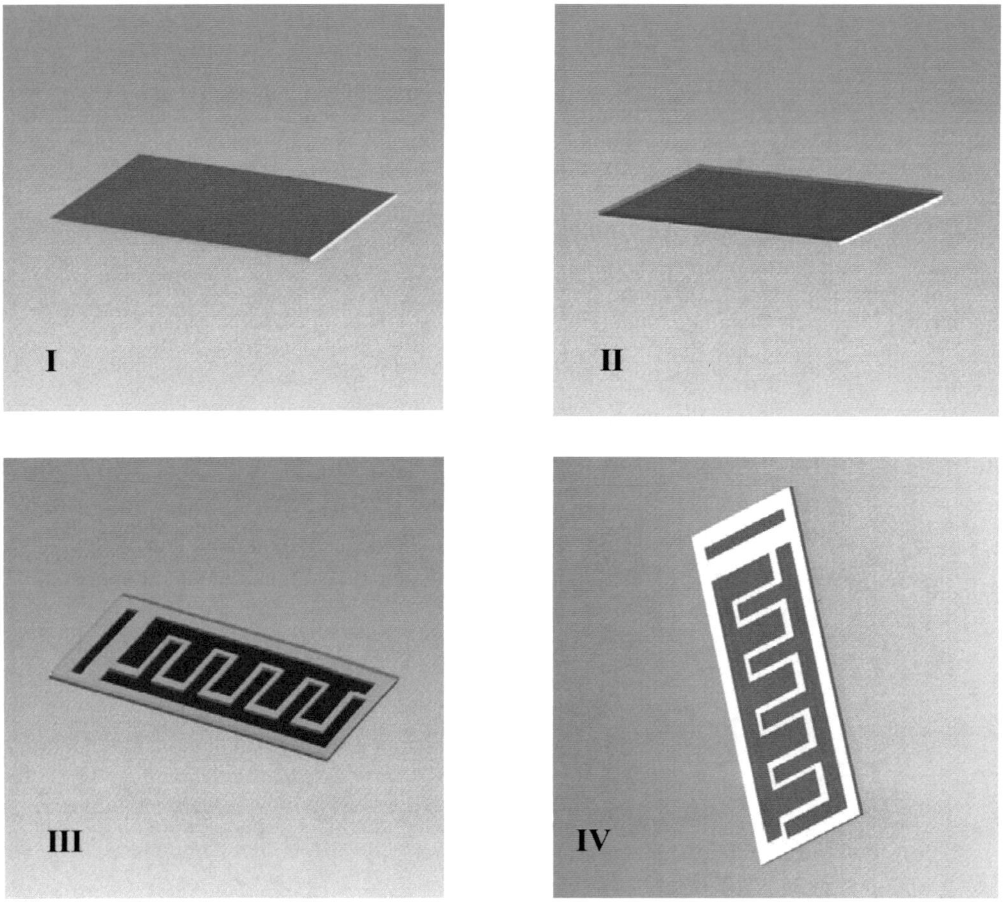

Figure 2. Process Flow for Maskless Lithographic PCBMEMS

2.3Electrochemical measurement

All experiments were performed with a PalmSens (Palm Instruments BV, Electrochemical Sensor Interfaces, Netherlands), which is a hand-held battery powered potentiostat instrument for use with electrochemical sensors or electrochemical cells. The PalmSens is connected to a Pocket PC (Compaq, USA) to specify the parameters of the measurement, to display the measured curves and calculate the results of the measurements. Osteryoung square-wave voltammograms and amperometric method are applied in these measurements. All experiments were performed at room temperature

3. Results and Discussions

The following results/discussion demonstrate the electrochemical detection capability of a three-electrode electrochemical planar sensor, developed on LCP substrate, using the maskless lithographic PCB/Laminate technique.

3.1 Detection of Nitro group

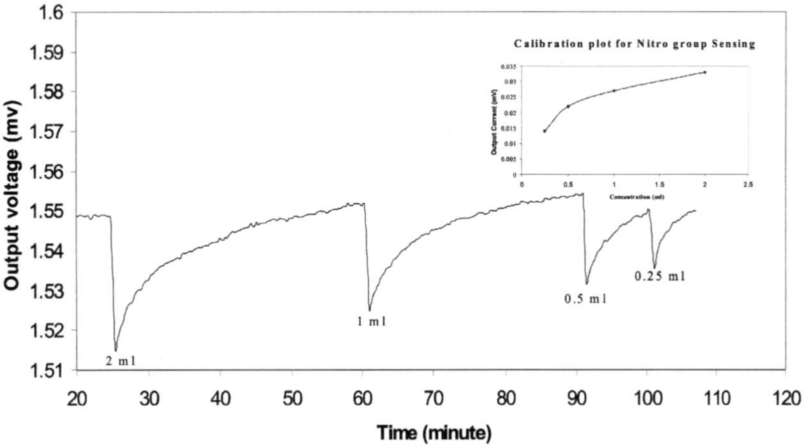

Figure 3. Amperometric detection of NO_2 gas

3.2 Detection of Hydrazine

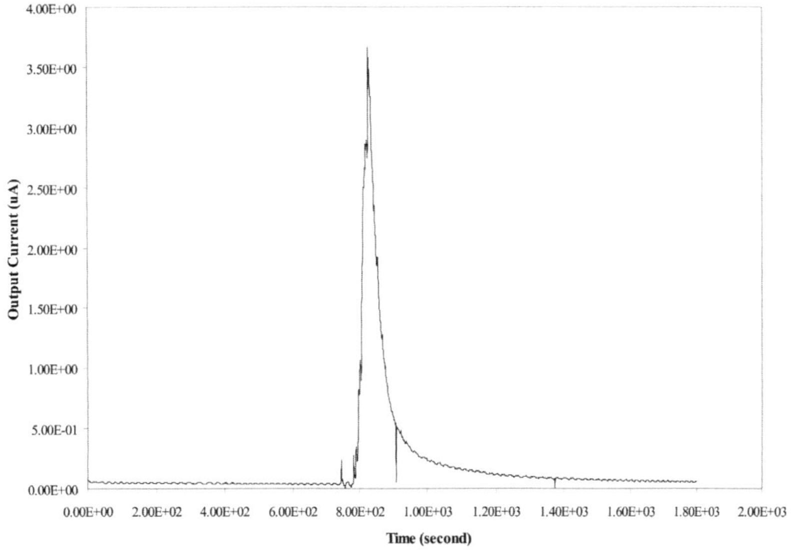

Figure 4. Amperometric detection of hydrazine gas.

3.3 Detection of TNT

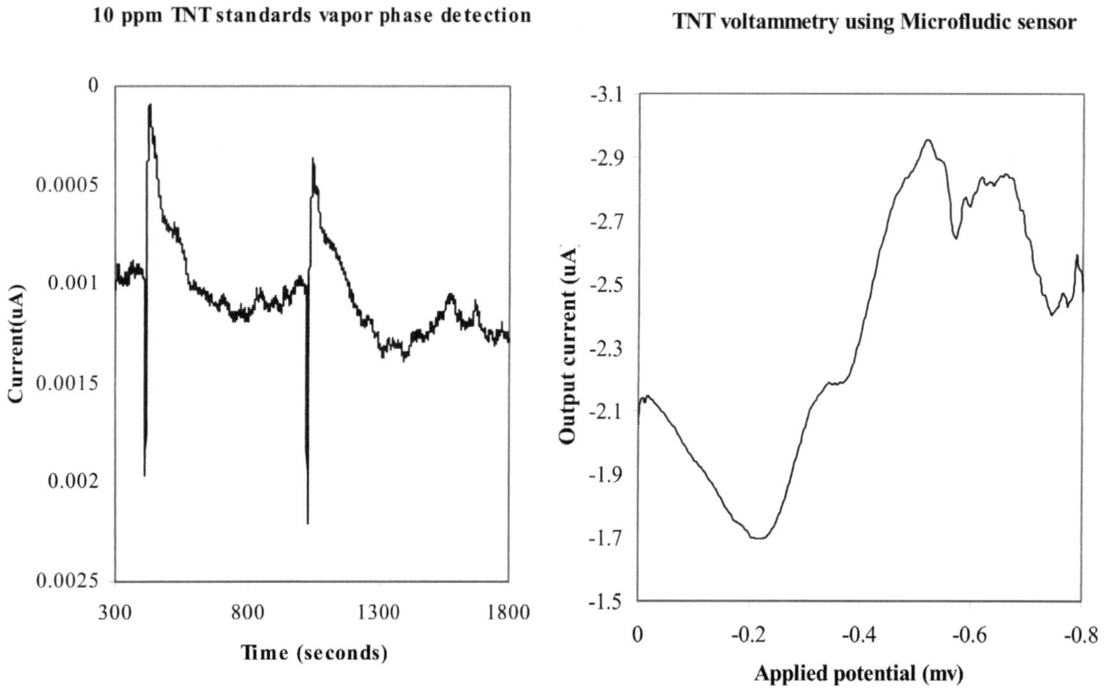

Figure 5. Amperometric and voltammetry measurement for TNT vapor.

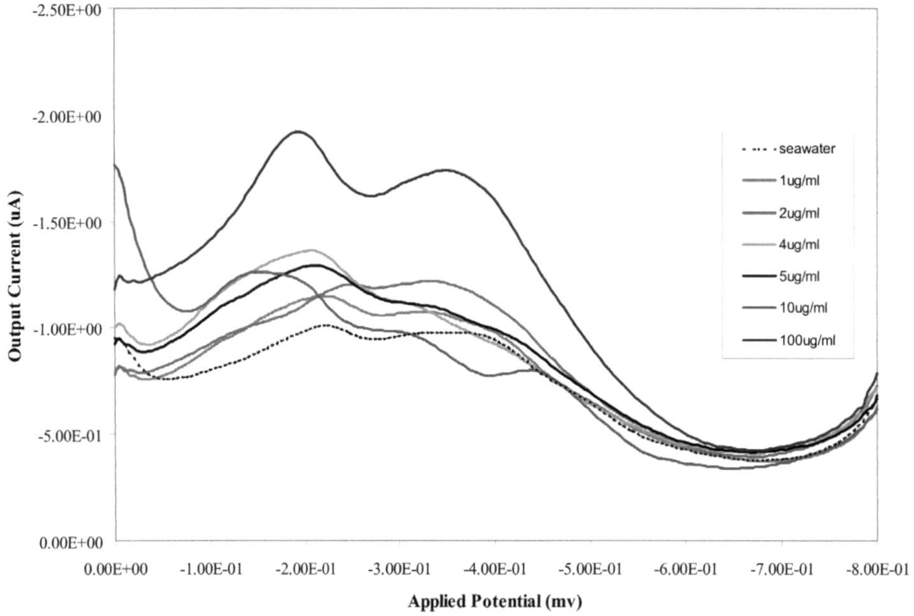

Figure 6. Voltammetry measurement for TNT in sea-water.

Distribution A. Approved for Public Release; distribution unlimited

4. Conclusions

Fabrication of PCBMEMS planar electrochemical sensor on Liquid Crystal Polymer has been developed. The direct write technologies for photoimageable processing of laminate materials allows computer to print capability and permits rapid changes of electrode designs. The fabricated MEMS electrochemical sensor has demonstrated the ability to detect explosive residues in the field. For many systems of use in harsh environments the attributes of the LCP and polyimide are preferred and the PCBMEMS route can provide the most versatile approach to enable micro total analysis systems.

5. Acknowledgements

Instrumentation support from J. Kolesar is greatly acknowledged.

6. References

1. P. Hrncirova, F. Opekar, K. Stulik, An amperometric solid-state NO_2 sensor with a solid polymer electrolyte and a reticulated vitreous carbon indicator electrode, Sens.Acuators,B 69 (2000) 199-204.
2. F. Opekar, K. Stulik, Electrochemical sensors with solid polymer electrolyte, Anal. Chim. Acta 385 (1999) 151-162.
3. J. Langmaier, F Opekar, Z.Samec, Amperometric solid-state NO_2 sensor based on plasticized PVC matrix containing a hydrophobic electrolyte, Sens. Actuators, B 41 (1997) 1-6.
4. J. Wang, Reticulated vitreous carbon – a new versatile electrode material, Electrochim. Acta 26 (1981) 1721-1726.
5. IPI Explosive Chemistry, http://www.islandgroup.com/ExplosiveChemistry.html
6. K.Jayaraj and B.Farrell, Liquid crystal polymers and their role in electronic packaging, Advancing Microelectrron 25 (1998), 15-18.
7. L.Chorosinski, Low cost, light weight, inflatable antenna array development using chip on flexiable membranc\es for space-based radar applications, Northrop Grumman Co.Report, 2000.
8. X.Wang, J.Engel and C.Liu, Liquid crystal polymer (LCP) for MEMS: process an dapplications, J. Micromech. Microeng. 13 (2003), p628-633.
9. C.G.L.Khoo, B.Brox, R.Norrhede and F.H.J.Mauer, Effect of copper lamination on the rheological and copper adhesion properties of a thermotropic liquid crystalline polymer used in PCB applications, IEEE Trans. Compon.Packag.Manuf.Technol. 20 (1997), 219-226.
10. E.C.Culbertson, A new laminate material for high performance PCBs: liquid crystal polymer copper clad films, Proc.45th Electronic Components and Technology Conf. 520-523.

11. P.Hrncirova, F.Opekar and K.Stulik, An amperometric solid-state NO2 sensor with a solid polymer electrolyte and a reticulated vitreous carbon indicator electrode, Sens.Actuators, B 69 (2000) 199-204.
12. J. Wang, Remote electrochemical sensors for monitoring inorganic and organic pollutants, trends, Anal. Chim. Acta. 16 (1997), 84-88.

Highly Selective Methanation in Microreactors

O. Görke, P. Pfeifer, K. Schubert,

Institute for Micro Process Engineering (IMVT), Forschungszentrum Karlsruhe GmbH, Hermann-von-Helmholtz Platz 1, 76344 Eggenstein-Leopoldshafen, Germany, www.fzk.de/imvt-en

Prepared for Presentation at 2005 Spring National Meeting / IMRET 8 / April 10-14 / Poster Session

Unpublished

Introduction

In the past, methanation has received considerable attention for natural gas substitution. The substitution process involves the partial oxidation of coal to CO and subsequent recombination of CO with H_2 to form methane [1]. The methanation reaction is also commonly applicable in ammonia plants to remove carbon oxides from inlet streams which otherwise poison the ammonia synthesis catalyst. In the present, selective methanation became also attractive in order to reduce the CO content to less than 100 ppm in hydrogen-rich reforming gases for fuel cell applications [2,3]. In the present study the results of the selective methanation will be presented for two catalysts, namely Ru/Al_2O_3 and Ru/SiO_2. These catalysts have been successfully applied to microstructured metal foils. The coated foils were stacked in a clamping device, the microchannel reactor, and tested for selectivity and conversion.

Experimental

The microchannel reactor consists of 27 stainless steel foils with etched microchannels and a clamping device. The foils were coated with the catalysts on the microstructured side and were subsequently stacked in the clamping device. The etched foils and the clamping device are forming the microchannel reactor (Fig. 1).

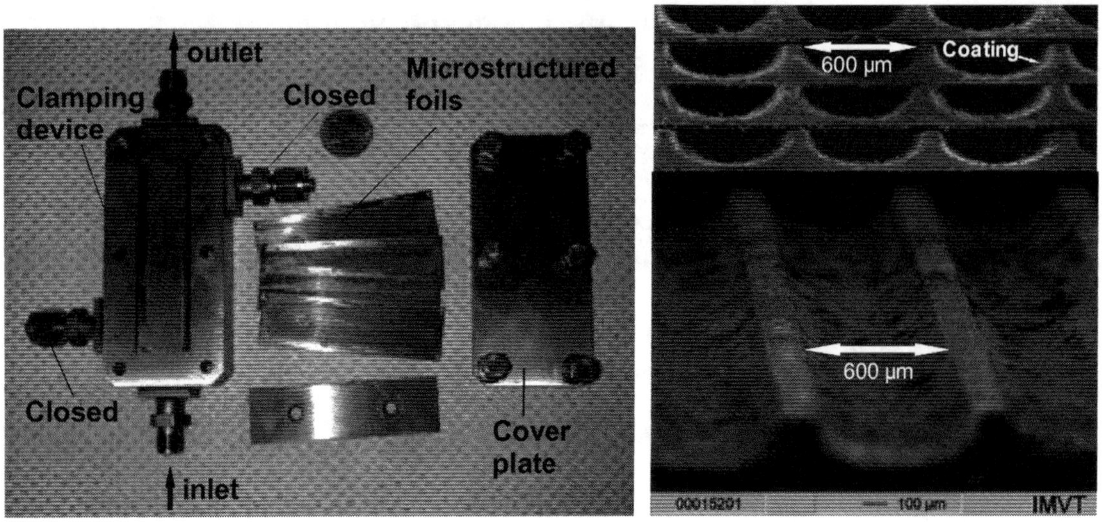

Fig.1: Microchannel reactor for methanation experiments with clamping device and coated foils (left); SEM of a microstructured foil coated with Ru/SiO_2 and stacked foils (right).

Because other methods to prepare the desired catalyst are not applicable to completely manufactured microchannel reactors, it was decided to use a two step catalyst preparation method: Sol-gel followed by wet impregnation (see also [4]). The main results of the catalyst characterization are summarized in table 1:

Table 1: substrate, specific mass, composition, surface enlargement and mean pore diameter of the catalysts applied to the microstructured foils

Catalyst	substrate	Catalyst mass/ Coated surface of stack (mg/cm^2)	Composition	Surface enlargement ratio (m^2m^{-2})	Mean pore diameter (nm)	Layer thickness (µm)
Ru/Al_2O_3	Stainless steel	0,54	55,7 % Al_2O_3	526	7	0,5 – 2
Ru/SiO_2	Stainless steel	0,60	19,8 % SiO_2	470	7	0,5 - 2

The total inlet gas flow $\dot{V}_{in}$ measured by the mass flow controllers was varied between 0.22 and 0.66 l/min (STP). The temperature range was from 100°C to 380°C. A mixture of 25 vol.% H_2, 1 vol.% CO, and 1 vol.% O_2 and 73 vol.% N_2 was applied to the catalyst at three standardized residence times τ (177 ms, 251 ms and 474 ms for at 250°C) of the gas mixture in the microchannel volume (V_R = 3222 mm^3), keeping into account that the washcoat layer has an average thickness of 1 µm.

Results and discussion

The CO-conversions plotted as function of reaction temperature for the two catalysts tested reveal that the Ru/SiO_2-catalyst has the higher activity (Fig. 3). At 285 °C and 177 ms standard residence time, the CO-conversion reaches a maximum value of 95 %. The

Ru/Al_2O_3-catalyst reaches that value of CO-conversion only at 360 °C. The difference between the CH_4-selectivities is considerable. The Ru/SiO_2-catalyst exhibits its highest CH_4-selectivity of only 82 % at a temperature of 305 °C, whereas a selectivity of 99 % is reached by the Ru/Al_2O_3-catalyst at 340 °C. Analysis of the product gas revealed that CO, not reacting to CH_4 was converted into CO_2 by total oxidation. Up to a temperature of 140 °C O_2 is completely converted to CO_2 and H_2O.

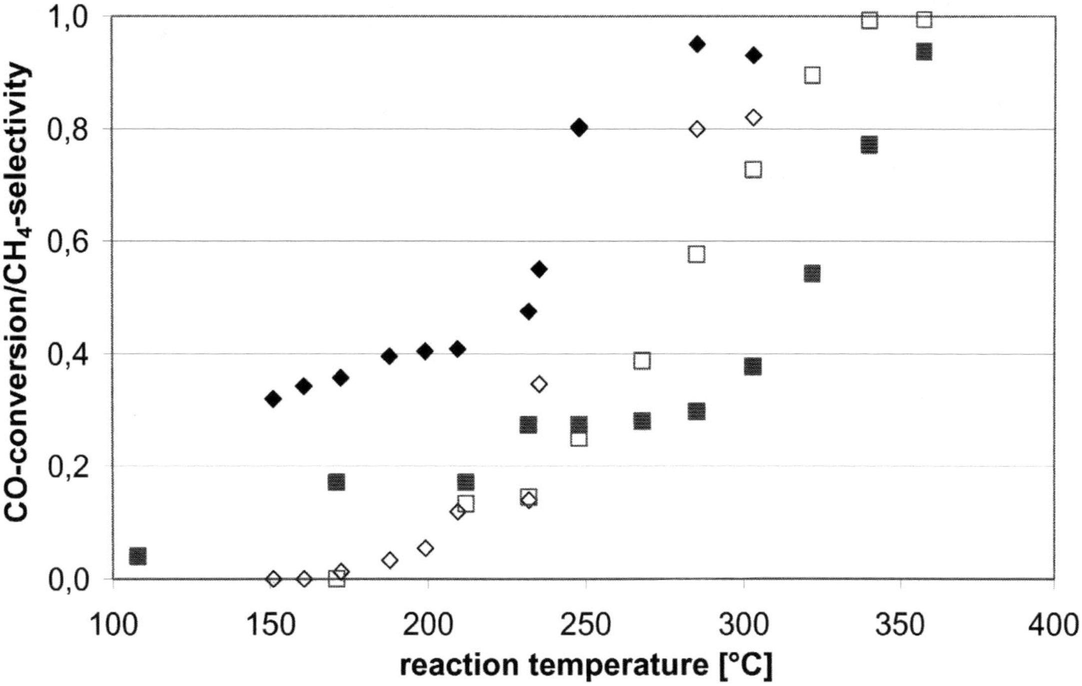

Fig 2: CO conversion X and CH_4-selectivities S of the methanation as a function of temperature for the two catalysts ((◆) X(CO) for Ru/SiO_2; (■) X(CO) for Ru/Al_2O_3 I; (◇) S(CH_4) for Ru/SiO_2 (□) S(CH_4) for Ru/Al_2O_3). Standard residence time of 177 ms. Inlet composition of model gas: $[CO]_{in} = [O_2]_{in}$ = 1 vol.%; $[H_2]_{in}$ = 25 vol.%; $[N_2]$ = 73 vol.%)

It can be demonstrated that the CH_4 space time yield of a gas mixture without H_2O is much higher compared to the CH_4 space time yield for a gas mixture with 20 vol.% H_2O. The CH_4 space time yield of the water free methanation shows an S-shaped curve, the one with H_2O exhibits a constantly rising curve.

To test the competitive methanation of CO and CO_2, experiments were performed with Ru/SiO_2 and different CO_2 contents. The CH_4 space time yields increase with rising temperature for all gas mixtures. The gas mixture containing no CO_2 and only 1 vol.% CO exhibits the highest CH_4 space time yield for temperatures below 240°C. The gas mixture that contains 4.5 vol.% CO_2 but no CO reveals the highest CH_4 space time yield at temperatures from 250 °C to 300 °C. In contrast to both measurements, the gas mixture with 4.5 vol.% CO_2 and 1 vol.% CO exhibits the lowest CH_4 space time yield in the entire temperature range.

To compare the CO-oxidation and CH_4 space time yield by CO_2 methanation with the CH_4 space time yield by CO-methanation, each of these reactions were conducted separately.

By comparing the CH_4 space time yield from CO with the amount of converted CO per volume, some interesting relations can be observed. The CH_4 space time yields of the CO_2-methanation (fig.3: ●,■) rise strongly with increasing temperatures. For temperatures below 260 °C the CO_2-methanation rate of the gas mixture containing 1 vol.% O_2 (fig.3: ●) is obviously lower than without O_2 (fig.3: ■). The converted CO per volume (fig.4: Δ) rises with increasing temperatures until complete conversion at 200 °C is reached. The CH_4 space time yield by CO methanation with addition of 1 vol.% O_2 to the gas mixture (fig. 3:◆) exhibits the fastest rising below 200 °C, but above that temperature the value remains constant for a CO-conversion of more than 99 % and a CH_4-selectivity > 98 %.

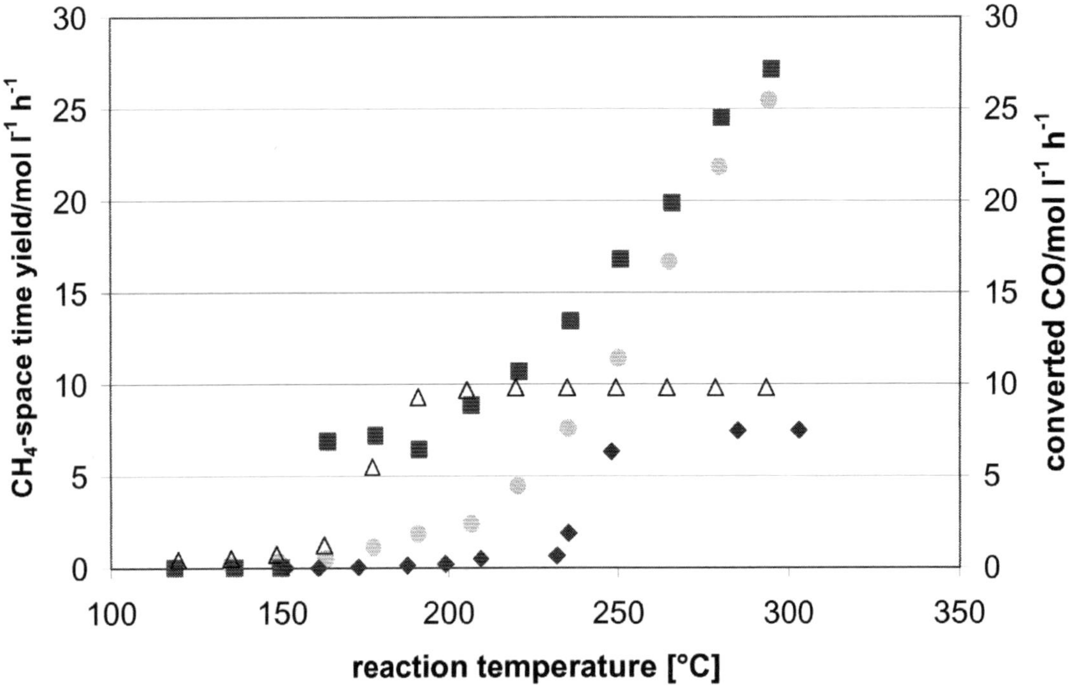

Fig. 4: CH_4 space time yield by CO_2 methanation; inlet gas composition:
(●): $[CO_2]_{in}$= 4.5 vol.%; $[O_2]_{in}$ = 1 vol.%; $[H_2]_{in}$ = 25 vol.%; $[N_2]$: balance
(■): $[CO_2]_{in}$= 4.5 vol.%; $[O_2]_{in}$ = 0 vol.%; $[H_2]_{in}$ = 25 vol.%; $[N_2]$: balance
CH_4 space time yield by CO methanation; inlet gas composition:
(◆) $[CO]_{in}$= 1 vol.%; $[O_2]_{in}$ = 1 vol.%; $[H_2]_{in}$ = 25 vol.%; $[N_2]$: balance
Converted CO per volume and time; inlet gas composition:
(Δ)$[CO]_{in}$ = 1 vol.-%; $[O_2]_{in}$ = 1 vol.%; $[H_2]_{in}$ = 0 vol.%); $[N_2]$: balance
all experiments performed at a standard residence time of 177 ms

Conclusions

A Ru/Al_2O_3 and Ru/SiO_2 catalyst were successfully applied to microreactors to reduce the CO content by methanation in a model gas mixture that contains the major product components of a natural gas reformer or of partial oxidizer used for hydrogen production. It could be shown, that the CO methanation rate with oxygen is higher than without oxygen addition at temperatures between 140 °C and 200 °C. For temperatures from 200°C to 320 °C the H_2 oxidation rate increases more than the CO methanation rate and consumes most of the O_2. At temperatures up to 250 °C obviously the CO oxidation rate rises and so the formed CO_2 can react "directly" to CH_4. This increases the entire CO-methanation rate again. If CO (in a mixture of CO and CO_2) has to be converted over a Ru/SiO_2-catalyst by methanation a sufficient amount of O_2 has to be added and temperature has to be controlled precisely. Because of the dimension and the heat transfer coefficients of the microchannel reactor, the latter demand can be met very easily. Temperature ranges can be controlled exactly, which is important to maximize the ratio of CO- to CO_2 methanation rate. The CO methanation space time yield clearly decreases with H_2O in the gas flow compared to the space time yield for same conditions but without water. The experiments demonstrate that the microchannel reactor is an excellent tool for studying the reaction network of methanation of CO in presence of oxygen, CO_2 and hydrogen without heat transfer limitation.

References

[1] K.O. Xavier; R. Sreekala; K.K.A. Rashid; K.K.M. Yusuff; B. Sen; Catal. today 49 (1999) 17-21.

[2] H. Igarashi; T. Fujino; M. Watanabe; J. Electroanal. Chem. 391 (1995) 119.

[3] H.F. Oetjen; V.M. Schmidt; U. Stimming; F. Trila; J. Electrochem. Soc. 143 (1996) 3838.

[4] K. Haas-Santo, O. Görke, P. Pfeifer, K. Schubert; Chimia - Chimia Report, 56 (2002), 605-610.

Fast Temperature Cycling of Catalytic Reactions using Microreactors

M. Luther[1,2,*], J. J. Brandner[1], K. Schubert[1], L. Kiwi-Minsker[2], A. Renken[2],

[1)] Institute for Micro Process Engineering, Karlsruhe Research Center, D-76344 Eggenstein-Leopoldshafen, Germany

[2)] Institute of Chemical Sciences and Engineering, Laboratory of Chemical Reaction Engineering, Swiss Federal Institute of Technology, CH-1015 Lausanne, Switzerland

Prepared for presentation at the 2005 AIChE Spring National Meeting (IMRET 8 – 8th International Conference on Microreaction Technology), April 10-14, Atlanta, GA, USA
Unpublished

First theoretical studies discussing forced temperature variations to increase the performance of heterogeneous catalytic reactors appeared in the late 1960th. Experimental studies realized in the late 1970th failed to confirm the predicted effects[1]. This can be explained by the high thermal inertia of the experimental devices used, allowing only slow temperature changes and therefore leading to cycling periods in the order of an hour. The appearance of microreactors (an example is illustrated in Fig. 1) in the last decade providing a lower thermal inertia, thus allowing temperature changes of several tens of degrees within seconds [2], has woke up the interest toward this field of research.

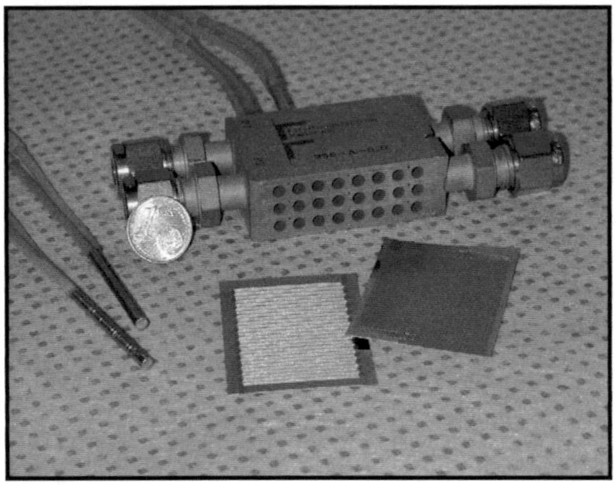

Fig. 1 Microreactor used for Fast Temperature Cycling experiments

The objective of forced temperature oscillations is to generate higher yield by increasing the mean reaction rate to a value which is not reachable under steady state conditions. The points in Figure 2 are representing three different "mean reaction rates", depending on the way to calculate it. One observes that the mean reaction rate calculated by taking the value at low and high temperature is higher than the reaction rate at the mean temperature. This is trivial and comes directly from the exponential dependence of the reaction rate with respect to the temperature. One can only conclude that the Fast Temperature Cycling of a reaction produce a non-trivial effect on the reaction rate if the

* to whom correspondence should be addressed. e-mail: martin.luther@imvt.fzk.de

mean reaction rate obtained under the cycling conditions is higher than the mean reaction rate calculated between the rate at low and high temperature.

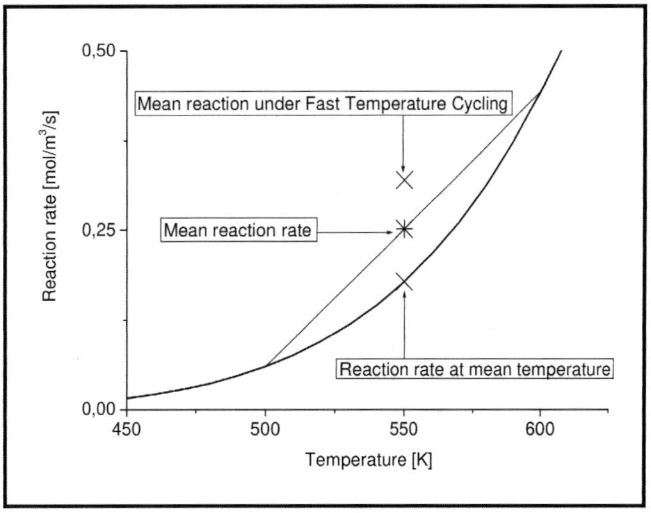

Fig. 2 Three different interpretation of the mean reaction rate

The simplest way to show the non-trivial effect of fast temperature variations on a heterogeneously catalyzed reaction is to consider a monomolecular reaction as an example and to simulate the transient behavior under unsteady state thermal conditions. For this simple reaction simulation, three different steps are taken into account: adsorption of the reactant to the catalyst surface, desorption of the reactant from the catalyst surface and reaction on the surface to form the product, which desorbs instantaneously. If the rate of desorption of the reactant from the surface is low compared to the time constant of the temperature changes, and the difference of surface coverage at high and low temperature is high enough to be of importance, a non-trivial increase of the instantaneous surface reaction rate can be expected.

In Figure 3, a simulation of the instantaneous reaction rate under periodic temperature variation is shown for a general case in comparison to the maximum rate under steady state conditions

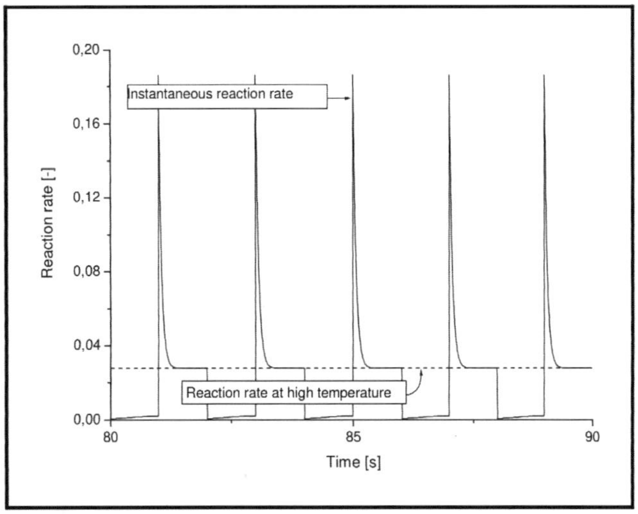

Fig. 3 Predicted instantaneous reaction rate compared to the maximal rate at high temperature

The non-trivial increase in the mean reaction rate is due to the peak obtained for the instantaneous reaction rate at the beginning of the high temperature half-period. During a short time, both the reaction rate constant and the reactant surface coverage are high (what never happens during the steady state) allowing a transitional higher instantaneous reaction rate.

For a more complicated reaction like a bimolecular reaction, the principles described above remain the same. It is the coupling of the surface reaction rate and the rate of the desorption/adsorption processes which can lead to a non-trivial increase in the reaction rate compared to the thermal steady state.

Due to the lack of theoretical and empirical background, it is not easy to select reactions which may be influenced by the applications of fast temperature changes. Some general schemes are listed by Neer[3] and Rouge[4]. The work realized by Brandner[2] shows in a qualitative way that for the reaction of CO with O_2 on a Pt/Al_2O_3 catalyst the application of Fast Temperature Cycling improves the conversion of CO.

Whereas the reaction mechanism of the CO oxidation is relatively well known, reliable kinetic parameters for the individual steps are missing. In general, kinetic constants determined under thermal steady state conditions can not be applied under thermal unsteady state conditions without any restrictions [5].

In the poster, the effects of Fast Temperature Cycling on the CO oxidation reaction are presented using simulations and experimental measurements. The simulations realized with realistic kinetic parameters show that the gain in performance can reach several tens of percent but strongly depends on the choice of the kinetic parameters, the frequency of the variation and the heating and cooling rate.

By using a fast online analysis method, it is possible to follow the variation of the concentration of the reaction product(s) at the reactor outlet when the reaction temperature is rapidly varied. In Figure 4 the temperature of the reactor and the concentration of CO_2 measured by FTIR spectroscopy are shown:

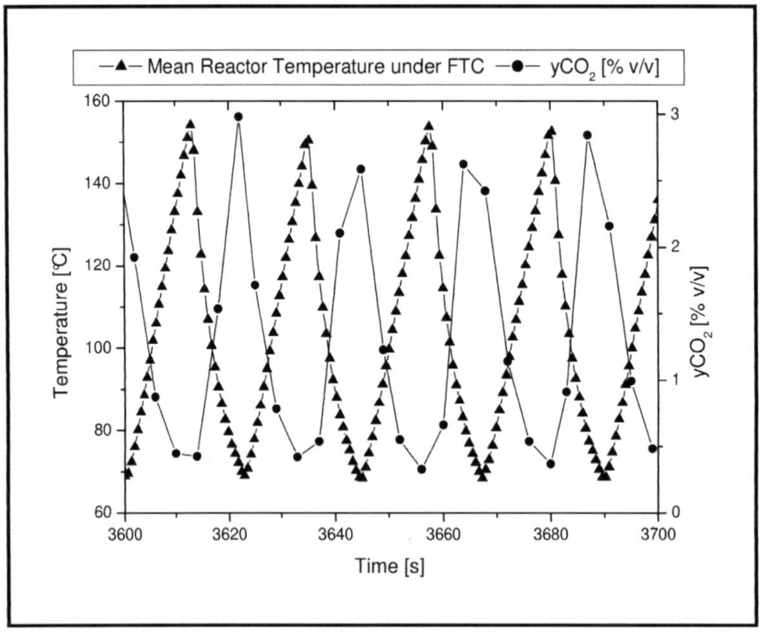

Fig. 4 Plot of the temperature of the reactor and the concentration of CO_2 versus relative time.

It can be seen that the sample rate for the CO_2 concentration measurement has to be increased in order to get a more regular and reproducible curve.

Since both the reactor temperature and the reaction product(s) concentration can be measured online, they can be correlated together and related to the time. Realizing this correlation is not trivial. Often a phase difference between these two measurements appears, as illustrated in the Figure 4. This phase difference occurs on one hand from the time that the gas needs to reach the detector, which is not situated directly at the reactor outlet, and on the other hand from the fact that the measurement of the concentration is not instantaneous but with a time delay of approximately 0.5 seconds.

The curves obtained experimentally are compared with the simulated curves. This comparison gives an indication about the validity of the theoretical model and allows to propose a qualitative explanation for the effects of the Fast Temperature Cycling.

---

[1] P.L. Silveston, R. R. Hudgins; Periodic temperature forcing of catalytic reactions; Chemical Engineering Science, 59 (2004), pp. 4043-4063

[2] Jürgen J. Brandner; Entwicklung von Mikrostrukturreaktoren zum thermisch instationären Betrieb chemischer Reaktionen; Thesis work, (2003); IMVT, Forschungszentrum Karlsruhe

[3] F. van Neer; Forced Oscillations in Heterogeneous Catalysis; Thesis Work (1999); Department of Chemical Engineering, Faculty of Chemistry, University of Amsterdam

[4] André Rouge; Periodic Operation of a Microreactor for Heterogeneously catalyzed Reactions: the Dehydration of Isopropanol; Thesis work (2001); Chemistry Department, EPF Lausanne

[5] A. Renken, R. Doepper, M. Schmid; Application of non-stationary and Periodic Processes in the Study of Heterogeneous Catalytic Kinetics; Dechema-Monographs, vol. 120, VCH Verlagsgesellschaft (1989), pp. 273-290.

A Fundamental Three-Phase Reactor Characterization Study in a Thin-Film Catalyzed Multifunctional Structure

X. Ouyang, D. Schweich, V. Meille, I. Pitault, C. de Bellefon

Three-phase industrial dearomatization processes are conventionally operated in trickle-bed or slurry reactors under high pressure. Recent research efforts to examine alternative reactor technologies for better performance of these processes have been extensive resulted from the tightening legislations on aromatic by-products in the refining industry [1]. Reactors with structured catalyst are attractive for these processes mainly due to the efficient gas-liquid-solid contact, leading to enhanced mass transfer and better catalyst utilization [1]. However, structured catalyst in monolith mode has significant drawback for exothermic reactions due to its poor heat dissipation. Dramatic improvement in gas-liquid interfacial area and heat transfer rate can be achieved with microreaction technology [2]. Demonstration of gas-liquid-solid catalysis in a micro-structured component up to 45 bar and with long residence times has also been published [3]. But microchannel applications in industrial scale-up are still hindered by the plugging and fouling of microchannels, cost and availability of microreactor modules for large-scale commercial reactions, etc. [4]

In this study, we adopt a multifunctional meso-structure [5], which is originally commercialized for heat exchange applications, to study three-phase toluene hydrogenation as a model reaction. A thin-film Pt/Al_2O_3 wall catalyst is washcoated as the solid catalyst phase. The well-proven heat transfer efficiency of this structure [5], coupled with the advantages of thin-film structured catalyst, can lead to great productivity improvement compared to conventional three-phase reactors. This paper focuses on examining this meso-structure with classical three-phase reactor characterization methods (*e.g.*, flow distribution, flow regimes, gas-liquid holdup, residence time distribution, mass/heat transfer issues), followed by discussion of advantages and limitations of the multifunctional structure.

The Compabloc® plate (Figure 1) has channels with ca. 1-cm feature size and a typical inclination angle of 45°. The Compabloc® has wide industrial applications for condensation and re-boiling and is served as a prototype reactor in this work. The gas and liquid components flow either in co-current or counter-current fashion, similar to a trickle-bed reactor operation. With each two plates assembled interlaced, the corrugation features are divided into small crosslinked regions promoting fluid exchange. Compared to trickle-bed reactors, this structure thus has large radial dispersion coefficient, leading to more homogeneous reactant distribution. In addition, these crosslinked regions also enhance fluid turbulence, resulting in a large effective gas/liquid contact area and consequently the volumetric mass transfer coefficient.

Hydrodynamic aspects of the multiphase flow in the structure, including flow distribution, flow regimes, liquid holdup, residence time distribution, gas/liquid contact area, and gas-side and liquid-side mass transfer coefficients, are under investigation. A mechanistic model needs to be constructed in order to correlate these hydrodynamic parameters to structure geometric features and flow characterization. Ultimately, this goal of this model is to adapt transport and flow behavior to different reaction kinetics, and thus significantly improve reaction productivity, selectivity and catalyst utilization.

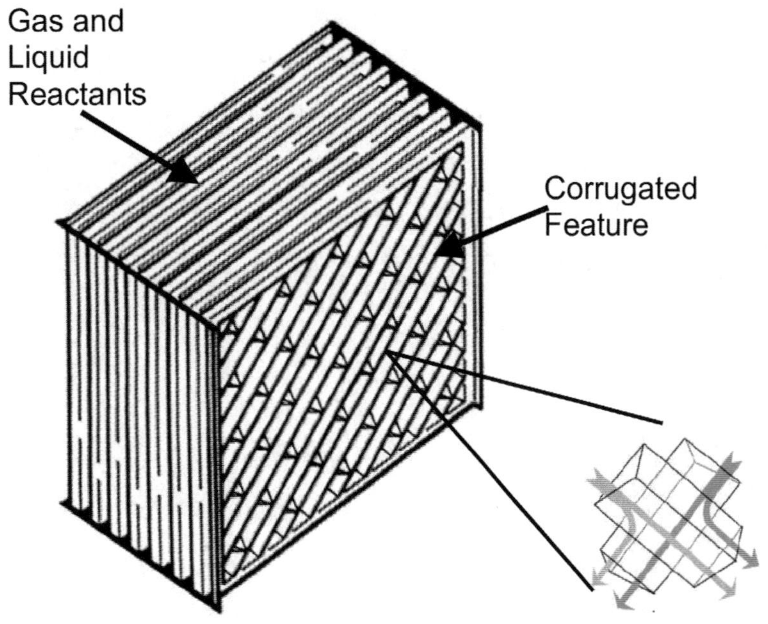

Figure 1. Compabloc® multifunctional structure from ALFA LAVAL™.

Mechanistic models exist for similar structures (e.g., KATAPAK®, MELLAPAK®) [6,7,8,9,10]. In principle, these models were constructed based on the interaction of falling liquid films with the gas phase by taking account of the flow and physical properties and geometric features. In addition, a general approach approximating the corrugated structure as a series of isolated inclined channels was used [11]. The feasibilities of these models on the Compabloc® structure over a wide range of liquid flow are examined with both theoretical and experimental analysis.

Finally, we will discuss a design methodology that incorporates fluid dynamics, transport phenomena, and kinetic-transport coupling with the combination of mechanistic modeling, CFD modeling and experimental methods.

Reference:

1. W. Liu, *AIChE J.* 48 (2002) 1519-1532.
2. M.W. Losey, M.A. Schmidt, K.F. Jensen, *Ind. Eng. Chem. Res.* 40 (2001) 2555-2562.
3. R. Abdallah, V. Meille, J. Shaw, D. Wenn, C. de Bellefon, *Chem. Commun.* 372 (2004) 372-373.
4. H. Löwe, V. Hessel, *Proceedings of the 6th International Conference on Microreaction Technology* (2002) 3-4.
5. P. Arvidsson, *Hydrocarb. Eng.* 8 (2003) 41-43.
6. Brunazzi et al., *Chem Eng. Technol.* 18 (1995) 248-255.
7. E. Brunazzi, A. Paglianti, *AIChE J.* 43 (1997) 317-327.
8. E. Brunazzi, A. Paglianti, *Ind. Eng. Chem. Res.* 36 (1997) 3792-3799.
9. J.A. Rocha, J.L. Bravo, J.R. Fair, *Ind. Eng. Chem. Res.* 32 (1993) 641-651.
10. J.A. Rocha, J.L. Bravo, J.R. Fair, *Ind. Eng. Chem. Res.* 35 (1996) 1660-1667.
11. M.G. Shi, A. Mersmann, *Ger. Chem. Eng.* 8 (1985) 87-96.

Preparing Mono-dispersed Mixed Metal Oxide in a Micro Fluidic Reactor

Jae Han Kim[b], Won Choon Choi[a], Hee Young Kim[a], Yong Kang[b] and Yong-Ki Park[a*]

[a] *Advanced Chemical Technology Division, Korea Research Institute of Chemical Technology, Yusung P.O Box 107, Daejon 305-600, Korea (Fax +82-42-860-7590, ykpark@krict.re.krT)*

[b] *Department of Chemical Engineering, Chungnam National University,Yusung-gu, Goong-dong 220 Daejon 305-764, Korea*

Micro technology is currently being considered as an important area in chemical process, because the micro reactor can be advantageously used, compared to the conventional chemical reactor. And if a large number of micro reactors are arrayed together, the required amount of products can be obtained [1]. In this study, we tried to control shape, dimension and composition of $Ni_xCo_yMn_zO_4$ mixed oxide used as an active cathode material of Li secondary battery by precipitating these metal precursors in a micro fluidic reactor.

The micro fluidic reactor was fabricated by inserting a micro capillary tube in a mm scale glass tube as shown in Fig.1. Metal precursor solution containing 2M of Ni, Co and Mn sulfates was introduced to the inner micro capillary tube while flowing 2-Propanol/Butanol mixture solvent which is insoluble with metal precursor solution through the outer glass tube.

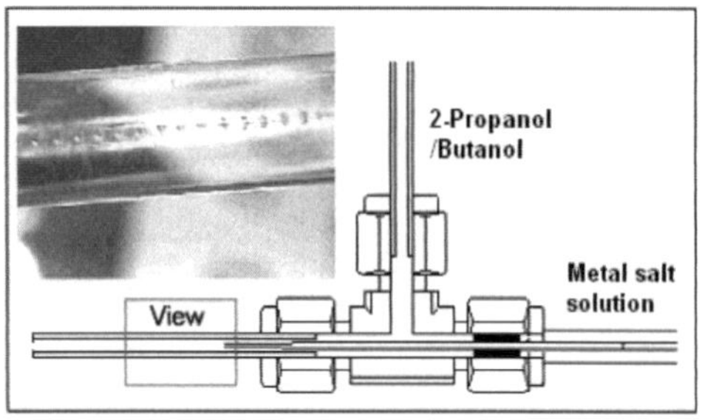

Fig. 1. Schematic drawing of micro fluidic reactor and photo-image of produced spherical micro droplets.

At the exit of the inner capillary tube, a different phase of Ni/Co/Mn precursor solution in micro scale was formed and then uniform and micro spherical droplets were formed spontaneously after a certain period of time. However, as shown in Fig. 2, the size and shape of micro droplets were sensitively influenced by the ratio of flow rate and viscosity of two solutions as well as the viscosity and flow rate themselves. That is there exist an optimum flow condition exist for a given geometry of micro fluidic reactor and property of solutions. As the flow rate increased and diameter of inner capillary tube decreased, the smaller size of sphere was obtained. The chemical property of solvent also played an important role for the formation of stable micro sphere particle. The solvent should have enough solubility to leach out water from the aqueous phase of micro droplet while maintaining two separate phases. By adjusting the ratio of Butanol/2-Propanol, the solubility of solvent was controlled.

The chemical composition of the produced micro spherical particles was analyzed by EDS because the ratio and uniformity of each metal component is very important to be applied as a cathode material of Li battery. The final micro crystallite has the composition of $Ni_{0.38}Co_{0.31}Mn_{0.31}$ which is quite close to the precursor composition of 0.4 Ni: 0.3 Co: 0.3 Mn. These results imply that the micro fluidic reactor in this study can be applicable successfully to the synthesis of micro spherical particles where the shape, size and uniformity are important like a cathode material of Li secondary battery [2].

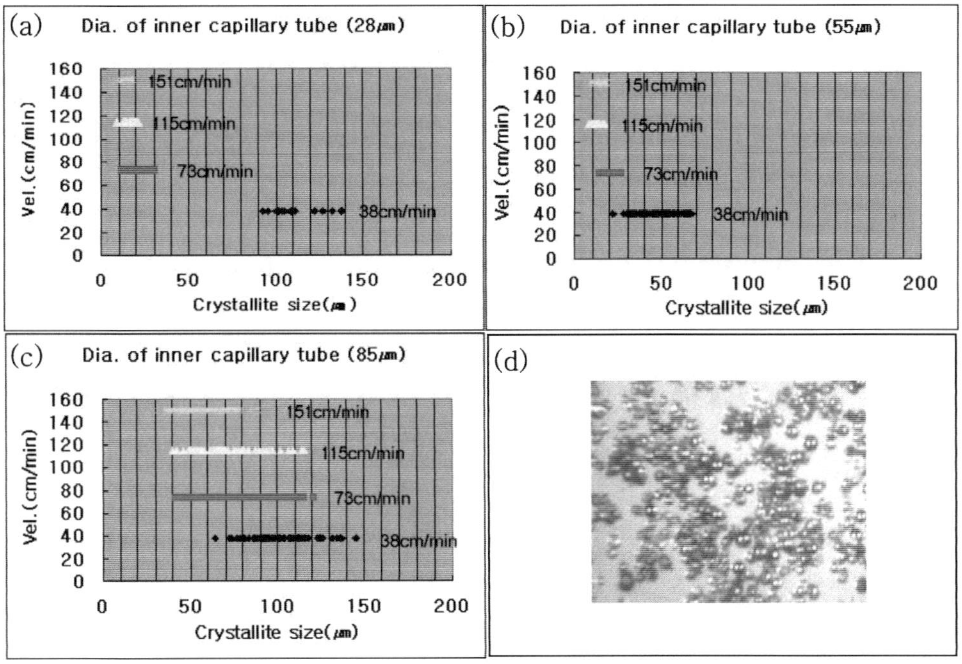

Fig. 2. Size distribution of micro sphere (a~c) and photo-image (d) of $Ni_{0.38}Co_{0.31}Mn_{0.31}$ crystallites depending on the size of inner micro capillary.

Reference

[1] K. Jähnisch, V. Hessel, H. Löwe and M. Baerns, Angew. Chemi. Int. Ed. 43 (2004) 406.
[2] C.-H. Lu and S. K. Saha, Materials Science and Engineering B 79 (2001) 247.

Measurement of Three-Dimensional Distributions Inside Microchannels

T. Hinouchi, M. Kawano, M. Satou, H. Koyama and K. Isozaki

Advanced Technology Research Center, Corporate Research and Development Headquarters, Yokogawa Electric Corporation, 2-9-32 Nakacho, Musashino-shi, Tokyo, 180-8750, Japan

1. Introduction

Microreactors are getting more and more attention for their unique properties. For example, large surface-to-volume ratio promotes heat exchange and small characteristic length gives laminar flow profile and quick diffusion mixing. To make use of these features for reactor, the size and shape of microreactors are often optimized by means of computer simulation.

Developing microreactors, however, faces an experimental problem: it is difficult to measure physical properties such as temperature and flow rate in microchannel, especially in a three dimensional manner.

In case of measurement of flow late, Santiago et al. first succeed to apply Particle Image Velocimetry (PIV) technique to micro fluidic device [1]. Their method, named to Micro PIV, uses hundreds nm diameter flow-tracer, epifluorescent microscope and intensified CCD camera to record high-resolution particle-image fields. According to their method, particles outside of watching plane have lower image intensities because of defocusing and can be filtered out by thresholding the image intensity. This worked well and they achieved 1.5um in-depth resolution. However, their technique is difficult to apply for other measurements, because it doesn't make use of another information; fluorescent intensity.

Fluorescent intensity is a useful measuring probe because each fluorescent dyes have their unique susceptibility to circumstances such as temperature, pH etc. If measurement is carried out in multi-color with multi fluorescent dyes, multiple properties can be measured at the same time.

In order to use this measurement principle in microchannel with high in-depth resolution, we needed to cut off fluorescence from unfocused area. In this paper, we employed confocal optical system for above-mentioned purpose. With this system, we measured local temperature rise at liquid-liquid interface when exothermic reaction occurred at interface.

Our method is powerful not only to develop microreactors but also to research reaction kinetics.

2. Experimental

2.1. Equipment setup

Experimental setup is shown in Fig. 1 schematically. Our system utilize fluorescent microscope (BX51, Olympus Co.), 640 x 480 x 8bits CCD camera (C7190, Hamamatsu Photonics Co.) for recording fluorescent image, Confocal Scanning Unit (CSU22, Yokogawa Electric Co.) for cutting off fluorescence from unfocused area, and piezo-electric actuator and its control system (RT-3D, Yokogawa Electric Co.) for moving objective lens to change focusing depth.

As a microreactor, we used a glass-made microchip (ICC-SY05, Institute of Microchemical Technology). Channel width and depth is about 100um and 45um, respectively and Channel cross section is almost semi-circle whose width and depth are about 100um and 45um, respectively. Reagent ware pushed into microchip by syringe pumps (model 210, KD Scientific Inc.).

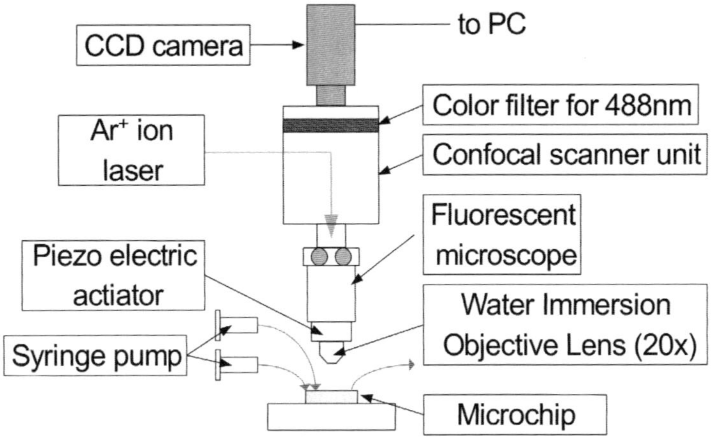

Fig. 1. Schematic diagram of the experimental setups

2.2. Reaction and Reagent preparation

We use a dilution of conc. $CaCl_2$ aqueous solution with water as a model reaction. This is highly exothermic reaction and occurs at liquid-liquid interface between water and $CaCl_2$ solution.

Aqueous solution of $CaCl_2$ was prepared by dissolving CaCl2•2H2O (purchased from Kanto Kagaku Co.) into de-ionized water to the concentration of 5.4 mol/dm^3. Fluorescein (purchased from Kanto Kagaku Co.) was added to both $CaCl_2$ solution and water at the same concentration of 20 mg/dm^3 for measurement probe.

3. Results and discussion

3.1. Temperature dependence of fluorescent intensity

First, we measured temperature dependence of fluorescent intensity of fluorescein for both $CaCl_2$ solution and water. The results are shown in figure 2. Fluorescein in $CaCl_2$ solution and fluorescein in water display different tendency: fluorescent intensity of fluorescein in $CaCl_2$ solution is enhanced by temperature rise, on the other hand, that of in water is weaken. Therefore, we used different calibration curve for respective reagent.

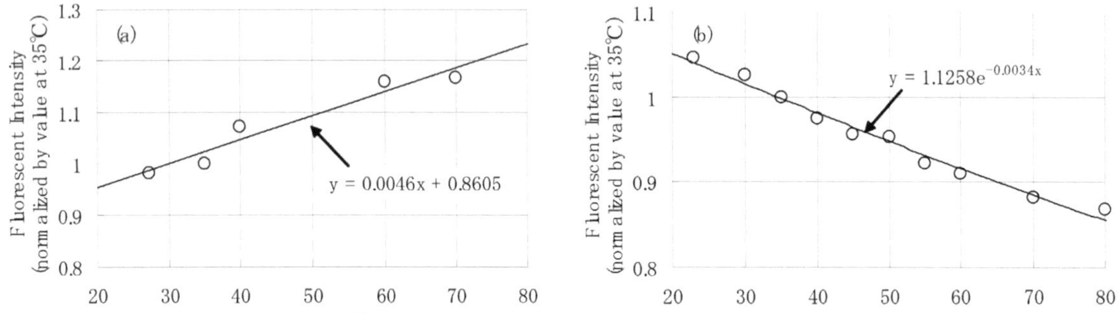

Fig. 2. Temperature dependence of fluorescent intensity

(a) Fluorescein (20 mg/dm^3) in conc. CaCl2 solution (5 mol/dm^3)

(b) Fluorescein (20 mg/dm^3) in water

3.2. Space resolution of measurement system

In-plane space resolution was 0.7 um, which was calculated by pixel size of

CCD camera (14 um x 14 um) and magnification of objective lens (x20).

In-depth space resolution was determined in following way. First, microchannel was filled with suspension of 1um diameter tracer beads. Then, shifting focus position from bottom to top, images of beads were obtained. Since each bead was observed in depth range about 5um, in-depth resolution was concluded to be about 5um.

3.3. Measurement of temperature

One of the obtained fluorescence image is shown in fig.3, which is top view of microchannel at center depth. $CaCl_2$ solution and water were supplied from upper and lower inlet, respectively. Watching near the interface, we noticed that CaCl2 solution side became bright and water side became dark. This phenomenon was unchanged by exchanging inlets or tilting the microchip. We assumed that this is caused by temperature rise with heat generation by dilution of conc. $CaCl_2$ solution. Remember that fluorescence in $CaCl_2$ solution is enhanced by temperature rise and that in water is weaken.

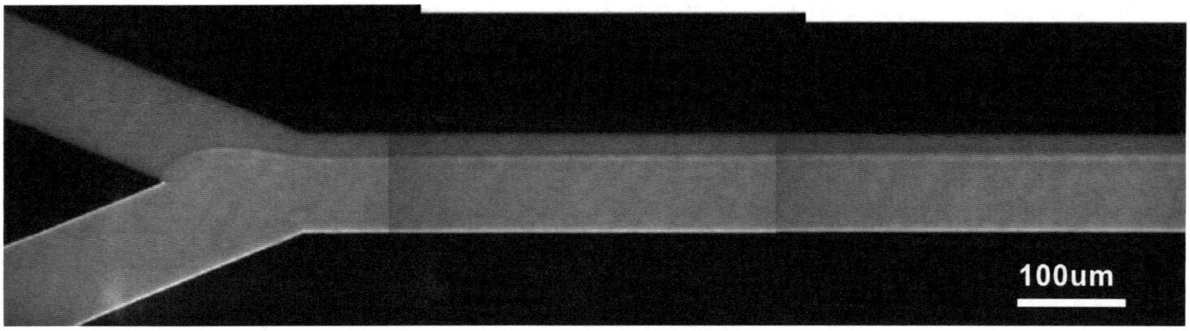

Fig. 3. Fluorescent image in microchannel (top view)

In Figure 4, we present a result of temperature calculation. Interface is also shown in the same. Temperature rise came out to be about 50 °C at interface. In Figure 4(a), as move to downstream, half bandwidth of temperature rise area became larger, but distribution shape was maintained. This indicates that heat was transferred by only diffusion.

In addition, Temperature decreased in water side. This is because in water side, $CaCl_2$ concentration arose, i.e. dissolution rather than dilution was occurred. This nature of dilution was made clear for the first time by watching liquid-liquid interface in micro scale with our measurement system.

Temperature distributions in cross section are also shown in figure 4 (b). Semicircle-like channel shape can be distinguished. Measurement was done in all area of channel cross section, however, signal from deep position was weak and calculation error became large. Further improvement is needed.

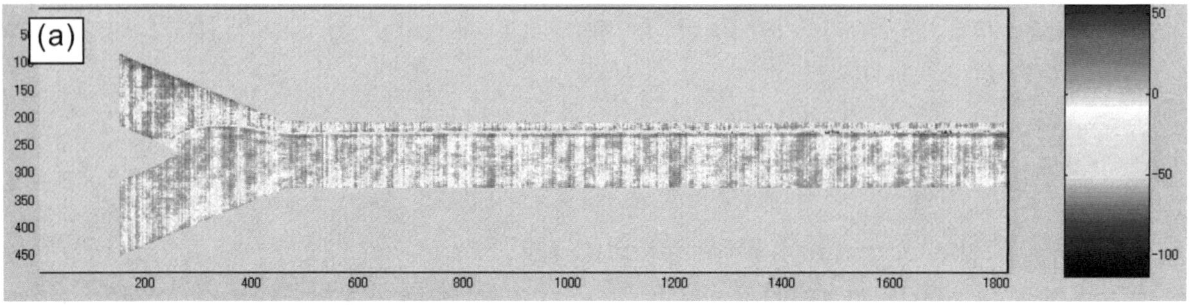

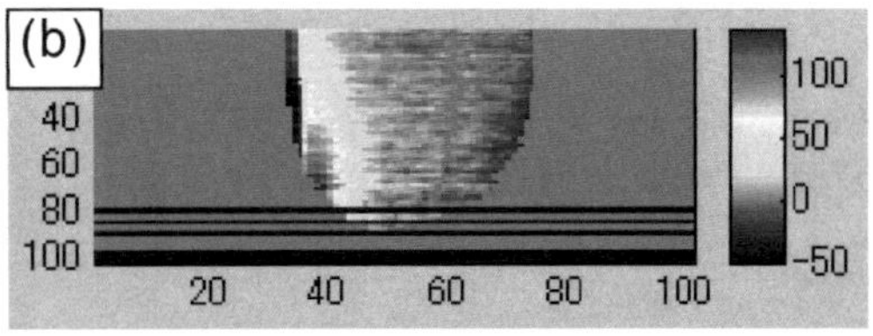

Fig. 4. Calculated temperature distribution

(a) Top view

(b) Cross section at channel merging position

4. Conclusion

We developed a method to measure temperature distribution in microchannel in three dimensionally. With this method, we measured local temperature rising when exothermic reaction such as dilution of conc. $CaCl_2$ solution occurred at interface.

We presented only the result of temperature measurement, however, this method is applicable to other measurements, such as pH, concentration of dissolved oxygen and all which have influence on fluorescent intensity.

In addition, our method is of great use for study for reaction mechanism, for laminar flow in microchannel gives new reaction field where micro scale phenomena are not destroyed by turbulent flow.

5. Notice

All other company and product names mentioned in this paper are trade names, trademarks or registered trademarks of their respective companies.

References

[1] J. G. Santiago, S. T. Wereley, C. D. Meinhart, D. J. Beede and R. J. Adrian, *Experiments in Fluids* **25** (1998), 316-319.

The Use of Input-Response Measurements as a Basis for Modelling the Residence Time Distribution of Microreactors

D. Boskovic, S. Loebbecke

Fraunhofer-Institute for Chemical Technology ICT
P.O. Box 1240, 76318 Pfinztal, Germany

Introduction

Residence Time Distribution (RTD) plays a decisive role in chemical process engineering and reaction technology. Depending on the reaction kinetics and thermodynamics RTD might significantly influence the yield and selectivity of chemical processes.
Compared to macroscopic processes, a much more precise measurement of RTD is required for an optimal processing of chemical reactions on a microscale, i.e. in microreaction technology.
Therefore, in this work an experimental method is described for measuring the RTD of microfluidic devices such as micromixers and microreactors by applying an input-response technique. Moreover, different possibilities for the processing of measured data and the modelling of the RTD are discussed.

Experimental Approach

The RTD is experimentally obtained by using an input-response technique. A dye tracer is injected into the constant flow of a pure solvent (here: water) and immediately monitored by a two-point spectroscopic method at the inlet and the outlet of the reactor. To make sure that the flow inside of the capillaries is not disturbed by additional devices, special spectroscopic flow cells have been developed. The flow-cells with the attached optical fibres were pulled over transparent PFA-capillaries and the concentration of the dye tracer was measured directly inside of the capillary (Fig. 1).

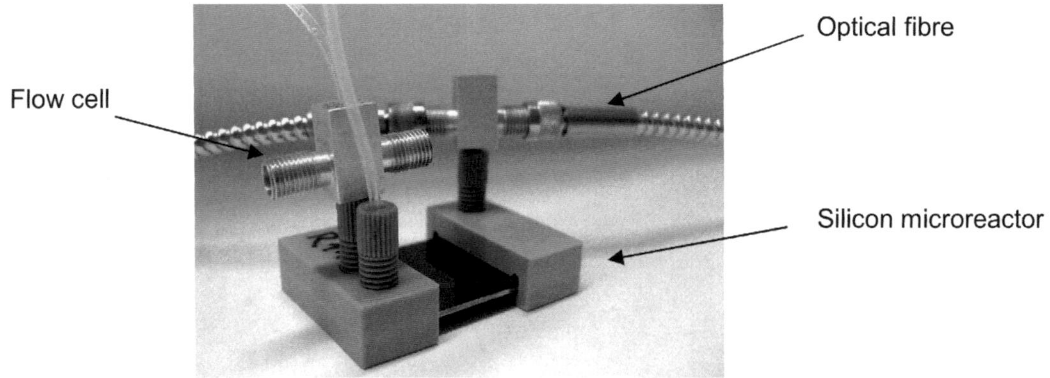

Fig. 1. Silicon microreactor with attached UV-Vis-flow cells

To ensure identical flow the pure solvent and the tracer solution are fed via the same pump (Fig. 2).

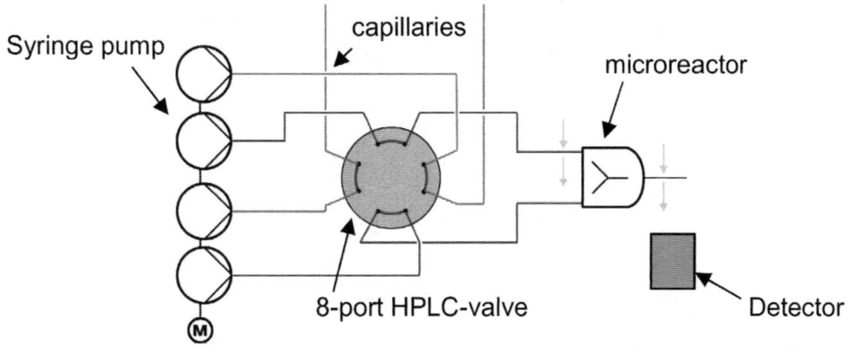

Fig. 2. Simplified scheme of the experimental setup

By using a fast cycling HPLC valve both step change and impulse experiments can be carried out which deliver F- and E-RTD-curves of a certain microfluidic device. (Fig. 3). The experimental method was optimized in order to eliminate systematic errors caused for example by absorption of the dye on the microfluidic channels, and thus to achieve a high repeatability of the measured RTD. The method can now be applied to different micro-devices for direct comparison of the RTD (Fig. 3).

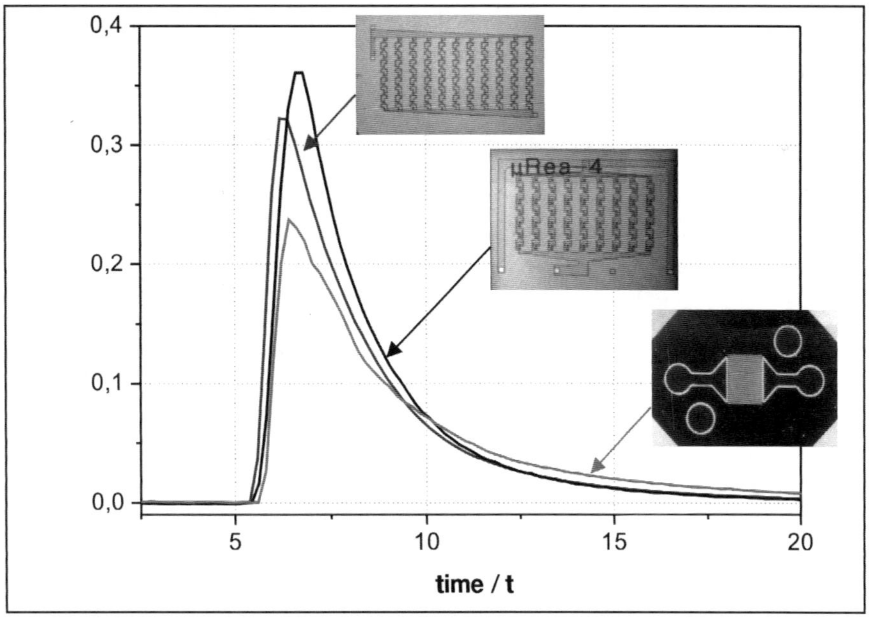

Fig. 3. Comparison of the RTD of different micro devices

Data processing

Since the measured RTD is a cumulative value of the microfluidic device itself and its connected periphery (e.g. capillaries), their individual contributions have to be separated by mathematical deconvolution methods to obtain the characteristic RTD behaviour of the microfluidic device only. Due to the fact that direct deconvolution leads to proper results only by using very simplified or idealized systems, it is suitable to choose another approach. There are several methods applicable which are all based on the convolution of the measured input signal with a model and a subsequent fitting to the output signal. Therefore it is necessary to derive models for describing the RTD-behaviour of microfluidic devices.

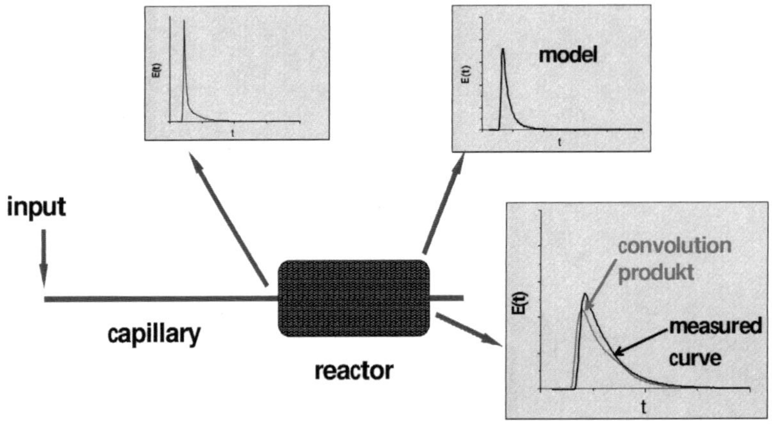

Fig. 4. Method of fitting the convolution product with the measured output signal

The convolution can be carried out in the frequency domain, Laplace domain or time domain. This is useful since analytical solutions of the models in the time domain are often difficult or impossible to achieve.

Modelling

Most of the one-dimensional models often used for macroscopic reactors like the Dispersion Model or the Tanks-in-Series Model are not applicable for microfluidic devices. Even if the dispersion coefficient for laminar flow $D=D_m+(u^2*d^2)/192*D_m$ is used as proposed by Aris [1] the dispersion Model is not applicable because of the strong impact of the laminar velocity profile.

A very promising way of modelling is the the application of 2-dimensional models based on the convective diffusion equation:

$$\frac{\partial c_T}{\partial t} + \upsilon_z(r)\frac{\partial c_T}{\partial z} = D_z \frac{\partial^2 c_T}{\partial z^2}$$

with the analytical solution [2]:

$$E(t) = \frac{1}{2\upsilon^2 t^3}\left\{\sqrt{\frac{D_z t}{\pi}}\left(L\exp\left(\frac{-L^2}{4D_z t}\right) - (L+\upsilon t)\exp\left(\frac{-(L-2\upsilon t)^2}{4D_z t}\right)\right) + \frac{1}{2}\left(L^2 + D_z t\right)\left(\operatorname{erf}\left(\frac{L}{2\sqrt{D_z t}}\right) - \operatorname{erf}\left(\frac{L-2\upsilon t}{2\sqrt{D_z t}}\right)\right)\right\}$$

with: υ...... average velocity
L...... length
D_z...... axial diffusion coefficient
t....... time
E(t)....residence time density function

Thereby, molecular diffusion in axial direction accompanied by convective dispersion in a fully developed laminar flow can be considered. First modeling results show a good approximation of the measured RTD curves (Fig. 6). However, the model equations have to be further refined concerning physical aspects and boundary conditions.

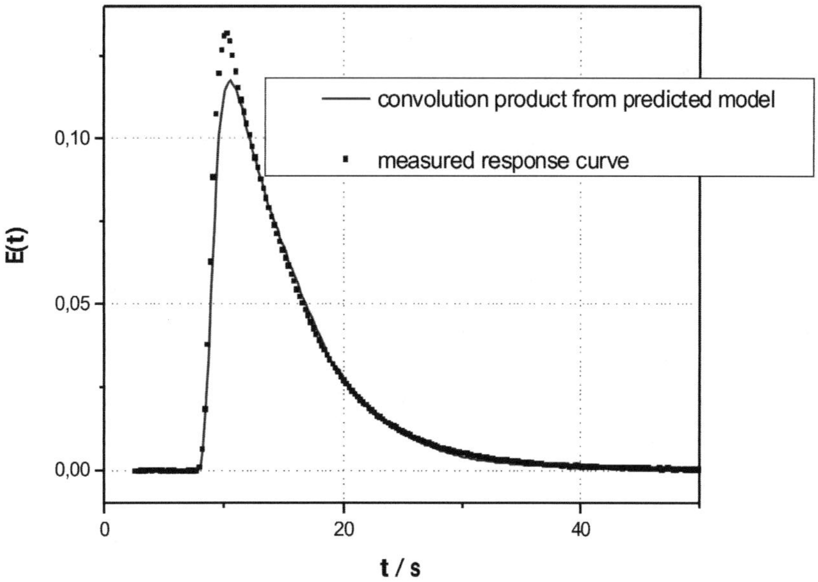

Fig. 5. Fit of the RTD-curve with the convective diffusion model

Results

An input-response technique for the measurement of the RTD was successfully applied to microfluidic devices. The RTD of different devices can be obtained with a high accuracy and good repeatability.
Models for the flow behaviour can be approximated on the basis of the measured curves. Two-dimensional dispersion models were used for an adequate description of the RTD of microfluidic devices.

References

[1] Aris, R., Proc. Roy. Soc.,1956, A256, 67
[2] Platzer B., Steffani, K., Große, S., Chem. Ing. Tech., 1999, 71, 795

PrOx in a Microreactor: Modeling of Reaction Kinetics and Effect of Heat Transfer Limitations

X. Ouyang,[1] P. Ho,[2] and R.S. Besser[3]

[1] *CNRS/CPE Lyon*
43 bd du 11 Nov 1918, BP 2077
Villeurbanne 69616
France

[2] *Reaction Design, Inc.*
6440 Lusk Boulevard, Suite D-205
San Diego, CA 92121

[3] *New Jersey Center for Microchemical Systems*
Department of Chemical, Biomedical and Materials Engineering
Stevens Institute of Technology
Castle Point on Hudson
Hoboken, NJ 07030

ABSTRACT

Silicon microreactors with Pt/Al_2O_3 thin-film wall catalyst were adopted for CO preferential oxidation (PrOx). In order to quantitatively study the intrinsic kinetics and effect of heat transfer limitations, two modeling tools were used. A microkinetic reaction model was used for the study of reaction pathways and the analysis of intermediate surface reactions which are difficult to trace experimentally. Principal kinetic mechanisms, including the rate-limiting steps, CO_2 formation pathways, and the effect of H_2 are discussed with these modeling results as a guide. This tool was used to assist the understanding of critical kinetic mechanisms, as well as increasing the efficiency of the project by eliminating certain experiments and generating fruitful new experimental directions through simulation. Secondly, a quasi-3D non-isothermal reactor model using the finite difference method was developed to understand differences in performance of the microreactor and mini packed-bed lab reactors (m-PBR's) with respect to the CO conversion vs. temperature characteristic. As a result, we discovered that the significant heat transfer limitations in m-PBR's favor the reverse water-gas-shift (r-WGS) side reaction, thus causing a much narrower range of permissible operating temperature compared to the microreactor. Accordingly, the efficient heat removal of the microchannel/thin-film catalyst system eliminates temperature gradients and prevents the onset of the r-WGS reaction. In conclusion, the microkinetic reaction model and the non-isothermal reactor model, coupled with our microreactor experiments, proved to be essential tools for a comprehensive understanding of reaction kinetics and reactor heat transport limitations with an ultimate view toward effective reactor design.

INTRODUCTION

The deep removal of CO in a H_2-rich stream is a critical step in on-site H_2 generation applications from hydrocarbon fuels. In order to prevent poisoning of the fuel cell electrodes, the CO concentration needs to be reduced from ~1% to below 10 ppm (Equation 1) while conversion of H_2 is minimized (Equation 2). Among a variety of catalyst candidates, Pt/Al_2O_3

has been the most studied PrOx catalyst because of its high CO conversion and stability at moderate temperatures (~200 °C) [1]. Despite the fact that PrOx reaction mechanisms for this catalyst system have been extensively investigated, a kinetic model at the molecular level that includes a thorough analysis of all intermediate species is still lacking. Most PrOx studies have been based on minimized packed-bed reactors (m-PBR's), assuming negligible axial and radial temperature gradient in the reactor. However, qualitative studies indicated significant heat transfer resistance for these devices due to the high reaction exothermicity [2]. It is thus imperative to carry out a comprehensive PrOx study using models that consider both the catalytic behavior and transport limitations.

$$CO+\frac{1}{2}O_2 \xrightarrow{\Delta H=-67kcal/mol} CO_2 \quad (1)$$

$$H_2+\frac{1}{2}O_2 \xrightarrow{\Delta H=-58kcal/mol} H_2O \quad (2)$$

$$CO_2+H_2 \xrightarrow{\Delta H=9.8kcal/mol} CO+H_2O \quad (3)$$

In our research, we conducted a two-fold modeling study on PrOx with a microkinetic simulation tool and a non-isothermal reactor model. A thin-film catalyzed microreactor, which has previously been shown to have a short mass and heat transfer resistance length [3], was used as the experimental tool and a guide for the modeling study. The CHEMKIN® software was used to explore the intrinsic reaction mechanism and to provide insight into the processes occurring in the reactor. Secondly, the effect of reactor heat transfer limitations was studied with a quasi-3D non-isothermal reactor model, comparing the microreactor with packed-bed reactors of different sizes.

EXPERIMENTAL

Thin-Film Catalyzed Microreactor

The kinetic data for PrOx was collected with the silicon microreactors used in our research [2].The Pt/Al_2O_3 catalyst with 2 wt% platinum was synthesized using a sol-gel technique [4]. The catalyst precursor was then selectively deposited in the microchannel. Multiple layers were deposited by repeating the procedures followed by calcination.

Microkinetic Reaction Simulation with CHEMKIN®

The CHEMKIN® software [5], an integrated software package for modeling complex chemical kinetic processes at the molecular level, was used to explore the intrinsic reaction mechanism and to provide insight into the processes occurring in the reactor. A reaction mechanism describing the detailed gas phase and surface chemical kinetics was constructed by extracting the relevant reactions from previously published work [6]. and The rate parameters for the oxygen dissociative adsorption reaction were altered in order to better match the experimental results. The model includes 8 adsorption reactions, 8 desorption reactions, and 12 surface reactions involving 9 gas-phase species (N_2, O, O_2, CO_2, H, OH, CO, H_2, H_2O) and 8 surface species (Pt(s), O(s), H(s), OH(s), H_2O(s), C(s), CO(s), CO_2 (s)). The CHEMKIN® software facilitates the analysis of reaction importance by providing information on rates of production of various species from each reaction, along with measures of the sensitivity of the solution to each reaction rate. Such an analysis resulted in the reaction network in Figure 1.

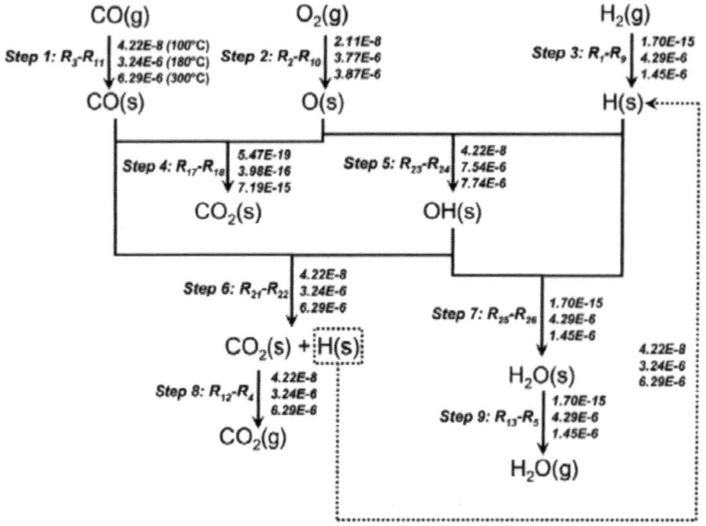

Figure 1. Elementary reaction network shows the pathways involving CO_2, H_2O and CO. Steady-state reaction rates (mol/s) in the whole microreactor at 100°C, 180°C and 300°C are also shown. WHSV = 1500 hr^{-1}, F(reformate) = 5.0 Ncm^3/min, F(air) = 0.5 Ncm^3/min.

Non-isothermal Reactor Model

The core of the non-isothermal reactor modeling is the construction of mole balance and energy balance equations. The majority of the reported PrOx kinetic studies has focused solely on the CO oxidation [7,8,9]. However, the inclusion of rate expressions for the coupled H_2 oxidation (Equation 2) and r-WGS (Equation 3) reactions is necessary for accurate representation of PrOx reaction behavior. Despite the importance of these other reactions, only few studies in the literature have addressed kinetic expressions for all three PrOx reactions [5]. Based on our previous work [10] and the PrOx kinetic studies in the literature [11], the kinetic expressions for PrOx reactions were formed as in Equations 4 – 9.

$$r_1^{'} = \frac{k_1 P_{CO} P_{O_2}}{(1 + K_1 P_{CO})^2} \tag{4}$$

$$r_2^{'} = k_2 P_{O_2}^{\beta 1} P_{H_2}^{\beta 2} P_{CO}^{\beta 3} \tag{5}$$

$$r_3^{'} = k_3 (\frac{P_{CO2} P_{H_2}}{K_3} - P_{CO} P_{H_2O}) \tag{6}$$

$$k_i = A_i \exp(-\frac{E_i}{RT}),\ i = 1,2,3 \tag{7}$$

$$K_1 = K_1^0 \exp(\frac{\Delta H_{ads}}{RT}) \tag{8}$$

$$K_3 = \exp(\frac{4577.8}{T} - 4.33) \quad (9)$$

$$\left\{ k\left(\frac{1}{r}\frac{\partial T}{\partial r} + \frac{\partial^2 T}{\partial r^2} + \frac{\partial^2 T}{\partial l^2} \right) \right\} - \left\{ \sum_{i=1}^{5} [H_i^{\circ}(T_R) \frac{dF_i}{dV}] \right\} = 0 \quad (10)$$

A multiple nonlinear regression analysis with the Marquardt method [12] was selected to evaluate the independent variables (A_1, A_2, A_3, E_1, E_2, E_3, β_1, β_2, β_3, K_1^0, ΔH_{ads}) in the kinetic expressions.

After the derivation of the mole balance equations, the energy balance equations were built as in Equations 10. The first bracketed term is the heat removal in axial and radial directions and the second is the heat generation from the three reactions in PrOx. Cylindrical geometry was used to approximate both the microchannel and the packed bed. Finally, the Gauss-Seidel iteration method [13] was used to solve the mole balance and energy balance equations simultaneously.

RESULTS AND DISCUSSION

Microkinetic Reaction Mechanism with CHEMKIN®

In order to verify the applicability and to facilitate the understanding of the CHEMKIN® model, the simulation results were compared with well-known PrOx mechanism studies in the literature.

The identification of O_2 adsorption as the rate limiting step at low temperatures has been substantiated experimentally in various reactor studies [7,14,15,16]. In our model, O_2 adsorption (step 2 in Figure 1) was also identified as the rate limiting step before CO light-off. Among the major adsorption reactions (i.e., adsorptions of CO, H_2, and O_2), CO adsorption has the highest sticking coefficient along with a relatively slow desorption rate. As a result, as shown in Figure 2, CO(s) dominates the catalyst surface, severely limiting the O_2 adsorption and consequently the CO(s) oxidation. Elevated temperatures increase the CO desorption rate, which is essential to open enough Pt sites for O_2 adsorption and the onset of CO(s) oxidation.

Various PrOx studies found a strong enhancement of CO oxidation by either H_2 or H_2O [7,14,15,17] addition, while it was later determined [18] that the presence of a hydroxyl surface species leads to a more efficient CO oxidation pathway than the direct CO oxidation on the Pt surface. The intrinsic mechanisms of H_2 on PrOx reaction was also studied with CHEMKIN®. First, as shown in Figure 1, there are two pathways for CO_2 formation: step 4 and steps 5 and 6. As indicated by the higher net-rate values, the major path for CO_2(s) production is through the formation of the OH(s) species (step 5) and the subsequent oxidation of CO(s) by OH(s) (step 6), rather than the direct oxidation of CO(s) (step 4). Further, simulation results based on reaction feeds with and without H_2 indicate a large increase in CO conversion in the presence of H_2, as shown in Figure 4.

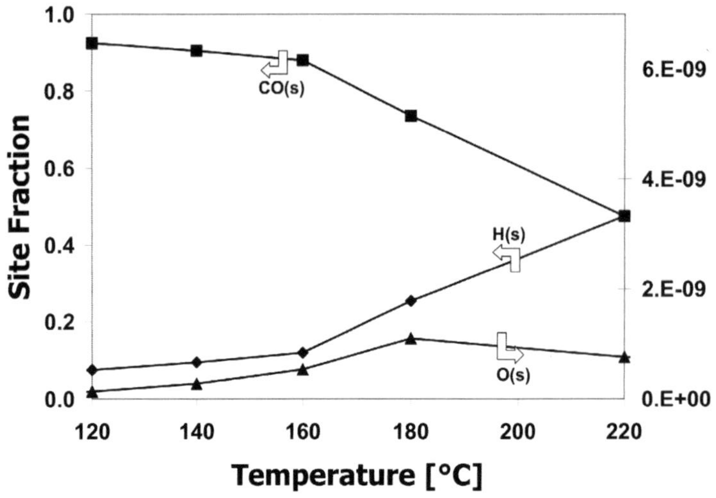

Figure 2. The temperature dependence of the site fractions of CO(s), H(s) and O(s) calculated by the CHEMKIN® model.

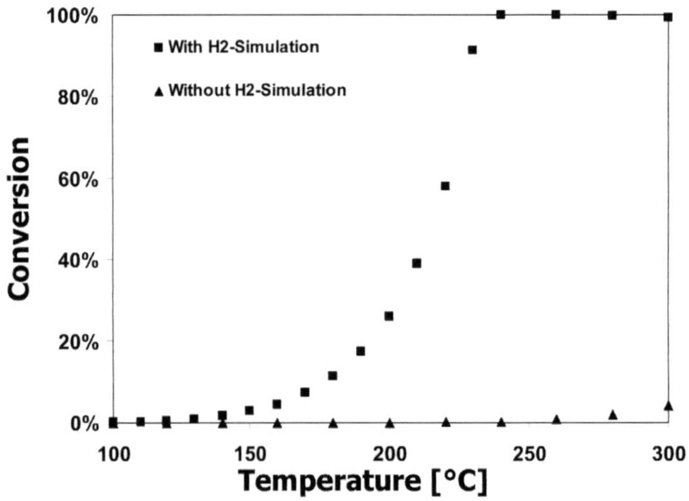

Figure 3. The addition of H_2 significantly increases CO conversion as calculated by the CHEMKIN® model.

After this analysis of the PrOx reaction mechanism, a wide range of simulations were done with the CHEMKIN® model to identify new regions of parameter space to explore experimentally that would offer higher efficiency.

Effect of Reactor Heat Transfer Limitations

Our calculation with Mear's criterion determined that the heat conduction distance (t_{cat}) of the thin-film catalyst ($5x10^{-6}$ m) is orders of magnitude smaller than that for the packed-bed catalyst (~$2x10^{-3}$ m). Due to the quadratic dependence of heat transfer resistance on the t_{cat}, the Damköhler's number (*Da*) of the m-PBR with $2x10^{-3}$ m radius (2-mm m-PBR) is more than 5 orders of magnitude higher than that for the microreactor. The Mear's criterion then suggested severe heat transport limitations for the m-PBR's for the highly exothermic PrOx reaction. Further, experimental studies comparing the CO conversion temperature dependence of the microreactor and m-PBR's in the literature also disclosed large difference caused by reactor heat transfer limitations [3]. Following the above analysis, the quantitative study of reactor heat transfer limitations was thus realized through the non-isothermal reactor modeling with the microreactor, the 2-mm m-PBR, and the 4-mm m-PBR.

The modeling results of the microreactor with $5x10^{-6}$ m catalyst film shows essentially isothermal temperature distribution in the thin-film catalyst of the microreactor, even at the highest operating temperature (300°C), supporting our qualitative analysis with Mear's criterion. On the other hand, the results for the 2-mm m-PBR show significant temperature gradients, as indicated by Figure 4. The temperature gradients become more dramatic and move upstream with higher isothermal wall temperatures, which suggests that the large accumulated reaction heat caused by inefficient heat transfer within the reactor.

Figure 5 plots the 2-mm m-PBR modeling results of O_2, CO, H_2 conversions and CO_2 selectivity along the reactor. With the wall temperature increases, full conversion of O_2 (with CO and H_2) takes place in shorter reactor lengths, implies a higher density of heat accumulation and consequently larger temperature gradients in the region with most oxidation activity (Figures 4).

Further understandings of the modeling results can be given as follows. Due to the low O_2 reaction activity (Figure 5a) at low temperatures (e.g., T_w = 120°C), the temperature gradient is negligible in both axial and radial directions, as shown in Figure 4a. At 180°C, O_2 reaction activity is mild up to $l = 2x10^{-2}$ m (Figure 5b), which leads to small temperature gradients in this region (Figure 4b). However, as also indicated by Figure 5b, the sudden increase of both CO and H_2 oxidation activity between $l = 2x10^{-2}$~$2.5x10^{-2}$ m causes a dramatic increase of accumulated reaction heat, leading to temperatures much higher internally than at the wall (Figures 4b). With further increase of the wall temperature, more severe temperature gradients are developed close to the entrance of the reactor (Figure 4c) since full O_2 conversion is reached at a length of only 10^{-3} m (Figure 5c).

Figure 6 plots the modeling results of CO conversion at different wall temperatures for the microreactor and 2-mm and 4-mm m-PBR's. As discussed above, the microreactor has uniform temperature distribution, resulting in a CO conversion curve coinciding with the result for ideal isothermal operation. On the contrary, the results for the m-PBR's show different characteristic curve essentially caused by the temperature gradients and the resulting different local reaction activities for all three Prox reactions (Equations 1-3). Particularly at high temperatures, the drop in the CO conversion curves for the m-PBR's is caused by strong r-WGS conversion activated by high local temperatures in these reactors. This phenomenon are seen to intensify as the radial thermal resistance increases (4-mm vs. 2-mm).

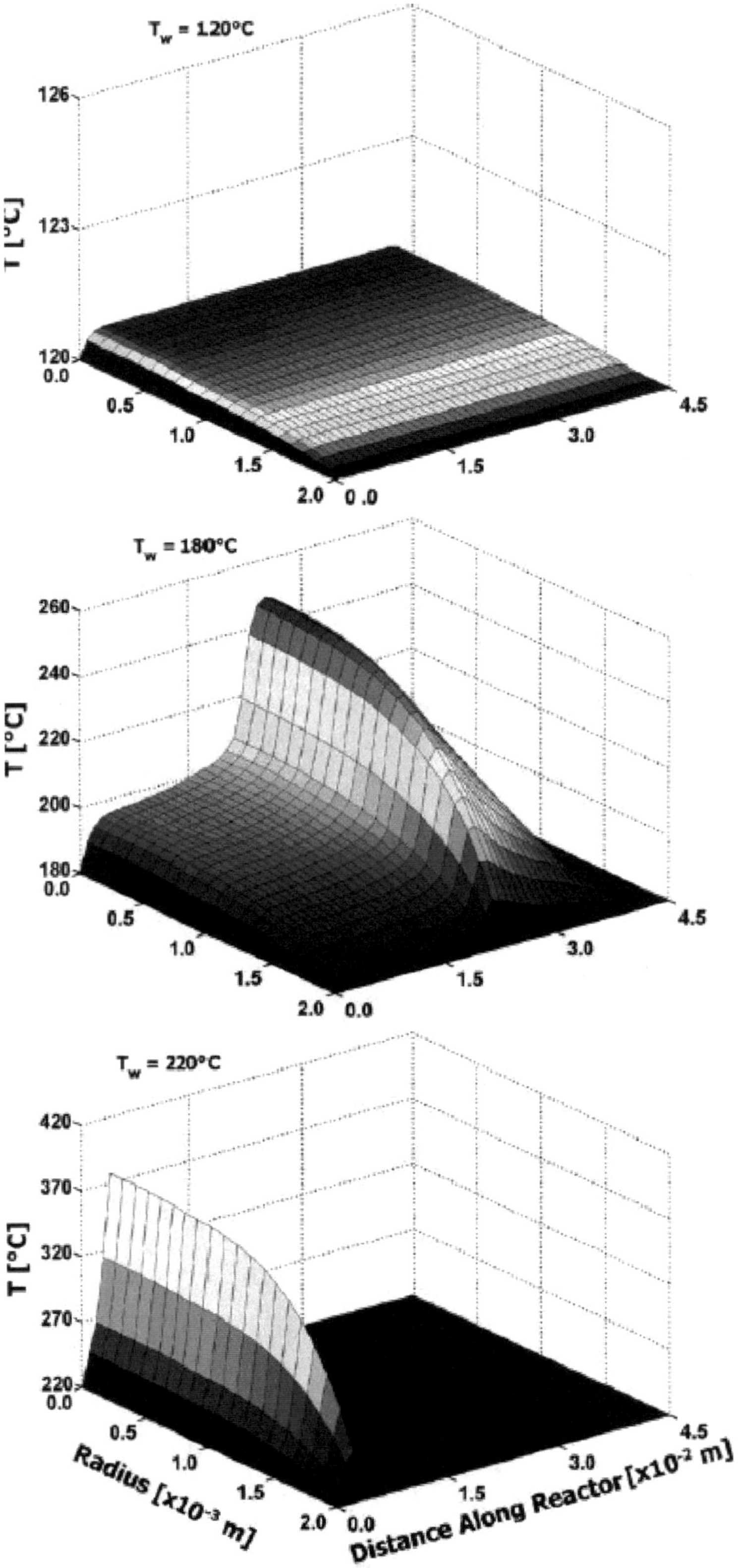

Figure 4. Plots of reactor temperature distribution for the 2-mm m-PBR with T_w = 120°C, 180°C, 220°C. $WHSV$ = 1500 hr^{-1}.

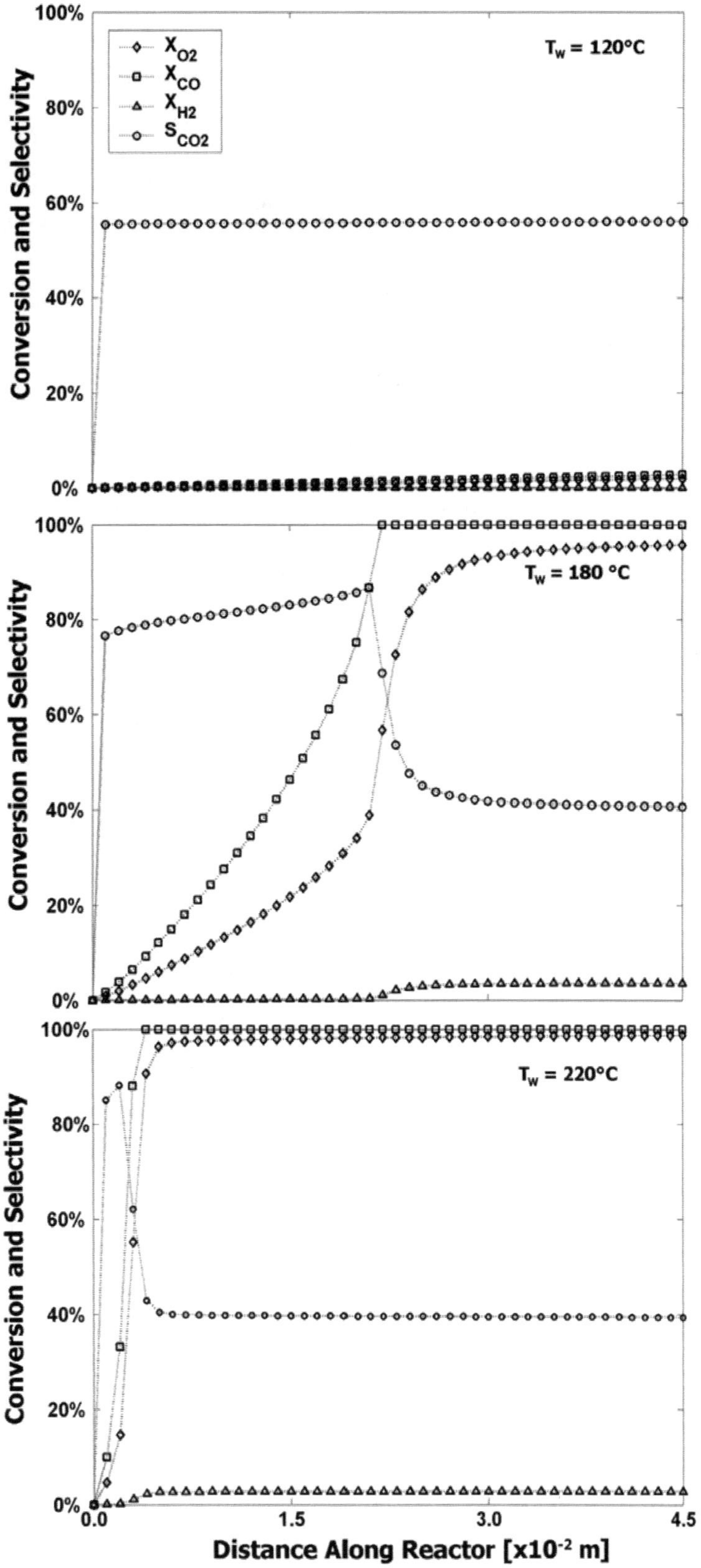

Figure 5. O_2, CO, H_2 conversion and CO_2 selectivity along the reactor length of 2-mm m-PBR with T_w = 120°C, 180°C, 220°C. *WHSV* = 1500 hr^{-1}.

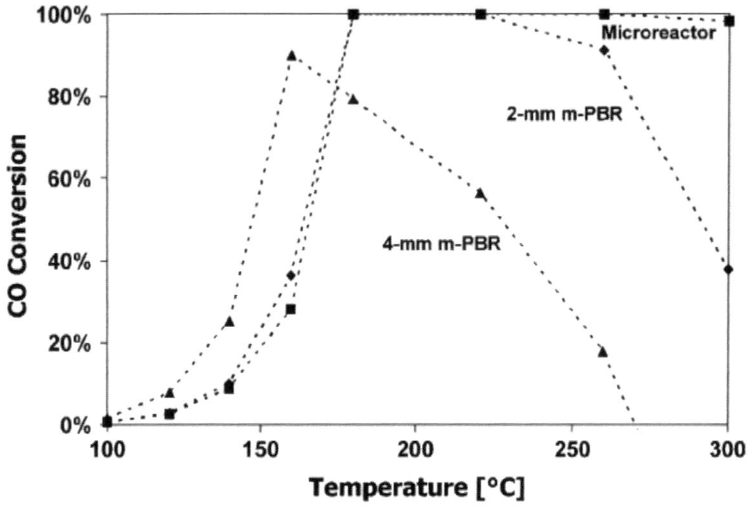

Figure 6. CO conversion vs. different reactor wall temperatures for the microreactor, 2-mm and 4-mm m-PBR's. *WHSV*: 1500 hr^{-1}.

CONCLUSIONS

This paper demonstrated the application of two reactor modeling tools, CHEMKIN® and the non-isothermal reactor model, to study PrOx in a thin-film catalyzed reactor. Critical reaction mechanisms were verified in the CHEMKIN® simulations, which provided further understanding of these mechanisms through analysis of elementary reaction steps. With the non-isothermal reactor model, divergence of the PrOx reaction behavior between a microreactor with thin-film catalyst and m-PBR's were quantified and delineated. Finally, significant advantages of the thin-film catalyzed microreactor were demonstrated for kinetic studies of the highly exothermic PrOx reaction system.

NOMENCLUTURE

Greek letter

β_i	reaction order for rate expression of H_2 oxidation (*i*: O_2, H_2, CO)

letters

A_i	pre-exponential factor in kinetic expressions
Da	Damköhler's number for heat transport
E_i	activation energy for kinetic expressions (J mol^{-1})
ΔH_{ads}	adsorption energy (J mol^{-1})
k_i	specific reaction rate for r_1', r_2' and r_3' (*i* = 1, 2 or 3)
K_1, K_1^0	adsorption equilibrium constant for CO
K_3	equilibrium constant of the r-WGS reaction
l	reactor axial position, *l* = 0 for the entrance of the reactor (m)
l_{Tmax}	the axial position where the temperature (at *r* = 0) is the highest (m)
L	total reactor length (m)
P_i	partial pressure of reactant species *i* (*i*: CO, O_2, CO_2, H_2, H_2O) (Pa)
r	reactor radial position, *r* = 0 for the center of the reactor (m)

$r_1^{'}$	reaction rate of CO in CO oxidation, moles of reactant per gram of active metal of catalyst per second (mol kg^{-1} s^{-1})
$r_2^{'}$	reaction rate of H_2 in H_2 oxidation, moles of reactant per gram of active metal of catalyst per second (mol kg^{-1} s^{-1})
$r_3^{'}$	reaction rate of CO_2 in r-WGS oxidation, moles of reactant per gram of active metal of catalyst per second (mol kg^{-1} s^{-1})
R	reactor radius (m)
t_{cat}	effective interparticle heat conduction distance: catalyst thickness for the microreactor or reactor inner radius for m-PBR (m)
$T_{w,}$	reactor inner wall temperature (°C)
WHSV	weight hourly space velocity, defined as mole of total reactant flow per mole of precious metal in the catalyst per hour (hr^{-1})

REFERENCE:

1. C. Song, *Catal. Today* 77 (2002) 17–49.
2. H. Surangalikar, X. Ouyang, R.S. Besser., *Chem. Eng. J.* 93 (2003) 217-224.
3. X. Ouyang, L. Bednarova, P. Ho, R.S. Besser, *AIChE J.*, accepted for publication (2004).
4. S. Zhao, R.S. Besser, in Proceedings of the 6th International Conference on Microreaction Technology, New Orleans, LA (2002) 289-296.
5. R.J. Kee, F.M. Rupley, J.A. Miller *et al.*, *Chemkin Collection, Release 3.7.1*, San Diego, Calif, Reaction Design, Inc (2003).
6 D.K. Zerkle, M.D. Allendorf, M. Wolf, O.Deutschmann, *J. Catal.* 196 (2000)18-39.
7. M.J. Kahlich, H.A. Gasteiger, R.J. Behm, *J. Catal.* 171 (1997) 93–105.
8. R.H. Nibbelke, M.A.J. Campman, J.H.B.J. Hoebink, G.B. Marin, J. Catal. 171 (1997) 358-373.
9. D.H. Kim, M.S. Lim, *Appl. Catal. A: Gen.* 224 (2002) 27-38.
10. X. Ouyang, P. Ho, H. Chen, W.C. Shin, L. Bednarova, W. Lee, R.S. Besser, S. Pau, C.S. Pai, J.A. Taylor, W.M. Mansfield, in AIChE 2003 Annual Meeting Proceedings (2003).
11. Y. Choi, H.G. Stenger, *J. Power Sources* 129 (2004) 246-254.
12. A. Constantinides, N. Mostoufi, Numerical Methods for Chemical Engineers with MATLAB Applications, Prentice Hall Inc., Upper Saddle River, NJ, 1999.
13. J.D. Hoffman, Numerical Methods for Engineers and Scientists, second ed., Part 3, McGraw-Hill, Inc., 1992.
14. S.H. Oh, R.M. Sinkevitch, *J. Catal.* 142 (1993) 254-262.
15. G. Avgouropoulos, T. Ioannides, Ch. Papadopoulou, *et al.*, *Catal. Today* 75 (2002) 157-167.
16. A. Manasilp, E. Gulari, *Appl. Catal. B: Env.* 37 (2002) 17–25.
17. R.H. Nibbelke, M.A.J. Campman, J.H.B.J. Hoebink, G.B. Marin, *J. Catal.* 171 (1997) 358-373.
18. M.M. Schubert, H.A. Gasteiger, R.J. Behm, *J Catal.* 172 (1997) 256-258.

132r Experimental and Numerical Evaluation of Micro Heat Exchanger Using Catalytic Combustion

Hee-Yeon Kim, In-Soo Ryu, Kwang-Sup Song, Nam-Jo Jeong, Sang-Phil Yu, Seung-Jae Lee

Catalytic micro heat exchangers were fabricated with stainless steel plates. Microchannels Having 0.3-0.5 mm of width, 0.2 mm of depth and 20 mm of length were formed on 40´40 0.3 mm^3 of metal plates by photolithography and chemical etching. The metal plates with microchannels were stacked and then brazed to seal up the gaps between the metal plates. Alumina to support palladium catalyst was deposited on the microchannels of the brazed metal plates by sol-gel process. The coated alumina was present in 5-10 μm of thickness, which was not uniform in the alumina layers.

The fabricated micro heat exchangers were evaluated in the efficiency of heat exchange with or without catalytic combustion of LPG. In the heat exchange without catalytic combustion, heat exchange efficiency increased with decreasing Reynolds number of heating air and increasing that of cooling air. This result suggested that the heat exchanger without catalytic combustion could be operated at high Reynolds numbers of heating and cooling air in order to obtain high heat transfer rate as well as high efficiency. On the other hand, in the heat exchange with catalytic combustion, there was a region to operate the micro heat exchanger, where the stable LPG combustion was obtained by adjusting the inlet concentration of LPG and the Reynolds number of the cooling air. The heat exchange efficiency of the heat exchange with catalytic combustion was lowered at high inlet concentration of LPG. The experimental results was compared with those in numerical simulation.

Multi-layered Composite Nanocrystals Prepared in a Micro-fluidic Reactor

Masato Uehara, Hiroyuki Nakamura and Hideaki Maeda

Micro-space Chemistry Laboratory, National Institute of Advanced Industrial Science and Technology (AIST)

807-1 Shuku, Tosu, Saga 841-0052, Japan

Abstract

Preparation of uniform-sized nanocrystals in a micro-fluidic system was reported by some researchers. Semiconductor nanocrystals such as CdSe have attractive luminescence properties. However, in order to obtain short fluorescence wavelength, the particle size must be smaller and the crystallinity would become lower, as a result that the fluorescence intensity would be lower. In this work, we have prepared ZnS/CdSe/ZnS multi-layered composite nanocrystals in a micro-fluidic reactor. The composite nanocrystals have a structure which sandwiched CdSe layer by ZnS core and outer shell. The composite nanocrystals have emitted blue fluorescence more strongly than the two-layered nanocrystals (ZnS-capped CdSe nanocrystals). It was thought the CdSe layer was high-crystallized uniformly on ZnS core nanocrystals by using micro reactor system, consequently blue fluorescence became stronger.

Introduction

The photoluminescence (PL) properties of the isolated semiconductor nanocrystals (NCs) are attractive for light emitting diode and fluorescence bio-molecules tag[1, 2]. These materials emit the variable light, ZnO, CdS and ZnSe for UV and blue, CdSe for visible. Red and near-infrared lights are obtained from CdTe, GaAs, InP and so on[3-5].Recently, the synthetic technique of colloidal NCs has greatly been improved, especially II-VI groups such as CdSe and CdS[3-5]. The structure control of NCs, such as size-tuning, alloying and coating(core/shell structure) were carried out and many papers were reported about the relationship between NCs structure and optical properties[3-7].

Now, synthesis of NCs in a micro-fluidic system was reported by some researchers[7-9]. The use of micro fluid system is effective to a uniform synthesis of NCs, because the system brings out the uniform and accurate control of heating condition, such as temperature and time. We have synthesized some NCs by a simple micro fluid system which consists of syringe, capillary and oil bath[7]. The reaction solution is pushed into the capillary in a constant rate by a syringe pump. A part of capillary is dipped in an oil bath. The solution was heated in the dipped part of capillary. The heating time calculated from the solution flow rate and the heated volume of the capillary. We have applied this system to the coating of NCs surfaces. The optical of NCs are influenced by surface and inner defects significantly. The PL quantum yield is improved by surface capping by the different material which have wider energy gap. We coated CdSe NCs by ZnS in the micro fluidic system[9]. The quantum yields have been more improved and the FWHM of PL peak was kept, compared with batch system. These suggested that this system is very useful for the uniform coating and modification of NCs.

In this work, we have applied the micro-fluidic reactor to synthesis multilayer NCs. The synthesis was carried out by ion-exchange process, in all papers about multilayer NCs [10-12]. This process has limitation of materials which consist of same cation or anion, such as CdS and HgS. We tried the synthesis in CdSe and ZnS system. We synthesized the ZnS/CdSe/ZnS (core/shell/shell) composite NC by this reactor. The composite NCs have a structure which sandwiched CdSe layer by ZnS core and outer shell. The wavelength of PL emission of NCs is tunable by the particle size. However, in order to obtain shorter wavelength, the particle size must be smaller and the crystallinity would become lower, as a result that the PL intensity would be lower. We tried to deposit CdSe on the ZnS NC surface for improvement of CdSe crystallinity. CdSe and ZnS have no same ions and the lattice mismatch was large 9%. However, on some atomic layers level, the epitaxial deposite might achieve such as ZnS and CdS [10]. Furthermore, it would be achieved by application of micro-fluidic reactor which could uniformly coat for NCs.

Experimental procedure

Firstly, we prepared two raw material solutions. Zinc diethyldithiocarbamate was dissolved to the mixture of tri-octylphosphine (TOP) and octadecene (ODE) with a small amount of oleic acid. This solution (solution A) was used for synthesis ZnS core NCs and ZnS shell layer. For the raw material solution of CdSe shell layer (solution B), cadmium acetate was dissolved in ODE with oleic acid at 180°C. The solution was mixed with Selenium TOP solution.

For the synthesis of ZnS core particles, the solution A was loaded into a glass syringe connected to a glass capillary. The solution was pushed into the capillary in a constant rate (10 L/min) by a syringe pump. The 80 cm of the capillary was dipped in a 240°C oil bath. The solution was heated in the dipped part of capillary. The heating time calculated 150sec from the solution flow rate and the heated volume of the capillary. The products were precipitated with alcohol and further isolated by centrifugation, and subsequently re-dispersed in ODE. This clear ZnS colloidal solution was mixed with the solution B. This solution was pushed into the same micro fluidic system for CdSe-coating of ZnS NCs. The reaction temperature and time were 180-240°C and 30sec-120sec. These products were washed in the same processing as the above-mentioned. Furthermore, these ODE solutions of ZnS/CdSe (core/shell) NCs were mixed with the solution A and ZnS-coated by the same micro fluidic system. Thus, we synthesized the ZnS/CdSe/ZnS (core/shell/shell) multilayered composites NCs.

A UV-1600 UV-visible spectrophotometer (Shimadzu Corp.) and FP-6600 spectrofluorometer (Jasco, Inc.) were used to determine the absorption and photoluminescence emission spectra. S-5200 FE-SEM (Hitachi High-Tech. Corp.) and EDX system (EDAX Inc.) and Tecnai FE-TEM (FEI Corp.) were used to observe the particle size and determine the elements of particles.

Result and discussion

The CdSe NC has a high-luminescence and the light-tuning can be brought out by size-control. However, the diameter must be smaller than 2.5nm for blue luminescence (below 490nm) from CdSe.[13] It is difficult to keep high-crystallinity, as a result that the fluorescence intensity is lower. So, we tried to deposit CdSe on the ZnS NC surface. The ZnS NCs were synthesized by pyrolysis of Zinc diethyldithiocarbamate. The size of ZnS NCs was about uniformly 2nm. CdSe-deposition was carried out on these ZnS NCs at 240°C. ZnS/CdSe/ZnS core/shell/shell NCs were synthesized by the futher coating by ZnS. Figure 1 shows the PL emission spectra of ZnS/CdSe/ZnS NCs for blue-light emission. It is also shown that of CdSe/ZnS core/shell NCs which have no ZnS core, for comparison. The Intensity of ZnS/CdSe/ZnS was higher than CdSe/ZnS. The CdSe was deposited on the ZnS NCs and the blue-light could be emitted highly. The quantum yield was determined about 50%. We were careful of the deposition of CdSe on the ZnS surface. If the CdSe was freely deposited not on the ZnS surface, the particle diameter was twice or larger than the layer thickness which deposit on the ZnS. In this case, other PL emission was not observed, and the FWHM of PL peak was about 80nm. Therefore, the CdSe deposition would be comparatively uniform.

The synthesis of multilayer NCs were carried out by ion-exchange process, in all papers [10-12]. This process has limitation of materials which consists of same cation or anion. However, in this case, we could demonstrate a variety for target materials, such as CdSe and ZnS. This micro-fluidic system would be applied to the synthesis of multilayer NCs.

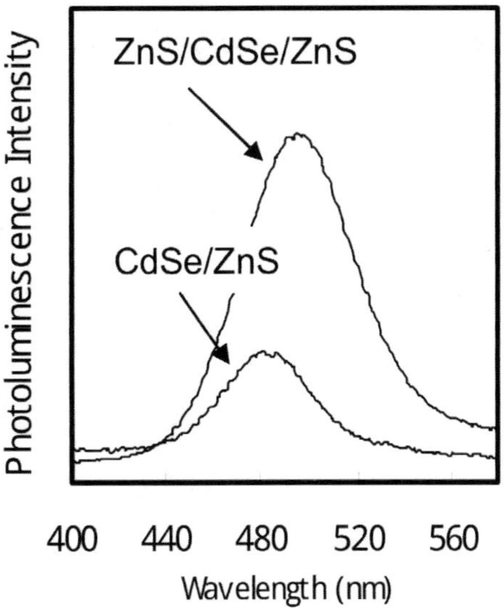

Figure 1 Photoluminescence spectra of ZnS/CdSe/ZnS core/shell/shell and CdSe/ZnS core/shell NCs for blue-light emission.

Figure2 shows the STEM image of the CdSe-coated NCs (ZnS/CdSe core/shell NCs). We observed at low accelerating voltage 30KV by FE-SEM equipment and we could get the high-contrast images. The spherical particles were observed and the diameter was 4-5nm. The diameter of ZnS core particles was about 2nm and it was confirmed to have grown by the coating processing. The EDX analysis about one particle was carried out. The signal of four elements Zn, S, Cd and Se could be detected from one particle, whereas other signals of C, O, Cu and Si from machine and micro-grid were also detected. This has suggested the success of CdSe-coating.

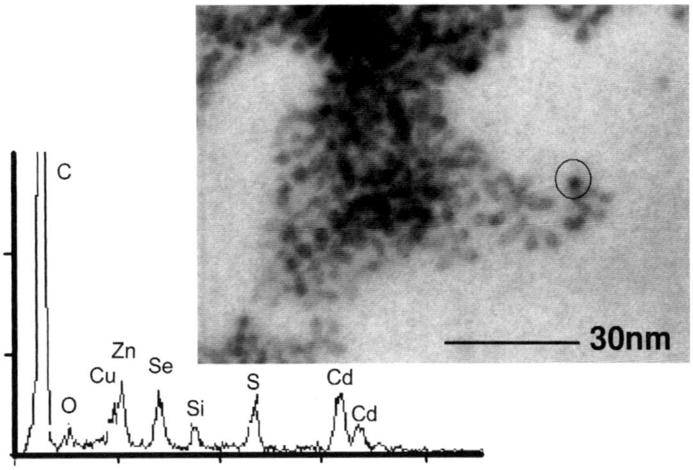

Figure 2 STEM image and EDX spectrum of the CdSe-coated ZnS NCs (ZnS/CdSe core/shell NCs).

The image of ZnS/CdSe/ZnS core/shell/shell NCs were also shown in Fig.3. The particle shape was spherical and the size was about 6nm. The thick ness of CdSe layer is calculated about 1-1.3nm from EM images. This is thinner than calculated value[13]. The confinement of exciton is not complete and it may exude to ZnS phases. According to EDX analysis, the NCs consisted of four elements Zn, S, Cd and Se. In the HRTEM observation, the lattice images of NCs were observed and it was confirmed that the particles were not amorphous but crystal. However, the layered structure, such as interfaces of ZnS and CdSe could not be observed. It was caused by the fineness of this composite structure or the epitaxial relation of each layer. Anyway, the more fine investigation has been needed.

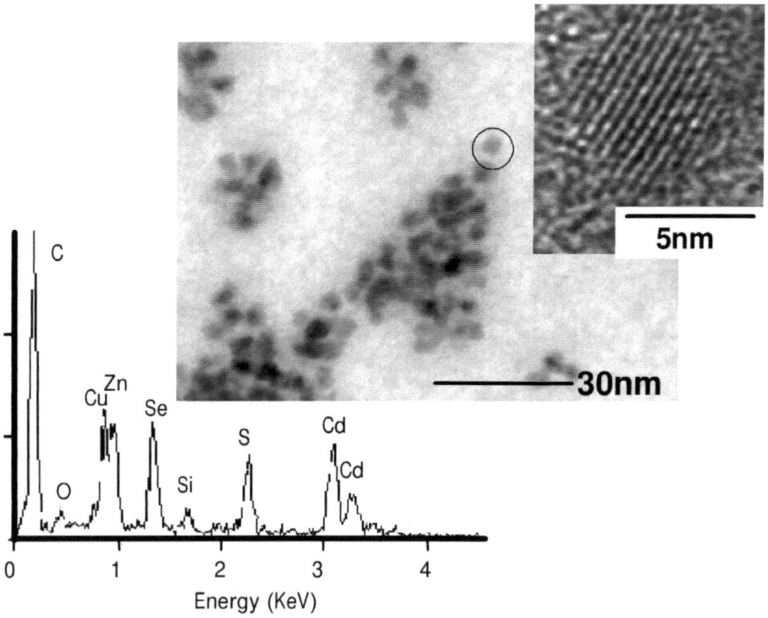

Figure 3 STEM and HRTEM images, and EDX spectrum of the ZnS/CdSe/ZnS core/shell/shell NCs.

We have tried light-tuning by control of CdSe layer thickness. The thickness could be controlled by the flow rate of the raw material solution in the CdSe deposition process. The PL spectra were shown in Fig.4. These depositions were carried out in the slow flow rates at 180°C. The residence time was 75-120 sec. The peaks shifted from 490 to 530nm. The wavelength was shorter with the flow rate. The FWHM of PL peaks were kept about 80nm and CdSe-coating would be carried out comparative homogeneously.

Thus, we have succeeded the synthesis of multilayer NCs and light-tuning by a micro-fluidic reactor. This has suggested that the uniform and accurate controllability of the micro-fluidic system are useful for the modification of NCs. This system have variety than conventional ion-exchange process[10-12]. The micro fluidic system would be applied to the synthesis of NCs which have special structure.

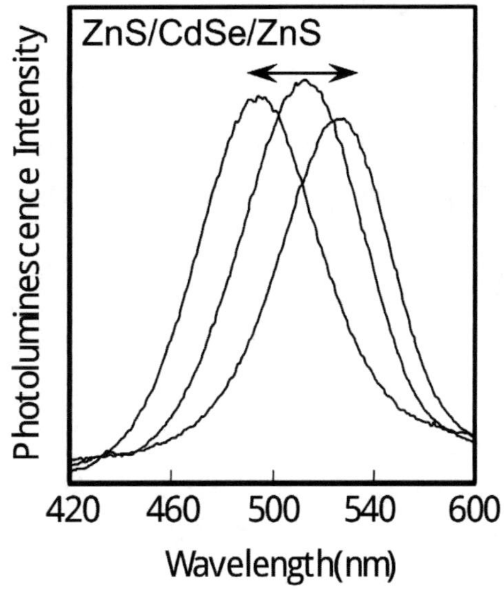

Figure 4 Photoluminescence spectra of light-tunable ZnS/CdSe/ZnS core/shell/shell NCs.

Summary

We have prepared ZnS/CdSe/ZnS multi-layered composite nanocrystals in a micro-fluidic reactor. The size was about 6nm. The composite nanocrystals have emitted blue fluorescence more strongly than the two-layered nanocrystals (ZnS-capped CdSe nanocrystals). It was thought the CdSe layer was high-crystallized uniformly on ZnS core nanocrystals by using micro reactor system. The light-tuning could be brought out by the control of flow rate in the CdSe deposition process. We could demonstrate the variety of this system for the modification of NCs.

Acknowledgement

TEM analysis in this work was supported by "Nanotechnology Support Project" of the Ministry of Education, Culture, Sports and Technology (MEXT), Japan.

References

[1] A. Henglein, *Chem. Rev.*, 1989, 115, 1861-1873.
[2] W. C. W. Chan, D. J. Maxwell, X. Gao, R. E. Bailey, M. Han, S. Nie, *Anal. Biotechnol.*, 2002, 13, 40-46.
[3] A. P. Alivisatos, *Science*, 1996, 271, 933-937.
[4] C. B. Murray, D. J. Norris, M. G. Bawendi, *J. Am. Chem. Soc.*, 1993, 115, 8706-8715.
[5] X. G. Peng, M. C. Schlamp, A. V. Kadavanich, A. P. Alivisatos, *J. Am. Chem. Soc.*, 1997, 119, 7019-7029.
[6] X. Zhong, M. Y. Han, Z. Dong, T. J. White, W. Knoll, *J. Am.Chem. Soc.* 2003, 125, 8589 - 8594.
[7] H. Nakamura, Y. Yamaguchi, M. Miyazaki, H. Maeda, M. Uehara, P. Mulvaney, *Chem. Commun.*, 2002, 2844-2845.
[8] E. M. Chan, R. A. Mathies, A. P. Alivisatos, *Nano Lett.*, 2003, 3, 199-201.
[9] H. Wang, X. Li, M. Uehara, Y. Yamaguchi, H. Nakamura, M. Miyazaki, H. Shimizu and H. Maeda, *Chem. Commun.*, 2004, 48-49.
[10] R. B. Little, M. A. El-Sayed, G. W. Braynt and S. Burke, *J. Chem. Phys.*, 2001, 114, 1813-1822.
[11] A.L. Rogach, M.T. Harrison, S.V. Kershaw, A. Kornowski, M.G. Burt, A. Eychmüller, H. Weller, Phys. Stat. Sol. (b), 2004, 224, 153–158.
[12] A. Eychmilller, A. Mews and H. Weller, Chem. Phys. Lett.,1993, 203, 59-62.
[13]V. Albe, C. Jouanin, and D. Bertho, Phys. Rev. B, 1998, 58, 4713-4720

GASEOUS FLOWS IN MICROCHANNELS

Aguru Yamamoto, Shinsuke Mori**, and Masaaki Suzuki***

** The Research Association of Micro Chemical Process Technology (MCPT), Japan*

*** Department of Chemical Engineering, Tokyo Institute of Technology, Japan*

INTRODUCTION

Microreactors have been developed in advantageous industrial fields. Many researches aim at creating new devices and fundamentals of fluid dynamics are also still discussed [1]. A feature of microreactors is that every microchannel in microreactors does not have circular crosssection. Typical shape of crosssection of microchannels is shown in **Table 1**. Shape of crosssection of a microchannel in a microreactor depends on fabrication method. Microchannels fabricated on glass plate by using standard isotropic wet etching technique, which is used to integrate some devices, have D-shaped crosssection. It is also characteristics that aspect ratio of crosssection can be far from one. Since laminar flow is usual in microchannels, shape of crosssection of a microchannel is important.

Table. 1 Crosssection of typical microchannels.

Crosssection	Example
Circle	Capillary tube
Rectangle	Microchannel in glass chip, piled glass
Isosceles Trapezoid	Microchannel in <100> Silicon wafer, wet etching
D-Shape	Microchannel in glass chip, isotropic wet etching

In microscale fluid dynamics of gaseous flow, the rarefaction effect is important. The slip between a fluid and a solid wall is originated by the rarefaction effect, and have been studied widely in the field of rarefied gas dynamics. As fabrication technology have been developed, the rarefaction effect can appear easily, even under an atmospheric pressure. The rarefaction effect is reported for rectangular crosssection [2-4] and trapezoidal crosssection [5,6] as well as circular crosssection. Though D-shape crosssectional microchannels have been used in several microdevices [7], fundamental study of fluid dynamics is lacked even in continuum flow region.

In this study, we carried out experiments and simulations for pressure-driven gaseous flows through a microchannel and studied characteristics of microflows. Macroscopic flow resistance of D-shape crosssectional microchannel was obtained and then the rarefaction effect was investigated in detail.

EXPERIMENTS

We obtained flowrate of pressure-driven flow through various microchannel, in experiments. Since it is hard to measure directly flowrate through a microchannel by using commercial flowmeter, flowrate is determined by pressure changes of inlet and outlet of a microchannel. Schematic diagram of experimental setup and a microchip with a microchannel is shown in **Fig. 1**. Pressure chambers were placed at microchannel inlet and outlet. These two chambers are closed after adjusting pressures to respective initial pressures. Time evolutions of pressures of those closed chambers were measured and flowrate was determined. The microchip is typical chip made of pyrex glass. In the chip microchannels are etched isotropically. Therefore their crosssections are D-shape. The microchannels used in this experiment are shown in **Table. 2**.

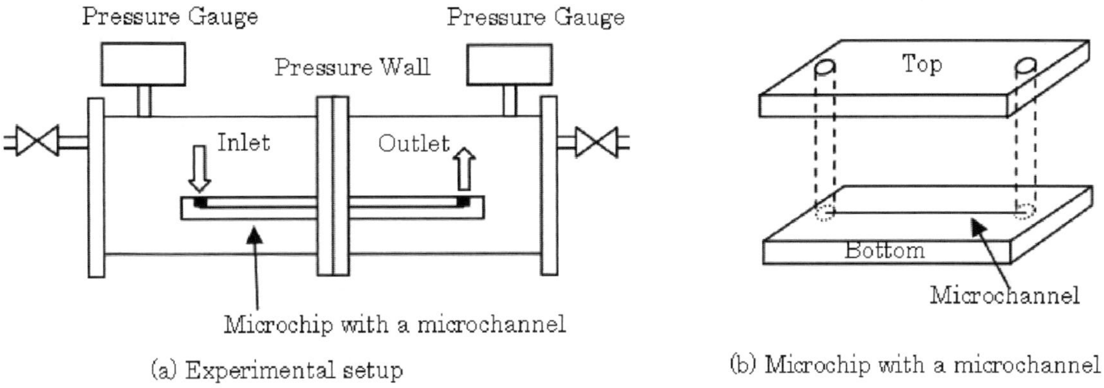

Fig. 1 Schematic diagram of experimental setup and a microchip with a microchannel.

Table 2 Crosssection of microchannels used in this study.

Type	A	B	C
Crosssection (SEM)	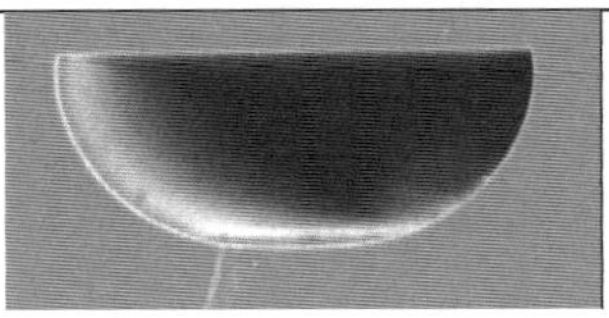	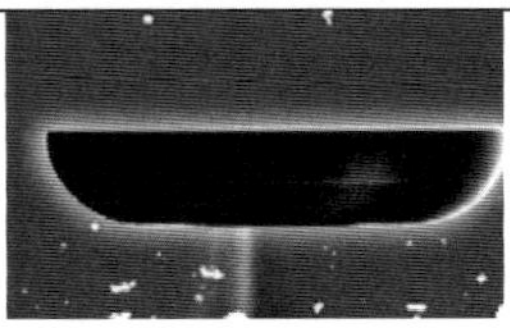	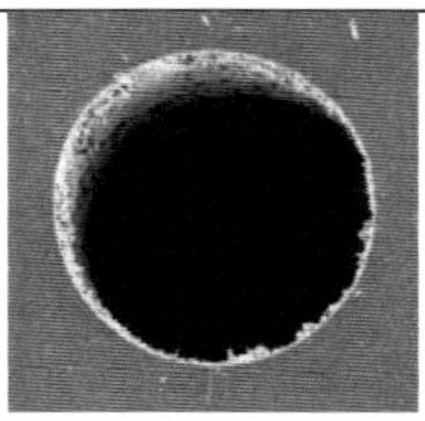
Size	114μm×50μm (×60mm)	29μm×6μm (×60mm)	ϕ 500μm (×60mm)

SIMULATIONS

Fluid through a microchannel is almost laminar and computer simulation is effective tool. Simulations for typical imcompressive fluid in a microchannel of D-shaped crosssection (**Fig. 2**) are carried out. Simulated results are arranged by using the equivalent diameter. We define the equivalent diameter as diameter of circle whose flow resistance equals D-shape.

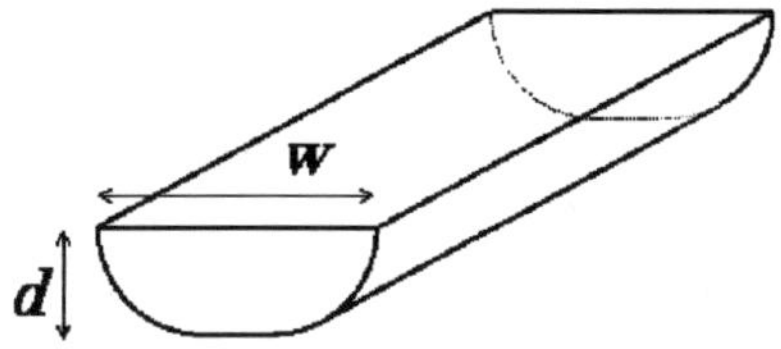

Fig. 2 Simulated microchannel of D-shaped crosssection.

RESULTS AND DISCUSSION

A key non-dimensional parameter for gas micro flows is the Knudsen number Kn, which is defined as the ratio of the mean free path over a characteristic geometry length. Rarefaction effects become more important as the Knudsen number increases. A classification of the different flow regimes is shown in **Fig. 3**.

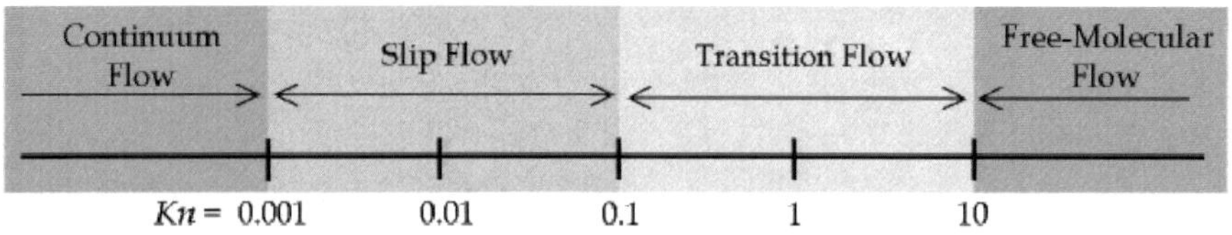

Fig. 3 A classification of the different flow regimes.

Experimental results are compared with theoretical values of the continuum flow regime (**Fig. 4**) and the free-molecular flow regime (**Fig. 5**). Viscosity of gas is determined by using kinetic gas theory concepts. It is confirmed that the determination is good.

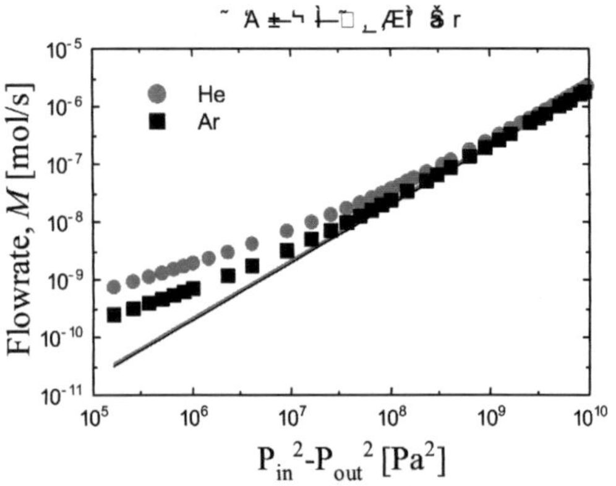

Fig. 4 Flowrate through a Type-A microchannel (the continuum flow regime).

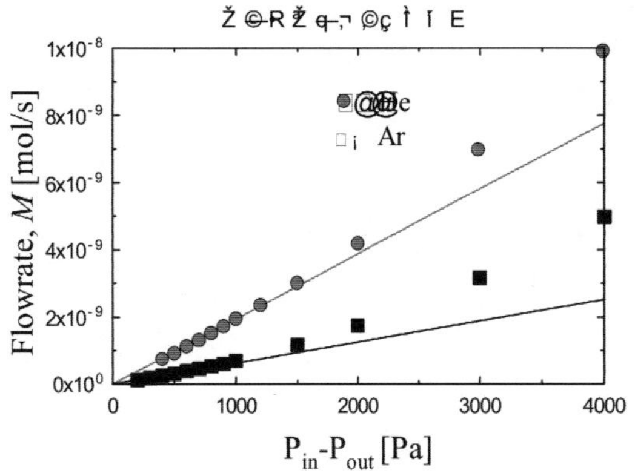

Fig. 5 Flowrate through a Type-A microchannel (the free-molecular flow regime).

In the continuum flow regime, computer simulation is effective. The basic characteristics of D-shape crosssectional microchannels are obtained by using computer simulations. Simulated results are shown in **Fig. 6**, where the equivalent diameter denoted by d_c divided by d is plotted as a function of aspect ratio. In **Fig. 6**, experimental results are also plotted and agree well with simulated results.

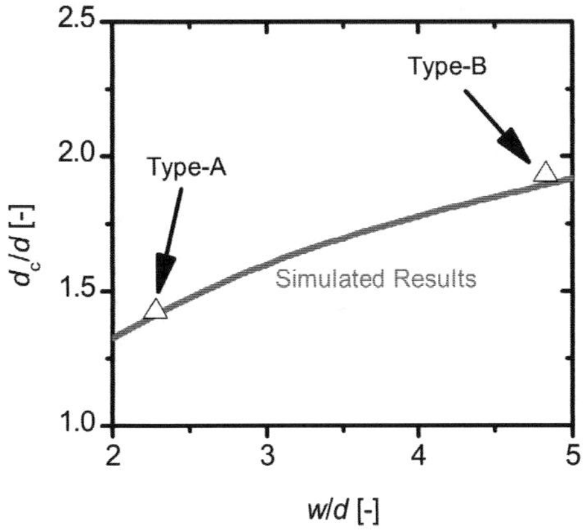

Fig. 6 The equivalent diameter of D-shape crosssectional microchannels.

In the slip flow regime and the transition flow regime, flowrates are corrected by using Kn. In this study, the correction is derived from experimental results. Experimental results are shown in **Fig. 7**, where it is observed that there is no difference in He and Ar. The accommodation coefficient σ used for the correction is determined as 0.8.

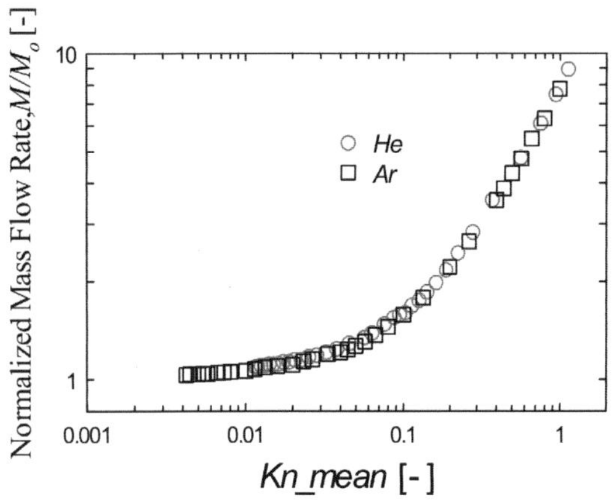

Fig. 7 Normalized flowrate as a function of Kn (Type-A).

A characteristic geometry length in Kn is determined by the comparison of results of Type-A, Type-B, and Type-C microchannel. For Type-A microchannel the equivalent diameter is good, and for Type-B microchannel the shorter length is good.

CONCLUSION

We carried out experiments and simulations for pressure-driven gaseous flows through microchannels having various kinds of crosssection and studied characteristics of microflows. Equivalent diameters of D-shaped crosssectional microchannels are obtained. It is confirmed that simulated results agree well with experimental results in continuum flow regime. We obtain the accommodation coefficient of 0.8, which is used in the correction for the slip flow regime and the transition flow regime. A characteristic geometry length used in Kn is determined at two D-shaped crosssectional microchannels.

ACKNOWLEDGEMENTS

We thank Professor C. Kuroda of Tokyo Institute of Technology, Japan. This work was supported by the New Energy and Industrial Technology Development Organization (NEDO), Japan, and performed in Department of Chemical Engineering, Tokyo Institute of Technology, Japan.

REFERENCES

[1] Choondal B. Sobhan and Suresh V. Garimella, A Comparative Analysis of Studies on Heat Transfer and Fluid Flow in Microchannels, Microscale Thermophysical Engineering, vol. 5, pp. 293-311, 2001.

[2] G. L. Morini, M. Spiga, Slip Flow in Rectangular Microtubes, Microscale Thermophys Eng, vol. 2, pp. 273-282, 1998.
[3] Cécile Aubert Stéphane Colin, High-order Boundary Conditions for Gaseous Flows in Rectangular Microducts, Microscale Thermophys Eng, vol. 5, pp. 41-54, 2001.
[4] S-S Hsieh, H-H Tsai, C-Y Lin, C-F Huang, C-M Chien, Gas Flow in a Long Microchannel, Int J Heat Mass Transfer, vol. 47, pp. 3877-3887, 2004.
[5] H. Y. Wu, and Ping Cheng, Friction Factors in Smooth Trapezoidal Silicon Microchannels with Different Aspect Ratios, Int J Heat Mass Transfer, vol. 46, pp. 2519-2525, 2003.
[6] Qu Weilin, Gh. Mohiuddin Mala, and Li Dongqing, Pressure-Driven Water Flows in Trapezoidal Silicon Microchannels, Int J Heat Mass Transfer, vol. 43, pp. 353-364, 2003.
[7] Kiichi Sato, Akihide Hibara, Manabu Tokeshi, Hideaki Hisamotoa, Takehiko Kitamori, Microchip-based chemical and biochemical analysis systems, Advanced Drug Delivery Reviews, vol. 55, pp. 379-391, 2003.

Optimal design approach for microreactors with desired temperature, concentration and residence time distributions

Osamu Tonomura[1], Manabu Kano[1], Shinji Hasebe[1], Masaru Noda[2]

[1] *Department of Chemical Engineering, Kyoto University, Nishikyo-ku, Kyoto 615-8510, Japan*
[2] *Nara Institute of Science and Technology, Ikoma, Nara 630-0192, Nara*
E-mail: tonomura@cheme.kyoto-u.ac.jp (O. Tonomura)

Prepared for Presentation at the 2005 AIChE Spring National Meeting
IMRET 8 – 8th International Conference on Microreaction Technology
April 10-14, Atlanta, GA, USA
Paper No. [132u] – Poster Session

Abstract

In microreactors, enhanced heat transfer can handle highly exothermic reactions due to the removal of the generated heat and the suppression of hot-spot formation. Such well-known characteristics of microreactors have been reported by many researchers. However, in most of the conventional design approaches, the cross-sectional area of each microchannel was assumed to be constant, and the channel design under that condition sometimes caused the unnecessary increase of pressure drop. In this research, we propose a systematic thermo-fluid design approach for a plate-fin microreactor where the size and shape are optimized so as to realize the desired temperature and concentration profiles and residence time distributions. The proposed approach is applied to a parallel competing reaction system. It is revealed that the optimally designed microchannel shape with varying width along the axial position is very effective in the realization of the desired reaction conditions. It can be concluded that the proposed design strategy and method have potential for being widely applied to the design problems of various types of microreactors and reactions.

Keywords: Microreactors; Thermo-fluid design approach; Optimal shape design

Introduction

Recently, extensive studies have been carried out on micro chemical process technologies. In particular, analysis of chemical reactions and transport phenomena in micro-space has been performed energetically. However, most microreactors are designed and fabricated by trial and error -- there is no systematic procedure for developing microreactors. Therefore, it takes much time and effort to reach the optimal design. Unless this situation is improved, the design period cannot be shortened. Recent advances in computational fluid dynamics (CFD) enable us to know both flow and temperature distribution in a microreactor precisely without conducting any experiments. Tonomura et al. (2004) proposed the CFD-based optimization method for design of plate-fin microdevices. However, it is not practical to apply CFD simulations directly to the optimal design problem of microreactors, since they require too much computational time and effort. In this research, we propose a new systematic approach based on a simplified thermo-fluid model to efficiently design microreactors with constraints related to flow distribution, temperature profile, pressure drop, product yield and size boundary conditions. The validity of the proposed design method is assessed through a case study.

Proposed optimal design approach[2,3]

The proposed design approach mainly consists of two stages (flowchart, Figure 1). In the first stage, the optimal microchannel shape is derived by using a thermal compartment model: the "thermal design stage." In the second stage, the optimal manifold shape is obtained by using a pressure drop compartment model: the "fluid design stage." In the thermal design stage, a microreactor having parallel channels is divided into an adequate number of thermal compartments, each of which is modeled as a lumped parameter system. The heat transfer among the compartments is also modeled. Microchannel width is optimized as a function of the longitudinal position by taking into account the constraints related to thermal balance, pressure drop, reaction yield, temperature, and reactor dimensions. By changing the channel width, that is, by controlling residence time of fluid at each longitudinal position, a desired temperature distribution is realized. In the fluid design stage, a pressure drop compartment model is used to estimate the pressure distribution over the microreactor, including the inlet and outlet manifolds. In the pressure drop compartment model, the relation between the inlet and outlet pressures is described as a function of the design parameters and flow velocity. Through the fluid design, the shapes of the inlet and outlet manifolds and the number of parallel microchannels are determined so as to optimize a given performance index such as the minimization of the total residence time in the microreactor. Many constraints related to the pressure and mass balances, flow uniformity, pressure drop, total flow rate and reactor dimensions are considered in this design stage.

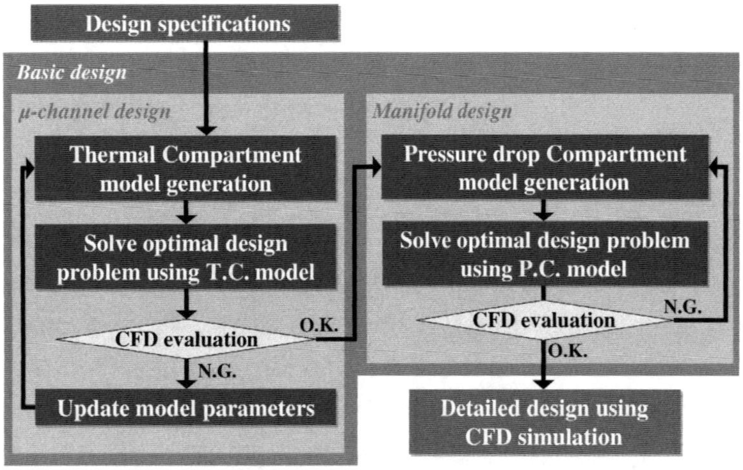

Figure 1. Proposed design approach.

Case Study

The proposed approach is applied to a microreactor design problem with the desired temperature and residence time distributions. Denbigh reactions [4] as shown in Figure 2 are taken up in this case study and have wide applications to a host of real reacting systems. Y is the desired product, X is an intermediate and P and Q are the products of side reactions. These reaction rate equations are all of first order and do not involve complex mathematics. The temperature-dependent rate constant k_i in each reaction is presented by Arrhenius' law:

$$k_i = k_i^* \exp\left(-E_i/(RT)\right)$$

where k_i^* is called the frequency or pre-exponential factor and E_i is called the activation energy of the reaction. Temperature dependency of Denbigh reactions is determined by the activation energy. A high temperature favors the reaction of higher activation energy, and a low temperature favors the reaction of lower activation energy. If $E_1 > E_2$, $E_3 > E_4$, the ideal temperature profile is always an isothermal one located on the level of maximum allowable temperature $T_{\max}$. This is shown by the sketch (I) in Figure 2.

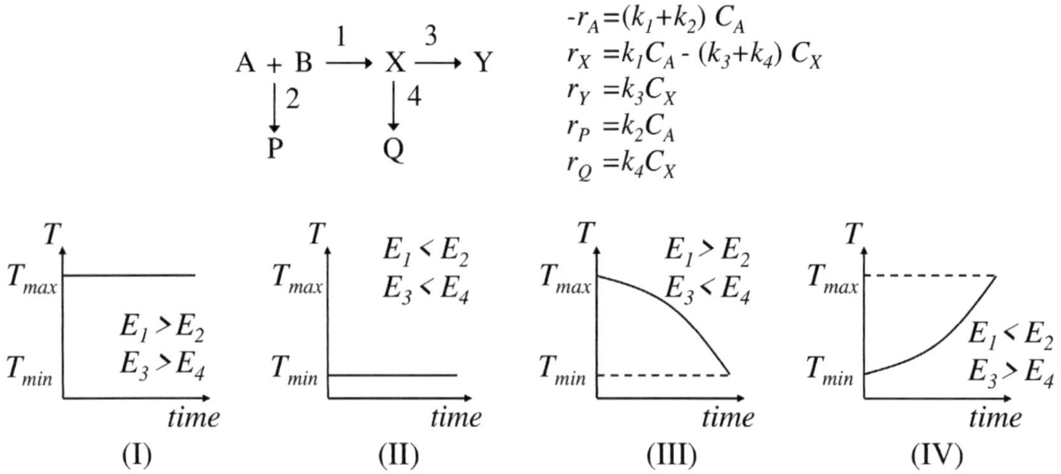

Figure 2. Temperature dependency of Denbigh reactions.

The objective of the microreactor design in this case study is to realize uniform temperature and residence time distributions, which insure the maximum yield of the desired product Y. Table 1 shows the main design conditions and assumptions in this case study. The reactants and products have the same physical properties as water (298 K) has. The design region is defined as a rectangular glass whose dimension is 1000 μm in thickness and 1000 μm in width. To calculate the temperature distribution, the design region is divided into 100 inner thermal compartments representing flow in the microchannel and 100 outer thermal compartments representing the wall. The objective function of the thermal design problem is minimizing the integrated temperature difference along the microchannel.

Figure 3 illustrates the thermal design results. The solid line in the left graph shows the optimal channel width along the microchannel and dotted line shows the channel width which is optimized as constant value. As shown in the right figure, the optimal shape of microchannel is effective to keep temperature distribution uniform along the microchannel. On the other hand, the reaction temperature in the conventional straight channel decreases gradually. In comparison to the optimal design result, the yields of P and Q in the conventional design increase by 21 and 6 %, and the amount of feed in the conventional design decrease by 6 %. In addition to the thermal design problem, the fluid design problem is solved to determine the optimal manifold shape, which can realize the flow equipartition in all parallel microchannels and avoid deterioration in the reactor performance.

Table 1. Design conditions and assumptions.

Objective	Minimize $\int (T(x)-T_{\max})dx$
Optimization variables	Channel width $W(x)$, Channel length L, Initial concentration of A C_{A0}, Inlet temp. T_{in}
Subject to	E_1=10, E_2=7, E_3=20, E_4=2 [kJ/mol], k_1^*=1, k_2^*=0.1, k_3^*=10, k_4^*=0.15 [1/s], $C_{X0}=C_{Y0}=C_{P0}=C_{Q0}$= 0 [mol/m^3], v_{in}= 1 [mm/s] T_{max}=700, T_{min}=400, $T_{surrounding}$=600 (Const.) [K], Reaction heat $\Delta H_{r,i}$=1000 [kJ/mol], Yield of Y $Y_Y \geqq 0.6$, Size limits

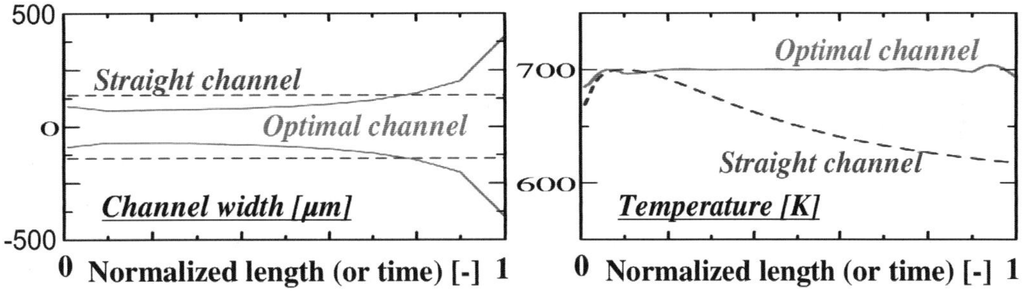

Figure 3. Thermal design results.

Conclusions

A new optimal shape design method based on simplified thermo-fluid models is proposed to efficiently design microreactors under constraints related to temperature and residence time distributions, concentration gradient, pressure drop, and production specifications. The proposed shape design method is applied to a case study, and it is revealed that the optimally designed microchannels with varying width are effective for realizing a desired reaction conditions.

Acknowledgments

This research was partially supported by the New Energy and Industrial Technology Development Organization (NEDO), and the Project of Micro-Chemical Technology for Production, Analysis and Measurement Systems.

References

[1] O. Tonomura, T. Shotaro, M. Noda, M. Kano, and S. Hasebe (2004) *Chem. Eng. J.* **101**, 397-402.

[2] O.Tonomura, M. Noda, M. Kano, and S. Hasebe (2004) *the 10th Asian Pacific Confederation of Chemical Engineering (APCChE), 17-21 October, Kitakyushu, Japan*, 4B-07.

[3] M. Noda, O.Tonomura, M. Kano, and S. Hasebe (2004) *the 10th Asian Pacific Confederation of Chemical Engineering (APCChE), 17-21 October, Kitakyushu, Japan*, 4B-08.

[4] A. Burghardt and J. Skrzypek (1974) *Chem. Eng. Sci.* **29**, 1311-1315.

Influence of Solubility/Mixing Effects on Chemical Reactions in Microfluidic Systems

Takeshi Honda,[1] Masaya Miyazaki,[1,2] Kenichi Yamashita,[1] Yoshiko Yamaguchi,[1] Masato Uehara,[1] Hiroyuki Nakamura,[1] and Hideaki Maeda*[1,2]*

[1]*Micro-Space Chemistry Laboratory, National Institute of Advanced Industrial Science and Technology (AIST), Tosu, Saga 841-0052, Japan*
[2]*Department of Molecular and Materials Sciences, Interdisciplinary Graduate School of Engineering Sciences, Kyushu University, Kasuga, Fukuoka 816–8580, Japan*

Abstract

Microchannel reactors have continued to attract attention in the fields of analytical and synthetic chemistry. An important feature of microchannel systems is their superior controllability of fluidic systems, which cannot be created in the batchwise reaction system with larger reaction apparatus. We have been demonstrated that secondary flow occurs at the curving structure of microchannel. Such secondary flow disrupts interfacial structure and enhances mixing of two miscible solutions. In this study, we examined the effects of such a microfluidic system on chemical reactions using amino acid substitution as a model reaction. Substitution of phenylalanine in a microreaction system using separate solutions was more efficient than in a batchwise reaction or in microchannel reaction using a homogeneous solution in DMF-water system. In acetnitrile-H_2O system, such enhancement was diminished. Substitution of other amino acids showed that this enhancement is caused by localization of hydrophobic amino acids at the organic solution–H_2O interface. Using the rapid Michael addition reaction of the SH group of cysteine to a maleimide group, we also demonstrated that such a reaction involving hydrophilic amino acids was diminished in this microreaction system. These results show that the microreaction system is a novel apparatus for regulating chemical reactions, depending on the structure of the reactant molecules, by controlling mixing of two different solutions.

Introduction

Microchannel reactors have attracted attention in synthetic and analytical chemistry fields.[1–4] One important feature of microchannel systems is their superior controllability of fluidic systems, which cannot be created in larger reaction apparatus.[5–7] This feature might be useful to control solvent effects on chemical reactions in a novel mode, which cannot be achieved in a bulk reaction.[8,9] Here we show that the chemical reaction can be regulated differentially in a microchannel by controlling formation of an interface of two different miscible solvents using amino acid substitution as a model reaction. Although the reaction rate did not differ in bulk reaction apparatus, dramatic acceleration of the reaction rate was achieved within the microchannel reaction system by accumulation of molecules at the interface, depending on the structure of amino acid side chains. This result demonstrates the possibility of microreactor as a new class of efficient reaction device, not simply depending on mixing.

In a microchannel system, fluid forms a laminar flow in straight portions; therefore the mixing of different solutions depends on the residence time within the microchannel. In such a simple system, the reaction rate correlates with the mixing of two solutions. We have found that inertial force induces the secondary flow at the turn structure.[10,11] Such secondary flow disrupts the solution interface and enhances mixing of the two solutions.[11] Therefore, the reaction rate might be affected by enhancement of secondary flow. We performed a simple reaction of phenylalanine with an active ester using a microreactor, as shown in Figure 1, and Michael addition of SH group to maleimide moiety (Fig. 2), to examine this hypothesis.

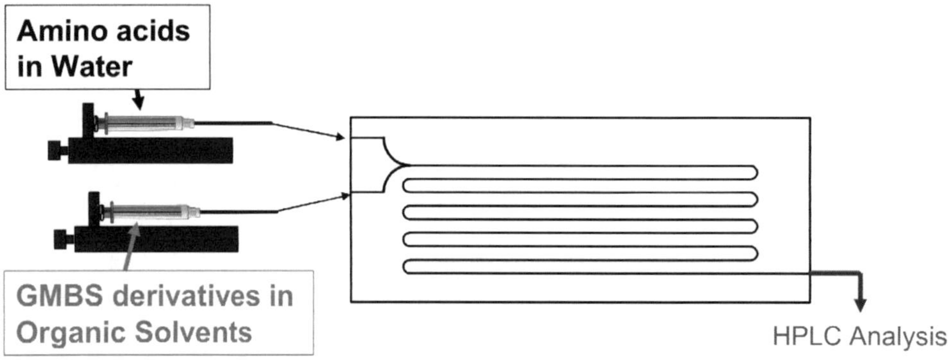

Fig. 1. The microchannel reaction system used in this study. Microchannel was fabricated as described in "Experimental" section.

Results and Discussion

First, we performed a simple condensation reaction of amino group with active ester using a DMF solution of N-(4-maleimidobutyloxy)succinimide (GMBS) and an aqueous solution of phenylalanine as a model reaction. According to previous results, the reaction might proceed even at a high flow rate, which yielded shorter reaction time by enhancing the mixture of the two solutions by the secondary flow. [11] Results are shown in Figure 2a. The reaction yield exceeded 90% at a lower flow rate. This result suggests that reaction efficiency was dramatically improved over that of bulk reaction, which required more than one day for completion by microchannel reaction. However, the reaction yield decreased sharply with a raised flow rate; no reaction occurred at high flow rates. This result is counterintuitive. Even at high flow rates, mixing of two solutions was completed within the microchannel, and therefore should give some product. To address whether mixing of two solutions affects the reaction yield or not, we simply loaded premixed solution to the microreactor. As shown in Figure 2b, the reaction did not proceed. Therefore, mixing of two solutions did not affect the enhancement of reaction in the microchannel.

Scheme 1

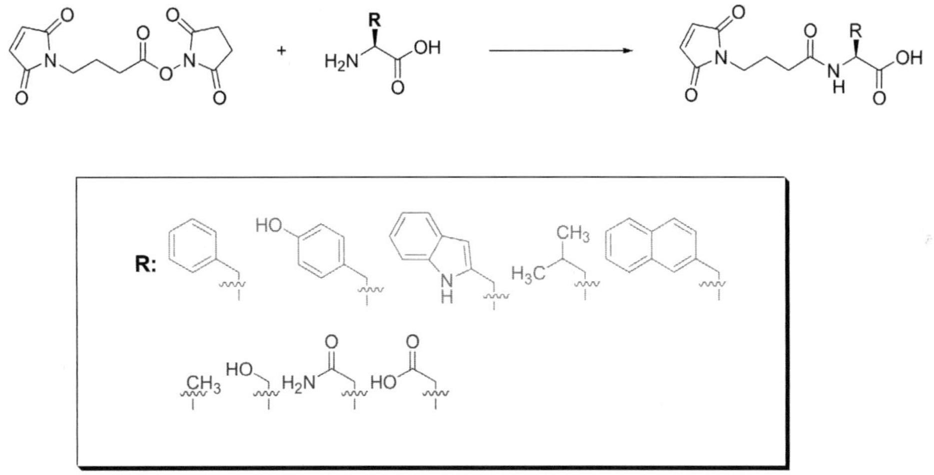

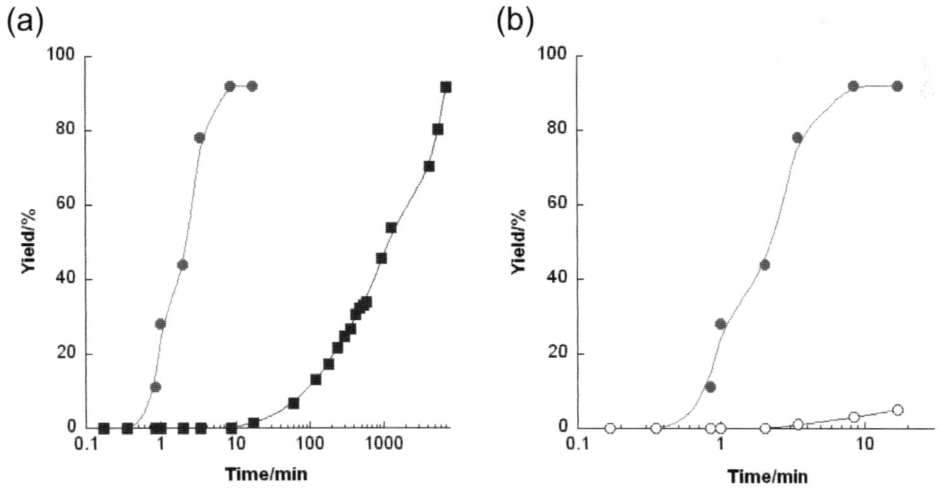

Fig. 2. Reaction of phenylalanine with GMBS in DMF. (a) Comparison of bulk (■) and microchannel (●) reactions. (b) Reaction yield in the microchannel using separate solutions (●) and premixed solution (○).

What enhances reaction efficiency in a microchannel? We have assumed that formation of an interface between two solutions is a key factor. Amino acids tethering hydrophobic side chains, such as phenylalanine, showed higher solubility in the mixture of DMF and water than in water alone.[12] These amino acids might accumulate or absorb to the interface of DMF and water in a microchannel. Local concentration of such amino acids at the interface becomes much higher; therefore the reaction efficiency improved dramatically at lower flow rates that maintain the interface in the microchannel. At high flow rates, although the interface was maintained in straight portions, the residence time was not sufficient to accumulate the molecules. The secondary flow at the curve disrupts the interface. Consequently, such enhancement did not occur.

We performed the following experiments to verify this inference. First, we performed the same reaction using acetonitrile as a solvent. In the presence of acetonitrile, phenylalanine was less soluble than in the case of DMF. Therefore, one would assume that the effect would be weakened. Results are shown in Figure 3. The reaction in the bulk system showed almost the same reaction kinetics, meaning that changing organic solvent (Figure 3a) induced no solvent effect. On the other hand, the reaction yield of an acetonitrile-water system was lower than that in DMF-water system in microchannel reaction (Figure 3b). Because the solvent effect was not observed in the bulk system, a difference of reaction yield in microchannel should result from the difference in accumulation of amino acids at the interface.

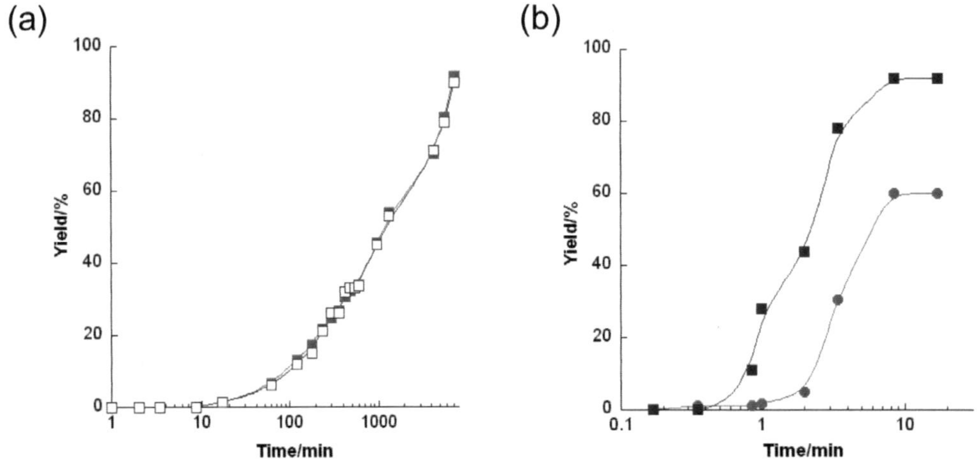

Fig. 3. Reaction of phenylalanine with GMBS in acetonitrile. (a) Comparison of yield of bulk reactions (□) with that in DMF (■). (b) Reaction yield in the microchannel using acetonitrile (●) and DMF (■) as the solvent.

Next, we examined the effect of a side chain of amino acids: Ala, Ser, Asn, Asp, Leu,

Tyr, Trp, and Naphtylalanine (Nal). The reaction rate was not different for each amino acid in a batchwise reaction (data not shown). The reaction in the microchannel was conducted as in the case of Phe. Results by amino acid having bulky side chains, Leu, Tyr, Trp, and Nal showed reaction with GMBS in the microchannel in order of hydrophobicity of its side chains. On the other hand, amino acids with smaller or hydrophilic side chains (Ala, Ser, Asn, and Asp) did not react in the microchannel. Thus, the structure of amino acids seemed to strongly influence the reaction rate in a microfluidic system.

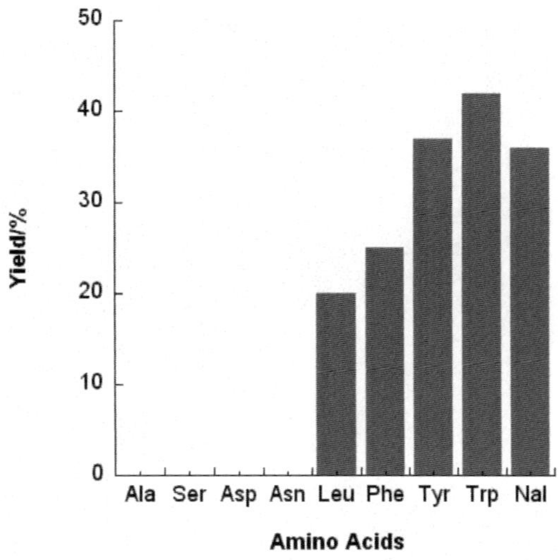

Fig. 4. Effect of amino acid side chain structure on reaction with GMBS. Each bar shows reaction yields at a flow rate of 15 μl/min.

We also examined reaction of maleimide with thiol group of cystein using DMF and water system (Scheme 2) to support this speculation. This reaction proceeds rapidly, even in a batchwise system. Free cystein shows higher solubility in H_2O alone than with presence of DMF; also, it might localize to the area that does not contain DMF in a microreactor. Therefore, the reaction should give a lower yield when the active mixing effect is weakened at a lowered flow rate, although reaction time would increase. The reaction was performed using H_2O solution of cystein and DMF solution of *N*-(4-maleimidobutyl) phenylalanine (Fig. 5). At high flow rates, the reaction proceeds rapidly, as in the batchwise system, in all cases. As expected, the reaction yield dramatically decreased at slower flow rates in the case of free cystein. Thus, it can be said that hydrophobicity of molecules enables both improved accessibility of amino acids to DMF-water mixture and higher reaction yields. Amino acids with hydrophilic side chain do not prefer presence of DMF and localize to the area not containing DMF. Therefore, reactivity of such amino acids was decreased in the microfluidic system and gave lower yields than that of hydrophobic molecule.

Scheme 2

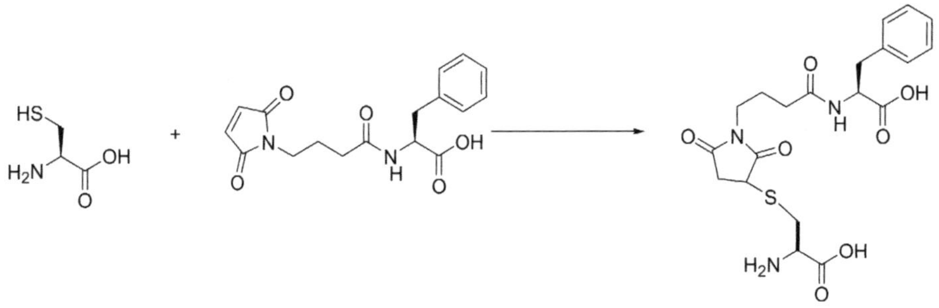

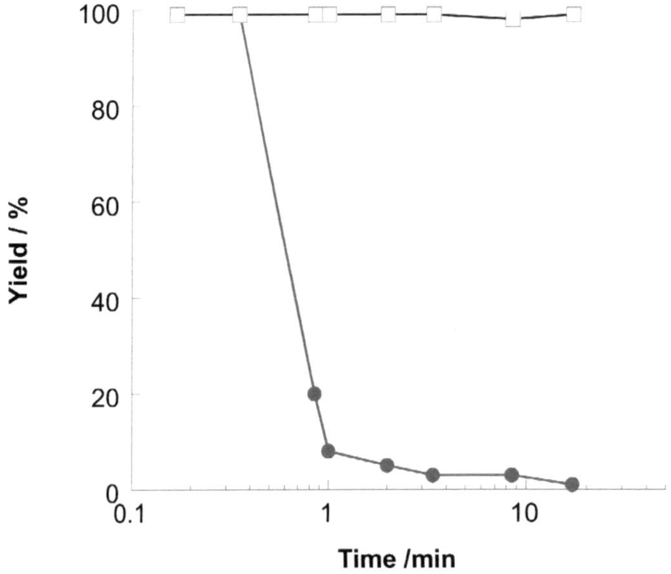

Fig. 5. Reaction of cystein with N-(4-maleimidobutyloxy) phenylalanine in batchwise (□) and microreactor (●). The reaction was performed in the same microreactor as in Fig. 2.

These results support our inference. Hydrophobic amino acids accumulate at the interface in a microchannel. Such amino acids yielded improved reaction efficiency when the interface was maintained. Disruption of the interface by secondary flow inhibits the accumulation. Thereby, the reaction efficiency was not improved. On the other hand, hydrophilic amino acids did not absorb to the interface and did not demonstrate improved reaction efficiency. These amino acids might localize in areas that do not contain organic solvents, rather than at the interface. Moreover, aside from organic solvent-water systems, this method might be applicable to other differential solvent systems such as solvents with different polarity, pH, and ionic strength. Although detailed mechanisms require verification, it should be possible to design an efficient reaction system by choosing appropriate differential solvents for each reactant.

In summary, we demonstrated that reactivity of amino acids in a substitution reaction can be regulated by microfluidics. The enhancement did not depend on the mixing. Formation of a differential interface is the key. A molecule that can accumulate or absorb to the interface yields dramatically improved reaction efficiency. These findings might be useful for further development of efficient microchannel reaction systems.

Experimental Section

General. Ceramic plates were obtained from Mitsui Mining Material Co. (Tokyo, Japan). Other materials for fabrication of microreactors were commercially available. Amino acid derivatives were purchased from Calbiochem-Novabiochem AG (Laufelfingen, Switzerland) or Wako Pure Chemical Inds., Ltd. (Osaka, Japan) and reagents for organic reactions were purchased from Tokyo Kasei Kogyo Co. (Tokyo, Japan) or Sigma-Aldrich (St. Louis, MO). Water was freshly prepared by Milli-Q (Millipore, Medford, MA), and DMF was distilled prior to use. The entire solution was filtered by a PTFE filter (0.45 μm) prior to use, and charged into 1 ml Hamilton Gastight Syringes (Hamilton Co., Reno, NV). The solutions were supplied to the microreactor using a KDS 230 syringe pump with a parallel syringe holder (KD Scientific Inc., New Hope, PA). Thereby, two solutions were charged at the equal volume per time and flow rates indicated in figures were calculated from total of two solutions. HPLC analysis was performed using Wakosil C18AR column (3.0 × 250 mm) using linear gradient of 0.05% TFA-acetonitrile over 60 min at a flow rate of 0.5 ml/min at 30°C. Each compound was confirmed with LC-MS (ABI Mariner ESI-TOF system).

Microreactor Fabrication. A square microchannel structure (200 μm wide × 200 μm deep × 50 cm long) was made mechanically on a ceramic plate (3 cm × 7 cm, 1 mm tick) using a Robodrill α-T21iDs (Fanuc Ltd., Yamanashi, Japan) equipped with a flat-end microdrill (Hitachi Koki Co., Ltd., Tokyo, Japan). Microchannel structure was confirmed by laser microscope analysis using a Keyence VK-8510 microscope (Keyence Co., Osaka, Japan). The ceramic plate was covered with a glass plate by heating; then inlet and outlet tubes were connected. The microreactor structure is shown in Figure 1.

Reaction of Amino Acids with GMBS in the microchannel. A 1 mM solution of GMBS in DMF or acetonitrile and a 1 mM aqueous solution of amino acids and triethylamine (1 equivalent to amino acid) were prepared separately. The solutions were filtered and degassed prior to use. Each solution was loaded into the microreactor by syringe pumping, as shown in Figure 1. The eluted solution was subjected directly to HPLC analysis. Reaction yields were calculated from the peak area calibrated with standard compound prepared in large scale. For the reaction using homogeneous solution, each solution was mixed and charged from both inlets. Figures 2, 3 and 4 show results.

Reaction of Cysteine derivatives with N-(4-Maleimidobutyloxy)phenylalanine. The maleimide derivative, *N*-(4-maleimidobutyloxy)phenylalanine, was prepared by reaction of GMBS with phenylalanine, and purified by preparative HPLC. A 1 mM solution of *N*-(4-maleimidobutyloxy)phenylalanine in DMF and a 1 mM aqueous solution of cysteine derivatives and triethylamine were prepared separately. The solutions were filtered and degassed prior to use. Each solution was charged into microreactor by syringe pumping, as shown in Fig. 1. The reaction was terminated by adding ammonium hydroxide solution at the end of channel; the recovered solution was subjected to HPLC analysis. The reaction yields were calculated from the peak area calibrated with standard compound prepared in large scale. Results are shown in Fig. 5.

References

[1] S. J. Hasewell, R. J. Middleton, B. O'Sullivan, V. Skelton, P. Watts, P. Styring, *Chem. Commun*. **2001**, 391–398.

[2] S. Taghavi-Moghadam, A. Kleemann, K. G. Golbig, *Org. Process Res. & Dev.* 2001, **5**, 652–658.

[3] M. Krishnan, V. Namasivayam, R. Lin, R. Pal, M. A. Burns, *Curr. Opin. Biotechnol*. 2001, **12**, 92–98.

[4] T. Chovan, A. Guttman, *Trends Biotechno*l. 2002, **20**, 116–122.

[5] N. L. Jeon, S. K. W. Dertinger, D. T. Chiu, I. S. Choi, A. D. Stroock, G. M. Whitesides, *Langmuir*, 2000, **16**, 8311–8316.

[6] A. D. Stroock, S. K. W. Dertinger, A. Ajdari, I. Mezic, H. A. Stone, G. M. Whitesides, *Science*, 2002, **295**, 647–651.

[7] B. Zhao, N. O. L. Viernes, J. S. Moore, D. J. Beebe, *J. Am. Chem. Soc.* 2002, **124**, 5284–5285.

[8] P. J. A. Kenis, R. F. Ismagilov, G. M. Whitesides, *Science*, 1999, **285**, 83–85.

[9] R. Ferrigno, A. D. Stroock, T. D. Clark, M. Mayer, G. M. Whitesides, *J. Am. Chem. Soc.* 2002, **124**, 12930–12931.

[10] H. Kawazumi, A. Tashiro, K. Ogino, H. Maeda, *Lab. Chip* 2002, **2**, 8–10.

[11] Y. Yamaguchi, F. Takagi, K. Yamashita, H. Maeda, K. Sotowa, K. Kusakabe, Y. Yamasaki, S. Morooka, *AIChE J.*, 2004, **50,** 1530-1535.

Partial chemical modification of a microchannel and stabilization of water-oil phase separation

Yoshiko Yamaguchi[1], Kazuya Ogino[2], Fuminori Takagi[2], Takeshi Honda[1], Kenichi Yamashita[1], Masaya Miyazaki[1,3], Hiroyuki Nakamura[1] and Hideaki Maeda[1,3]

[1]National Institute of Advanced Industrial Science and Technology, Saga, Japan
[2]Graduate School of Science and Engineering, Saga University, Saga, Japan
[3]Graduate School of Engineering Sciences, Kyushu University, Fukuoka, Japan

Abstract

A microchannel which enables water-oil phase separation was developed using silicon and glass plates. Si surface was hydrophobized using alkanethiol solution. Stabilization of water-oil phase liquid and separation performance using this microdevice were evaluated by experiments. It was found that hexane phase can be separated from aqueous phase when the fluid velocity is relatively small. This device is useful for effective extraction of chemical compounds through the water-oil interface.

Introduction

Microreactor systems, which have great potential to realize large-item-small-scale production and to lessen the impact that chemical process has on the environment, have attracted much attention in recent years. The large surface-per-volume ratio of interface area between fluids in microchannels gives an advantage to some processes such as product extraction and reaction through the liquid/liquid interfaces. Some examples of highly efficient extraction and reaction system have been reported (Hisamoto et al., 2001; Zhao et al., 2003; Maruyama et al., 2003). In a separation method using organic and aqueous phases, stability of the liquid/liquid interface is an important key. One way to achieve interface stability is to control hydrophilic and hydrophobic character of the microchannel inner wall. Partial chemical modification of the glass surface by alkyl compounds is one of the major methods in making the wall surface hydrophobic (Hibara, et al., 2002, Maruyama et al., 2003). However, fluidic behavior of the alkyl compound solution for partial modification should be severely controlled in the microchannel, in order to modify the proper region. Moreover, bonding of the top and bottom glass plates is generally difficult.

In this study, a silicon-glass microreactor was devised which can provide effective stabilization of the interface and at the same time address the problem of bonding difficulty in a glass-glass microreactor. Unlike many other microreactors with a vertical interface, our reactor provides a horizontal interface. Following bonding a glass plate above and a silicon plate below, surface modification is applied to make the silicon plate hydrophobic. This microreactor offers several advantages. First, a horizontal interface type microreactor can easily supply wide interface area, in contrast, it is generally difficult to create a wide vertical interface area because of the difficulty in fabricating deep channels by conventional methods. Second, the methods used in bonding and partial modification of silicon-glass plate are simple. Following anodic bonding of the glass plate and the silicon plate constituting Au-coated microchannel, we simply inject alkanethiol solution into the microchannel. Self-assembled monolayer (SAM) of alkanethiol is produced only on the surface of Au film. This easy method enables partial modification of microchannels without special fluid control. In this way, a horizontal interface-type microreactor was developed and separation performance of water/hexane liquid was evaluated.

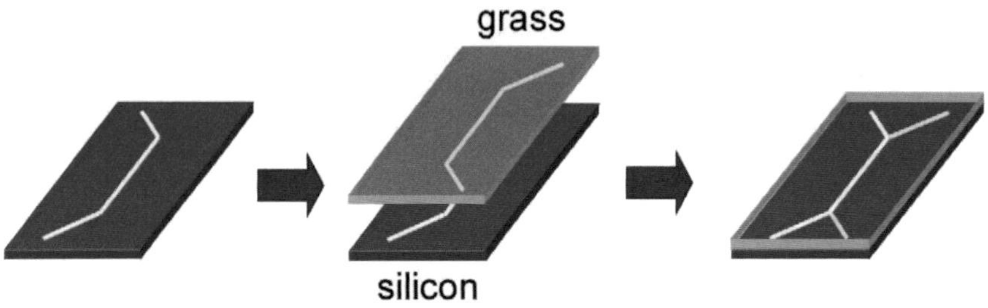

Fig. 1 A partially modified silicon-glass microreactor.

Experiments

The width and length of a Pyrex glass and a silicon plate used were 3 cm and 7 cm, respectively. A microchannel was fabricated mechanically on a silicon plate, after application of a photoresist on the silicon plate in order to allow coating only of the inner wall of the microchannel with Au film by magnetron sputtering. The width and depth of the microchannel were 200 μm and 100 μm, respectively. The same microchannel was also fabricated on a Pyrex glass plate. After evaporatation of Ti, the silicon plate was coated with Au by magnetron sputtering. The photoresist on the silicon plate was removed, followed by anodic bonding to the glass plate. Alkanethiol in ethanol solution of (1:9 vol) was injected into the silicon-glass microchannel. In order to test the hydrophobicity, other silicon flat plates with alkyl group of different lengths were prepared. Contact angles were measured for the nine kinds of alkanethiol SAM plates.

Separation performance of water and hexane was evaluated using the fabricated microreactor as follows. Pure water and hexane solution of 10 mmol/L L-Phe-OBzl were injected into the microreactor. Whether the liquid collected from the hydrophilic outlet involves L-Phe-OBzl was checked by measuring the absorbance at 250 nm.

Results and Discussion

Fig. 2 shows the result of contact angle measurement of nine kinds SAMs of alkanethiol on the silicon plate. The surface of the microchannel becomes more hydrophobic as the length of alkyl group increases. It is considered that our method is efficient in creating a hydrophobic surface.

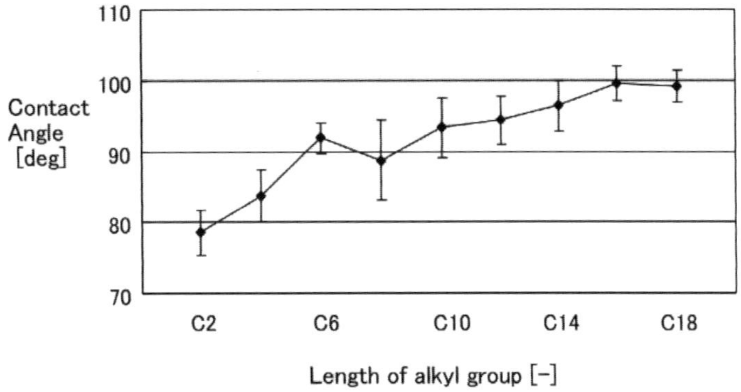

Fig. 2 Effect of hydrophobization by alkyl group on contact angles of water.

Separation performance of water and hexane was evaluated in two microreactors; a partially hydrophobized microreactor by 1-octadecantiol, and a non-hydrophobized microreactor for comparison. As shown in Fig. 3a, the

L-Phe-OBzl was detected from the liquid collected from the outlet of the glass side of the non-modified microreactor,. Apparently, the microreactor was not modified, hence, no separation of hexane phase occurred. On the other hand, the liquid from the hydrophilic outlet of partially modified microreactor did not contain L-Phe-OBzl under water velocity of 0.5 to 2 μl/min (Fig. 3b).The result proves that indeed hexane phase separated from the aqueous phase. A stable two-layer laminar flow reaching the separation point was observed by confocal fluorescence microscopy. On the other hand, it was observed that liquid from the hydrophobic outlet occasionally contained a little water.

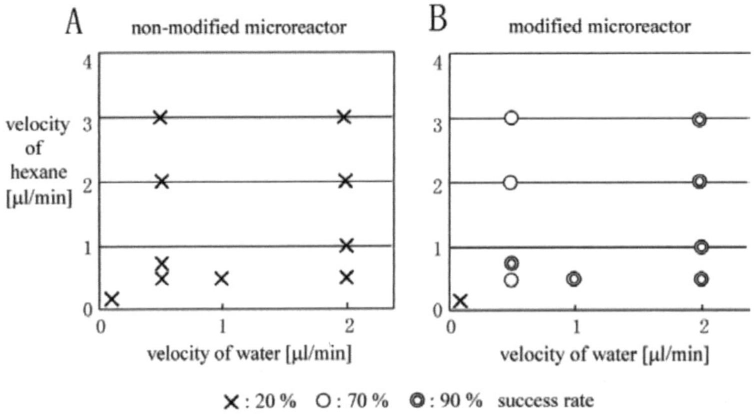

Fig. 3 Evaluation of water separation from hexane phase by partially modified microreactor. (A) silicon-glass microreactor without surface modification. (B) microreactor with hydrophobized silicon plate by the method as mentioned in the Experiments part. Each cross, open circle, and double circle indicates hexane/water separation success rate in more than five -time measurements.

Conclusions

A microreactor which enables stable two-layer laminar flow of aqueous and organic liquid was developed by anodic bonding of silicon and glass, and partial hydrophobization of the silicon surface. It was possible to separate hexane from water phase by more than 90% success rate when the fluid velocity was relatively small. Our results demonstrate a microreactor which offers considerable potential in separation and refinement of extraction products.

Literature Cited

Hisamoto, H., et al., "Fast and High Conversion Phase-transfer Synthesis Exploiting the Liquid-Liquid Interface Formed in a Microchannel Chip," Chem. Commun., (24), 2662 (2001).

Maruyama, T., et al., "Enzymatic degradation of p-chlorophenol in a two-phase flow microchannel system," Lab On a Chip, 3, 308-312 (2003).

Hibara, A., et al., "Stabilization of Liquid Interface and Control of Two-Phase Confluence and Separation in Glass Microchips by Utilizing Octadecylsilane Modification of Microchannels," Anal. Chem., 74, 1724-1728 (2002).

Zhao, B., et al., "Control and Applications of Immiscible Liquids in Microchannels," JACS, 124, 5284-5285 (2002).

A Comparison of Passive Mixing Microchannels

Kristin Hecht
Department of Chemical and Fuels Engineering

Justin Bishop
Jeremy Goeckeritz
Department of Electrical and Computer Engineering

University of Utah
Salt Lake City, Utah

Abstract

Due to the small scale of microscale devices, Reynolds numbers are typically under 100, which imply laminar flow. Consequently, turbulence is not an available mixing mechanism and diffusion dominates mass transport. As a result, researchers have developed various novel designs to enhance mixing. These can be categorized into active and passive devices. Active devices exert some form of control over the flow field through moving parts or varied pressure gradients. While active mixers produce impressive results, they are usually offset by increased difficulty in fabrication and integration into the rest of the system. Solutions based on passive mixers are preferred as they are easy to fabricate and highly compatible with applications.

There is no consensus on the best method to mix fluids in integrated microscale applications. Many authors have proposed channel designs that enhance mixing for microscale applications, but to this author's knowledge, there has been no standard created that would allow comparison of the advantages of these different designs. Here, a method for comparison of passive mixing channels is presented.

Standardized channel layouts were created to allow comparison among different channel designs and FLUENT was used to model mixing in the channels. All of the parameters and dimensions were controlled so that channel geometry was the dominant source of improvement in mixing profiles. FLUENT was used to model mixing in channels. Four channel shapes were examined in great detail. They are shown in Table 1, below: straight, zigzag, square wave, and serpentine.

TABLE I. CHANNEL GEOMETRIES

Straight mixer 	Zigzag mixer 
Square wave mixer 	Serpentine mixer 

Each channel had two inlets with dimensions of 50 μm by 100 μm that combined to form a main channel that is 100 μm by 100 μm. Liquid water and ethanol were used at equal volumetric flow rates to represent two fluids mixing. At various Reynolds numbers, channels were evaluated for pressure drop and extent of mixing at a distance 2 millimeters downstream from the intersection of the fluids. The extent of mixing was determined according to:

$$Mix = \frac{\sum_{i=1}^{n} 1 - \left| \frac{x_i - \bar{x}}{\bar{x}} \right|}{n} \qquad (1)$$

where x_i = volume fraction water in a cell

$\bar{x}$ = average volume fraction water, 0.5 for equal volumetric flow rates of water and ethanol

n = number of cells

According to the simulations, straight, zigzag, serpentine, and square wave channels have similar pressure drops, and zigzag, serpentine, and square wave channels show improvement in extent of mixing over the standard y-mixer, with the serpentine and square wave channels having the highest extents of mixing. At lower flow rates, altered channel geometries offered little improvement over a straight channel, but at moderate and high flow rates, significant differences in the extent of mixing were observed. The serpentine and square wave channels demonstrated the highest extents of mixing and also the highest pressure drops. This result remained while the flow rate increased, but the differences in pressure drop increased.

Experiments to validate the simulations have been designed. For the experiments, PDMS channels are made using common microfabrication technologies. SU-8 is patterned via lithography and used as a mold for PDMS channels. The PDMS channels are oxygen bonded to glass slides. Syringe ports and pumps are used to feed two liquids to the channel. Physically obtaining a quantitative evaluation of mixing is achieved by injecting CY-5 into one of the inlets. The dye is excited using an evanescent field created by endfire coupling a 635 nm diode laser into the glass substrate, and the fluorescence created by the CY-5 within the evanescent field is measured with a CCD camera.

A standardized method for comparing the effect of channel geometry on mixing of fluids for microscale devices has been proposed. Simulations have been performed to compare straight, zigzag, serpentine, and square wave channel geometries. Square wave and serpentine geometries offer better mixing at higher flow rates, but they also have higher pressure drops. These extent of mixing enhancement and pressure drop depend on flow rate. A method to verify these simulations experimentally has been designed. The information from these studies may be useful in determining the most appropriate channel geometry for use in specific microfluidic applications where mixing is important to device operation and a passive mixer, such as a microchannel, is the preferred method.

Acknowledgements

This work is made possible through a grant from the National Science Foundation Integrated Graduate Education Research Training (IGERT) program.

Continuous Polymerization of PTFE in a Microreactor System

Katsuya Ueno, Nobuyuki Kasahara, Kazuhiko Yamada and Shin Tatematsu

Asahi Glass Co., Ltd. Research Center
1150 Hazawa-cho, Kanagawa-ku, Yokohama-shi
Kanagawa, 221-8755, Japan

Katsuya-ueno@agc.co.jp

The development of microreactor (MR) systems is an innovative trend in the chemical engineering industry as these MR are capable of performing chemical reactions under aggressive conditions that conventional systems are in the face of difficulties. MR systems are expected to have a large potential to produce new and innovative products in this situation.

We achieved to synthesize polytetrafluoroethylene (PTFE) safely and continuously by applying MR to polymerization of tetrafluoroethylene (TFE), as opposed to the conventional batch process. In this presentation (1) the performance characterization of the MR with respect to co-current two-phase flow patterns and (2) the solution and emulsion polymerization of PTFE in MR will be discussed. Many flow patterns were observed at various gas and liquid flow conditions (Fig.1), and they greatly depended on the shape and inner diameter of the channel as expected. The gas absorption rates were also evaluated for each flow pattern.

By applying the flow pattern and gas absorption test results to polymerization, it was found that the yield and polymerization rate were increased at higher gas absorption rate. Both solution and emulsion polymerization methods were performed in this study.

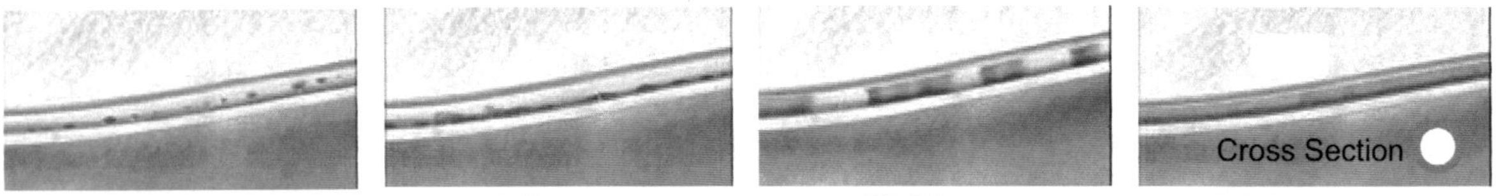

i) Dispersed Droplet   ii) Slug   iii) Plug   iv) Annular

Fig.1 Flow patterns of gas (N_2) and liquid (colored water)

132z Layer-by-Layer Self-Assembly of Nano- and Micro-particles as a Catalyst Integration Method for Microreactor Applications

Hongwei Qiu, Woo Lee

Commercially available silica spheres in the size range of about 20 nm to 2 µm were initially used as our model catalyst particles for exploring electrostatic layer-by-layer self-assembly, as a novel method of incorporating commercial catalyst powder into a thin-film catalyst structure as an integral part of microreactor fabrication. For the negatively charged SiO_2 surface, this method basically consisted of depositing polyelectrolyte multilayers by alternately dipping the substrate into a polycation solution and a polyanion solution. The final polymer surface was made to be positively charged since the silica particles are negatively charged. Substrates that were coated with the polymer film was dipped into silica colloidal suspension and then rinsed in a cleaning solution to electrostatically attach only one layer of the silica particles. The silica particles were immobilized by applying a 1 µm thick sol-gel alumina layer that contained pores with the size of about 3 nm. The overall procedure was refined by systematically studying the effects of pH, ionic strength, and salt and/or buffer concentrations of the polyelectrolyte solutions, the silica colloidal suspension, and the rinsing solutions. The effects of silica particle size on surface coverage and uniformity were assessed. Also, the self-assembly behavior of the model spheres was compared to that observed for ball-milled commercial catalyst powder to assess the practicality of this self-assembly concept.

Application of Controlled Thermal Expansion in Microlamination for the Economical Production of Bulk Microchannel Systems

Christoph Pluess and Brian K. Paul

Department of Industrial and Manufacturing Engineering, Oregon State University, Corvallis, Oregon 97331-2407, USA

Abstract

Diffusion bonding has been widely used within microlamination architectures for the fabrication of Micro Energy and Chemical Systems (MECS). MECS are microsystems with the ability to process bulk amounts of fluid within highly-parallel microchannel arrays capable of accelerated heat and mass transfer. Thus far, diffusion bonding of the microchannel arrays is commonly done in a vacuum hot press system. The use of the hot press greatly restricts the production rate due to vacuum pump-down time and heating-up and cool-down periods. Furthermore, larger substrates are gaining interest in the system design of MECS devices and it is not apparent that uniaxial pressing within a hydraulic vacuum hot press will provide the bonding pressure uniformity necessary for large substrate bonding. This paper presents a novel fabrication approach for the high-volume thermal bonding of MECS devices with the use of controlled thermal expansion. A thermal bonding fixture based on the principle of differential thermal expansion was developed with focus on controlling the bonding pressure magnitude, the pressure timing and its sensitivity. The application of such a fixture within a conveyorized furnace system could be the key to a continuous thermal bonding approach for the mass production of MECS devices.

Introduction

Today, the function of many bulk microfluidic devices has been validated in the laboratory. Future development of bulk microfluidic devices will focus on the integration of different unit operations at low-cost and high production volume. The full commercial potential of bulk microfluidic systems will be realized through the integration of multiple, highly-paralleled unit operations into one single device. The innovative design of a thermal expansion bonding unit presented in this paper could be a key technology for the mass production and commercialization of microlaminated devices. The use of thermal expansion as a driving force for the pressure application during a microlamination approach was first suggested by Pacific Northwest National Laboratory (PNNL) in 1999.[0]

Theoretical Concept and Pilot Study

The general concept of a bonding fixture based on the principle of differential thermal expansion for the application of bonding pressure consists primarily of a frame, two bonding platens and, in the simplest case, an expansion block interposed between them as visualized in Figure 1. The rigid frame is composed of a low thermal expansion material (α_{f1}). The expansion block has a significantly higher coefficient of thermal expansion (α_{e4}) than the frame of the fixture. All things being equal, the height of the expansion block is directly proportional to the amount of clamping pressure to be delivered. Between the expansion block and the frame are placed the bonding platens between which the laminae are aligned and stacked. In this case, the bonding platens serve to prevent solid state diffusion to the fixture parts and may be made of graphite or some other suitable material. As a guideline, the coefficients of thermal expansion (CTE) should preferably differ at least by a factor of two to guarantee functionality of the fixture.

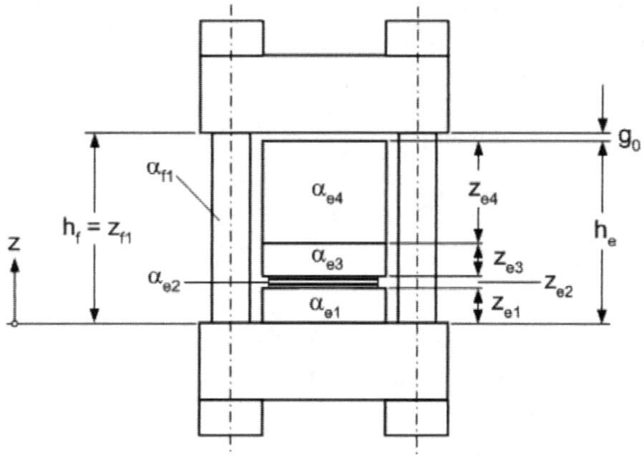

Figure 1: ΔCTE Fixture Concept

When the bonding unit is heated up to the bonding temperature, the expansion block and the platens inside the frame expand relatively to the frame by the difference in the sum of their coefficients of thermal expansion scaled by the product of the height of the expansion block and the change in temperature. An initial gap (g_0) can be designed into the fixture assembly to scale and time the application of the bonding pressure. As soon as the initial gap is consumed (i.e. inner parts come into contact with the frame), compression is applied to the laminae and increases with increasing temperature. An initial test fixture according to Figure 2 was developed to demonstrate the feasibility of bonding laminae by the use of differential thermal expansion. Prior to the experimental pilot test the generation of bonding pressure was quantified by a finite element model. The model has shown a bonding pressure of 7 MPa at a temperature of 800°C between the bonding platens. The expansion block was shaped conical to uniformly distribute the pressure to the graphite platens. Highly expanding stainless steel 321 was selected as a material for the expansion cone. The core piece was built of a lower expanding Inconel™ nickel-alloy.

Five donut-shaped copper (alloy 110) layers were bonded successfully with this test fixture and proved the feasibility of diffusion bonding with the use of thermal expansion.

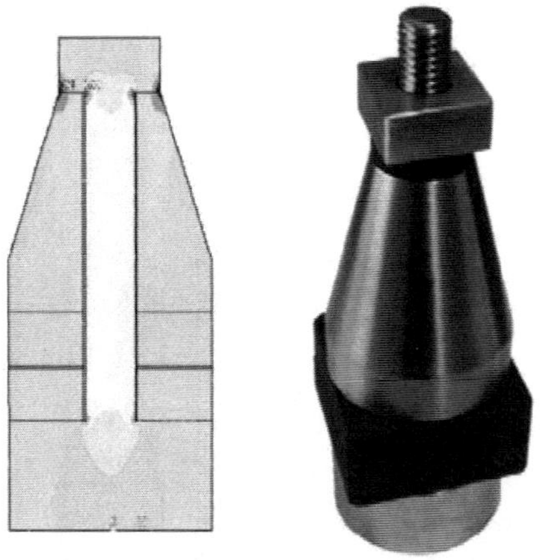

Figure 2: FE-Model and Initial Test Fixture for Bonding Pilots

Novel Fixture Design to Control Bonding Conditions

Experimental and theoretical studies of ΔCTE-bonding fixtures have shown that a desired bonding pressure due to differential thermal expansion is rather difficult to control and a small change either in temperature or initial gap size can result in an enormous variation of bonding pressure (± 100%). This is due to the fact that the bonding pressure is regulated by the modulus of elasticity of the material. To avoid warpage leading to flow maldistribution within bulk microfluidic devices, it is necessary to: 1) dampen the amount of pressure caused by thermal expansion so to minimize pressure variations due to thermal fluctuations or gap adjustments; and 2) avoid putting pressure onto the stack until just before the bonding temperature. The second requirement extends from the difference in thermal expansion between the laminae and the bonding platen in the lateral dimension. One way to control the timing of applying bonding pressure, an initial gap adjustment is necessary to offset the point of contact until just before the final bonding temperature. However, to be effective, the pressure must be applied as a step function with low pressure sensitivity.

A more complex design of a thermal expansion bonding fixture claims the implementation of high-temperature spring elements to decrease the pressure sensitivity of the bonding fixture. The resulting bonding pressure is due to the amount of active spring compression related to the substrate area. By introducing spring elements into the fixture design, the amount of thermal expansion will always be consumed by the springs and the magnitude of the bonding pressure is no longer dependent on the material properties of the fixture platens. The application of disc springs is considered as favorable, since disc springs work with minimal compression by high load forces. Besides disc springs are available in various materials and alloys some of which, for instance, are capable of working at high temperatures. Taking this concept a step further, the spring elements can even be preloaded to the desired final pressure level as visualized in Figure 3. The springs are held by screws in the base plate of the bonding unit in the pre-compressed state. The appropriate amount of preload force can be applied with a hydraulic press and the screws snugged during the preload procedure to secure the amount of spring compression. During a bonding cycle, as soon as the expansion block comes in contact with the frame of the fixture, the preloaded force is transmitted into the laminae stack applying the pre-adjusted pressure. The timing of the force release can be controlled by a set screw in the fixture frame. The initial gap has to be compensated so that contact is made a few degrees below the final bonding temperature to guarantee release of the pre-loaded force to the laminae.

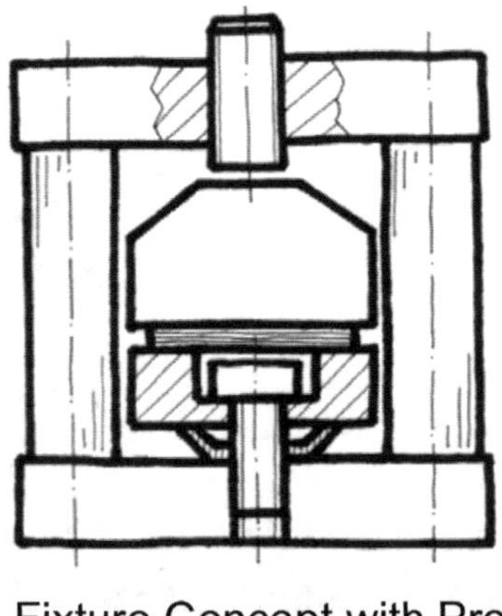

Figure 3 : ΔCTE Fixture Concept with Preload Mechanism

Fluctuations in temperature follow differential thermal expansion behavior of the fixture modulated by the spring constant which can be modified through the design of the spring. Consequently, a thermal expansion fixture with spring elements will yield minimal bonding pressure sensitivity. However, the presented concept involves some complications in the selection of appropriate materials and components to allow the fixture application at high temperatures without diffusion bonding structural components together or reaching limits of the maximal service temperature.

To provide a more fundamental understanding of this fixture concept, a finite element model was created in ANSYS as shown in Figure 4. The model was used to study theoretically the behavior of the differential thermal expansion unit and prove its feasibility. The theoretical assessment of the resulting bonding pressure, timing of the pressure and its sensitivity depending on changes in temperature was of interest to design and optimize a prototype fixture for experimental studies. The prototype fixture is composed of multiple high temperature materials selected depending on their thermal expansion potential. The base and the top of the thermal expansion fixture are made of molybdenum connected by four ceramic standoffs to guarantee a rigid and low expanding fixture frame. The test articles are placed between isotropic graphite ISO-63 bonding platens and aligned by three tungsten alignment pins. On top of the graphite bonding platen the engagement block is placed and centered. The engagement block is fabricated out of high temperature stainless steel 321 with a high coefficient of thermal expansion to drive the main displacement of the load cell sitting on top of it. The load cell consists of two molybdenum platens with four integrated Inconel™ high temperature disc springs. The device is held together by four high-temperature fasteners and is mainly responsible for controlling the magnitude of the bonding pressure. The initial gap between the load cell top and the fixture frame is adjusted with a fine threaded set screw in the top plate of the fixture frame to control the timing of the bonding pressure. The set screw is made of graphite to prevent diffusion bonding to the molybdenum top plate.

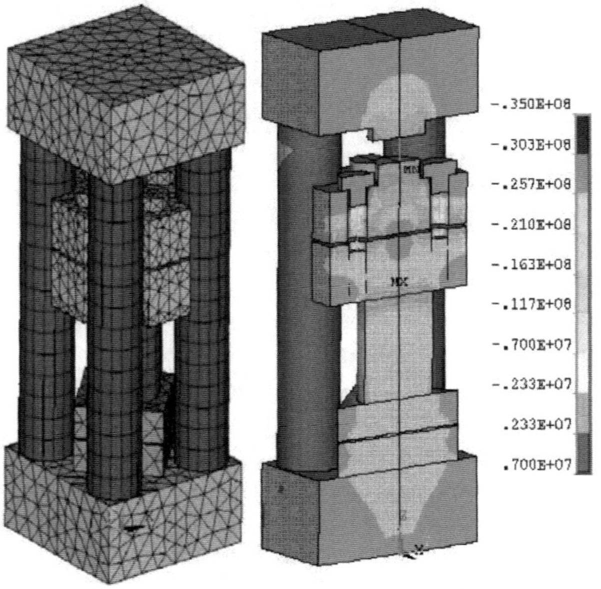

Figure 4 : FE-Model of the ΔCTE Fixture Prototype

The gap size can be adjusted by fitting a shim or assembly of shims according to the calculated gap size. Favorably is the use of a feeler gauge with a variety of precise shims for the adjustment of the initial gap and therefore for the timing of bonding pressure.

Based on the initial gap size the timing of the pressure can be controlled as shown in Figure 5. As soon as the load cell gets into contact with the fixture frame the bonding pressure kicks in with the pre-loaded force. The remaining portion of the target bonding pressure will be provided by the additional compression of the load cell depending on the difference between contact and bonding temperature. The slope of the additional pressure increase equals the pressure sensitivity of the fixture and is mainly dependent on the total spring constant of the load cell and the differential thermal expansion behavior. The finite element simulations reach a final pressure of 4.07 MPa which is only 1.75% off from the target bonding pressure of 4 MPa. The finite element model validated the concept of the ΔCTE fixture design. Consequently, a prototype fixture was built as shown in Figure 6.

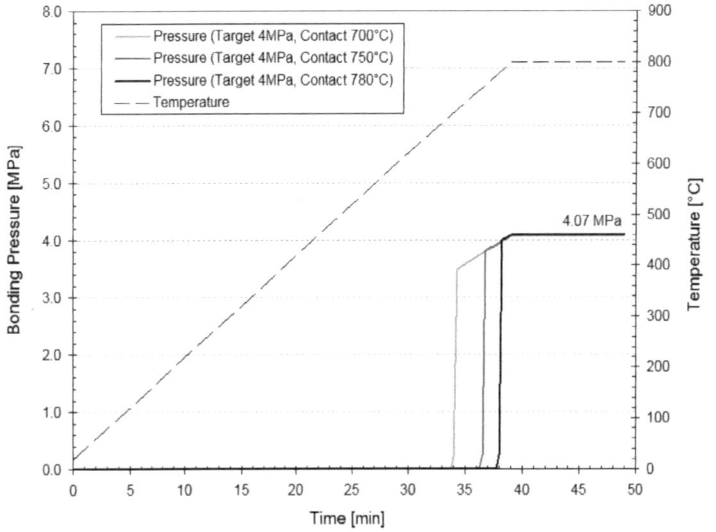

Figure 5 : Theoretical Pressure Engagement of FE-Model ΔCTE Fixture

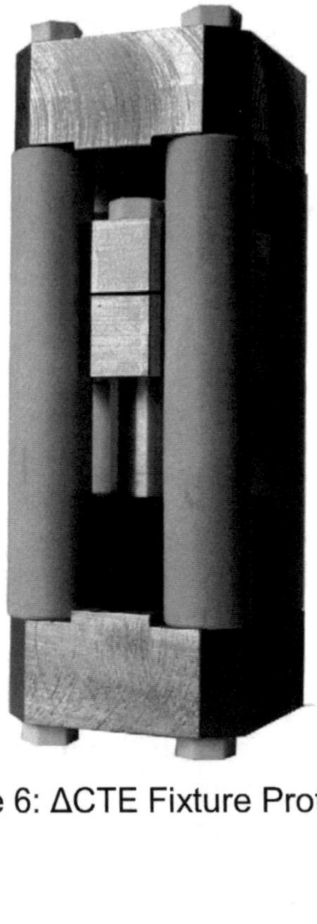

Figure 6: ΔCTE Fixture Prototype

Experimental Results

Different experiments have been conducted to validate the functionality of the proposed bonding fixture. Fuji pressure sensitive measurement film was used to evaluate pressure uniformity, magnitude and timing in the low temperature regime. The behavior of fin warpage was used as a reference for the calibration of the bonding fixture (pressure timing). The test articles (cross-section in Figure 7) were also used for the metallurgical assessment of the bond line. The void fractions of samples bonded in the ΔCTE fixture (Figure 8) were compared with samples bonded within the hot press (Figure 9) under similar conditions. Finally, an analysis of variance (ANOVA) was performed to compare samples bonded in the thermal expansion fixture and samples bonded within the hot press to determine if the differential thermal expansion fixture was capable and repeatable. After processing the 32 experimental runs the fin warpage was measured on the 64 test articles by scanning the channel fin with a Dektak[3] surface profiler. Since each test article had two fins a total of 128 measurements were taken. A multifactor analysis of variance (ANOVA) was performed to decompose the variability of the measured fin warpage into contributions due to the various factors (mode, temperature, pressure, time) and their interactions. The analysis of variance has shown that there is no statistical significant difference between samples processed in the hot press or within the fixture on the basis of a p-value of 0.92 for the experimental factor mode.

Figure 7: ΔCTE Bonded Sample Showing Minimal Fin Warpage

Figure 8: Bond Line ΔCTE Fixture (800°C/6MPa/30min)

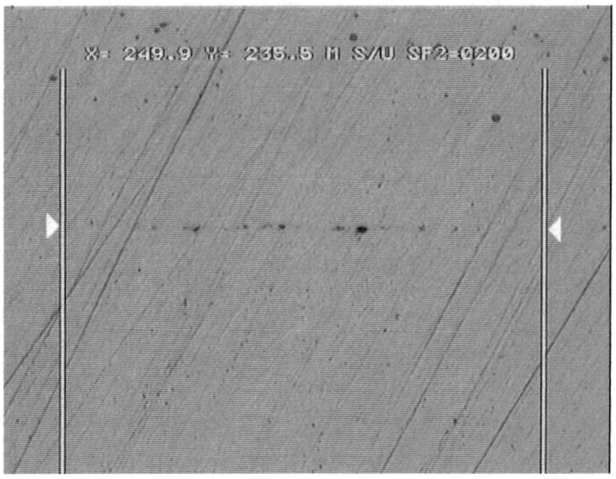

Figure 9: Bond Line Hot Press (800°C/6MPa/30min)

Conclusions

A new and unique diffusion bonding fixture capable of operating within a continuous furnace has been presented in this paper. It was shown that the use of disc springs can significantly reduce the pressure sensitivity of the bonding fixture compared with prior efforts to simply put a highly expanding material between a lower expanding frame construction. The feasibility of the final device concept was theoretically validated by a finite element model. Experimental investigations with a fixture prototype have proven the functionality of the proposed bonding device. Based on this study it can be concluded that the diffusion bonding of devices within a ΔCTE fixture is not significantly different from processing devices in a hot press system. This is significant especially in light of the fact that the ΔCTE fixture has to engage prior to the bonding temperature which could lead to warpage.

The validation of a ΔCTE bonding fixture opens the possibility for the high-volume microlamination of MECS devices within a continuous furnace system which will reduce costs associated with diffusion bonding (or ultimately any thermal bonding) cycles. A major advantage of this fixture is the ability to decouple the pressure magnitude from the process temperature through the concept of preloading and force storage within a load cell. Initial gap settings are set independent of the pressure magnitude and, therefore, any level of bonding pressure can be applied at any level of bonding temperature by design. This flexibility makes the investigated ΔCTE bonding device not only interesting for diffusion bonding, but also for any thermal bonding process. Major advantages also exist for the use of a ΔCTE fixture over existing vacuum hot press platforms. In addition to better cycle times associated with continuous furnace processing, a second major advantage of the fixture is that it does not require dynamic loading at the bonding temperature. Loading of the bonding pressure in the ΔCTE fixture is more gradual. This may become important with heightened efforts to increase layer-to-layer registration precision and reduce fin warpage.

References

P. M. Martin, D. W. Matson, W. D. Bennett, D. C. Steward, C. C. Bonham, „Laminated Ceramic Components for Micro Fluidic Applications“, Proceedings of SPIE, 1999, v. 3877, p. 95-100

Commercial Scale Microchannel Technology Methodology and Capabilities: Demonstrating Sufficient Flow Distribution

Anna Lee Tonkovich[1], Sean Fitzgerald[1] & Ravi Arora[1]
[1] Velocys, Inc. 7950 Corporate Blvd., Plain City, OH (614) 733-3300

Abstract

Microchannel reactor technology is being commercialized based on an economy-of-mass manufacturing versus the conventional economy-of-scale paradigm. Modular commercial reactor assemblies, which represent a portion of the total plant capacity, are being developed with the promise of generating both capital and operating cost savings for applications including steam reforming, oxidation, dehydrogenation, and others.

Scaling microchannel systems to commercial capacities (tens to tens of thousands of tons per year) requires new methodologies. It has been widely reported that microchannel systems increase scale by 'numbering-up' - adding parallel channels with equivalent hydrodynamics to achieve the desired reactor capacity. This approach reduces conventional scale-up risk, because flow regimes, hydrodynamics and transport effects remain the same regardless of scale. However, numbering up microchannels requires sufficient flow distribution to permit robust process operation during both steady-state and transient operation.

A critical element in the design of a commercial microchannel reactor is flow distribution. Manifold designs with minimal flow maldistribution over many thousands of parallel microchannels have been developed. These designs have been theoretically modeled and experimentally validated. While the flow maldistribution is quite low, perfect uniformity is not possible. The design methodology has proven sufficient to meet both performance requirements and manufacturing constraints.

Introduction

In recent years there has been intense industrial and academic interest toward developing microscale devices for chemical processing. A recent review of microscale reactors, containing 236 citations, has been provided by Gavrilidis [1]. Microscale chemical processors, which are characterized by fluid channel dimensions of 5 mm, or less, can provide unique advantages due to short heat and mass transfer distances, and, in some instances, different flow characteristics.

Although these devices offer many advantages, new designs and differing flow characteristics within these devices create challenges for controlling flow, particularly flow through manifolds that distribute flow to many connecting channels. In particular, the small channel dimensions that give rise to improved heat and mass transport often impose a laminar flow regime, which lowers flow resistance compared to transition and turbulent regimes. Thus, a large microchannel processing device can have relatively small pressure drops at high overall flow rate, adding to the flow distribution challenge due to low flow resistance.

The methodology described in this paper develops a manifolding solution that achieves sufficient flow distribution with minimal pressure drop. A flow distribution device with seventy-two microchannels in parallel spanning more than 40 cm was built and tested to validate the design methodology.

Background

Microchannel technology reactors need an effective manifold designs to achieve sufficient flow distribution without incurring much additional pressure drop while also minimizing cost and volume. The work of large scale manifold design [2-6] and compact heat exchangers [7-8] provides the background and initial concepts for manifold design.

Recent work in microchannel manifold has been done for low Reynolds number flow regimes that utilize changes in cross-sectional area of the header and footer manifolds [9] to balance flow. Golbig [10,11] devised a means of flow distribution which varied the dimensions of the connecting microchannels to vary the connecting channel flow resistance as a means of equalizing the residence time in microchannel arrays using fully developed Fanning friction factors. Use of fractal channel distribution has also been discussed as a means of distributing flows [12,13].

Model Development

To design a micro-manifold system for distributing flow in process channels, a flow resistance network comprising of momentum balances and mass continuity was developed akin to electrical circuit analysis. Figure 1 shows a schematic of micro-manifold and connecting channels network. The figure also provides examples of locations in the flow resistance network at which pressures and flow rates are calculated to determine the flow distribution.

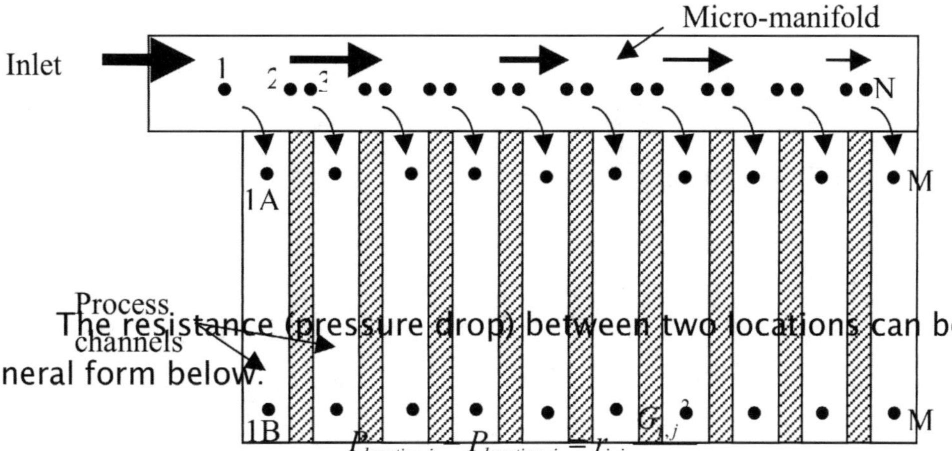

Figure 1: Schematic of flow resistance network

The resistance (pressure drop) between two locations can be expressed in the general form below:

$$P_{location,i} - P_{location,j} = r_{i,j}\frac{G_{i,j}^2}{2\rho_{i,j}}$$

where

$r_{i,j}$ [kg/m-s^2] = Flow resistance between location i and j

$G_{i,j}$ [kg/m^2/s] = mass flux rate between location i and j

$P_{location,\,i}$ [Pa] = Pressure at location i

$P_{location,\,j}$ [Pa] = Pressure at location j

$\rho_{i,j}$ [kg/m^3] = Density of fluid between location i and j

The resistance function ($r_{i,j}$) represents flow resistance such as friction losses, cross-sectional area changes, momentum compensation and other losses. The resistance is a function of mass flux rate, geometry, molar composition changes, and temperature changes among others.

In micro-manifolds, the hydraulic diameters are on the same order of the connecting channels (5 mm or less), much smaller than many the conventional pipe or duct based manifold systems. Due to the micro-manifold having small hydraulic diameters, fairly large specific velocities or mass flux rates can be in the turbulent, transition or laminar flow regimes, which affect flow distribution in unexpected ways. For example, the turning losses and momentum compensation coefficients can be substantially higher for micrchannels than for pipes.

Flow distribution is preferred to be as uniform as possible. However, a small amount of flow maldistribution is acceptable and does not negatively impact overall device performance. The efficacy of a micro-manifold design is measured by uniformity of flow distribution in the connecting channel which is measured as "Quality Factor" defined below:

$$Q_1 = \frac{m_{max} - m_{min}}{m_{max}} \times 100\% \quad (1)$$

where

m_{max} [kg/sec] = maximum connecting channel mass flow rate

m_{min} [kg/sec] = minimum connecting channel mass flow rate

Experimental

To validate the flow distribution model a device, shown in Figure 2, was evaluated. The large square section in the lower right hand corner is the macro manifold volume in which accepts flow from a 0.5 inches diameter tube. From the macromanifold box fluid enters six parallel sub-manifolds, which distribute flow to the individual microchannels. The connections between the parallel sub-manifolds consists of three small rectangular openings, called gates, that bridge over a solid wall in the top plate and enter a larger rectangular opening. Each open rectangle allows flow into four downstream microchannels. The sub-manifold width varies from sub-manifold to sub-manifold, and the gate openings vary in each sub-manifold to regulate the flow to the seventy-two channels, which all exit into a common environment.

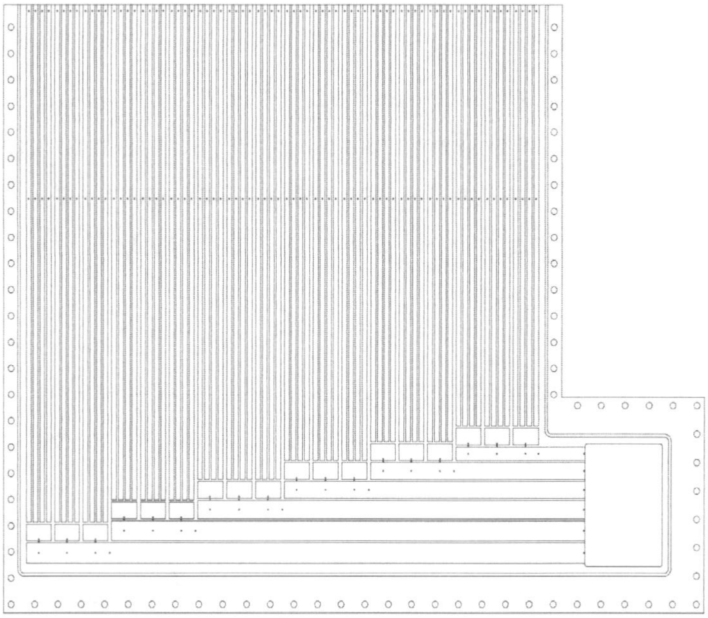

Figure 2. Drawing of the flow distribution device: Top plate only

The microchannels were machined into 12.7 mm aluminum plates. When a thin gasket material was added the heights of the channels were about 1.1 mm tall. The seventy-two channel matrix has channel widths of 40 mm and a minimum of 292 mm long. The channels are separated by a 1.5 mm thick wall, except for every fourth separation walls of 3.1 mm thick. The overall width of the microchannel matrix is 428 mm. The width dimensions of the sub-manifolds and gates are listed in Table 1. The sub-manifold numbers correspond to those in Figure 2.

Table 1. Dimensions of the sub-manifold microchannels

Sub-manifold number	Sub-manifold width (mm)	Width of Gates (mm)		
		#1	#2	#3
1	10.16	6.86	6.35	6.22
2	12.70	6.91	6.48	6.38
3	12.70	8.94	8.38	8.26
4	13.97	9.91	9.22	9.09
5	15.24	9.35	8.86	8.69
6	15.24	14.73	11.18	10.92

To measure local static pressure and pressure changes, tubes were mounted flush with the internal geometry surface through wire EDM created holes, and held in place with adhesive. There are two pressure taps in each downstream microchannel and eight taps in every sub-manifold: One in the manifold adjacent to each gate, one just inside the gate, one just upstream of the first gate and one at the sub-manifold connection to the macro manifold. Each pressure tap was connected to a plastic tube that was either plugged or attached to a pressure sensor. Sensotec differential gages of 6.89 kPa and 34.47 kPa were used to measure pressure with an accuracy of 0.0068 kPa and 0.0344 kPa, respectively. Additional thermowells were added to the system to measure the metal temperature.

Air was metered through the device at rates from 30 to 300 SLPM, which evaluated the flow rates in the sub-manifolds ranging between laminar to transition to turbulent flow.

Results and Discussion

The flow distribution hardware with 72 parallel channels was demonstrated under many operating conditions. Figure 3 shows that the predicted values of flow per channel were in close agreement with the measured values. The model showed excellent agreement with the experimental channel mass flow rates, with an average relative residual between the model and experiment range from 4.5% to 9%.

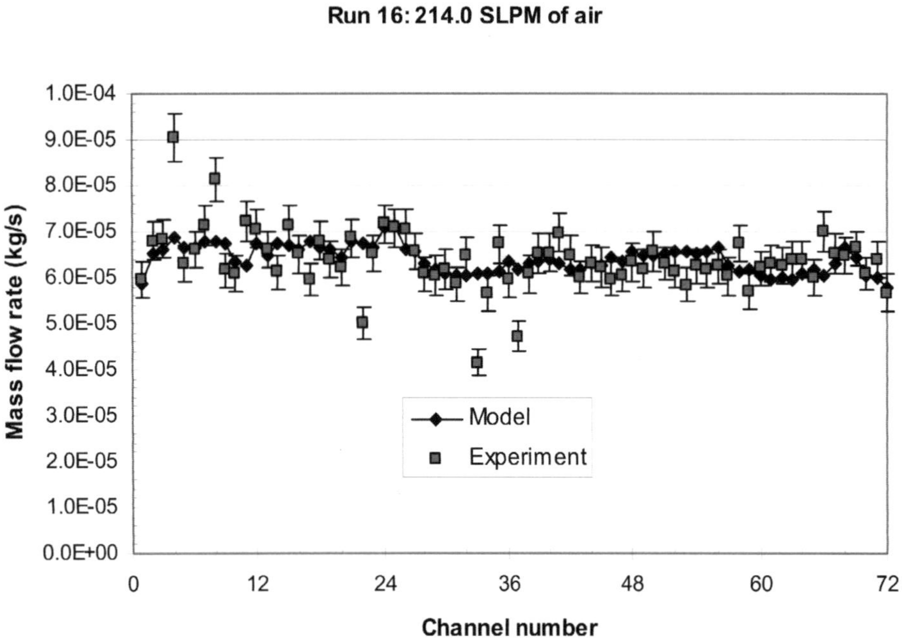

Figure 3. Channel mass flow rates, predicted and experimental results.

There were a few outlier experimental points that were preferentially biased to one side of the device. A post-operation evaluation of the device revealed that the gaskets partially failed on one-side of the device, resulting in some channels having a larger effective cross sectional area and others a smaller cross sectional opening. On the other half of the device (channels 37 to 72 as operated), the channel cross sectional openings were very close to the designed values.

Figure 4 shows overall agreement between the model and experiments, where the measured hardware pressure drop had excellent agreement with the predicted value, from the entrance of the macro manifold to the outlet

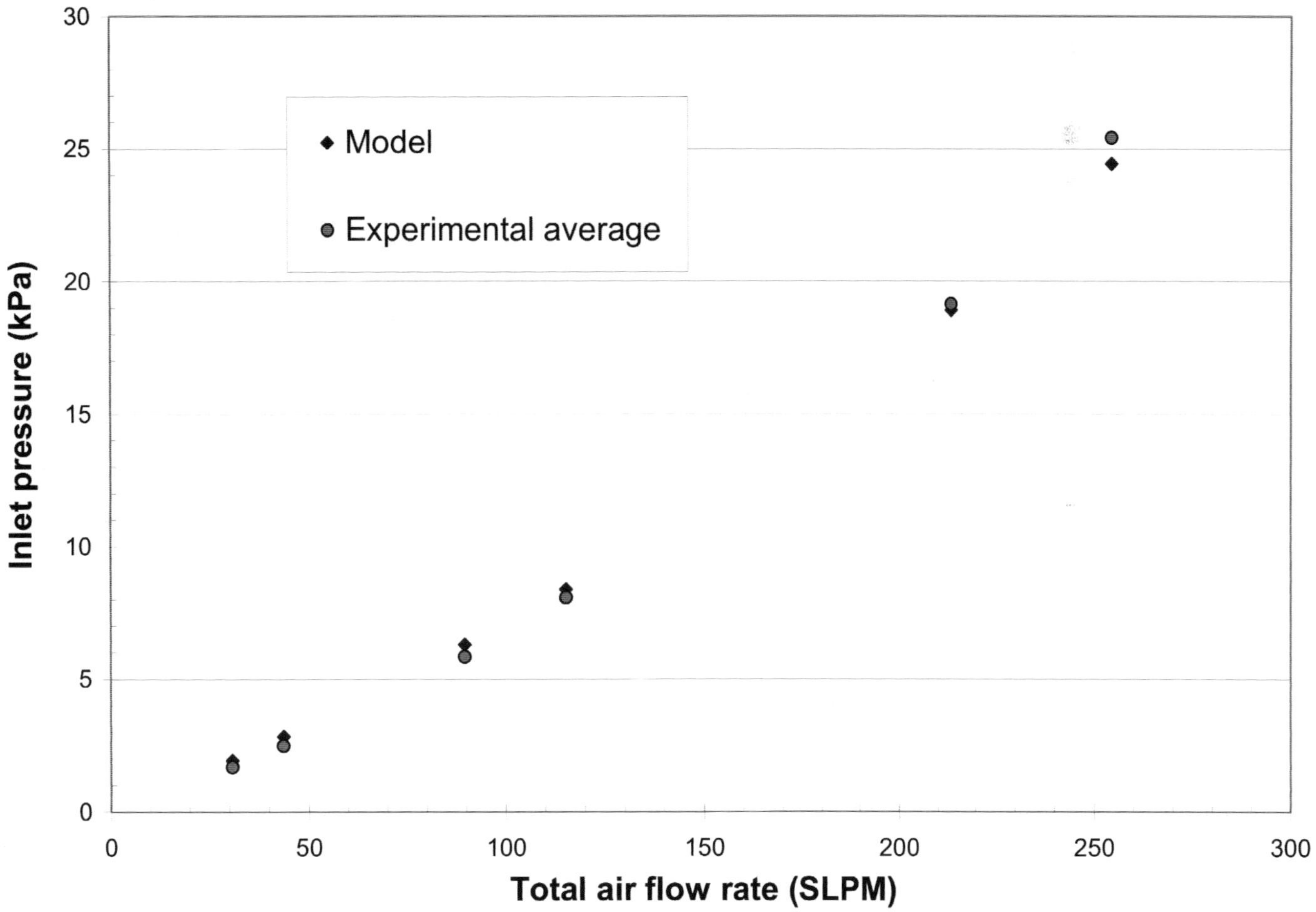

Figure 4. The overall pressure drop for test device versus air flow rate

The quality factor for the reported data was 21%. For all experiments, the quality factor was less than 40%. These quality factors were sufficient for operating a reactor. The design can be further tailored to improve the flow distribution with the aid of the validated model.

Conclusions

A methodology to create acceptable flow distribution within arrays of tens, hundred, and thousands of microchannels has been developed, modeled, and experimentally validated. Flow distribution is achieved by creating nearly identical pressure drop circuits from the inlet to outlet for all connecting channels. Each channel is designed with nearly identical cross section, and the inlets or outlets to each channel is tailored to add a slight amount of additional pressure drop as required to equalize the pressure drop circuit. The measured flow rates were within 4.5% to 9% of the experimental flow rates, which is in the range of the experimental uncertainty of 7%. The pressure drop predicted by the one dimensional model matches the experimental values well.

Acknowledgements

The authors gratefully acknowledge financial support for this work from Velocys, Battelle, and Conoco-Philips. The authors further acknowledge Dr. Jim Kimble for his thoughtful inspiration, Prashant Sholapurmath for his careful attention to experimental detail, and Dr. Frank Rosenberg for his careful patience in navigating complex phenomena.

References

[1] Gavriilides, A., Angeli, P., E. Cao, K. Yeong, Y. Wan, "Technology And Applications Of Microengineered Reactors," Trans. IChemE, Vol. 80, Part A, pp.3-30 (Jan. 2002)

[2] Fried, E. and I.E. Idelchik, "Flow Resistance. A Design Guide for Engineers", Hemisphere, 1989

[3] Pigford,R.L. M. Ashraf, and Y.D. Miron "Flow distribution in piping manifolds", Ind. Eng. Chem. Res., pp. 463-471, 1983

[4] Acrivos,A, B.D. Babcock and R.L. Pigford, "Flow distribution in manifolds", Chem. Eng. Sci., pp. 112-124, 1959

[5] Haerter, A.A., "Flow distribution and pressure change along slotted or branched ducts", ASHRAE J., pp. 47-59, 1963

[6] Bajura, R.A., "A Model for Flow Distribution in Manifolds", J. Eng. Power Trans. ASME, pp. 7-12, 1971

[7] Kays, W.M. and A.L. London, "Compact Heat Exchangers", 3rd ed., Krieger, 1984

[8] London, A.L., "Compact Heat Exchangers - Design Methodology", from Low Reynolds Number Flow Heat Exchangers, Proc. 4th NATO ASI on Heat Transfer, Ankara, Turkey, Jul. 13-14, pp. 814-845, 1981

[9] Commenge, J. M., L. Falk, J.P. Corriou; M. Matlosz, "Optimal design for flow uniformity in microchannel reactors", AIChE J., v. 48(2), pp. 345-358, 2002.

[10] Golbig, K., S. Taghavi-Moghadam, P. Born, "CYTOS-Technology - Microreaction in Practical Sense", 6th International Conference on Microreaction Technology, March 10-14, 2002, pp. 131-134

[11] Golbig, K., V. Autze, P. Born, C. Drescher,"Enhanced fluid flow in a stacked plate microreactor", Patent WO 03/043730 A1

[12] Kearney, M., " Process Intensification Using Engineered Fluid Transporting Fractals", Process Intensificatiion, Topical Conference Proceedings, AIChE Spring National Meeting, New Orleans, 2003, pp. 134-139

[13] Kearney, M., "Fractal Cascade As An Alternative to Inter-Fluid Turbulence", U.S. Patent 5,938,333, August 17, 1999

133c Modular Lab Tool for Microreaction Processes: The FAMOS System

Dusan Boskovic, Juergen Antes, Slobodan Panic, Stefan Loebbecke, Tobias Tuercke

Microreaction technology is currently on the threshold of a more widespread application in industry and research. However, a broad application will only be achieved as soon as suitable equipment will be made available that allows the users to test the potential of microreaction technology intensively for their own applications.

In contrast to the past years when research focused on the application of individually designed microreactors for specific chemical problems, today the development of versatile applicable microreaction systems moves into the centre of interest. In particular, modular concepts for microreaction systems based on a "toolkit concept" are intensively discussed.

In general, a flexible modular setup of microreaction processes, interchangeable modules, and a uniform assembling technique are indisputable benefits of a toolkit concept. However, the deciding factor for a successful microreaction toolkit concept is its acceptance by the (industrial) user. Hence, the toolkit philosophy is only accepted as long as modularization and singularization of microreaction processes result in improved functionality, easy handling and a broad applicability.

The "FAMOS concept" presented here consists of standardized microfluidic modules with suitable fluidic, analytical and electronic interfaces. The individual modules have been designed for very different types of reactions and experimental conditions, and are thus made of different materials like silicon, ceramics, metals and plastics.

A crucial issue of the toolkit development was to ensure an appropiate degree of modularity by avoiding needless singularization of unit operations and other toolkit components. For example, microstructured mixers or heat exchangers are not designed as individual toolkit modules any more. They are rather integrated into functional reaction modules which contain microstructures for mixing, heat exchange and retention time within one module to ensure best performance for a specific reaction type such as single or multiple phase reactions, mixing-sensitive reactions, thermal gas phase reactions, heterogenous catalysis, etc.

Microstructures are exclusively used to achieve a functional benefit for the chemical process. Otherwise, macroscopic operability dominates the toolkit concept. As a consequence, temperature control, online analysis as well as measurement and control techniques are realized by avoiding needless miniaturization. Hence, the user is able to integrate the microreaction toolkit into his familiar laboratory environment.

The developed microreaction system is based on a planar and symmetric design. Hexagonal microfluidic modules can be mounted and connected on a baseplate with suitable slots to set up microreaction processes. All standardized modules are of identical shape but may differ in height, internal microfluidic structure and material.

The fluidic modules can be easily interconnected via PTFE tubes or other types of capillaries at the underside of the baseplate. Pressure-tight sealing is conventionally achieved by flanged tubings and clamping claws which press the modules into the slots of the baseplate.

Up to six standardized fluid ports are located in the angles of each hexagonal module allowing a flexible set-up of the microreaction process, since modules can be easily turned in their baseplate slot by 60, 120 or 180 degrees. Moreover, multiple connections of modules and a compact set-up of more complex, multistage chemical processes are also possible.

This construction ensures defined thermal conditions over the entire microreaction process. A temperature bath, on which the baseplate is mounted, is connected with an external cryostat to control the temperature of the interconnections, tubings and fittings at the bottom side of the baseplate. Surprisingly, until now temperature control of the interconnections and tubings has been often neglected in microreaction set-ups, although their hold-up can be several times the hold-up of the microreactor itself.

In addition, the microfluidic modules can be individually cooled or heated by coolants or heating oils which are passed through the heat exchange microstructures of the modules (by using fluid ports on their upside).

To make possible that the user is able to run experiments under accurate thermal conditions (e.g. isothermal processing) temperature measurement and control is an inherent part of the toolkit concept. Temperatures can be measured at two independent positions in each fluidic module by integrated miniaturized Pt100 thermoresistors. The two temperature values can be read out at each slot on the baseplate and displayed in realtime.

Besides measurement of temperature values analytical information about the ongoing process can be also obtained online either IR spectroscopically via optical interfaces which are integrated into the reaction modules or by using separate analytical modules (such as flow-through UV/Vis/NIR cuvettes) mounted on the baseplate.

The toolkit concept is complemented by a simulation software which provides set-up specific calculations to design and optimize microreaction processes on basis of the toolkit hardware.

(The Fraunhofer Alliance for Modular Microreaction System FAMOS is a common development of six Fraunhofer Institutes, see also: www-microreaction-technology.info)

133d Evaporation of Hydrogen Peroxide in a Microstructured Falling Film

Elias J Klemm, Georg Markowz, Guhan Mathivanan, Johannes Albrecht, Steffen Schirrmeister

Introduction

Organic oxidation processes which use hydrogen peroxide (H2O2) as an oxidant are usually homogeneously or heterogeneously catalyzed liquid phase processes. Either an organic solvent is used or a complex multiphase system has to be handled. A heterogeneously catalyzed gas phase reaction with hydrogen peroxide vapour has several advantages like lower separation effort (solvent and/or catalyst) and higher reaction rates (faster mass transport in gas phase and higher temperatures). A crucial requirement and an extreme challenge is a safe and stable evaporation of hydrogen peroxide. In the frame of the DEMiS® project (Demonstration Project for Evaluation of Microreaction Technology in Industrial Systems) hydrogen peroxide evaporation has been developed from lab scale to pilot scale, i.e. from 10-100 g/h to 1 – 10 kg/h per evaporator module. By arranging modules in a stack, evaporator units with capacities of some metric tons per hour are possible. In the present contribution we report about lab experiments which have been the data basis for the design of the pilot scale evaporator.

Microstructured falling film

Necessary requirements for a safe and stable evaporation of hydrogen peroxide are partial evaporation of a suitably stabilized hydrogen peroxide, avoiding overheating and minimizing liquid hold-up. A microstructured falling film fulfils all the mentioned requirements. The microstructured lab scale evaporator plate possesses an area of about 200 cm2. The choice of the material and the pickling and passivating procedure are important factors which have to be optimized.

Experimental setup

A rotary gear pump was used to dose the aqueous hydrogen peroxide solution through a preheater to the top of the falling film evaporator, where it is distributed across the width of the microstructured evaporator plate. A heating medium was conducted in the core of the evaporator module. At the bottom of the evaporator a carrier gas (nitrogen) was introduced and by overflowing the film it transported the evaporated hydrogen peroxide to the top of the evaporator where it is withdrawn. For the purpose of analysis the vapour is condensed and the concentration of hydrogen peroxide determined by permanganometric titration. The non-condensable outlet stream of the condenser which consisted of nitrogen and oxygen was analyzed by an electrochemical oxygen sensor. This signal allows the on-line monitoring of the decomposition during evaporation. The liquid was collected at the bottom of the falling film and the concentration of hydrogen peroxide was determined by titration. The O- and H-balances reached almost 100 %. An observation window in the bottom part of the evaporator allowed visual control of the film and temperature sensors allowed measurement of the vapour temperatures in the upper part.

Experimental results

The design of experiments considered the following main influencing variables: geometry of microstructures, quality and concentration of hydrogen peroxide feed, flow rates of hydrogen peroxide feed and carrier gas, and temperature. The objective was to reach maximum hydrogen peroxide evaporation per unit of area at minimum decomposition and minimum demand of carrier gas. Here only influence of nitrogen flow rate is discussed and other influencies will be shown and discussed during oral presentation.

As said before, the main objective was to reach maximum hydrogen peroxide evaporation per unit of area at minimum decomposition and minimum demand of carrier gas. Unfortunately, relative decomposition increases and evaporation rate decreases when carrier gas flow is reduced. On the other hand, these tendencies can be compensated or even attenuated by minimizing residence time of the vapor through reducing gas channel width and by increasing temperature.Relative decomposition is lower than 10 % and can be reduced to 1-3 %, when optimum operation conditions are selected.

Gas-Liquid Mass Transfer in a Microstructured Falling Film Reactor

S. Claudel, C. Nikitine, C. Boyer, P. Font

Institut Français du Pétrole, BP3, 69390 Vernaison, France

Prepared for Presentation at
IMRET 8 - 8th International Conference on Microreaction Technology
AIChE Spring National Meeting
Atlanta
April 10-14, 2005

Unpublished

Abstract

Micro-structured reactors can be advantageously used as testing tools for catalyst or solvent screening at low scale before scale-up of industrial gas/liquid reactor or contactor. As a matter of fact, with these micro-systems, hydrodynamics can be controlled, and mass and heat transfer intensified. Of interest are the falling film micro-reactors, where a liquid film falls under gravity over plate with micro-channels at counter-current with gas flow. These reactors make it possible to precisely control gas/liquid interfacial area, liquid film wettability, confinement and thickness and are therefore expected to yield faster gas/liquid mass transfer rates.

So far, very few experimental data on mass transfer parameter k_L or k_La is available for gas/liquid mass transfer in micro-structured falling film reactor. In order to start new applications development, fundamental information should be obtained on mass transfer in gas-liquid micro-reactor. Two gas-liquid chemical reactions and one physical adsorption system were studied to determine mass transfer parameters and for parameter sensibility (concentration of gas and liquid, gas velocity, ...), when using 3 types of plates in a falling film micro-reactor. The reactor is used at counter-current with gas flow-rates between 50 $NL.h^{-1}$ and 95 $NL.h^{-1}$, and liquid flow-rates between 0,05 $L.h^{-1}$ and 1,5 $L.h^{-1}$. Channel hydraulic diameter varies from 226 to 907 µm.

The system CO_2/NaOH is used to predict both the interfacial area, *a*, and the mass transfer coefficient on gas side k_G operating in rapid chemical reactional mode. The second one, CO_2/NaClO/Na_2CO_3 is used to predict simultaneously both the interfacial area, *a*, and the mass transfer coefficient on liquid side k_L, operating in intermediate chemical reactional mode. Physical absorption of CO_2 in water has also been measured to determine the overall mass transfer coefficient k_La.

Measured interfacial area, a, lies between 180 and 200 $m^2.m^{-3}$ of reactor that corresponds to a range of 6000 to 9000 $m^2.m^{-3}$ of liquid. Gas mass transfer appears to be much higher than liquid one and mass transfer coefficient k_L ranges significantly between 4.10^{-4} $m.s^{-1}$ and 10^{-3} $m.s^{-1}$, depending on the plate geometry and liquid flow rate. Since the geometrical interfacial area is almost the same for the three plates, the overall mass transfer coefficient k_La is found to be changing with the channel width. For a given plate, flow-rates has no effect on the interfacial area and little effect on the mass transfer coefficient. The values of k_L are closed to highest values that can be obtained on classical structured packing column. The measured values of interfacial area *a* are very high when referred to liquid volume. When referred to reactor volume, *a* values are of the same order as those obtained on structured packing column but it has to be noted that, in this micro-reactor, the volume left for gas flow appears to be oversized. This results will be used for designing a testing tool for catalyst or solvent screening.

1. Introduction

In the development of new catalytic processes, micro-reactors appear to be attractive new tools for test and screen new catalysts species or solvent phases. In these systems, it is possible to control both and independently hydrodynamics and heat transfers of the reactant phases that are flowing through the micro channels.

Since during reactor scale down the ratio of feed rate over catalyst volume is kept constant, this induces very small flow rates when working at small pilot scale. It becomes therefore quite important to control flow structure to ensure in particular a perfect wetting of catalyst phase by liquid flow. For exothermic or endothermic reactions, the reaction temperature has to be properly controlled to perform kinetic measurement at isothermal conditions.

Among the tools already developed, a falling film gas/liquid micro-reactor commercialized by Institute of Microtechnology Mainz (IMM) has been chosen. The objective of such a tool is to create and control falling film flow over micro-structured plates in order to enhance mass transport phenomena in gas-liquid flows with low back mixing. By using a second plate for heating or cooling flow at the rear of the reaction plate, it is possible to enhance heat transfer and control the reaction temperature.

Hessel et al. (2000) have performed mass transfer experiments on a falling film micro-reactor with the CO_2/NaOH system. But no results are presented for interfacial area, a, nor for the mass transfer coefficients k_G or k_L. The purpose of this article is thus to determine the interfacial area, a and the mass transfer coefficients k_G and k_L and k_La. Several diphasic systems are considered to examine gas/liquid absorption with or without chemical reaction.

2. Experimental Study

2.1. Description of the falling film micro-reactor

Figure 1 shows the falling film micro-reactor disassembled into different elements.

From the left to right hand side of Figure 1, the elements are the following :

- The reactor housing with two connections used to introduce liquid reactants at the top and to extract liquid products at the bottom. The copper plate, is a structured plate used for cooling or heating fluid flow.
- The reaction plate with parallel micro-channels for the liquid film.
- Two gaskets to avoid flow leakage around the structured plates
- The reactor upper housing with connections, on the front side, for the gas inlet and outlet and an assembled inspection glass window. In this system, gas and liquid are in contact at the level of the micro-structured plate.

Figure 1. Falling film micro-reactor

This study deals with the characterization of the gas/liquid contact efficiency using different reacting systems in gas/liquid flow configuration. Liquid phase is distributed in micro-channels and it flows by gravity.

The range of the micro-reactor operating conditions are listed on Table 1. For each micro-structured plate, a maximum liquid flow rate is defined. This value corresponds to the limit from which liquid flows over the micro-channel.

Table 1. Operating conditions ranges for the micro-reactor

Operating conditions	Range with G/L flows	
Liquid flow rate ($L.h^{-1}$)	Plate A	0 – 0.05
	Plate B	0 – 0.4
	Plate C	0 – 1.5
Gas flow rate ($NL.h^{-1}$)		0 – 100
Pressure (bar)		1 – 20
Temperature (°C)		0 – 180

Three different reaction plates with different channel geometry (number and hydraulic diameter) have been tested. To calculate the exact hydraulic diameter, an image of the channels cross section has been acquired with an optical microscope. Images showed that the channels section is not rectangular but partially circular or semi-elliptic. The images have been analyzed to determine the exact channel perimeter.

When the cross section is shown to be partially circular (Figure 2), the curvature radius R and angle θ are deduced from image analysis. These image analysis allows to accurately evaluate the different hydraulic diameters for the three plates (see Table 2) when the channel is full of liquid.

a) b)

Figure 2. Channel cross section of reaction plate A : a) parameters used for hydraulic diameter determination, b) image of channels cross-section acquired with an optical microscope and used for measuring R and θ

Hydraulic diameter $\varnothing_{hyd}$ is calculated with the following equation (Blevins, 1992) :

$$\varnothing_{hyd} = \frac{2R\left(\theta - \sin\theta\right)}{\theta} \quad (1)$$

Table 2. Reaction plates characteristics considering a partially-circular or semi-elliptic channel section : channel number n, channel height h, channel width *l*, channel length L, channel hydraulic diameter $\varnothing_{hyd}$ (full of liquid)

Plate	n (-)	*l* (μm)	h (μm)	L (mm)	$\varnothing_{hyd}$ (μm)	R (μm)	θ (rad)
A	64	328	97	66.4	226	187	2.13
B	32	645	226	66.4	523	343	2.44
C	16	1200	400	66.4	907	650	2.35

2.2. Experimental procedure

To characterize the gas/liquid mass transfer of micro-channels, three types of experiments have been performed with different gas/liquid systems :

- *Absorption with chemical reaction in rapid reaction regime* : the system CO_2/NaOH is used to predict interfacial area a and mass transfer coefficient on gas side k_G.
- *Absorption with chemical reaction in intermediate reaction regime* : The system CO_2/NaClO/Na_2CO_3 is used to predict interfacial area a and mass transfer coefficient on liquid side k_L.
- *Physical absorption* : The system CO_2/H_2O is used to determine the overall mass transfer coefficient k_La. This last system is used to check results obtained with the previous methods.

➢ ***Absorption with chemical reaction in rapid reaction mode***

The system CO_2/NaOH is operating in rapid chemical reaction mode. During CO_2 absorption in caustic soda, carbonate ions are produced.

$$CO_2 + 2OH^- \longrightarrow CO_3^{2-} + H_2O \qquad (2)$$

Operating conditions are chosen to be in rapid regime :
- Non reversible pseudo-first order reaction
- Hatta number $Ha \geq 5$
- $Ha \ll Ei$ (instantaneous acceleration factor)

For each plate, experiments are performed with liquid flow rate lower than the maximum value specified in Table 1 and gas velocity is fixed at 10 $cm.s^{-1}$. To respect the criteria of chemical rapid regime, concentration of caustic soda is 1 $mol.L^{-1}$ and CO_2 concentration remain in range described in Table 3.

Table 3. CO_2 concentration range to obtain rapid regime

Plate	C_{CO2} % vol.
A	0 - 30 %
B	0 - 20 %
C	0 - 15 %

Once operating in rapid reaction regime, the CO_2 flux transferred in liquid phase is directly depending on interfacial area and mass transfer coefficient on gas side :

$$\frac{p_{CO_2}}{\Phi} = \frac{1}{a}\left[\frac{1}{k_G} + \frac{H}{\sqrt{D_{CO_2} k_2 C_{NaOH}}}\right] \tag{3}$$

The value of absorbed flux is calculated with this equation :

$$\Phi = \frac{C_{CO_2,\,dissolved} \cdot Q_L}{V_{reactor}} \tag{4}$$

The quantity of CO_2 dissolved is determined by using a potentiometric method. Liquid sample is dosed with sulfuric acid H_2SO_4.

The concentration of CO_2 dissolved is the carbonate ions concentration in the liquid. To minimize uncertainties, CO_2 concentration is calculated using the values of HCO_3^- and CO_3^{2-} concentrations measured at the two pH jumps :

$$C_{CO_2,\,dissolved} = \frac{1}{2}\left(C_{CO_3^{2-},\,final} + C_{HCO_3^-,\,final}\right) \tag{5}$$

Pohorecki and Moniuk (1988) determined the kinetic constant of CO_2/NaOH system as a function of temperature and soda concentration in liquid phase.

$$\log(k_2) = 0.221 C_{OH^-} - 0.016\left(C_{OH^-}\right)^2 + \log k^{\infty} \tag{6}$$

At infinite dilution, the kinetic constant is described by following correlation :

$$\log k^{\infty} = 11.895 - \frac{2382}{T} \tag{7}$$

By varying caustic soda concentration (between 1 and 2 mol.L^{-1}), it is possible to determine the mass transfer coefficient on gas side k_G and the interfacial area a with the method of Danckwerts (1966 and 1970). When P_{CO2} is plotted against NaOH concentration, k_G and a can be deduced from curve slope and interception with P_{CO2} axis at zero concentration of soda.

➢ ***Absorption in intermediate chemical reaction mode***

The system CO_2/NaClO/Na_2CO_3 is operating in intermediate chemical reaction mode. In that case, the reaction described below produces hydrogenocarbonate ions.

$$CO_3^{2-} + CO_2 + H_2O \longrightarrow 2HCO_3^- \tag{8}$$

The criteria to be in intermediate regime are the following:
- Non reversible pseudo-first order reaction
- $0.3 < Ha < 3$
- No limitation by gas transfer

The presence of hypochlorite ions is necessary to accelerate reaction kinetics and reach in the desired chemical regime. Those ions have a role of catalyst and kinetic constant depends only on the catalyst concentration.

As previously, experiments are performed with the liquid flow rate lower than the maximum values specified in Table 1. Gas velocity is fixed at 10 cm.s^{-1}. To reach intermediate regime, CO_2 concentration is fixed at 10 %, carbonate concentration is 0.5 mol.L^{-1} and the concentration of hypochlorite ions varies between 0.1 and 0.5 mol.L^{-1}.

In the intermediate chemical reaction mode, the expression of absorption flux is given by the following equation :

$$\Phi = a\, C_{CO_2,i} \sqrt{D_{CO_2} k_2 C_{ClO^-} + k_L^2} \qquad (9)$$

Absorbed flux Φ is calculated with equation (4) as previously. Concerning the quantity of CO_2 dissolved, it is always obtained by a sulfuric acid dosage. The concentration is calculated with formula (10) :

$$C_{CO_2,\,dissolved} = C_{CO_3^{2-},\,initial} - C_{CO_3^{2-},\,final} \qquad (10)$$

The k_2 constant has been evaluated by Benadda (1994) : k_2 = 1450 L.mol^{-1}.s^{-1} at 293 K. Calculation of CO_2 molecular diffusion coefficient is described in annex.
The curve which represents Φ^2 versus C_{ClO^-} gives a straight line (Danckwerts, 1966 and 1970). The slope and the intercept of this curve allows to calculate the mass transfer coefficient on liquid side k_L and interfacial area a.

➢ ***Physical absorption***

Physical absorption of CO_2 in water is the third type of experiments performed. Operating conditions are not far from the previous ones. Liquid flow rate is maximum for micro-channels. Gas velocity is equal to 10 cm.s^{-1}. And CO_2 concentration is fixed to 20 %.

When assuming that liquid flow is a plug flow regime, the expression of absorption flux is given by :

$$\Phi = k_L\, a \left(C_{CO_2,\,interface} - C_{CO_2,\,bulk}\right) \qquad (11)$$

Since the CO_2 desorption from liquid phase is quasi-instantaneous, the flux of CO_2 is estimated using CO_2 concentration measurement in gas phase with infra-red analyzer.

3. Results and discussion

3.1. Experiments with system CO_2/NaClO/Na_2CO_3

Experiments were performed for the plates A, B and C. Different concentration of hypochlorite in a constant CO_3^{2-} solution are used at different liquid flow rates (Table 4). The gas phase flow rate is kept constant (corresponding to a superficial velocity of 10 cm.s^{-1}) during these experiments and CO_2 volumetric concentration is fixed at 10 %. It can be noted that for each plate, higher is the flow rate, lower is residence time and the conversion and higher is uncertainty on the CO_2 flux determination (since it is deduced from a concentration difference with order of magnitude close to those of concentration measurement). For this reason, it was not possible to exploit results obtained with plate A.

Table 4. Liquid experimental conditions

Plate	Flow rate ($L.h^{-1}$)	$C_{CO_3^{2-}}$ ($mol.L^{-1}$)	C_{ClO^-} ($mol.L^{-1}$)
A	0.05	0.5	0.1 – 0.5
B	0.2 - 0.4	0.5	0.1 - 0.5
C	0.5 - 1.0 - 1.5	0.5	0.1 - 0.5

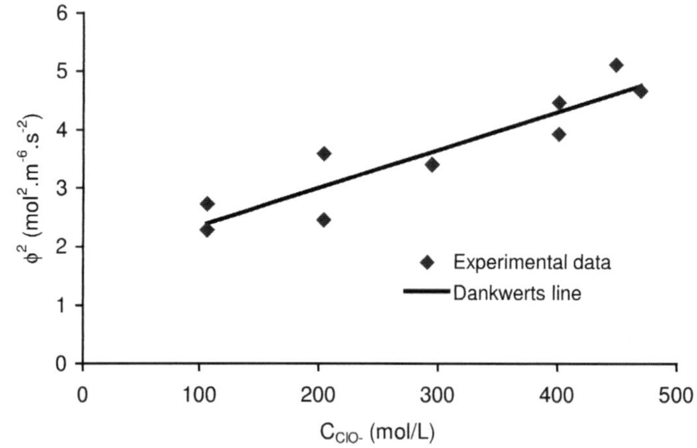

Figure 3. Variation of Φ^2 as a function of ClO^- concentration for the plate C with a liquid flow rate of 0.5 $L.h^{-1}$

For each experiment, the square of the mass flux is plotted as a function of the hypochlorite concentration (Figure 3). The interfacial area *a* and the mass transfer coefficient k_L can deduced from the fitted line. The measured values obtained with plates B and C are summarized in Table 5.

Table 5. Measured values for mass transfer coefficient on liquid side and interfacial area for plates B and C

Plate	Liquid flow rate ($L.h^{-1}$)	a ($m^2.m^{-3}$)	k_L ($m.s^{-1}$)	k_La (s^{-1})
B	0.2	183 ± 1	$4.81.10^{-4} \pm 0.12 .10^{-4}$	$8.80.10^{-2} \pm 0.18 .10^{-2}$
	0.4	189 ± 6	$4.78.10^{-4} \pm 1.15 .10^{-4}$	$9.03.10^{-2} \pm 0.77 .10^{-2}$
C	0.5	188 ± 4	$8.29.10^{-4} \pm 0.34 .10^{-4}$	0.156 ± 0.004
	1	187 ± 10	$10.80.10^{-4} \pm 0.90 .10^{-4}$	0.202 ± 0.007
	1.5	169 ± 17	$7.04.10^{-4} \pm 2.01 .10^{-4}$	0.119 ± 0.018

The interfacial area remains rather constant close to a value of 180 $m^2.m^{-3}$. The measured mass transfer coefficient on liquid side are varying between values of $4.7.10^{-4}$ and 10^{-3} $m.s^{-1}$ that are among highest values obtained on structured packing configuration.

For the whole experiments, Ha varies between 0.34 and 1.11 and $(H/k_LEi)/(1/k_G)$ between 57.11 and 98.42. The criteria of intermediate regime $0.3 < Ha < 3$ and $(H/k_LEi) >> (1/k_G)$ are thus validated.

3.2. Experiments with system CO_2/NaOH

Experiments were performed with this system on the plate B with 0.4 $L.h^{-1}$ as liquid flow-rate and 0.79 $mol.L^{-1}$ to 1.85 $mol.L^{-1}$ as NaOH concentration in liquid phase. The gas velocity was 10 $cm.s^{-1}$ with 10 % vol. of CO_2.

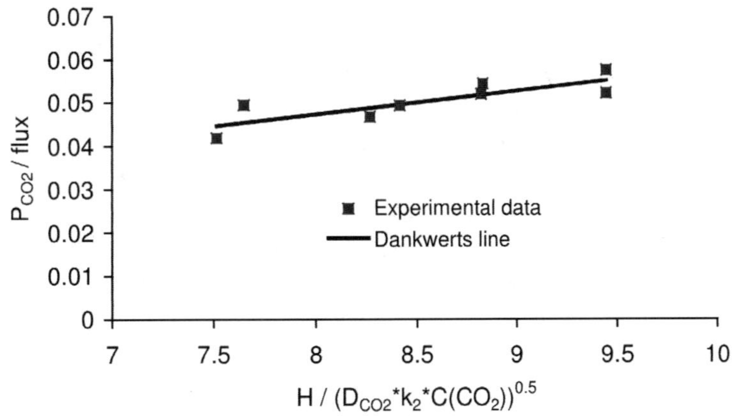

Figure 4. Variation of CO_2 partial pressure as a function of $\sqrt{D_{CO_2} k_2 C_{NaOH}}$ for the plate B and with a liquid flow rate of 0.4 $L.h^{-1}$

The analysis of Dankwerts line (Figure 4) allows to obtain the following mass transfer parameters on gas side, k_G and *a* for plate B (Table 6) :

Table 6. Measured values for mass transfer coefficient on gas side and interfacial area for plate B for liquid flow rate of 0.4 L/h

a	185 ± 24 $m^2.m^{-3}$
k_G	1.32 ± 0.36 $mol.m^{-2}.s^{-1}.bar^{-1}$
k_Ga	244 ± 98 $mol.m^{-3}.s^{-1}.bar^{-1}$

These results clearly show that mass transfer on gas phase is not the limiting phenomenon on whole mass transfer since mass transfer resistance on liquid side $\frac{H}{k_L E}$ is 4000 times higher than mass transfer resistance on gas side $\frac{1}{k_G}$. This results confirm that the hypothesis of chemical intermediate regime can be assumed for CO_2/NaClO/Na_2CO_3 system. It can also be noted that the measured value for interfacial area is very close to values obtained with system working in chemical intermediate mode.

For the systems CO_2/NaOH, Table 7 presents the calculated value of Hatta Number, Ha, and the instantaneous acceleration factor, Ei. Reactions of the two systems are non reversible and pseudo-first order.

The results show that conditions of rapid regime (Ha > 5 and Ha << Ei) are realized for the system studied.

Table 7. Values (Min/Max) of Ha, Ei and Ei/Ha for CO_2/NaOH system

	Min	Max
Ha	5.65	10.53
Ei	180.24	622.47
Ei/Ha	31.91	60.42

3.3. Experiments with CO_2/water system

Experiments performed with this system on the plate B with a liquid flow rate of 0.4 $L.h^{-1}$, gas velocity of 10 $cm.s^{-1}$ and 10 % of CO_2. Since a very low quantity of CO_2 has transferred in liquid phase during the liquid residence time, the accuracy on the measurement of a CO_2 concentration difference between inlet and outlet was too low to give an acceptable value for k_La.

3.4. Discussion of measured data

Interfacial area a

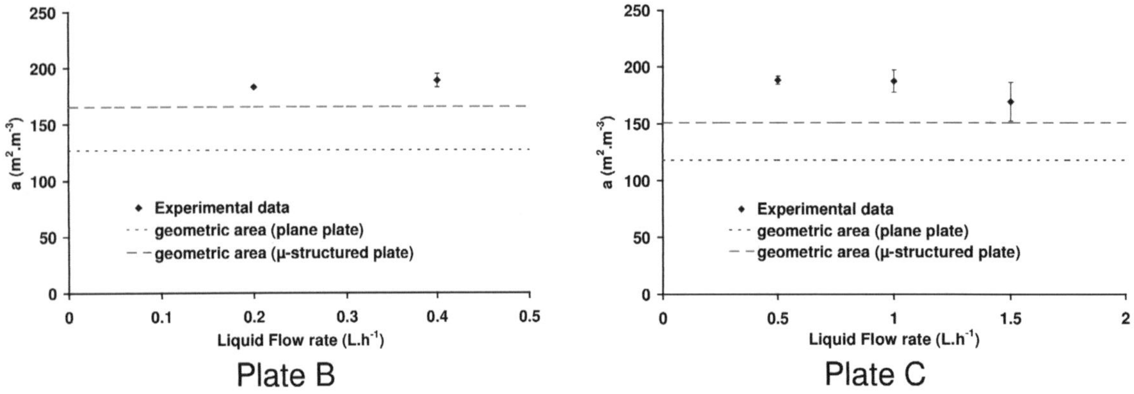

Figure 5. Comparison of measured and estimated interfacial area on plates B and C

The measured values of interfacial areas are compared to geometrical area with or without micro-channels (Figure 5). In all cases, the experimental data appears to be higher than geometrical area obtained considering the area developed on structured plate. The measured values are about 10 % higher for plate B and 25 % higher for plate C. One possible explanation may be the presence of waves on the interface. Ishigai et al (1972) distinguish five regimes of flow depending on the Reynolds number, Re for liquid film : Laminar, First Transition, Stable Wavy, Second Transition, Turbulent. The operating conditions indicate that experiment were performed in the First Transition regime (between Re = 9 and 42). The flow regime is near the first transition regime that implies possible presence of waves on liquid interface.

This values of interfacial areas appears not large enough compared to those obtained on classical structured packing column but could be easily increased by optimizing the reactor geometry. As a matter of fact the gas chamber height (5.8 mm) is quite larger compared to µ-channel height (< 400 µm). Since the mass transfer on gas side is not a limiting phenomenon, it is possible to reduce the gas chamber cross-sectional area and thus the interfacial area could be increase up to 1000 $m^2.m^{-3}$ of reactor by reducing the gas chamber from 5.8 mm to 1 mm (with liquid hold up increasing from 1 % up to 8 %).

Mass transfer coefficient k_L

Concerning the liquid side mass transfer coefficient k_L, the experimental values obtained from the chemical method are shown in Figure 6.

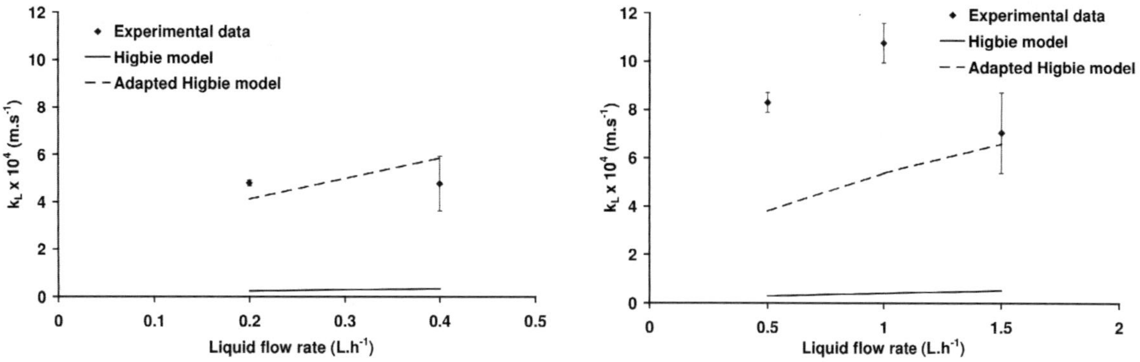

Plate B Plate C

Figure 6. Comparison of measured and estimated mass transfer coefficients on plates B and C

Experimental data for plates B and C are compared to two models, based on the penetration theory of Higbie (1935). The models make a correlation between the mass transfer coefficient k_L and the liquid exposure time τ :

$$k_L = 2\sqrt{\frac{D_{CO_2}}{\pi\,\tau}}$$

The exposure time τ corresponds to the liquid film renewal time, when a gas flow is in contact with the liquid falling film. It is the ratio of a characteristic length by a characteristic velocity. For the determination of the characteristic velocity, liquid velocity is calculated assuming micro-channel filled of liquid. Indeed, liquid film thickness profiles obtained by Yeong (2004) show that the micro-channel is nearly full of liquid at various flow rates. Consequently, the liquid velocity chosen is the ratio of the flow rate in a micro-channel by its section. For the classical Higbie model, the characteristic length is the channel length L. So, the exposure time τ is the total liquid residence time in the micro-reactor. This choice shows an important underestimation of k_L in Figure 6 by the continuous line. The measured values are about 10 times higher for plate B and 20 times higher for plate C. This result indicates that the liquid film renewal time is much more lower than the total liquid residence time over the plates.

As said previously, there is probably the presence of waves on the interface that induces local mixing in the film. In a second approach, it has been assumed that the renewal of the liquid surface occurs on a length, close to the channel height h. Local mixing structure is assumed to be homogeneously developed in channel height and length. The corresponding adapted Higbie model is plotted in Figure 6 by the dashed line and shows a rather good agreement between the simulated values and the experimental data, particularly for plate B. For plate C, a difference remains for the lowest liquid flow rates. It can be noted that the main difference appears for data points which were already characterized by highest interfacial area probably corresponding to higher wavy regime.

4. Conclusions

Gas-liquid mass transfer is measured in a micro-structured falling film reactor, using 3 different plates (hydraulic diameter range : [226 – 907] μm). It consists in determining transfer coefficients on gas and liquid sides k_L, k_G and the interfacial area a.
Different chemical reaction systems were implemented to determine those parameters. Absorption in rapid reaction mode with a CO_2/NaOH system permits to obtain a, k_G and k_Ga. The other parameters a, k_L and k_La are obtained using an absorption in intermediate reaction mode with a CO_2/NaClO/Na_2CO_3 system.

Values obtained for k_G and k_Ga are higher than those for k_L and k_La, showing there is no transfer limitation on gas side. Interfacial area measured for plate B and C is constant with liquid flow rate and superior of 10 to 25 % to the geometrical area developed on structured plate. It can be explained by the presence of waves on the surface of the liquid film.
The values of interfacial area are lower than those obtained on classical structured packing but could be enhanced up to 1000 $m^2.m^{-3}$ of reactor with a smaller gas chamber.

k_L values determined with a chemical method are varying between 4.10^{-4} and 10^{-3} $m.s^{-1}$ and correspond to highest values obtained on classical packing. Results match rather well with an adapted Hidgie model, based on the characteristic length equal to channel height and velocity of the liquid film.

5. Annex

Calculation of molecular diffusion coefficient of CO_2 in liquid phase

Molecular diffusivities depend on temperature and phases composition. To estimate the diffusion coefficient of CO_2 in the studied aqueous solutions, this coefficient is first determined in pure water. Then, diffusion coefficient is found with a viscosity correction.
The diffusion coefficient of CO_2 in pure water is obtained with the following correlation (Pohorecki and Moniuk, 1988) :

$$\log\left(D_{CO_2-H_2O}\right) = -8.1764 + \frac{712.52}{T} - \frac{259100}{T^2} \qquad (14)$$

To obtain, the diffusion coefficient of CO_2 in an electrolyte aqueous solution, the relation of Wilke and Chang (1955) is used :

$$\left[D_{CO_2-A^-}\mu_{A^-}\right] = \left[D_{CO_2-H_2O}\mu_{H_2O}\right]^n \qquad (15)$$

with A^-, anionic species in aqueous the solution.
Exponent n is 1 for CO_2/Na_2CO_3 system (Benadda, 1994) and n = 0.637 for CO_2-NaOH system (Dragan et al, 2000).

6. Notations

Latin alphabet :

a	interfacial area ($m^2.m^{-3}$)
C	concentration ($mol.L^{-1}$)
D	diffusion coefficient ($m^2.s^{-1}$)
E	acceleration factor
Ei	instantaneous acceleration factor
g	gravity constant ($m^2.s^{-1}$)
h	channel height (m)
H	Henry constant ($Pa.m^3.mol^{-1}$)
Ha	Hatta number (-)
k	kinetic constant ($m^3.mol^{-1}.s^{-1}$)
k_G	gas phase (gas-side) mass transfer coefficient ($m.s^{-1}$)
k_L	liquid phase (liquid-side) mass transfer coefficient ($mol.m^{-2}.s^{-1}.bar^{-1}$)
L	channel length (m)
l	channel width (m)
n	channels number (-)
P	partial pressure (Pa)
Q	flow rate ($m^3.s^{-1}$) or ($L.h^{-1}$)
R	channel curvature radius (m)
Re	Reynolds number (-)
T	temperature (K)
V	volume (m^3)

Greek alphabet :

$\varnothing_{hyd}$	hydraulic diameter (m)
μ	dynamic viscosity (Pa.s)
Φ	absorbed flux ($mol.m^{-3}.s^{-1}$)
τ	residence time (s)

Index :

G	gas
L	liquid

7. References

Benadda, B., *Contribution à l'étude du transfert de matière dans une colonne à garnissage - influence de la pression.* Thèse Doctorale, Université Claude Bernard – Lyon 1, 1994.

Blevins R.D., *Applied Fluid Dynamics Handbook*, Krieger Publishing Compagny Chap. 6, (1992), 38-123.

Danckwerts, P.V. and Sharma M.M., *Chem. Eng. Sci.*, 23 (1966), 559.

Danckwerts, P.V., *Gas-Liquid Reactions*, McGraw-Hill, New-York, 1970.

Dragan S., Cristea M., Simionescu, B.C., *Formation and properties of some polyelectrolyte complexes between strong complementary polyelectrolytes*, Revue Roumaine de Chimie, Vol. 45, N°., 7-8 (2000), 763-769.

Hessel V., Ehrfeld W., Herweck Th., Haverkamp V., Löwe H., Schiewe J., Wille Ch., Kern Th., Lutz N., *Gaz / Liquid Microreactors : Hydrodynamics and Mass transfer*, 4th Intal Conference on Microreaction Technology, 5-9 Mars, Atlanta, USA, (2000), 175-186.

Higbie R., Trans. Am. Inst. Chem. Engrs., Vol. 31, (1935), 365.

Ishigai S., Nakanisi S., Koizumi T., Oyaki Z., *Hydrodynamics and heat transfer of vertical falling films*, Bull. JSME, Vol. 15, N°83, (1972), 594.

Wilke C.R., Chang P., A.E.CH.E. Journal, 1 (1955), 264-270.

Yeong K.K., Gavriilidis A., Zapf R., Hessel V., *Experimental studies of nitrobenzene hydrogenation in a microstructured falling film reactor*, Chem. Eng. Sci., 59 (2004), 3491-3494.

Selective Condensation Reaction of Phenols and Formaldehyde Using a Micromixer

Noboru Daito[*], *Jun-ichi Yoshida*[**], *Kazuhiro Mae*[***]

**The Research Association of Micro Chemical Process Technology*
***Department of Synthetic Chemistry and Biological Chemistry, Graduate School of Engineering, Kyoto University*
****Department of Chemical Engineering, Kyoto University*

Prepared for Presentation at IMRET 8, Wed., April 13, 2005, Modularization and Multiscale Design

Unpublished

Abstract

The method of mixing should significantly affect the efficiency and selectivity of some chemical reactions. Some fast reactions are expected to be conducted in a highly controlled manner through extremely fast mixing and efficient heat transfer using microdevices. We studied the condensation reaction of hydroxybenzylalcohol (HBA) with phenol, to control the selectivity of bis-phenols (desired product) over tris-phenols (by-product) using various micromixers in a flow system. We attempted to terminate the main condensation reaction rapidly without raising the mole ratio of phenol to formaldehyde, in order to avoid the consecutive side reaction, by utilizing a micromixing method. We also compared the selectivity for bis-phenols under various phenol/HBA mole ratios with a batch system using a flask.

The selectivity for bis-phenols using a multilamination type micromixer (channel width 40 μm) was high at low mole ratios of phenol/HBA, and mole ratios only half of that used in the batch system were necessary to obtain the same selectivity. The flow rate, however, could not be increased to more than 6 mL/min, because of the high pressure loss. Furthermore, clogging of the microchannels occurred within 2 min. To overcome these problems, a new micromixer was tested, which has several channels converging on one point (8 channels of 200 μm width), developed by MCPT. The reaction was conducted steadily at a flow rate of more than 18 mL/min with low pressure loss and without clogging. The selectivity for bis-phenols was successfully increased to a level comparable to that achieved with a multilamination-type micromixer.

1. Introduction

Paint formulations have recently been shifting to more environmentally friendly products, such as nonsolvent and powdered formulations, to reduce the emission of volatile organic compounds (VOCs). As a result of this trend, the demand for bisphenol-F increases year by year, as a material for epoxy resin having superior heat resistance and low viscosity. Bisphenol-F is synthesized by the condensation reaction of phenol and formaldehyde with an acidic catalyst. Figure 1 shows the overall reaction scheme for bisphenol-F synthesis. In these reactions, the intermediate, hydroxybenzylalcohol (HBA), forms immediately and is then converted into bisphenol-F. At the same time, the condensation reaction forming higher-order condensates also proceeds, by the reaction of HBA and the bisphenol formed [1,2]. Thus, the bisphenol-F product includes high molecular weight, highly condensed compounds of formaldehyde and phenol, such as tris-phenols, as a result of the limitations of reaction control. Since these by-products significantly impair the desired low viscosity of bisphenol-F, the condensation reaction must be suppressed. To achieve this, the mole ratio of phenol to formaldehyde is fixed at an extremely high value, between 30 and 40, in the current process for the production of general-grade bisphenol-F (purity 90–94%). This value is more than 15 times larger than the theoretical mole ratio required. The excess feed causes problems such as increased expense for recovery of the unreacted phenol, and low productivity. There is a strong desire to improve this situation through development of an efficient and environmentally benign production method to increase the selectivity for bis-phenols using the theoretical mole ratio of reactants.

A lot of attention has been paid to micromixing operations as an expedient way of achieving highly efficient mixing. In addition, operation in the microspace enables the strict control of temperature and residence time. Proper use of these features would give efficient conditions for fast reactions, which are difficult to control by conventional operations in macroscale processes [3, 4]. From this viewpoint, we focused on the condensation reaction of HBA and phenol with an acid catalyst using various micromixers. We then compared the selectivity for bis-phenols under various phenol/formaldehyde mole ratios between the microsystem and a batch system using a flask, in order to clarify the validity of microreaction systems for increasing the selectivity for bis-phenols.

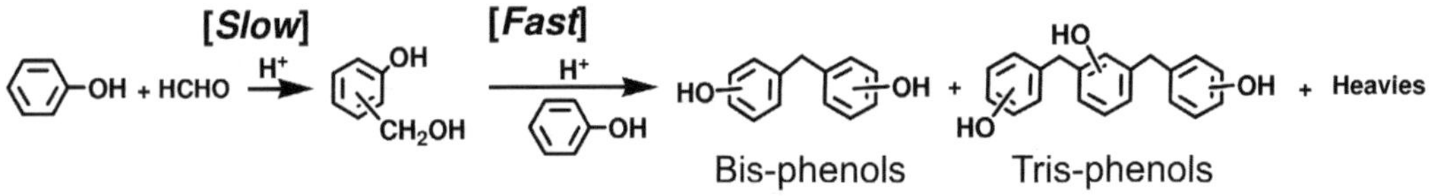

Figure 1 Reaction scheme of bisphenol-F synthesis

2. Experimental

2.1 Micromixers

Three types of mixers, a T-shaped mixer with 500 μm I.D., a multilamination type micromixer with 40 μm channels of Si (Institut für Mikrotechnik Maintz GmbH), and a K-M mixer were used. The K-M mixer is a new model developed by MCPT, consisting of three cylindrical components. As shown in Figure. 2, the two fluids are fed into the ring header on the inlet plate (1) to be uniformly distributed to the plural channels on the inner plate (2). The fluids then converge at the center of the inner plate (2). Mixing is promoted by both the diffusion and the shear rate at the center. Finally, the mixed fluid is discharged through the hole of the exit plate (3) [5, 6]. In this experiment, an inner plate having eight channels of 200 μm width and 200 μm depth was used.

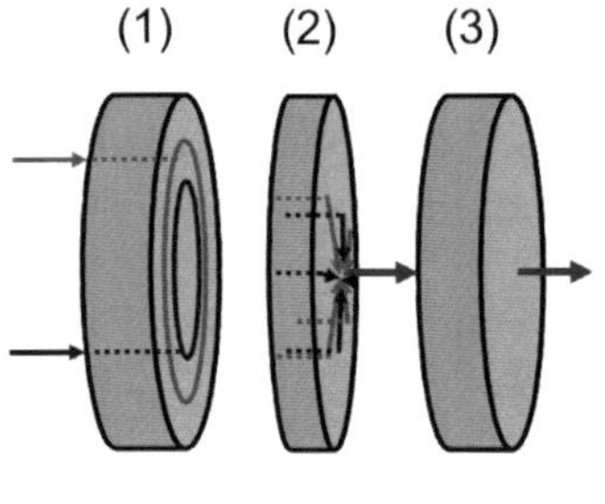

Figure 2 Structure of K-M mixer

2.2 Substrates

o-HBA (Wako Pure Chemical Industries Ltd., 98%) and the catalyst p-toluenesulfonic acid (PTS) (Wako, 99%) were dissolved in phenol (Wako, 99%), giving a HBA/phenol solution and a PTS/phenol solution. For one set of experiments the HBA solutions and the PTS solutions were prepared in same volume so that the mole ratios of phenol/HBA prepared were 10, 15 and 30, and the PTS concentration was 0.1 M after mixing. For a further set of experiments, the HBA solution and the PTS solution were prepared at a volume ratio of 14:1 with a mole ratio of phenol/HBA of 15, and a PTS concentration of 0.1 M after mixing.

2.3 Reaction and analysis

Figure 3 shows the experimental apparatus. The micromixers were soaked in an oil bath kept at 110 °C, as were the PTFE inlet tubes (50 cm, 1/16" O.D., 1 mm I.D.), to ensure that the solutions were preheated. Solution A (HBA/phenol) and Solution B (PTS/phenol) in plastic syringes, and the delivery tubes (1/16" O.D., 1 mm I.D.), were kept at 70 °C using cord heaters. The solutions were fed into the mixers at the same flow rate by syringe pump (HARVARD 11-IW), and the mixture continued to react in a 20 cm PTFE tube (1/16" O.D.,1 mm I.D.) subsequent to the mixer. In some experiments, the reaction was conducted under conditions where the flow rate ratio of Solution A to Solution B was 14. Bis-phenols and tris-phenols in the products were quantitatively measured by HPLC, using the internal standards method. Figure 4 shows the retention time for each compound in the UV spectrum. We measured the concentration of products based on this UV spectrum and calculated the selectivity for bis-phenols using Equation 1.

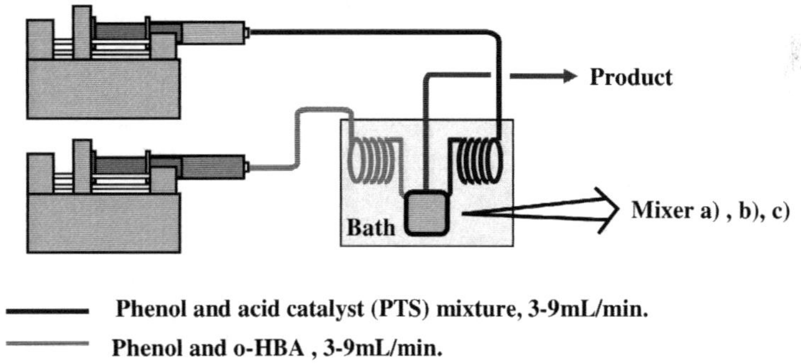

Figure 3 Reaction apparatus

$$\text{Bis-phenols selectivity} = \text{Bis-phenols} /(\text{Bis-phenols} + \text{Tris-phenols}) \quad (1)$$

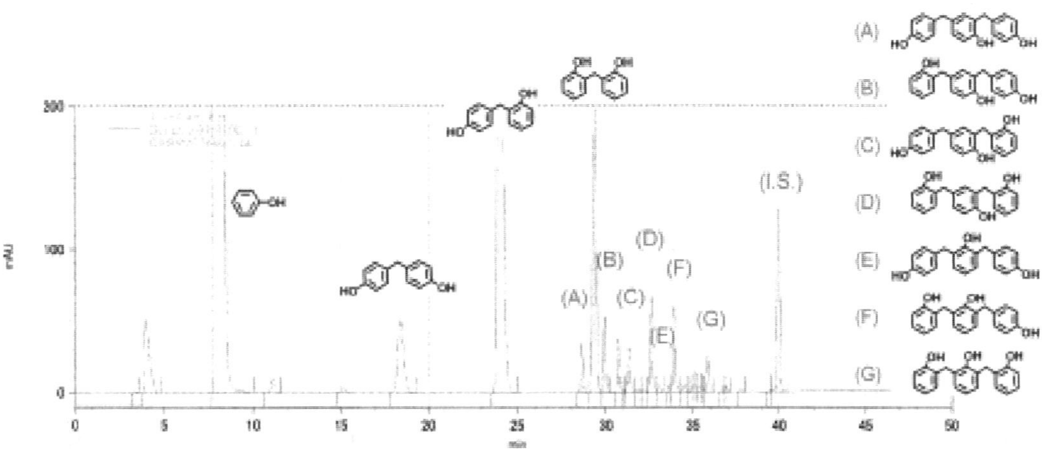

Figure 4 HPLC chromatogram: uv:254 nm, pseudocumene internal standard

3. Results and discussion

3.1 Comparison of mixer performance for bis-phenols selectivity

Table 1 summarizes the selectivity for bis-phenols under various conditions. The condensation of HBA was 100% in all the experiments. Initially, we compared the bis-phenols' selectivity under the same flow rate ratio of reactants. The selectivity achieved using the T-shaped mixer was not particularly high, even at high flow rates, with values comparable with those achieved by batch operation. We suspect that the mixing was insufficient in the T-shaped mixer, since the reactions of phenol and HBA are very fast. In contrast, the selectivity for bis-phenols using multilamination and K-M type micromixers was higher. Comparing the results using a phenol/HBA mole ratio of 15 in the K-M mixer, the selectivity for bis-phenols increased with the increase in the total flow rate. This suggests that the mixing of reactants is improved by higher flow rates.

Table 1 Experimental data for condensation of phenol and HBA using various micromixers

Entry	Mixer	Flow rate (mL/min.)		Phenol/HBA	Bis-phenol Selectivity
		Solution A	Solution B	(mol/mol)	(%)
1	T-Shaped	3	3	15	84.8
2	T-Shaped	9	9	15	87.2
3	Multilamination	3	3	20	91.7
4	K-M	3	3	15	89.7
5	K-M	9	9	15	90.3
6	K-M	9	9	10	86.1
7	K-M	14	1	15	90.0
-	Batch	-	-	15	85.6
-	Batch	-	-	30	91.8

Operability is one of the crucial factors to consider when developing an actual process. The flow rate was limited to a maximum of 6 mL/min when using the multilamination type micromixer, because of its high pressure loss. A further serious problem was clogging, caused within 2 min by the sticking of adhesive materials inside the exit slit. The K-M mixer was successfully operated at a flow rate of more than 18 mL/min with low pressure loss and without clogging. After the operation, no fouling was observed. The T-shaped mixer had less pressure loss and fewer problems with clogging, but the interface of the two fluids in laminar flow is too small to expect rapid mixing through diffusion. In the multilamination-type micromixer, the interface is large and the diffusion length is considerably shorter, and therefore instant mixing is accomplished, but the small cross-sectional surface increases pressure loss and induces clogging with viscous fluids. In contrast, dispersion of the fluid with kinetic energy and diffusion between the microfluxes are efficiently realized in the K-M mixer, by collision through the plural passages in a relatively opened space for mixing. This allows a high throughput with low pressure loss and simultaneously prevents clogging with viscous substances or precipitation of solids. Summarizing these discussions, the K-M mixer showed excellent performance for the production of bis-phenols at short reaction times.

3.2 Effect of phenol/HBA ratio on the selectivity for bis-phenols

The main concern in this study is to reduce the ratio of phenol to HBA while keeping the selectivity for bis-phenols. Hence, we performed the reaction of phenol with HBA at various mole ratios. Figure 5 shows the effect of the phenol/HBA ratio on the selectivity for bis-phenols. The selectivity increased with an increase in the phenol/HBA ratio. The selectivity using the K-M mixer was the highest at all the phenol/HBA ratios tested, and its advantage was greatest at low phenol/HBA ratios. Comparing the selectivity to the commercial process, it was found that the phenol/HBA ratio could be diminished by half when using the K-M mixer. This clearly shows that micromixing is useful for obtaining bis-phenols in high yields with economic advantages.

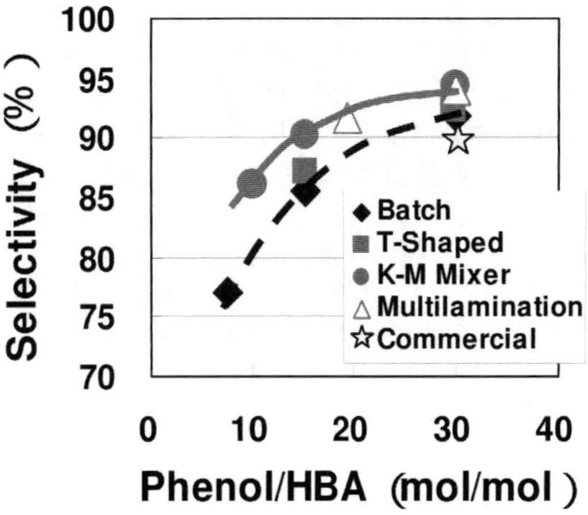

Figure 5 Influence of phenol/HBA mole ratio on the bis-phenols selectivities: the highest value of each mixer

Another restriction in the production of bis-phenols is the high flow rate of phenol required. In HBA synthesis, excessive phenol is required as a dilution agent to increase the HBA yield. Therefore, the HBA/phenol mixture was fed to the bis-phenols production section at extremely high flow rates. To achieve the production of bis-phenols economically, a small amount of phenol containing the acid catalyst should be mixed with a large amount of HBA/phenol mixture feed. The ideal flow rate of acid catalyst/phenol solution (Solution B) is 1/14th that of the HBA/phenol solution (Solution A). From this viewpoint, we performed the reaction under these conditions using the K-M mixer. As listed in Table 1, the selectivity for bis-phenols reached 90%, indicating that the K-M mixer has a high potential to produce bis-phenols under actual operation conditions.

3.3 Role of micromixing on the condensation reactions

As mentioned above, micromixing was effective in giving high selectivity for bis-phenols. To demonstrate this effect, we examined the yield of each product in the light of the reaction mechanism. Figure 6 shows the detailed reaction scheme for the condensation reaction of HBA and phenol with PTS catalyst. HBA reacts immediately with phenol to give 2,2'-bis-phenol and 2,4'-bis-phenol by mixing with the PTS catalyst. However, these bis-phenols react simultaneously with the remaining HBA to produce tris-phenols. To suppress the production of tris-phenols, we should instantly consume all the HBA in the formation of bis-phenols. If HBA remains, the consecutive condensation reactions proceed at a certain rate determined by the kinetics.

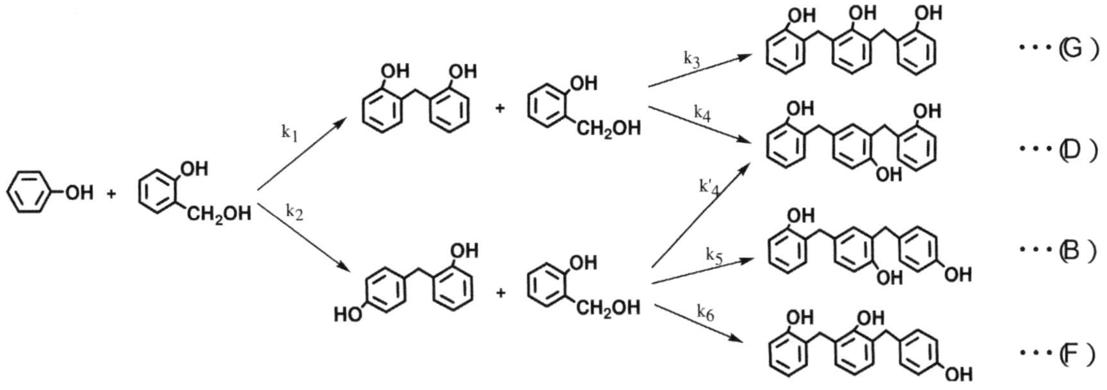

Figure 6 Detailed scheme for condensation reaction of phenol and HBA

Table 2 lists each product yield, based on HBA, using T-shaped and K-M mixers. Comparing the yields of bis-phenols, it was found that the formation rate of 2,4'-bis-phenol is faster than that of 2,2'-bis-phenol. As the mixing is not as efficient in the T-shaped mixer, the 2,4'-bis-phenols that formed had a chance to react with unreacted HBA forming tris-phenols. This led to a decrease in the yield of 2,2'-bis-phenol. In contrast, the mixing in the K-M mixer was so rapid that HBA effectively reacted with phenol and was consumed rapidly. This caused an increase in 2,2'-bis-phenol and a decrease in tris-phenols. From these observations, it was demonstrated that micromixing brought about effective consumption of HBA by phenol and suppressed the consecutive condensation reaction to tris-phenols.

Table 2 Difference in product yield by mixers. [% on HBA basis]

	4,4'-Bis	2,4'-Bis	2,2'-Bis	Tris (G)	Tris (D)	Tris (B)	Tris (F)
T-Shaped	0.09	55.46	31.65	4.21	5.89	1.57	1.13
K-M	0.09	55.35	34.86	3.15	4.65	0.91	0.99

4. Conclusion

We have proposed a new microreaction method for the efficient production of bis-phenols. Several experiments were performed using three types of mixers, and it was demonstrated that fast and uniform mixing, rapid heat removal and strict residence time control in the microspace realizes both the rapid termination of the main reaction and the simultaneous prevention of the consecutive side reactions. This system enables us to achieve high selectivity for bis-phenols, even at low phenol/formaldehyde mole ratios. Thus, the new production method for bisphenol-F detailed here is expected to lead to an advanced operation for a practical efficient process.

6.Acknowledgement

This research was supported financially by the Project of Micro Chemical Technology for Production, Analysis and Measurement System of NEDO, Japan.

References

[1] Martin, R.W.; The Chemistry of Phenolic Resins. New York, J. Wiley, 1956.
[2] Malhotra, H.C.; J. Appl. Polym. Sci. 20, 2461 (1976).
[3] Suga, S.; Nagaki, A.; Yoshida, J. Chem. Commun. 355 (2003).
[4] Suga, S.; Nagaki, A.; Tsutsui, Y.; Yoshida; J. Org. Lett. 5, 945 (2003).
[5] Nagasawa, H.; Aoki, N.; Mae, K.; Proceedings 3rd Workshop on Micro Chemical Plants, 42 (2005).
[6] Nagasawa, H.; Aoki, N.; Mae, K.; J. Chem. Eng. & Tech. (in press).

Continuous Hydrogenation Reactions in a Novel Gas-Liquid Flow Reactor Packed with Pd/C

Nungruethai Yoswathananont, Kohei Nitta, Yumi Nishiuchi, and Masaaki Sato

Department of Chemistry, Faculty of Arts and Sciences, Osaka Prefecture University, Gakuencho 1-1, Sakai, Osaka, 599-8531, Japan

Prepared for presentation at AIChE Spring National Meeting/ April 13, 2005/ IMRET 8/ 133 Modularization and Multiscale Design

Unpublished

Abstract

A novel gas-liquid-solid microflow system was applied to hydrogenation reactions of aldehyde and nitrile compounds. The microflow reactor, we present here, consisted of a stainless steel tube or a Teflon tube, of which the inner diameter was 1 mm and the length was varied from 5 to 25 cm, and a T-shaped mixer, where substrate solution and hydrogen gas were mixed together. Inside the tube, palladium on carbon catalyst (Pd/C) was packed firmly. Comparing the hydrogenation reactions by the microflow and conventional batch systems at 25°C and at elevated temperature (50-90°C) under the pressure of 0.5-2.5 MPa, the hydrogenation reactions by the microflow system revealed to be highly reactive and more efficient than those by the corresponding batch systems.

Keywords: microreactor, hydrogenation reaction, heterogeneous catalyst, microflow reactor

Introduction

Hydrogenation reactions play important roles in the chemical and pharmaceutical industries.[1] They are normally carried out in a large scale batch reactor to get products in large quantities for economical reasons. A representative example is a hydrogenation reaction of aromatic hydrocarbons. Benzene, for instance, is hydrogenised to cyclohexane, which is converted further to such as ε-caprolactum or adipic acid. Hydrogenation of phenol and its derivatives is also an important reaction since the products, cyclohexanone and its derivatives, are key materials for preparing liquid crystal compounds.[2] Nitriles are converted to the corresponding amines in the presence of Pd-alumina catalyst. In the case of adiponitrile, the hydrogenated product, hexamethylenediamine, which is an important industrial intermediate to prepare polyamides such as nylon, are produced more than 1,300,000 tons per year by the use of slurry reactor or trickle bed reactor.[3] These large scale reactors in industry, however, are thought to cause some decreases in the conversion and selectivity of the hydrogenation reactions compared with small scale reactors such as 100 ml flask or autoclave. In addition, these large scale operations need careful control to maintain uniform reaction without hotspots to avoid runaway reaction.

Over the recent years, "microreactor", which was defined as a miniaturized flow reactor with channel space in micrometer range, was gaining increasing interest since it was expected to open a new field in the chemical production.[4] Some chemical reactions in a stream through the small channel were reported to enable hazardous and/or exothermic reactions such as direct fluorination of organic compound with the elemental fluorine,[5] nitration of aromatic compounds,[6] and diazotization of aromatic amines[7] in high efficiencies and with minor risks. In addition, the stream through the small channel was reported to have advantages over multiphase hydrogenation reaction by increasing mass transfers between different phases.[8,9] Bellefon had reported an efficient high-throughput screening system for an asymmetric hydrogenation reaction with the use of a microreactor, where tiny hydrogen gas bubbles were generated in ethylene glycol/water with surfactant.[9]

In this work, we wish to report a novel gas-liquid-solid microflow system for hydrogenation reactions. In this microflow system, the hydrogen gas and the substrate solution are mixed at T-shaped mixer and the resulting plug flow is introduced to a stainless steel tube or a Teflon tube, where Pd/C is packed firmly. The hydrogen gas bubbles and the substrate solution travel through the micro or nano-scale channels between and through the Pd/C particles giving rise to a highly reactive and efficient hydrogenation system.

Experimental

All reagents were of analytical grade and were used without further purification. Pd/C (AER type, 5% wt Pd) was purchased from N. E. CHEMCAT Co.

The microflow reactor consisted of the stainless steel tube or the Teflon tube, 1 mm in inner diameter and 5-25 cm in length. The packing of the Pd/C into the tube was performed by filling a slurry solution of Pd/C in methanol at a flow rate of 0.18 mL/min for 12 hours. At both ends of the tube, two filters were placed as to hold the slurry catalyst inside.

The substrate solution was introduced to one inlet of the T-shaped mixer by a HPLC pump, while the hydrogen gas was introduced in the other inlet by controlling the pressure at around 2.5 MPa. The resulting mixture (a plug flow having almost identical volume of gas and liquid phases) was introduced to the inlet of the tube. The residence time in the tube (25 cm in length) was only 2 minutes. The solution coming out from the outlet of the tube was collected. The methanol solvent was removed by evaporation under reduced pressure and the product was determined by ^{1}H-NMR.

In the batch experiment, Pd/C was suspended in methanol solution containing the substrate, where the substrate concentration was the same as in the microflow experiment. The hydrogen gas was admitted to an autoclave (50 mL capacity) up to 2.5 MPa. After letting the reaction run for 2 minutes, the Pd/C was filtered off and the solvent was evaporated under reduced pressure to isolate the product. The yield of the product was determined by ^{1}H-NMR.

The reaction temperature was controlled by placing the reactor in a temperature controlled water bath or by heating with a ribbon heater wrapped round the reactor, and the temperature was monitored by a digital thermometer.

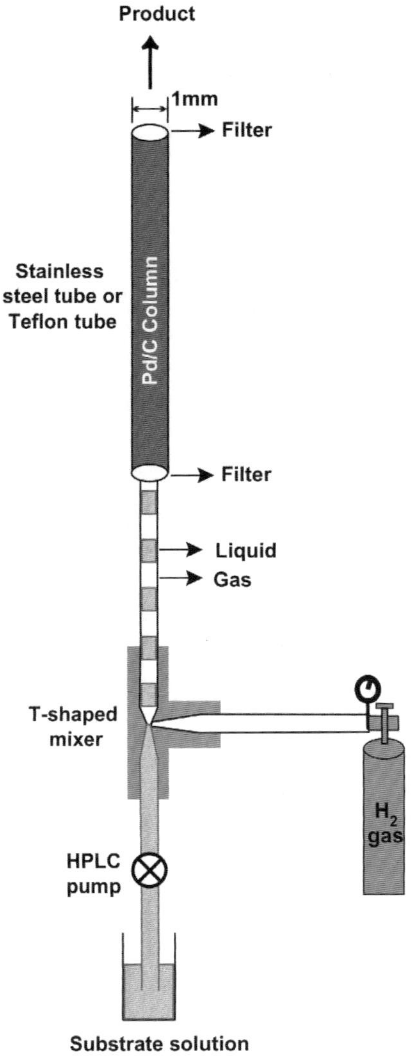

Figure 1 Schematic representation of the gas-liquid-solid microflow system for efficient hydrogenation reactions.

Results and Discussion

To evaluate the efficiency of this microflow system, hydrogenation reaction of 4-cyanobenzaldehyde (**I**) was chosen as a model reaction. Compound **I** had two functional groups, aldehyde and cyanide, which were hydrogenised individually. Hydrogenation reaction of compound **I** gave 4-cyanobenzyl alcohol (**II**), 4-hydroxymethyl-benzylamine (**III-1**), bis(4-hydroxymethylbenzyl)amine (**III-2**) and 4-methylbenzyl alcohol (**IV**) as shown in Scheme 1. The secondary amine, **III-2**, seemed to be formed by an intermolecular reaction of the **III-1** with the imine intermediate.[10] The hydrogenation reactions of compound **I** by the microflow and batch systems under corresponding conditions are summarized in Table 1.

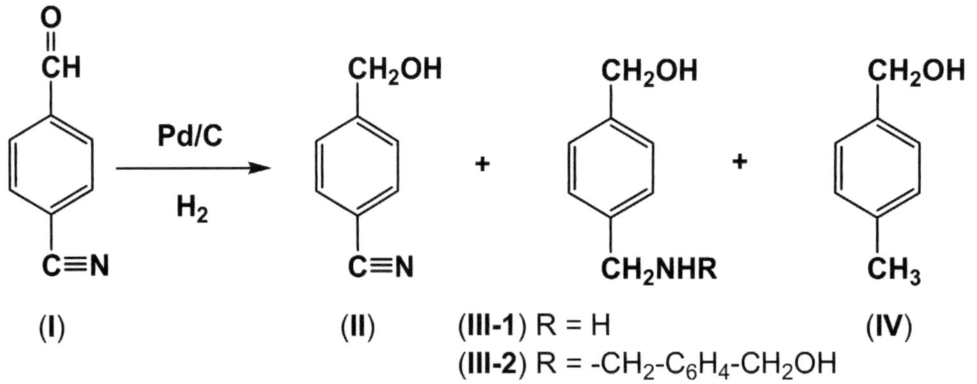

Scheme 1. Hydrogenation reaction of 4-cyanobenzaldehyde (**I**)

Table 1. Hydrogenation reactions of 4-cyanobenzaldehyde (**I**) in methanol (0.167 mol/L) by the microflow and batch systems at 2.5 MPa in 2 minutes

Temperature	Experimental	Compound (%)				
(°C)	system	(**I**)	(**II**)	(**III-1**)	(**III-2**)	(**IV**)
25	microflow	0	72	25	3	0
	batch	49	51	0	0	0
50	microflow	0	48	48	3	1
	batch	5	95	0	0	0
70	microflow	0	10	61	22	7
	batch	0	100	0	0	0
90	microflow	0	0	71	16	13
	batch	0	100	0	0	0

At 25°C, using the microflow reactor under 2.5 MPa, compound **I** was hydrogenated to give product **II** in 72%, product **III-1** and **III-2** in 25% and 3%, respectively. In the case of the batch system under the corresponding condition, product **II** was obtained in 51% as a sole product and 49% of the starting material remained intact. In the microflow system, increasing the reaction temperature up to 50 and 70°C, the hydrogenation reaction of the cyano group proceeded. At 90°C, isolated products were amino (**III-1**: 71% and **III-2**: 16%) and toluene derivatives (**IV**: 13%).[11] In the batch system even at 90°C, on the contrary, 100% of the starting substrate was converted to product **II** but neither product **III** nor **IV** was obtained. These results indicated that the flow system was powerful enough to hydrogenise the cyano group.

Table 2 shows the hydrogenation reactions of compound **I** in the microflow system where the substrate concentration was decreased from 0.167 mol/L to 0.033 mol/L. Here, the production of the secondary amine, **III-2**, was retarded completely since an intermolecular reaction between primary amine, **III-1**, and the imine intermediate was thought to be hardly taken place in this diluted condition. Interestingly, hydrogenation reaction at 90°C, product **IV** was obtained quantitatively.

Table 2. Hydrogenation reactions of 4-cyanobenzaldehyde (**I**) in methanol (0.033 mol/L) by the microflow system at 2.5 MPa for 2 minutes

Temperature	Compound (%)				
(°C)	(**I**)	(**II**)	(**III-1**)	(**III-2**)	(**IV**)
25	0	20	73	0	7
50	0	5	75	0	20
70	0	0	25	0	75
90	0	0	0	0	100

To demonstrate applicability of this efficient microflow system to other nitriles, hydrogenation reactions of benzonitrile and three benzonitrile derivatives have been studied. Table 3 summarizes yields of primary amines obtained by the microflow and the batch systems. Here, it is also obvious that the hydrogenation reaction by the microflow system is much more efficient than that by the batch system.

Conclusions

This microflow system gave pure product simply by introducing the substrate solution and the hydrogen gas through the microflow reactor packed with Pd/C. No Pd ion or metal were detected in the product solution (checked by atomic absorption spectroscopy), even at 90°C under 2.5 MPa. A purification process, separating the product from the catalyst, was not necessary since the catalyst was isolated firmly within the microflow reactor. As the structure of the microflow reactor was simple, as explained above, packing and renewing the catalyst were quite easy. In addition, the microflow system showed much higher reactivity and efficiency than the conventional batch system in the hydrogenation reactions of aldehydes and nitriles. Therefore, it was concluded that the microflow system, we demonstrated here, could be utilized in many multiphase, gas-liquid or gas-liquid-solid, reactions in future.

Table 3. Hydrogenation reactions of benzonitriles using the microflow and batch systems

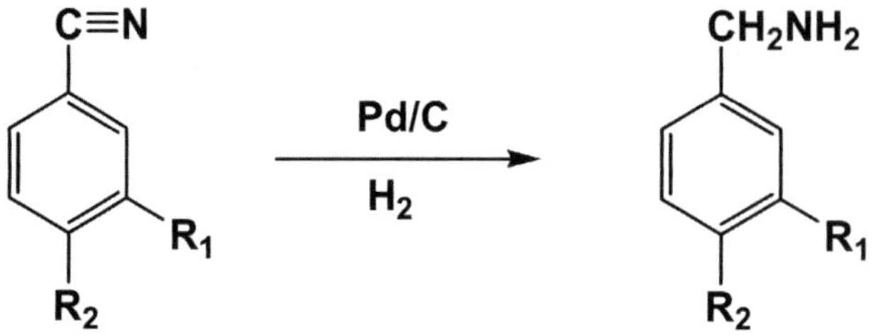

R_1	R_2	Temperature (°C)	Yield (%)	
			microflow	batch
H	H	25	60	2
		50	100	16
		90	100	7
Cl	H	25	81	21
		50	100	25
		90	100	16
H	Cl	25	100	0
		50	100	29
		90	100	18
H	CH_3	25	4	0
		50	21	0
		90	98	0

Acknowledgement

The authors wish to thank the Research Association of Micro Chemical Process Technology (MCPT), Japan and New Energy and Industrial Technology Development Organization (NEDO), Japan for financial support.

References and Notes

1) G. W. abalka and R. S. Verma, in *Comprehensive Organic Synthesis*, Eds. B. M. Trost and I. Fleming, Pergamon Press, Oxford, 1991, **Vol. 8**, p. 363; P. L. Mills and R. V. Chaudhari, *Catal. Today*, 1997, **37**, 367.
2) M. Wydra and H. Vogel, *Chemie Ingenieur Technik*, 2002, **74**, 800; W. Kiel, Ger. Offen., 1999, DE19727712; T. H. Coffield, A. H. Filbey, G. G. Ecke and A. J. Kolka, *J. Am. Chem. Soc.*, 1957, **79**, 5019.
3) S. Alini, A. Bottino, G. Capannelli, A. Comite, R. Di Felice and E. Frigo, *J. Mol. Cat. A-Chemical*, 2003, **206**, 363; R. Fischer, P. Bassler, H. Luyken, F. Ohlbach, J.-P. Melder,

M. Merger, A. Ansmann, A. Rehfinger and G. Voit, Ger. Offen., 2000, DE 19839346; G. Bartalini and M. Giuggioli, Ger. Offen., 1971, DE2053799; A. W. Larchar and H. S. Young, U.S. Patent, 1942, US2284525.

4) W. Ehrfeld, V. Hessel and H. Löwe, *Microreactors: New Technology for Modern Chemistry*, Wiley-VCH, Weinheim, 2000; W. Ehrfeld, V. Hessel and H. Lehr, *Microsyst. Technol. Chem. Life Sci.*, 1998, 233; K. F. Jensen, *Chem. Eng. Sci.*, 2001, **56**, 293; P. D. I. Fletcher, S. J. Haswell, E. Pombo-Villar, B. H. Warrington, P. Watts, S. Y. F. Wong and X. Zhang, *Tetrahedron*, 2002, **58**, 4735; S. J. Haswell and P. Watts, *Green Chemistry*, 2003, **5**, 240; H. Pennemann, P. Watts, S. J. Haswell, V. Hessel and H. Löwe, *Org. Process Res. Dev.*, 2004, **8**, 422.

5) R. D. Chambers and R. C. H. Spink, *Chem. Commun.*, 1999, 883; K. Jähnisch, M. Baerns, V. Hessel, W. Ehrfeld, V. Haverkamp, H. Löwe, C. Wille and A. Guber, *J. Fluorine Chem.*, 2000, **105**, 117; R. D. Chambers, D. Holling, R. C. H. Spink and G. Sandford, *Lab Chip*, 2001, **1**, 132.

6) J. R. Burns and C. Ramshaw, *Trans. Inst. Chem. Eng.*, 1999, **77**, 206; G. N. Doku, S. J. Haswell, T. McCreedy and G. M. Greenway, *Analyst*, 2001, **126**, 14; G. Dummann, U. Quitmann, L. Gröschel, D. W. Agar, O. Wörz and K. Morgenschweis, *Catal. Today*, 2003, **79-80**, 433.

7) H. Salimi-Moosavi, T. Tang and D. J. Harrison, *J. Am. Chem. Soc.*, 1997, **119**, 8716; R. C. R. Wootton, R. Fortt and A. J. de Mello, *Lab Chip*, 2001, **2**, 5; H. Hisamoto, T. Saito, M. Tokeshi, A. Hibara and T. Kitamori, *Chem. Commun.*, 2001, **126**, 14.

8) Wieβmeier and D. Hönicke, *Ind. Eng. Chem. Res.* 1996, **35**, 4412; M. W. Losey, R. J. Jackman, S. L. Firebaugh, M. A. Schmidt and K. F. Jensen, *J. Microelectromech. Syst.*, 2002, **11**, 709; J. Kobayashi, Y. Mori, K. Okamoto, R. Akiyama, M. Ueno, T. Kitamori and S. Kobayashi, *Science*, 2004, **304**, 1305.

9) C. de Bellefon, N. Tanchoux, S. Caravieilhes, P. Grenouillet and V. Hessel, *Angew. Chem. Int. Ed.*, 2000, **39**, 3442; C. de Bellefon, R. Abdallah, T. Lamouille, S. Caravieilhes and P. Grenouillet, *Chimia,* 2002, **56**, 621.

10) For reviews on the contamination of secondary amine in the catalytic hydrogenation of nitriles, see: P. Rylander, in *Catalytic Hydrogenation in Organic Syntheses*, Academic Press, New York, 1979, Ch 8; M. Freifelder, in *Catalytic Hydrogenation in Organic Synthsis. Procedure and Commentary*, Wiley-Interscience, New York, 1978, Ch 6; H. Greenfield, *Ind. Eng. Chem., Prod. Res. Dev.*, 1976, **15**, 156.

11) Toluene derivative can be produced by the hydrogenolysis of benzylamine in the hydrogenation reaction of benzonitrile using palladium-loaded pyridyl catalyst at 100°C and 2.9 MPa; M. B. Dines, L. Beach, P. J. DiGiacomo, M. Viejo, K. P. Callahan and C. Mesa, US Patent 4 384 981, 1983.

134a Applications of Microreactor in Pharmaceutical Development & Production

Yeung Chan

As pharmaceutical industry faces increasing challenges on many fronts in recent years, reducing the time to bring a new drug to market becomes an even important goal that every company in the industry is striving to achieve. Process intensification with the use of microreactor is one of the novel engineering approaches that have shown promising potentials for reaching such lofty objective. Key among the advantages offered by microreactor is the fast process development time and flexible response to changes in process parameters. Efficient heat and mass transfer rates allow not only the safe handling of highly hazardous process, but also accessing extreme chemistry or operating conditions that otherwise cannot be achieved. Another important characteristic of microreactor is the positive impact on pharmaceutical manufacturing. Process intensification creates tremendous cost savings associated with facilities, maintenance, flexible production capacity, and enhanced <i>Just-in-Time</i> manufacturing capability.

Using examples from some of our recent efforts, case studies are presented to demonstrate the successful application of microreactor technology respectively to handle: highly hazardous process and chemistry with problematic scale up. The characteristics of microreactor that are ideal to enable the expedient pharmaceutical process development with limited material will be particularly emphasized through the case studies. Besides the methodology taken to overcome the scientific challenges, our scale up approaches and results will also be discussed.

Finally, the role of microreactor and process intensification will be proposed in the context of FDA's 21st Century cGMP and PAT initiatives. At the conclusion, we will offer some of our experience and perspective regarding this paradigm shift to strengthen the science of manufacturing control.

134b A New Concept for the Measurement of Strong Exothermicities in Microreactors

Daniel Schifferdecker, Juergen Antes, Stefan Loebbecke

The scale-up of chemical reactions from the laboratory to production units is potentially dangerous. Therefore, thermal safety parameters such as reaction enthalpies and other relevant parameters have to be determined in advance. These parameters are usually measured with reaction calorimeters, such as the RC1 from Mettler-Toledo. However, commercially available reaction calorimeters require an unfavorable large test volume of about 0.3 – 2.0 l and thus a relatively huge amount of test substance. Hence, the safety analysis of fast exothermic reactions with such reaction calorimeters raise several problems. The control of the reactor temperature and a sufficiently rapid mixing are difficult to achieve. Furthermore, strong exothermicities in combination with short reaction times cause large numbers of secondary, consecutive and decomposition reactions due to hot spots and accumulated reaction heat. This may cause significant selectivity and safety problems.

To overcome the drawbacks of commercially available calorimeters, a new small reaction calorimeter with a hold-up of only 50 l have been developed. Such a device is of particular importance for the synthesis of new fine chemicals and pharmaceutical products, where typically only small amounts of test substance are available and time-to-market is crucial. Moreover, it is highly suitable for potentially hazardous or even explosible reactions due to its small hold-up.

The µl-calorimetric device consists of a microreactor made of glass and thermoelectric modules. The entire calorimeter is placed into a thermostated unit to control the reaction temperature. Pulsation-free syringe pumps are used for the control of the reactants flow rates. The entire system is computer controlled by using LabVIEW as automation software. The microreactor has been specially designed to achieve a homogenous flow distribution and an intensive mixing of the reactants.

The microreactor is embedded between two Seebeck elements, which are commercially available. The power of the Seebeck elements is recorded by a LabVIEW program every 0.3 s. The calorimetric measurement is performed by measuring continuously the heat flow caused by the chemical reaction from the microreactor through the Seebeck element to the Peltier element. The latter is regulated by a PID controller to generate a constant temperature difference between the microreactor and the lower Seebeck element. By integrating the voltage signals of the Seebeck element over the time the released reaction heat can be calculated. The huge surface-to-volume ratio of the microchannels ensures a fast transfer of the reaction heat.

The new µl calorimeter has a very small time constant of about 3 s which is a factor 20 smaller than that of conventional calorimeters. Hence, it is ideally suited to measure fast and highly exothermic reactions at isothermal conditions. The performance and the accuracy of the new device have been demonstrated on different strong exothermic examplary reactions, e.g.:

1. nitration of toluene with 100% nitric acid: the reaction enthalpy at 50 C was measured as 110 kJ/mol (literature: -115 to -135 kJ/mol).

2. dilution of concentrated sulphuric acid with water: the reaction enthalpy at 25 C was measured as -42 kJ/mol and is thus again in very good agreement with macroscopically measured data from literature (50-60 kJ/mol)

134c An Integrated Stacked Micro Fluidic Reactor System for Nanoparticle Synthesis

Challa Kumar, Josef Hormes, Jost Goettert, Paryog Datta, Willi Hempelmann, Yujun Song

Micro reactors are gaining an increasingly important role in specialized chemical production and process development.1 First successful experiments clearly indicate that they have also the potential to play a very important role in wet chemical synthesis of nanoparticles.2 This is mainly due to the fact that micro systems for chemical synthesis have a significant advantage in terms of increased mass and heat transfer.3 The reduction in size and the integration of multiple functions provides the potential for producing devices with capabilities that exceed those of conventional macroscopic systems. Polymer based micro reactors have the additional advantage of low initial fabrication cost in combination with the fabrication capability through inexpensive replication technologies such as embossing, casting, or photo- and injection molding. We have designed and fabricated a user-friendly computer controlled stacked polymeric micro reactor system for the synthesis of nanoparticles. The reactor system also can be utilized in general for wet chemical synthesis and process development of specialty chemicals.

This user-friendly system consists of three basic functional blocks for controlled flow from the chemical container (Inlet), a custom made, temperature controlled micro reactor stack, and an outlet unit to control and optimize all critical reaction parameters. The system operation is controlled by a computer. Based on the experimental parameters and conditions developed for the synthesis of Pd nanoparticles using a single polymeric micro reactor in our laboratory, 2 we are studying issues of scale-up production utilizing stacked micro reactor system. We will present details of the fabrication of stacked micro reactor system and its utility in scale-up production of nanoparticles as well as specialty chemicals.

1. Web page Ehrfeld Mikrotechnik, http://www.ehrfeld-mikrotechnik.com/.

2. Y. Song, Challa S.S.R.Kumar, and Josef Hormes, Synthesis of Pd nanoparticles using a continuous flow polymeric micro reactor, Journal of Nanoscience and nanotechnology, 4(7), 788-793, 2004.

3. Haswell, S. J.; Fletcher, P. D. I.; Greenway, G. M.; Skelton, V.; Styring, P.; Morgan, D. O.; Wong, S. Y.F.; Warrington, B. H. Micro-chemical reactors: the key to controlling chemistry. Special Publication - Royal Society of Chemistry 2000, 250(Automated Synthetic Methods for Speciality Chemicals), 25-33.

Catalytic Investigations in Microchannels at Reduced Pressure

A. Kölbl, P. Pfeifer, K. Schubert

Institute for Micro Process Engineering (IMVT), Forschungszentrum Karlsruhe GmbH, Hermann-von-Helmholtz Platz 1, 76344 Eggenstein-Leopoldshafen, Germany, http://www.fzk.de/imvt-en

Prepared for Presentation at 2005 Spring National Meeting /

IMRET 8 / April 10-14 / Oral Contribution

Unpublished

Introduction

Catalysis is of utmost importance in industrial chemistry. It is estimated, that 85-90 % of all industrial reactors are driven with a catalyst [1]; predominantly solid catalysts. Despite this, the detailed understanding of elementary steps in catalysis under process relevant conditions is in most cases insufficient.

Research on heterogeneous catalysis is traditionally divided in two operating fields of work. Surface scientist are investigating elementary steps of chemical reactions at well defined conditions (UHV and model catalysts), while chemical engineers are usually working at industry-like conditions (atmospheric/elevated pressure and empirically optimized catalyst samples).

Usually it is not possible to extrapolate the results of surface scientist to industrially relevant conditions. Successful extrapolation is limited to a few examples like ammonia synthesis on iron catalysts [2], the methanation of carbon monoxide on nickel catalysts [3] and the CO-oxidation on platinum metals [4]. These difficulties are often described with the terms pressure and materials gap [5].

In order to carry out experiments in the "pressure gap region" a new type of flow apparatus including a microstructured reactor is presented, which allows experiments from ambient to very low pressures. Different dosing concepts to realise low reactant partial pressures in the reactor are discussed. It is shown and experimentally verified in the case of a microstructured reactor that the split of the generated feed stream before the reactor and partially dosing via capillaries allows realising smaller reactant partial pressures compared to classical methods like additional dosing of carrier gas [6].

Methanol steam reforming on a nanoparticle based PdZn catalyst has been investigated in a microstructured reactor by Pfeifer [7] in a classical flow type apparatus at total pressures of 3 to 6 bar. Based on experimental results and literature data on mechanistic steps on PdZn, he developed a kinetic model (equation 1), which describes the experimental results for following conditions: reaction temperature θ = 210-290 °C, partial pressure methanol $p_{Methanol}$ = 90-736 mbar and molar ratio water to methanol $n_{water} / n_{methanol}$ = 1.2-3.

$$-R_{methanol} \equiv r_s = \frac{k_s K_{Me,Ads} c_{Me}}{\left(1 + K_{Me,Ads} c_{Me} + \sqrt{\frac{c_{H2}}{K_{H2,Des}}}\right)^3} \qquad (1)$$

The present study aims at contributions to bridge the pressure gap on the example of methanol steam reforming as well as the application of the kinetic model to different experimental conditions.

Experimental Setup

Reactor Concept

Figure 1 shows the core of the test reactor (right) and the catalytically coated microstructured foils (left). Channels of 100 µm x 100 µm cross section have been used for this work. A reactor with a similar geometry has been used by Pfeifer [7] for high pressure measurements.

Figure 1: Core of a microstructured reactor, catalytically coated micromachined foils and SEM of the PdZn catalyst.

Dosing Concepts

In order to vary vapour partial pressures in a flow type reactor different methods can be applied. When dosing liquid reactants, usually the temperature in a saturator system is altered (Figure 2).

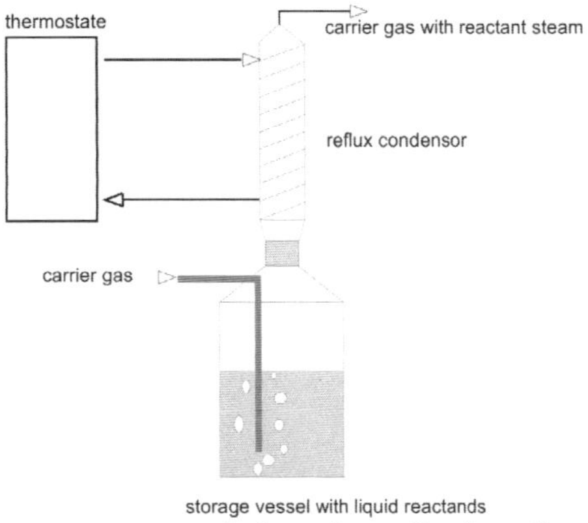

Figure 2: Reactant partial pressure variation by altering the temperature in a saturator system.

Applying this method for the present study on methanol steam reforming in a microchannel reactor, the variation in partial pressure of the reactants (water and methanol in a molar ratio of 1/1) was possible in a range between 25 and 200 mbar [6].

The partial pressures can further be decreased, when a lower total pressure in the reactor is generated by a pump downstream of the reactor (figure 3). This method is applicable both in classical fixed bed reactors and microstructured reactors.

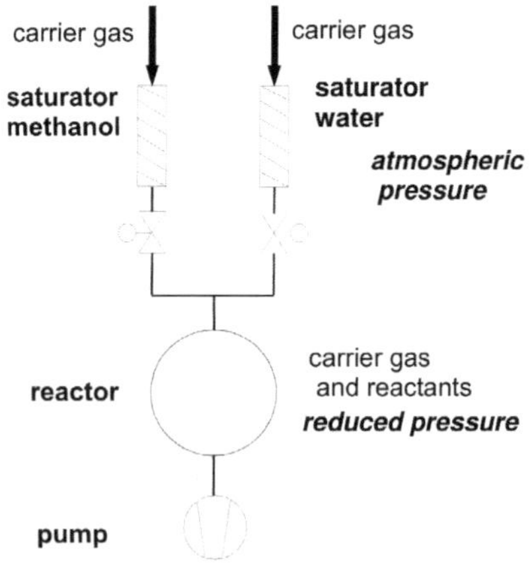

Figure 3: Reactant partial pressure variation by altering total pressure.

Applying this method for dosing water and methanol in the case of methanol steam reforming one is able to realise partial pressures between 5 and 200 mbar. Limiting for the lowest partial pressures available is the working range of the mass flow controllers.

Another classical method to go to low reactant partial pressures is the additional dosing of carrier gas (figure 4).

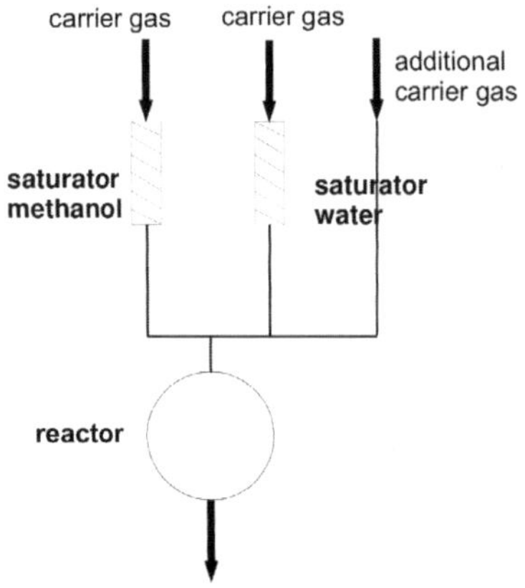

Figure 4: Reactant partial pressure variation by additional dosing of carrier gas.

Applying this method a mayor difference between microstructured reactors and classical fixed bed reactors is obvious. In the case of fixed bed reactors, the length of the fixed bed can be varied very flexible in order to keep residence time constant, while the length of the micromachined channels is restricted to reasonable dimensions. Consequently, additional dosing of carrier gas is a suitable method to get low reactant partial pressures in case of fixed bed reactors, but not for microstructured reactors. Applying this method in the present study, it would be possible to cover a partial pressure range between 5 and 200 mbar in the microreactor.

In order to get to significant lower reactant partial pressures a different method was applied (Figure 5).The split of the generated feed stream before the reactor and partially dosing via capillaries allows realising lower reactant partial pressures. Experiments have been conducted in a methanol partial pressure range between 0.6 and 200 mbar.

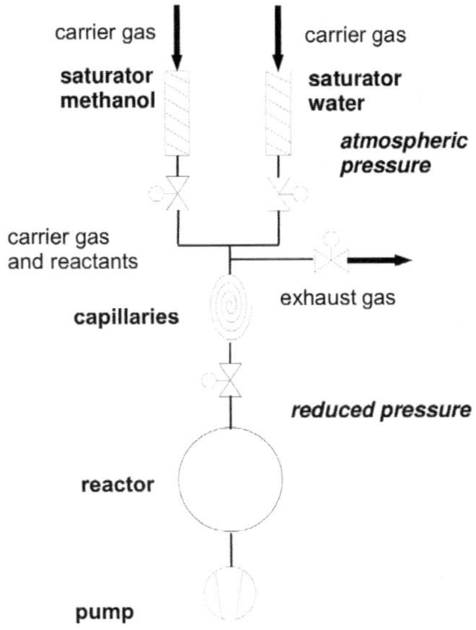

Figure 5: Reactant partial pressure variation by split of the generated feed stream.

Experimental Results

Temperature variation in a saturator system

Experiments have been carried out in a range of $p_{Methanol} = p_{water} = 25\text{-}200$ mbar, running the test rig at atmospheric pressure. The obtained conversions of methanol in dependence of the temperature are shown in figure 6. They are compared with the results of the kinetic model (equation 1). Model and experimental results are in good accordance.

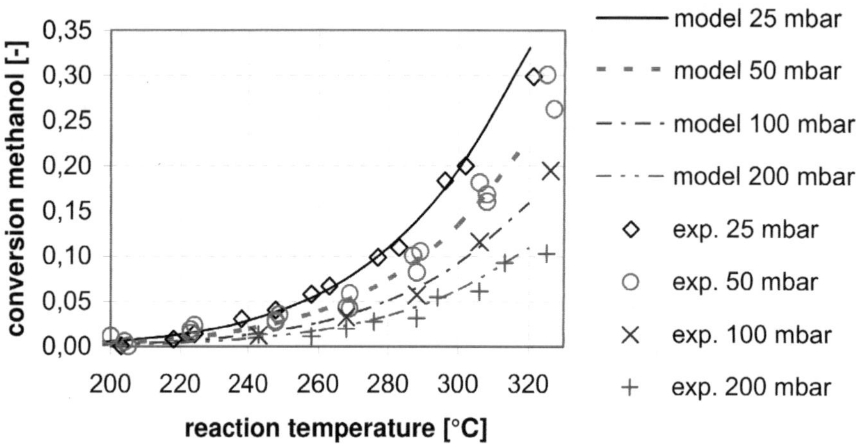

Figure 6: Reactant partial pressure variation by temperature variation in a saturator system.

Experimental results at reduced total pressure

In order to get to significant lower partial pressures experiments have been carried out in the vacuum range (30 and 500 mbar total pressure) using the dosing concept described in figure 5 (split of the generated feed stream). Figure 7 shows the obtained conversions of methanol over reaction temperature. The lines represent the predictions of the kinetic model (equation 1). Model and experimental results are in good accordance.

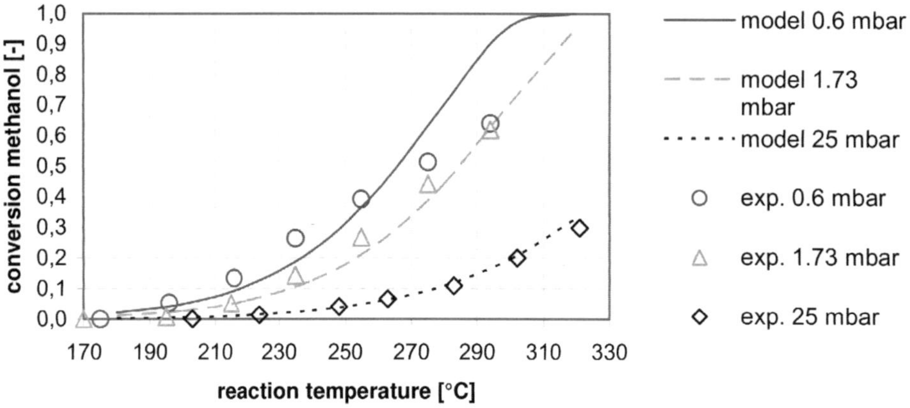

Figure 7: Experimentally obtained conversion degrees of methanol at reduced total pressure (split of the generated feed stream).

Application of the experimental data to the kinetic model

The obtained experimental results have been compared with the predictions of the kinetic model, which has been derived from high pressure measurements [7]. The application of the model to the presented reaction conditions is an extrapolation of the model parameter space with respect to molar ratio of water to methanol, reaction temperature and partial pressure of the compounds. Due to different deactivation of the catalyst observed at high [7] and low [6] pressure measurements, the surface rate constant k_S (equation 1) had to be corrected by a factor of 10^{-3} (200 mbar methanol partial pressure at a total pressure of 1 bar was the reference experiment). When applying this factor to the other measurements, it can be shown, that the model describes the experimental results within a deviation of +/- 20 % (figure 8).

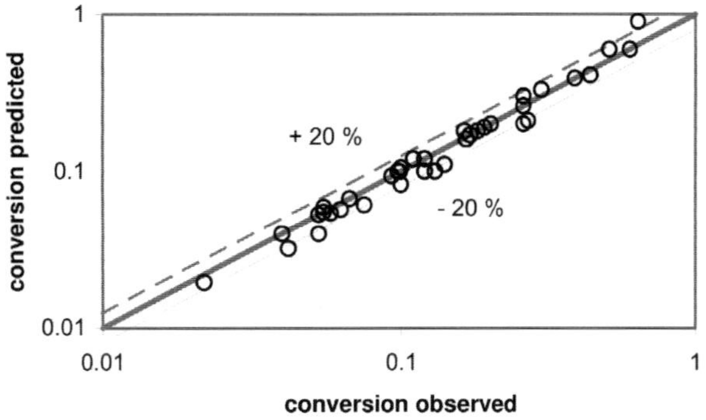

Figure 8: Parity plot for the experimental results given in figure 6 and 7.

Conclusion

Kinetic investigations in a microstructured reactor in the vacuum range have been carried out in order to overcome the pressure gap in heterogeneous catalysis on the example of methanol steam reforming. Experimental results are presented in a methanol partial pressure range between 0.6 and 200 mbar. These results are described by a kinetic model, which is derived from high pressure experiments, within a deviation of ± 20%. Together with high pressure measurements [7], the model covers a methanol pressure range of three decades (0.6-736 mbar), in which no change of the mechanism of the chemical reaction can be detected.

An alternative dosing system (split of the generated feed stream) has to be applied in case of a microstructured reactor in order to get to significant low pressures of methanol.

In order to fill out the experimental gap significant lower total pressures have to be realised in the microstructured reactor. Special care has to be taken to decide whether a change in microkinetic mechanism occurs when changing the pressure from technically relevant pressures to UHV conditions. Either way, an increase in conversion degree upon pressure decrease is expected due to an enhanced number of collisions of molecules with the catalytically coated wall. This effect has been predicted by Norberg et al. [8] and, for the dehydrogenation of cyclohexane experimentally verified by Jones et al. [9].

Acknowledgement

The authors thank the Deutsche Forschungsgemeinschaft for financial support

(Fi 816/1-1,2,3)).

Symbols

Θ	Reaction temperature	[°C]
$p_{methanol}$	Methanol partial pressure	[mbar]
p_{water}	Water partial pressure	[mbar]
n_{water} / $n_{methanol}$	Molar ratio water to methanol	[-]
$R_{methanol}$	Integral methanol consumption rate	[mol/($g_{catalysts}$)]
r_s	Surface reaction rate (volume specific)	[mol/m^3s]
k_s	Surface reaction constant (volume specific)	[mol/m^6s]
$K_{Me,Ads}$	Methanol adsorption equilibrium constant	[m^3/mol]
c_{Me}	Methanol concentration	[mol/m^3]
c_{H2}	Hydrogen concentration	[mol/m^3]
$K_{H2,des}$	Hydrogen equilibrium constant	[mol/m^3]

References

[1] J. Weitkamp, R. Gläser; "Katalyse", Winnacker Küchler, Chemische Technik, Vol. 1, Wiley-VCH, Weinheim, 2004, ISBN 3-527-30767-2.

[2] R. Schlögl; "Handbook of Heterogeneous Catalysis", Vol. IV, VCH Weinheim, 1997, ISBN 3-527-29212-8.

[3] R.D: Kelly, D.W. Goodman; Surf. Sci. 123(1982)L743.

[4] C.H.F. Peden in Surface Science of Catalysis, ACS Symposium Series, Vol. 482, Eds. D.J. Dwyer and F.M. Hoffmann; American Chemical Society, Washington, 1992.

[5] H.P. Bonzel; Phys. Bl. 32(1976)393.

[6] A. Kölbl, PhD thesis, University Erlangen-Nuremberg, 2004.

[7] P. Pfeifer, PhD thesis, University Erlangen-Nuremberg, 2003.

[8] P. Norberg, L.G. Peterson, L. Lundström; Vacuum, 45(1994)139.

[9] F. Jones, A. Kuppusamy, A. Zheng; in proc. of SPIE, the int. Society for optical engineering (1999)3877.

134e Rapid Optimization of Glycosylations in Microreactors

Daniel M Ratner, Edward R Murphy

Carbohydrates serve a wide array of biological purposes from mechanical supports and energy sources to cellular recognition and cell signaling. The organic synthesis of oligosaccharides through glycosylation has been an area of ongoing research for more than a century, and yet significant challenges remain.[1] Glycosylations are notoriously difficult due to the variety of conformations, sterics, and electronics of the reaction partners. This variety makes it difficult to predict the reactivity of the glycosylating agent with its corresponding nucleophile. The success of each step in the synthesis is dependent upon controlling a variety of reaction conditions from temperature and mass transfer to concentration and reaction time.[2]

Much of the effort of synthetic chemists in both academic and industrial settings is directed toward identifying the optimal conditions for a particular transformation. Traditionally, this bench top optimization requires commitments of both time and large quantities of valuable reagents. As a result, the availability of data on the behavior of a reaction is severely limited. The ability to find ideal reaction conditions quickly and efficiently would have a major impact on the practice and pace of research and development. Improved data collection can reduce waste and accelerate reaction optimization, while simultaneously removing a significant bottleneck from oligosaccharide synthesis and simplifying the process of obtaining reaction kinetics. We used a silicon microreactor in conjunction with automated HPLC analysis to rapidly scan the effects of temperature and flow rate on the formation of reaction products. The microreactor was designed with laminar mixing structures and an inline quench. This design enabled us to control both the order of mixing and confine the reaction to the silicon device. A novel packaging scheme was also developed to improve the performance of the microreactor when exposed to aggressive solvents such as methylene chloride and cryogenic temperatures down to -78°C. The two glycosylations examined with this reactor were the mannosylation of diisopropylidene galactose and 2,3,4-tri-O-benzyl-methyl mannoside by the glycosyl donor, mannosyl trichloroacetimidate. Both reactions used TMSOTf as an activator and reaction mixtures were quenched by triethylamine before exiting the reactor.

For this study, the effects of reaction temperature and residence time on product conversion were monitored. The reactor temperature was varied from -78°C to 20°C and residence times ranged from 27 to 214 seconds. The results showed clear relationships between reaction conditions and product formation. Additionally, we were able to observe the formation of orthoester, a significant byproduct and possible intermediate, at lower temperatures.

Unlike batch methods, which are challenged by the difficulty of handling microliter quantities of volatile solvents and the possibility of external contamination, this microreactor system serves to rapidly profile the effects of temperature and residence time on a given transformation. With a single preparation of reagents, 44 reactions were completed requiring just over 2mg of glycosylating agent for each reaction.

References

[1]J. Lehmann, in *Carbohydrates: Structure and Biology*, Thieme, Stuttgart, 1998, pp.1-37, 67-129.

[2] K. Toshima and K. Tatsua, *Chem. Rev.*, 1993, **93**, 1503.

Comparisons between a Thin-Film Catalyzed Microreactor and Conventional Lab Reactors for PrOx Kinetic Studies

X. Ouyang,[1] L. Bednarova,[2] H. Chen,[2] S. Pau,[3] N. Lawal,[2] W.Y. Lee,[2] R.S. Besser[2]

[1] CNRS/CPE Lyon
43 bd du 11 Nov 1918, BP 2077
Villeurbanne 69616
France

[2] Reaction Design, Inc.
6440 Lusk Boulevard, Suite D-205
San Diego, CA 92121

[3] New Jersey Nanotechnology Consortium
Lucent Technologies
Murray Hill, NJ 07974

ABSTRACT

In our study of CO preferential oxidation (PrOx) on Pt/Al_2O_3 with a thin-film catalyzed microreactor, systematic differences of kinetic study results and characteristic curves of CO conversion were identified between the microreactor and conventional minimized packed-bed lab reactors (m-PBR's). Evidence suggested that the difference in heat transport limitations is the origin of these differences. In this paper, these differences were examined systematically. First, based on existing criteria for mass and heat transport limitations, we found negligible mass transport resistance for both types of reactors within typical PrOx operating conditions, but strong heat transport limitations for m-PBR's. Further, a non-isothermal reactor model was built to quantitatively study the effect of the heat transport limitations on PrOx reaction behavior. Model results showed that the temperature gradients in m-PBR's favor the r-WGS reaction, thus causing a much narrower range of permissible operating temperature compared to the microreactor. Accordingly, a more direct access to PrOx reaction kinetics is permitted by the microreactor because of its excellent heat transfer efficiency.

INTRODUCTION

The reaction kinetics of PrOx on Pt/Al_2O_3 was extensively studied in the literature due to its significance in on-board fuel processing applications. However, most of these studies were carried out in m-PBR's. A comparison of catalytic activity studies from our Pt/Al_2O_3 thin-film catalyzed microchannel reactor and m-PBR's in the literature discovered significantly higher TOF's for the microreactor at temperatures above 200°C, as shown in Table 1. In addition, our study of CO conversion temperature characteristic showed large difference compared to other PrOx studies base on the same catalyst system, especially at higher temperatures as indicated by Figure 1. These differences implies possible influence of strong mass and/or heat transport limitations for the m-PBR's and a more direct access to PrOx reaction kinetics for the microreactor. A detailed examination of transport limitations is thus necessary. In this paper, classical criteria for mass and heat transport limitations are adopted to identify and compare the significance of these limitations, which is followed by 3D reactor modeling coupling both mole balance and energy balance equations to quantitatively study the effects of these limitations on PrOx reaction behavior.

Table 1. Comparison of the micoreactor and m-PBR's based on Pt/Al_2O_3 for PrOx with *TOF* calculations. D – catalyst dispersion.

References	Catalyst System	Temperature (°C)	Pressure (atm)	P_{CO} (Torr)	P_{O2} (Torr)	P_{H2} (Torr)	TOF (s^{-1})
This work	Pt/γ-Al_2O_3 (D=20%)	100°C	1	13,68	13,68	469,68	0,006
Sirijaruphan *et al.* [1]	Pt/γ-Al_2O_3 (D=45.4%)	90°C	1,8	13,68	13,68	342,00	0,009
This work	Pt/γ-Al_2O_3 (D=20%)	150°C	1	7,60	7,60	469,68	0,077
Kahlich *et al.* [4]	Pt/γ-Al_2O_3 (D=38%)	150°C	1	7,60	7,60	570,00	0,080
This work	Pt/γ-Al_2O_3 (D=20%)	200°C	1	7,60	7,60	469,68	1,266
Kahlich *et al.* [4]	Pt/γ-Al_2O_3 (D=38%)	200°C	1	7,60	7,60	570,00	0,710
Kim *et al.* [2]	Pt/γ-Al_2O_3 (D=100%)	200°C	1	7,60	7,60	380,00	0,330
This work	Pt/γ-Al_2O_3 (D=10%)	250°C	1	7,60	7,60	469,68	4,831
Kahlich *et al.* [4]	Pt/γ-Al_2O_3 (D=38%)	250°C	1	7,60	7,60	570,00	3,800

EXPERIMENTAL

PrOx reaction system

There are three reactions in PrOx on Pt/Al_2O_3 within typical PrOx reaction temperatures (160 ~ 220°C) as indicated by Equations 1-3. The oxidations of CO and O_2 are exothermic with similar reaction enthalpies while the r-WGS reaction is endothermic and strongly undesired due to its production of CO.

$$CO + \frac{1}{2}O_2 \xrightarrow{\Delta H = -67 kcal/mol} CO_2 \quad (1)$$

$$H_2 + \frac{1}{2}O_2 \xrightarrow{\Delta H = -58 kcal/mol} H_2O \quad (2)$$

$$CO_2 + H_2 \xrightarrow{\Delta H = 9.8 kcal/mol} CO + H_2O \quad (3)$$

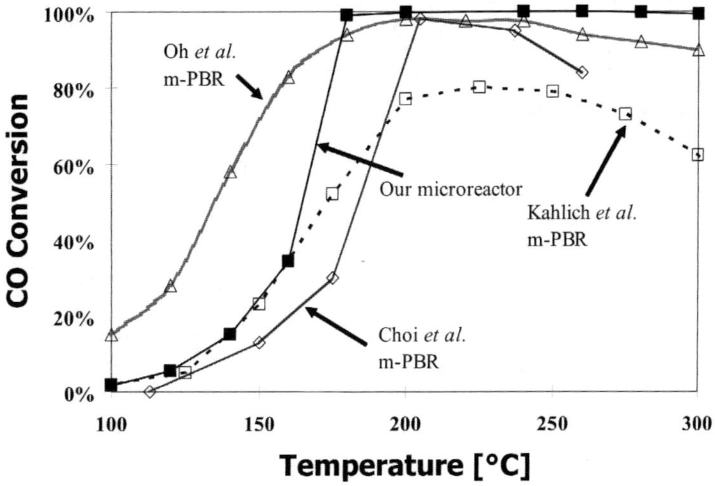

Figure 1. CO conversion vs. temperature. WHSV: our work – 1500 hr^{-1}, Kahlich et al. [4] – 1250 hr^{-1}, Choi et al. [5] – 200 hr^{-1}, Oh et al. [6] – 75 hr^{-1}.

Microreactor and m-PBR's

The microreactor used in this study was fabricated with well-known silicon micromachining processes, with geometric features of 500 μm (width) x 470 μm (depth) x 4.5 cm (length). Platinum supported on alumina was synthesized using a sol-gel technique and followed by multilayer selective-coating on the channel wall of the microreactor.

The m-PBR's are conventional tubular lab reactors for the acquisition of kinetic data. The parameters and kinetic data of the m-PBR's are taken from the literature, with the reactor inner radius of 2 mm or larger and catalyst particle diameter of ~100 μm [2,3,4,5,6].

Quasi-3D Non-isothermal Reactor Model

A finite difference method was adopted to calculate the 3D temperature distribution inside the reactor, which distinguishes the approach from previous 1D PrOx reactor modeling efforts [5]. The 3D modeling is necessary since both axial and radial temperature gradients in the reactor can be significant. Modeling was done for both the microreactor and m-PBR's with cylindrical finite-difference grids of various diameter approximating both the microchannel and the packed bed. With kinetic expressions and parameters determined from literature and our kinetic data, mole balance and energy balance equations were solved at each grid point, assuming identical boundary temperatures.

RESULTS AND DISCUSSION

Mass Transport Limitations

Both internal and external mass transport limitations are examined for the microreactor and the m-PBR's.

Estimation of the internal mass transport limitation is realized by the calculation of the Weisz-Prater (W-P) parameter (C_{WP}), which needs to be much smaller than 1 for negligible internal mass transfer resistance [7]. The internal mass transport is evaluated based on O_2 concentration as indicated in Equation 4.

$$C_{WP} = \eta\phi_1^2 = \frac{-r''_{O_2}(obs)\rho_c L_c^2}{D_e C_{O_2(s)}} \qquad (4)$$

The calculation results of C_{WP} show a significant difference for the two types of reactors, mainly due to the difference in the characteristic catalyst lengths (L_c). For the thin-film catalyst, L_c equals the catalyst thickness, which is normally ~ 5 µm, resulting $C_{WP} \ll 1$ even at the highest operating temperature. In comparison, L_c for the m-PBR (~ 50 µm) is ~ 10 times that of the microreactor, and the calculated C_{WP} can be ~ 100 times that for the microreactor due to its quadratic dependence on L_c. However, the estimated C_{WP} of the m-PBR no longer satisfies the $C_{WP} \ll 1$ criterion only for temperatures above 220°C.

External mass transport is negligible below 300°C for PrOx with m-PBR's because of their relatively small catalyst particle radii (~50 µm) and hence the small external mass transport distance for reactant gases, as reported previously [2,8]. However, it is necessary to address the possibility of external mass transport limitation for our microreactor due to its comparatively longer external mass transport distance, which is from the center of the microchannel to the wall (~250 µm). Two separate quantification methods different from above are applied to examine the external mass transfer resistance.

$$W_{CO} = k_{c-CO}(C_{CO(b)} - C_{CO(s)}) \qquad (5)$$

The first method examines the ratio of the surface and bulk gas–phase concentration for both CO and O_2 since the external mass transfer resistance can be considered negligible if both ratios are larger than 95% [9] (Equation 5). The mass transfer coefficient $k_{c\text{-}CO}$ is estimated with the correlations described in [10]. Figure 2 plots $C_{CO(s)}/C_{CO(b)}$ and $C_{O_2(s)}/C_{O_2(b)}$ at different temperatures. Both concentration ratios decrease slowly with temperature. As seen from this figure, these ratios are higher than 95% below 220°C and 240°C, respectively. Our typical PrOx operating temperature is ca. 180°C and both ratios are >98%, indicating the insignificance of external mass transport resistance. However, as temperature increases above 240°C, this advantage diminishes.

The second method evaluates the overall specific reaction rate (k_r) at different reactant flow rates. The assessment is based on a well-known method [11] for determining the limiting step between mass transport and surface reaction in a catalytic reactor. Since the change in flow rate results in different mass transport resistance, the calculated k_r would also change significantly if mass transport is limiting the reaction. On the other hand, if the change in the calculated k_r is small and without trend, it would imply that mass transport is not rate-limiting. An experiment was carried out with a microreactor running at 180°C with various reactant flow rates. The ratio of k_r and the average specific reaction rate ($k_{r(e)}$) are plotted against the ratio of $C_{O_2(b)}$ and $C_{CO(b)}$ in Figure 3. The calculated variation of k_r shows no dependence on the reactant flow, which further corroborates the conclusion that external mass transport is not rate-limiting in the thin-film catalyzed microchannel under typical operating conditions.

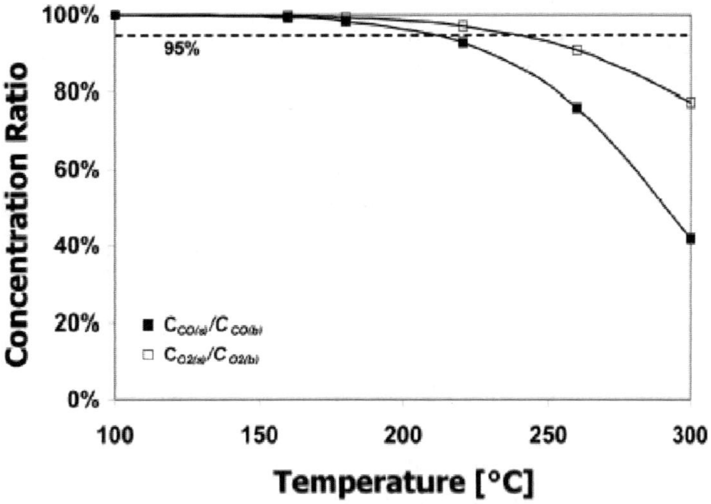

Figure 2. Calculated ratios of $C_{CO(s)}/C_{CO(b)}$ and $C_{O_2(s)}/C_{O_2(b)}$ from mass-transport analysis at different temperatures for the microreactor.

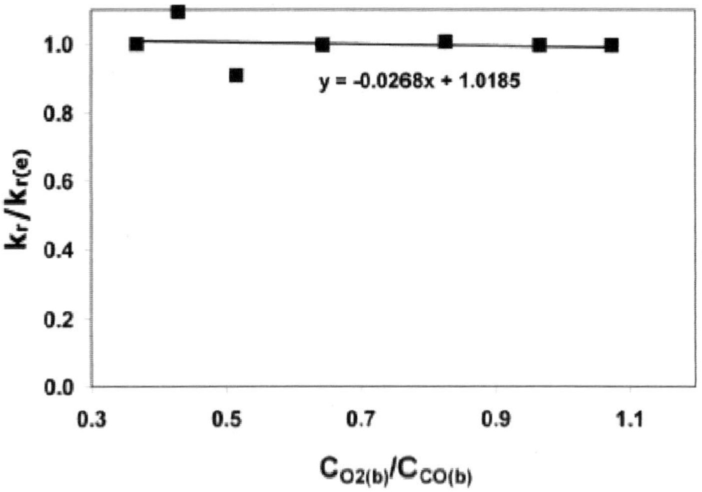

Figure 3. Ratios of experimental specific reaction rate (k_r) and average specific reaction rate ($k_{r(e)}$) at different $C_{O_2(b)}/C_{CO(b)}$.

Heat Transport Limitations

Heat transport inside the reactor can be categorized in three domains– intraparticle, interphase and interparticle transport. Mears [12] suggested that the magnitudes of the heat transport resistances are generally in the order of interparticle > interphase > intraparticle when the thermal Biot number (*Bi*) is small and the heat transfer resistance at the reactor wall are negligible. With the calculation of the *Bi* and the ratio of the reactor hydraulic diameter and catalyst particle size, as the criteria for the above assumptions proposed by Mears [13], interparticle heat transport are found to be dominant for both the microreactor and m-PBR's.

With the above justification, calculations of the *Damkőhler* numbers (D_a) with Equation 3 can provide a qualitative estimation of heat transport limitations, stating negligible heat transport resistance when the inequality is satisfied [12,13].

$$D_a = \frac{|-\Delta H(-r_{CO})t_{cat}^2|}{\lambda_e T_w} < 0.4\frac{RT_w}{E_a} \qquad (3)$$

In this calculation, the effective interparticle heat conduction distance (t_{cat}) is the determining factor. The t_{cat} is ~ 5 μm for the microreactor and about three orders of magnitude larger for the m-PBR's. The results show that at 180°C the left hand side for the microreactor is orders of magnitude smaller than the right hand side while the left hand side for m-PBR's are higher than the right hand side.

To summarize the above two sections, we found both types of reactors show negligible internal and mass transport resistance within typical PrOx operating temperatures (160 ~ 220ºC). On the other hand, within the same temperature range, the microreactor has negligible thermal gradient while the heat transport resistance of the m-PBR is significant for PrOx.

Non-isothermal Reactor Model

Based on the above qualitative analysis disclosing the significance of heat transfer in comparing the microreactor and the m-PBR's for PrOx, the non-isothermal reactor modeling was conducted neglecting mass transport resistance.

Essentially isothermal temperature distribution is shown in the modeling results for the microreactor for a wide temperature range (100 - 300°C). This result supports the above qualitative analysis with Mear's criterion. Also in agreement with the qualitative analysis, the modeling for the m-PBR's shows the development of large temperature gradients, both in axial and radial direction, as shown in Figure 4. At low wall temperature (T_w < 180°C), temperature differences in the reactor are small indicating slow rates of the exothermic oxidation reactions (i.e., Equations 1 and 2). At 180°C, phenomenal temperature gradients are developed in latter part of the reactor as a result of increased CO and H_2 oxidation activity. This phenomenon is intensified with further temperature increase (T_w ≥ 220°C) since the shorter the reactor length for full oxygen conversion (with higher temperature), the larger the heat accumulation within the corresponding length.

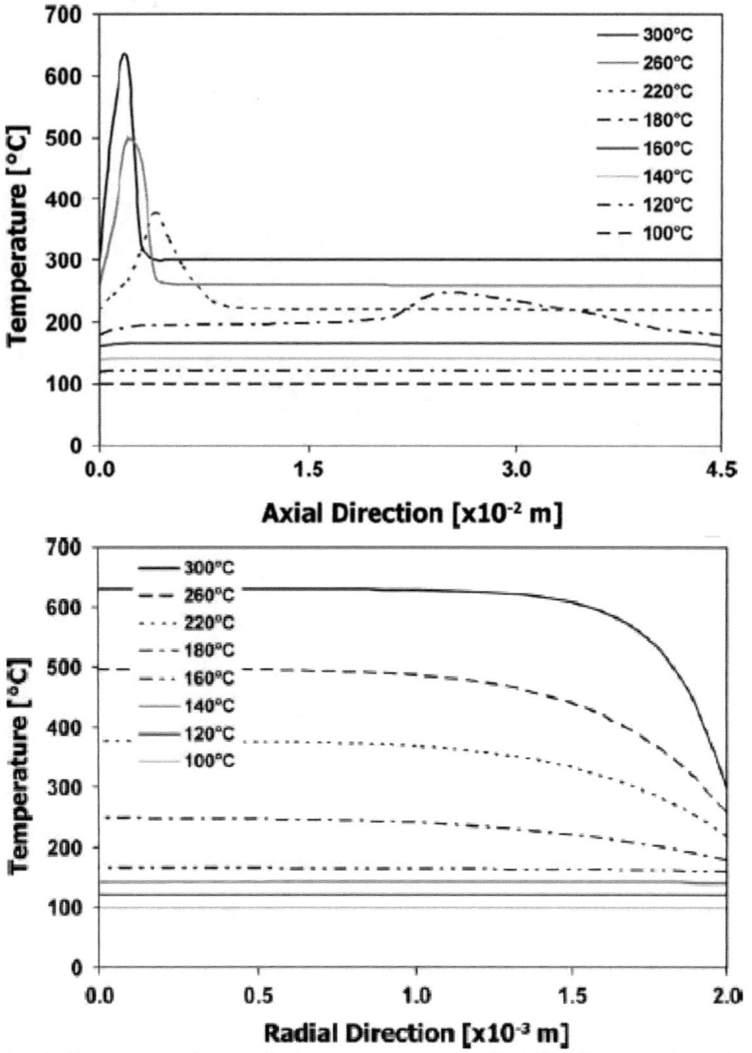

Figure 4. Modeling results of temperature distributions for the 2-mm m-PBR at different isothermal wall temperatures: (a) Temperature distribution in the radial direction at $l = l_{Tmax}$; (b) Temperature distribution in the axial direction at r = 0.0 m. *WHSV*: 1500 hr^{-1}.

In order to examine the effect of heat transport limitations on PrOx reaction activity, Figure 5 shows the modeling results of the average reaction rates of CO oxidation, H_2 oxidation and r-WGS reaction for the microreactor, the 2-mm and 4-mm m-PBR's. It clearly indicates that the m-PBR's has increasing r-WGS reaction rate with wall temperatures higher than 220°C. Consequently, these phenomena are more significant for the 4-mm than the 2-mm m-PBR.

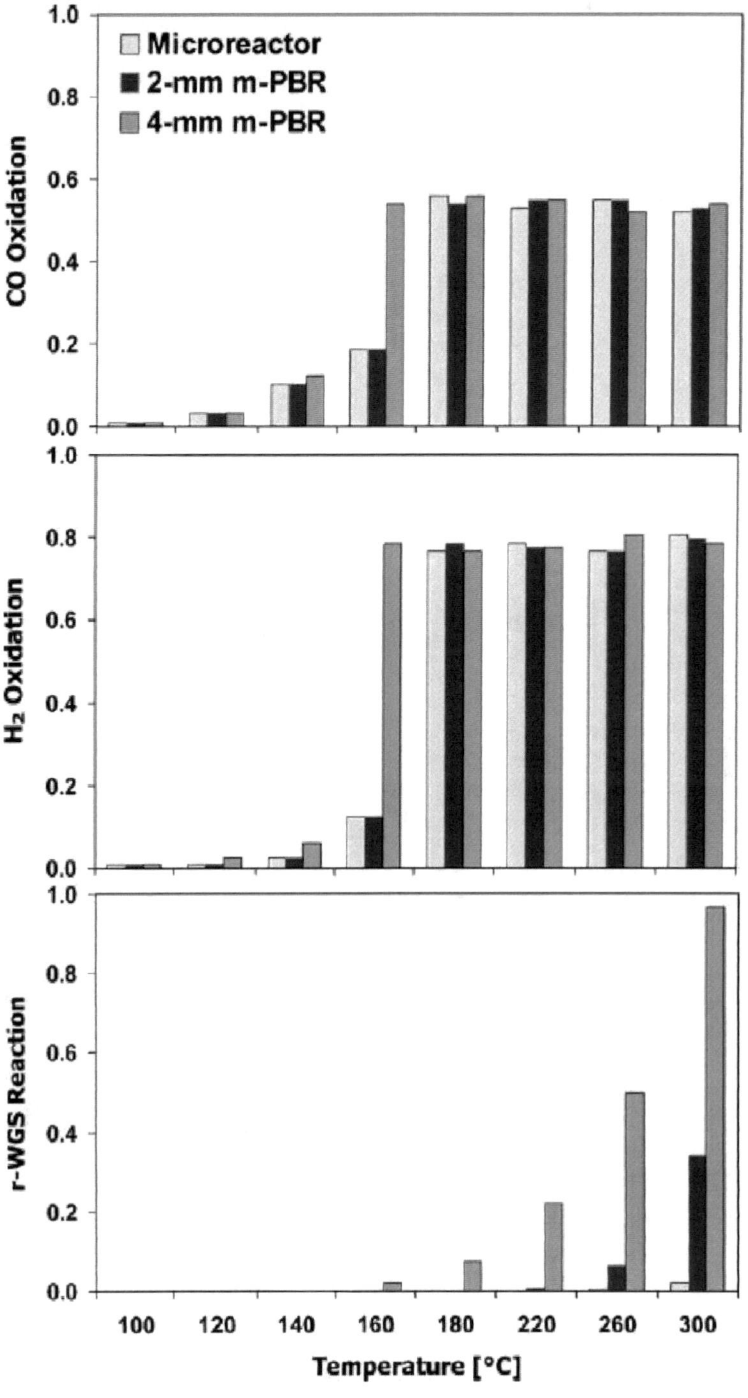

Figure 5. The average reaction rates (normalized) for CO oxidation, H_2 oxidation and the r-WGS reaction at T_w = 120°C, 180°C, 220°C for the microreactor, 2-mm and 4-mm m-PBR's.

Finally, Figure 6 plots the modeling results of CO conversion at different wall temperatures for the microreactor and 2-mm and 4-mm m-PBR's, all with the same *WHSV* (1500 hr^{-1}) and isothermal wall temperatures. The CO conversion curve for the microreactor essentially coincides with the result for ideal isothermal operation. On the contrary, the CO light-off curves for the m-PBR's shift to lower temperatures and CO conversion drops significantly at higher wall temperatures. The light-off shift to lower temperature is due to an increase in CO oxidation rate at local hot spots. With further temperature increase, however, the downfall of the CO conversion curve is caused by the strong r-WGS reaction activity. Especially for 4-mm m-PBR, the maximum CO conversion is only ~ 90% indicating severe hot spots at relatively low wall temperatures; the CO conversion drops below 0 above 260°C implying stronger r-WGS activity than CO oxidation activity. This result clearly shows the strong effect of reactor size, or more rigorously the t_{cat} of the reactor, on PrOx performance.

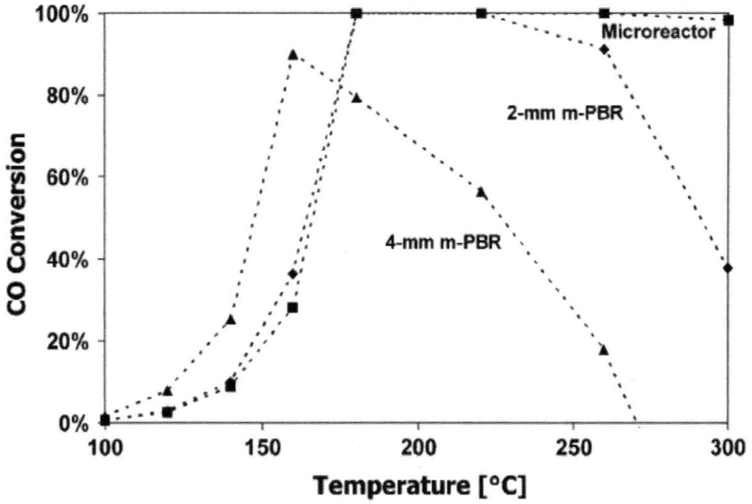

Figure 6. CO conversion vs. different reactor wall temperatures for the microreactor, 2-mm and 4-mm m-PBR's. *WHSV*: 1500 hr^{-1}.

CONCLUSIONS

In this paper, we carried out a study to systematically compare the thin-film catalyzed microreactor and the m-PBR's as kinetic tools for PrOx on Pt/Al_2O_3, provoked by their differences in TOF numbers and characteristic CO conversion behavior. The results of the qualitative analysis revealed significant differences of heat transport limitations essentially caused by their large difference of t_{cat}. Strong heat transfer limitation exists for the m-PBR's while heat transfer in the microreactor is efficient within large operating conditions. The quantitative study with the non-isothermal reactor modeling results further corroborated the above judgements and showed the inefficient heat transfer of the m-PBR's can lead to strong r-WGS activity and consequently large deviation of the CO conversion temperature characteristic compared to the isothermal microreactor.

NOMENCLATURE

Greek Symbols

ϕ_n	Thiele modulus of nth ($n \geq 0$) order reaction
η_i	internal effectiveness factor of catalyst

ρ_c	catalyst density (g/m^3)
ρ	gas density of the feed mixture (g/m^3)
μ	gas viscosity (g/m-s)
λ_e	effective thermal conductivity of the catalyst (W/m-K)

Symbols

D	catalyst dispersion
Da	Damköhler's number for heat transport
l	reactor axial position, l = 0 for the entrance of the reactor (m)
l_{Tmax}	the axial position where the temperature (at r = 0) is the highest (m)
k_{c-CO}	mass transfer coefficient of CO (m/s)
L	total reactor length (m)
P_i	partial pressure of reactant species *i* (*i*: CO, O_2, CO_2, H_2, H_2O) (Pa)
r	reactor radial position, r = 0 for the center of the reactor (m)
$-r_{CO}$	CO reaction rate (mol/m^3-s)
$-r''_{O_2}(obs)$	observed (actual) O_2 reaction rate at the entrance of the microchannel (mol/m^2-s)
R	reactor radius (m)
t_{cat}	effective interparticle heat conduction distance: catalyst thickness for the microreactor or reactor inner radius for m-PBR (m)
$T_{w,}$	reactor inner wall temperature (°C)
TOF	turnover frequency of CO oxidation on Pt (molecules/site-s)

WHSV weight hourly space velocity, defined as mole of total reactant flow per mole of precious metal in the catalyst per hour (hr^{-1})

REFERENCE:

1. A. Sirijaruphan, J.G. Goodwin Jr., R.W. Rice, *J Catal.* 224 (2004) 288-293.
2. D.H. Kim, M.S. Lim, Appl. Catal. A: Gen. 224 (2002) 27-38.
3. R.H. Venderbosch, W. Prins, W.P.M. van Swaajj, Chem. Eng. Sci. 53 (1998) 3355-3366.
4. M.J. Kahlich, H.A. Gasteiger, R.J. Behm, J. Catal. 171 (1997) 93–105.
5. Y. Choi, H.G. Stenger, J. Power Sources 129 (2004) 246-254.
6. S.H. Oh, R.M. Sinkevitch, J. Catal. 142 (1993) 254-262.
7. H.S. Fogler, Diffusion and reaction in porous catalysts. In: Elements of Chemical Reaction Engineering, 3rd ed., Prentice Hall, Englewood Cliffs, NJ (2000).
8. S.K. Ajmera, C. Delattre, M.A. Schmidt, K.F. Jensen, J. Catal. 209 (2002) 401-412.
9. Reactor and Catalyst Engineering, Delft University. Mass and heat transport effects - catalyst particles. Available at: http://www.dct.tudelft.nl/race/education/st316/chap04.pdf. Accessed October 25, 2003.
10. Arashi N, Hishinuma Y, Narato K, Nakajima F, Kuroda H. Mass transfer to a parallel-plate catalyst. Int Chem Eng. 1982;22:489-494.
11. Fogler HS. *Elements of Chemical Reaction Engineering*, 3rd ed., Prentice Hall, Englewood Cliffs, NJ (2000) 773.
12. D.E. Mears, *J. Catal.* 20 (1971)127-131.
13. D.E. Mears, *Ind Eng Chem Process Des Develop* 10 (1971) 541-547.

134g Polymerization in Microfluidic Devices: Numerical Simulations and Experimental Investigations

Christophe A Serra, Nicolas Sary

Microfluidic devices are now widely used in scientific fields such as biology, chemistry and chemical engineering. They proved themselves as powerful tools to study highly exothermic, fast and diffusion-controlled chemical reactions. Despites those great advantages they are scarcely used in polymer science. This communication presents a work aiming at addressing this issue.

We have studied the potentialities of several micromixers to control the polymerization. The micromixers are used either like microreactors or like devices to mix the reagents. The framework of this study was the synthesis of new functional polymer materials for high technologies such as electronic plastics for which polymers with a well defined macromolecular architecture are required. This work was divided in two parts: 1) numerical simulations and 2) experimental investigations.

Concerning the numerical simulations, we modelled the styrene free radial polymerization in two different interdigital multilamination micromixers as well as in a simple tee junction for comparison purpose. The simulations show that despite the heat released by the polymerization reaction, the thermal transfer in such microfluidic devices is high enough to ensure isothermal conditions. Moreover, one of the two micromixers allows a better control over the polymerization as polydispersity indexes are close to the theoretical value and smaller than that obtained with the tee junction or a lab balloon. Finally, one gets higher monomer conversion with this micromixer.

Concerning the experimental part, different kinds of related polymerization reactions were investigated 1) a homogeneous polymerization 2) a diffusion-controlled polymerization and 3) an interfacial homopolymer coupling reaction. The reagents were mixed either thanks to an interdigital multilamination micromixer or a simple tee junction. Thereafter, the polymerization was performed in a 0.9 mm ID tube reactor. The results were then compared with those obtained using a lab balloon. For the homogeneous polymerization, no differences were found between the micromixer and the tee junction. Indeed, the diffusion coefficients of the chemical species were so high that a perfect mixing is obtained in the first few inches of the tube reactor. However, the polydispersity indexes were found to be very close to the theoretical value of 1.5 where as for the lab balloon experiment it was found to be equal to 2. For the two other kinds of related polymerization reactions, the micromixer is by several order of magnitudes better in controlling the polymerization. The high contact area and the small reagent fluid strands thickness favour a lot the chemical reaction, leading to higher monomer conversions and smaller polydispersity indexes.

As a conclusion, both numerical simulations and experiments show that micromixers present strong capabilities to improve control over the polymerization and lead to better defined macromolecular architecture. Thus, microfluidic devices represent a new tool in polymer science.

Reaction rate enhancement of catalytic CO oxidation under forced thermal oscillations in microreactors with real-time gas detection

S. Jensen[1], J.L. Olsen[2], H. Hansen[2], O. Hansen[1], and U.J. Quaade[2]
[1] MIC - Department of Micro and Nanotechnology
[2] Department of Physics, Interdisciplinary Research Center for Catalysis (ICAT)
Technical University of Denmark, 2800 Kgs. Lyngby, Denmark

Abstract

We present microreactors with integrated mass spectrometer interfaces that allow real-time detection of the reaction products. The microreactors are made from silicon and fabricated using deep reactive ion etching. The low thermal mass of the reactor and the high temporal resolution of the detection system allow investigations of the effect of forced, fast thermal oscillations of the reaction temperature. CO oxidation is studied as an example of the use of the reactors, and using fast temperature cycling, reaction rates up to 40% higher than during steady state operation are observed.

1 Introduction

Catalytic chemical reactions performed under periodically varying operation conditions have shown increased reaction rates, improved selectivity, and reduced catalyst deactivation compared to reactions performed under steady-state conditions [1]. Most research has so far been directed towards studies of the effect of varying reactant concentrations, but it has been predicted that also forced oscillations of the reaction temperature can lead to increased reaction rates [2]. Investigations of the influence of temperature oscillations have been limited by the large thermal mass of conventional reactors, and it is only with the recent advent of microfabricated reactors (microreactors) with low thermal mass that it has become possible to investigate the effect of temperature oscillations with cycle times on the order of seconds [3, 4]. Here, we present microreactors with low thermal mass and integrated mass spectrometer interfaces that allow real-time detection of the reaction products. CO oxidation under fast temperature cycling is investigated as an example of the possible use of the microreactors, and reaction rates up to 40% higher than during steady state operation are observed.

2 Microreactor fabrication and characterization

To investigate the effects of temperature oscillations, we use microfabricated silicon reactors, since the multitude of available processing tools provides vast possibilities for integration of sensors and actuators, and enables parallel production of typically 20 reactors at a time.

The microreactors are fabricated using standard microfabrication processing technologies as UV photolithography and deep reactive ion etching (DRIE) [5]. The fabrication protocol is described in [6]. The use of DRIE allows fabrication of integrated short and narrow capillaries (30-50 μm long, 2-4 μm wide) for use as direct mass spectrometer interfaces [6]. Figure 1 (middle) shows SEM pictures of a fabricated capillary. The integrated capillaries are reproducible

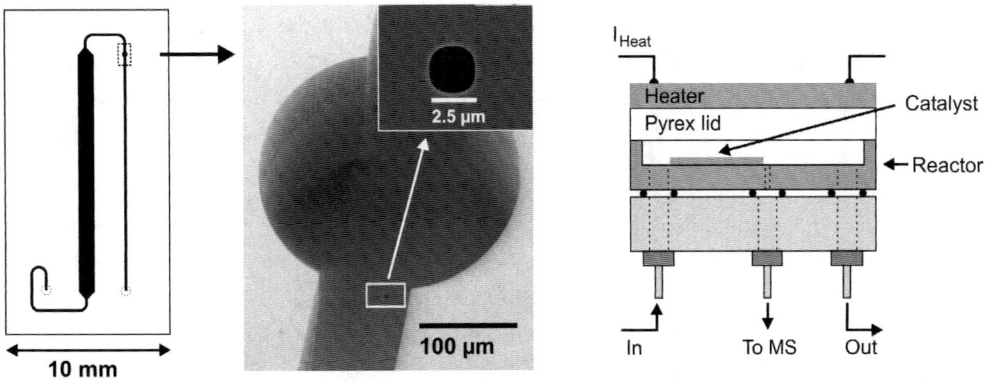

Figure 1: Left: overview of the microreactor chip. **Middle:** overview of the outlet channel and the mass spectrometer interface. **Right:** cross sectional view of the microreactor measurement setup. The microreactor is connected to the gas lines through an aluminium interface block, and the connection is sealed with Viton o-rings. Heating is provided by an external heater element, placed on top of the reactor. The temperature is recorded using a thermocouple placed underneath the reactor.

and provide several advantages over traditional external quartz capillaries [7], which suffer from limited reproducibility and limited knowledge about the exact orifice geometry. These limitations prevent a detailed understanding and modelling of the flow in such quartz capillaries, whereas the flow through the integrated capillaries has been shown to be highly predictable [8]. The short capillary length minimizes the dead volume and the response time, and all the involved species can be measured dynamically with real-time resolution.

For use in experiments, the microreactor is placed in a setup illustrated in Figure 1 (right). The setup comprises an aluminium interface block with gas line connections and a thermocouple for temperature measurements. The reactor is heated using an external resistive heater element connected to a computer-controlled power supply. With the current setup, the response to a heater current step pulse is a heating rate of 19 K/s on the rising edge, and a cooling rate of 10 K/s on the falling edge, as illustrated in Figure 2 (left). When oscillating the temperature,

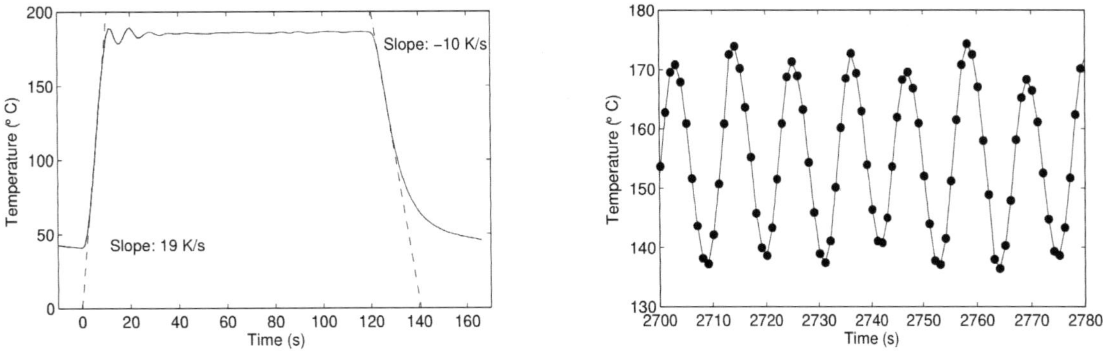

Figure 2: Left: the system temperature response to a square shaped heater current pulse. The maximal heating and cooling rates are indicated. **Right:** the temperature recorded at a cycle time of 11 s and an amplitude of 17 °C.

the amplitude and frequency cannot be chosen independently, due to the finite cooling rate. An example of a used oscillation with a cycle time of 11 s and an amplitude of 17 °C is shown in Figure 2 (right). Addition of recently developed integrated heaters and thermometers [9] is, together with thermal optimization of the reactor, expected to extend the parameter space that can be investigated to include sub-second regime cycle times and increased amplitudes.

3 Experiments

Many different types of catalytic reactions are expected to be sensitive to temperature oscillations [3], and as an example, we here investigate CO oxidation over a supported platinum catalyst. CO oxidation has proven sensitive to oscillatory forcing of the gas concentrations, and since oscillation of the surface temperature is bound to change the surface concentration of species, temperature oscillations are also expected to have an influence on the reaction.

Prior to the experiments, a porous support for the catalyst is realized by calcinating an AlOOH-containing gel, placed in the reactor using a syringe. The support is then impregnated with a platinum acetylacetonate solution, which, as the solvent evaporates during a heat treatment, leaves platinum particles behind in the support structure.

Before looking at the effects of temperature oscillations, we investigate possible adsorption delays in the system, which could obscure the detection of rapidly varying signals. A gas mixture of 90% He, 5% CO, 2.5% O_2, and 2.5% Ar was streamed through the reactor, while the temperature was increased from 40 °C to 175 °C over a period of 15 s. At increased temperatures CO and O_2 react to form CO_2, which we detect with the mass spectrometer. The gas lines leading to the mass spectrometer are heated to minimize adsorption delays.

Figure 3 (left) shows the mass spectrometer response as a function of time during the

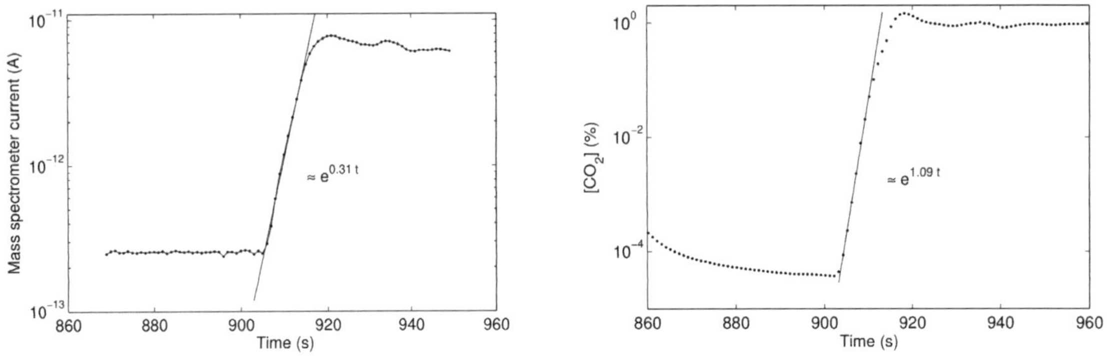

Figure 3: Left: the mass spectrometer response when the reactor is heated from 40 °C to 175 °C over a period of 15 s. The reponse is exponential with a time constant of approximately 3.2 s. **Right:** the calculated CO_2 production rate during the temperature step.

heating of the reactor. The response is exponential with a time constant of approximately 3.2 s.

Due to the small reactor dimensions and the low flow rates used, the reactor is diffusion mixed, and the output response to a step-increase of the inlet concentration is an exponential increase with a time constant of V/S, where V is the reactor volume and S the flow rate. With the used flow rate of 0.15 mL/min, and a reactor volume of 4.65 mm^3, this time constant is approximately 1.9 s. The discrepancy between this and the measured 3.2 s can be explained by the fact that the CO_2 concentration increase during the heating is not a perfect instantaneous step, but rather an exponential increase with a time constant of approximately 0.9 s, as illustrated in

Figure 3 (right), which shows the expected CO_2 production rate calculated from the temperature measured during the heating. This deviation from a perfect step is bound to make the output response slower than in the ideal case.

Having demonstrated real-time gas detection, the effects of forced temperature oscillations are now investigated using the same setup and gas composition as before. In Figure 4 (left), the CO_2 concentration in the outlet gas is plotted as a function of temperature for both

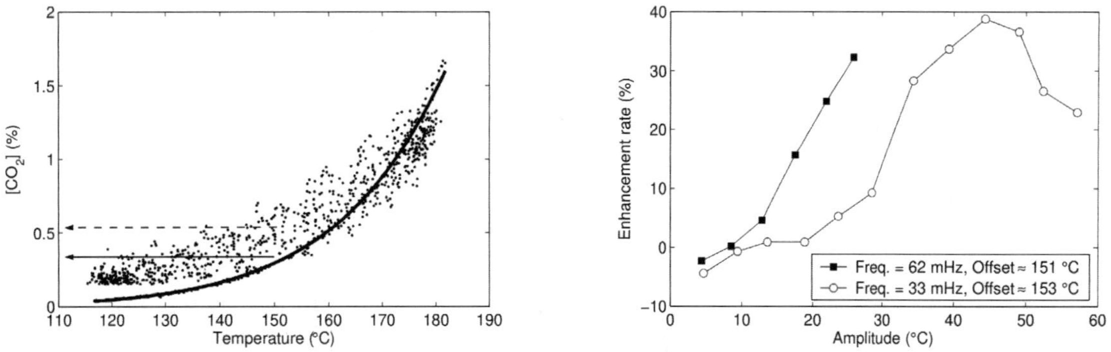

Figure 4: Left: the CO_2 concentration in steady state (solid line) and during oscillation (scattered data). At 150 °C, the average concentration during forced oscillations is 40% higher than the steady state concentration. **Right:** the rate enhancement as a function of oscillation amplitude for two different frequencies.

steady-state experiments and experiments using a forced temperature oscillation with an amplitude of 30 °C, an offset of 150 °C, and a cycle time of 20 s. Although the data from the experiments with forced temperature oscillations are somewhat scattered, they clearly seem to lie above the concentrations obtained in steady-state.

The reaction rate was studied as a function of oscillation amplitude, frequency, and offset, and some clear tendencies and substantial improvements over the measured steady-state rate were observed. As an example, the relative reaction rate enhancement is plotted as a function of oscillation amplitude in Figure 4 (right), and it is observed how it increases with oscillation amplitude and reaches a maximum of approximately 40% at a 40 °C amplitude, 150 °C offset, and 30 s cycle time. The reaction rate enhancement has also been observed to increase with increasing oscillation frequency, and the increase seems to appear at lower frequencies for larger amplitudes.

A model for the reaction rate enhancement has not yet been developed, but a possible mechanism could include a temperature induced break-up of large CO and O islands on the platinum surface similar to the one described for feed gas concentration variations [10], and/or changes in effective surface occupancies [2].

4 Conclusion

Microreactors with integrated mass spectrometer interfaces that allow real-time detection of the reaction products have been developed. The mass spectrometer interfaces provide fast and accurate gas sampling from the reactor volume without dilution effects. Using temperature cycling, with cycle times on the order of 10s of seconds, reaction rates up to 40% higher than during steady state operation have been achieved.

References

[1] P. Silveston, R.R. Hudgins, and A. Renken, "Periodic operation of catalytic reactors - introduction and overview", *Catal. Today* **25**, pp. 91-112 (1995).

[2] F.J.R. van Neer, A.J. Kodde, H. den Uil, and A. Bliek, "Understanding of resonance phenomena on a catalyst under forced concentration and temperature oscillations", *Can. J. Chem. Eng.* **74**, pp. 664-673 (1996).

[3] P.L. Silveston and R.R. Hudgins, "Periodic forcing of catalytic reactions", Chem. Eng. Sci. 59, pp. 4043-4053, (2004).

[4] J.J. Brandner, G. Emig, M.A. Liauw, and K. Schubert, "Fast temperature cycling in microstructure devices", *Chem. Eng. J.* **101**, pp. 217-224 (2004).

[5] A.M. Hynes, H. Ashraf, J.K. Bhardwaj, J. Hopkins, I. Johnston, J.N. Shepard, "Recent advances in silicon etching for MEMS using the ASE^{TM} process", *Sensors and Actuators* **74**, pp. 13-17 (1999).

[6] U.J. Quaade, S. Jensen, and O. Hansen, "Microsystem with integrated capillary leak to mass spectrometer for high sensitivity temperature programmed desorption", *Rev. Sci. Instrum.* **75**, pp. 3345-3347 (2004).

[7] B. Kasemo, "Quartz tube orifice leaks for local, fast-response gas sampling to mass spectrometers", *Rev. Sci. Instrum.* **50**, pp.1602-1604 (1979).

[8] U.J. Quaade, S. Jensen, and O. Hansen, "Fabrication and modelling of narrow capillaries for vacuum system gas inlets", *J. Appl. Phys.* **97** (2005).

[9] S. Jensen, U.J. Quaade, and O. Hansen, "High-temperature compatible nickel silicide heater and thermometer for catalytic chemical microreactors", in *Proc. 18th IEEE Int. Conference on Micro Electro Mechanical Systems (MEMS 2005)*, Miami, FL, USA, January 2005, pp. 463-466.

[10] X. Zhou, Y. Barshad, and E. Gulari, "CO oxidation on Pd/Al_2O_3: transient response and rate enhancement through concentration cycling", *Chem. Eng. Sci.* **41**, pp. 1277-1284 (1986).

Brominations in micro reactors using high and low p,T-processing

P. Löb, V. Hessel*, H. Löwe*, C. Hofmann*, K. Mazanek**, H. Klefenz***
**Institut für Mikrotechnik Mainz GmbH, Carl-Zeiss-Str. 18-20, D-55129 Mainz, Germany*
Tel.: +49 6131 990 377, E-mail: loeb@imm-mainz.de
***Europa Fachhochschule Fresenius, Idstein, Germany*

Abstract

Bromination reactions using elemental bromine in micro reactors have not been reported so far. To get a broad overview about the potential of micro chemical processing in the case of bromination, three different reactions covering different syntheses tasks were investigated: the bromination of toluene with competing side-chain and core substitution, the bromination of thiophene with competition of single and multiple substitution and the high-temperature (up to 230°C) side-chain bromination of *m*-nitrotoluene. All reactions were performed thermally without adding catalyst, irradiation and using solvents. Different micro mixers and tube sections were used depending on the specific required reaction conditions. Micro reactor processing thereby addresses the following aspects: transformation of batch in continuous processing, avoidance of solvents, safe processing, increase of space-time yields, and influence of process conditions on selectivity and yield and fast experimentation.

Introduction

This is, to our best knowledge, the first public report on performing bromination reactions using elemental bromine in micro reactors, while fluorination and chlorination reactions using the elemental halogens have already been studied by various groups (see [1]). In the case of fluorinations (e.g. [2-4]) and chlorinations (e.g. [5]), reported micro-reactor processings applied halogen always as gas and mostly specialized microstructured gas/liquid - contactors were used for performing the reaction resulting in concise flow patterns and phase conditions. In contrast, the brominations reported here were performed starting with the liquid bromine in standard off-the-shelf micro mixer/tube set-ups which are not multiphase speciality tools. Phase conditions here are thus quite complex through evaporated bromine and gaseous hydrogen bromide formed during the reaction. A special focus of the investigations was the targeted increase of space-time-yield. This was achieved by increased system pressure (up to 25 bar) and increased reaction temperature – both not feasible (at least not easily) with conventional equipment.

The bromination of thiophene is very fast in the ms-range, even at low temperatures (0 °C), demanding efficient mixing by the micro mixer [6]; the brominations of toluene and *m*-nitrotoluene however are rather sluggish, demanding hours-operation by batch processing and posing the need to set up an entirely novel experimental protocol, the so called high p,T-processing (p being pressure, T being temperature).

Experimental set-ups

For all the experiments micro mixer/tube set-ups were used built-up basically by connecting a micro mixer to a tube section. Two different micro mixers of IMM, the so-called triangular interdigital micro mixer (TIMM) [7] made of photostructureable glass and a so-called caterpillar micro mixer with the denomination CPMM-R600/12 made of PVDF [1] (more specifications are given in Table 1), and 1/8´´ tubes of different lengths and materials (PFA or Hastelloy C276) were thereby the basis for the different used combinations. Micro mixer and tube section were embedded in thermostated baths. Depending on reaction requirements, additional thermostated tube sections before the micro mixer were employed for pretempering of the reactants. Furthermore, it was possible to apply pressure to the system via a modified valve at the outlet of the tube section. Pressure within the experimental set-up was determined by a gauge built-in in the feeding line of the organic reactant. Temperature indications always refer to the temperature of the thermostated baths. Feeding of the reactants was accomplished by piston pumps with glass cylinders and Teflon or PVDF as sealing material.

Table 1. Micro mixer specifications

Triangular interdigital micro mixer (TIMM)		**Caterpillar micro mixer CPMM-R600/12 PVDF**	
Outer dimension with housing	106 mm x 56 mm x 25 mm	Outer dimension with housing	27 mm x 30 mm x 40 mm
Pressure stability	Up to 2 bar	Pressure stability	Up to 10 bar
Temperature stability	Up to about 200 °C (because of the Teflon tubing)	Temperature stability	-30 °C – 150 °C
Material in contact to reaction fluids	Glass, Teflon	Material in contact to reaction fluids	PVDF
Mixer feed channels: number; width; depth / width of separating walls	2 x 15; 60 μm; 150 μm / 50 μm	Mixing channel: cross section; length; number of mixing steps	600 μm x 600 μm; 14.4 mm; 12
Mixing chamber: length; initial width; height	8 mm; 3.25 mm; 150 μm		
Mixing channel: length; width; height	19.4 mm; 500 μm; 150 μm		

Bromination of toluene

Bromination of toluene was investigated with regard to product distribution and conversion in the temperature range of 0 °C to 120 °C [1, 8]. Focus thereby laid on competing formation of benzylbromide and the three bromotoluene isomers. Multi bromination was not aim of the reactions carried out and therefore the molar ratio toluene to bromine was set to 1. A PFA tube with an inner volume of 4.9 cm^3 and the triangular interdigital micro mixer, both embedded in the same thermostated bath, were used. The toluene was pretempered in the same bath, while the bromine was fed at room temperature. The system was open to atmospheric pressure. Two different total flow rates were applied (74.5 and 149.0 cm^3/h). For both investigated total flow rates, conversion of toluene and bromine increases with temperature up to 100 °C, while a decrease was observed going from 100 °C to 120 °C (Figure 1). Full conversion is achieved for a total flow rate of 74.5 cm^3/h at 80 °C and 100 °C, while at the higher flow rate the conversion was incomplete because of the reduced residence time. The decrease of conversion for 120 °C can also be attributed to a reduced residence time caused by evaporation of toluene. As expected only low selectivities of benzylbromide are obtained, decreasing with an increase in temperature. For temperature above 80 °C almost only core substituted products are formed. In no case multiple bromination products were detected.

Although the experiments performed represent only a first survey of toluene bromination in a continuously operated micro reactor set-up, these have already pointed out some very interesting aspects. The direct contacting of bromine and toluene without any use of solvents can be handled safely in a broad temperature range in the micro reactor set-up. The reaction velocity is high and therewith one important prerequisite for useful application of micro chemical processing is given. Assuming single/liquid phase condition in the reactor, the residence time was 3.9 min for the experiments leading to full conversion. Taking into account gas formation (assuming hydrogen bromide as ideal gas and a mean conversion of bromine of 50% in the reactor) the residence time reduces dramatically to only 3.0 s! Furthermore, the speed of experimentation is that high that work-up and analysis of the samples is now the bottleneck rather than sample generation.

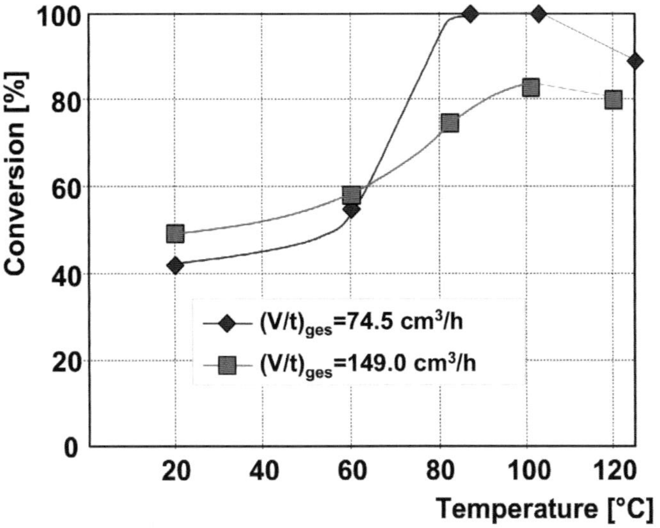

Figure 1. Conversion of toluene as function of temperature for two different flow rates.

Bromination of thiophene

Bromination of thiophene was investigated with regard to multiple bromination. The same experimental set-up as for the toluene bromination was used. The total flow rate was kept constant at 93.5 cm^3/h. Residence time was in the range of a few seconds when working at atmospheric pressure. In one set of experiments the molar ratio of bromine to thiophene was varied in the range from 1 to 5 for a temperature of 50 °C (Figure 2). High conversion of bromine was found despite the short residence time. Even at a molar ratio of 5, a conversion of 70% of bromine was achieved. Best yields (up to 80%) for the industrial fine-chemical intermediate 2,5-dibromothiophene (which is of importance as starting material for LEDs) were obtained for a molar ratio of 2. Furthermore, temperature was varied in another set of experiments between 0 °C and 60 °C for a molar ratio of 1. Complete conversion of the bromine was observed. The ratio between single and two-fold substituted thiophene is quite constant over the investigated temperature range. An increase of tri-fold substitution product is observed with higher temperatures. The bromination of thiophene is even faster than in the case of toluene. Already in the mixing chamber a discolouring of the reaction solution and the formation of hydrogen bromide gas can be observed.

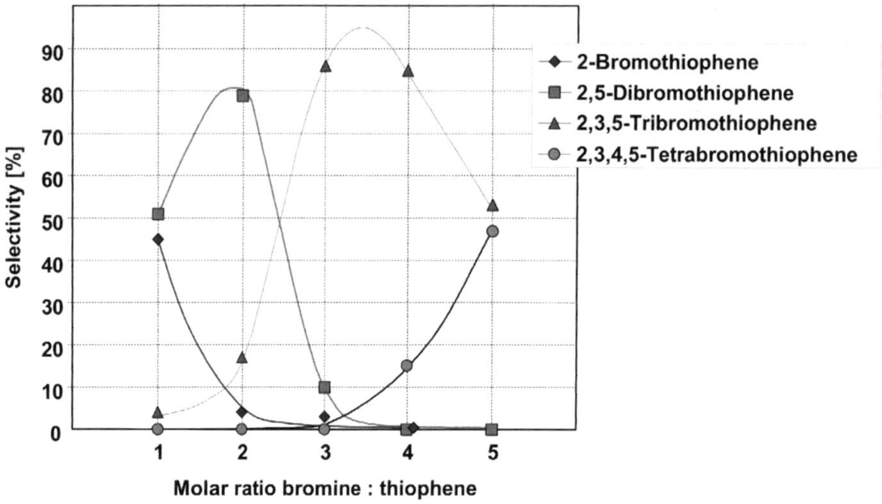

Figure 2. Product distribution for the thiophene bromination as a function of molar ratio bromine to thiophene at 50 °C.

Bromination of *m*-nitrotoluene

In a first set of experiments at atmospheric pressure and temperatures between 170 °C and 190 °C the monobromination of the side-chain was investigated with molar ratios bromine to nitrotoluene between 0.25 and 1. A tube section of PFA with an inner volume of 3.8 cm^3 embedded in a thermostated oil bath and the triangular interdigital micro mixer to contact the reactants at room temperature were used. Already in these experiments an increase of space-time yield by a factor of 18 can be achieved when switching from batch to micro reactor processing. The residence time was, assuming single/liquid phase condition, about 2 min. Further experiments were performed at molar ratios of 2 leading to a competition between mono- and disubstitution at the side-chain. The experimental set-up was changed to allow also experiments at higher pressures. The caterpillar micro mixer CPMM R600/12 PVDF and a 1/8´´ Hastelloy C276 tube (8.4 cm^3 inner volume) were used. A tremendous increase of turnover of nitrotoluene from about 40% to 95% was observed when increasing pressure to 15 bar (for a total flow rate of 3.3 cm^3/min corresponding to a residence time 2.6 min assuming single phase liquid conditions). A further increase of pressure to 25 bar led only to minor changes (Figure 3). Furthermore temperature (between 170 °C and 230 °C) and flow rate (between 1.5 and 6.5 ml/min) were varied which influenced significantly the selectivity of the reaction. With the selected continuous working set-up up to 17 experiments a day were performed, underlining the potential for an accelerated process development.

Conclusions

The direct contacting of bromine with toluene, thiophene and *m*-nitrotoluene without the use of solvents was realised in micro mixer/tube set-ups under controllable conditions even for pressures up to 25 bar and 230 °C. The experimental set-up is suited for changing different parameters quite fast, thereby allowing an accelerated process development. In all three cases high space-time yields were obtained. Especially, the bromination of nitrotoluene thereby underlines the acceleration of reaction by working under high pressure and high temperature conditions, not or at least not easily to realize in conventional equipment. Naturally, reactors can and should be further optimised, but the results of the described reaction examples already justify a broader use of this technology.

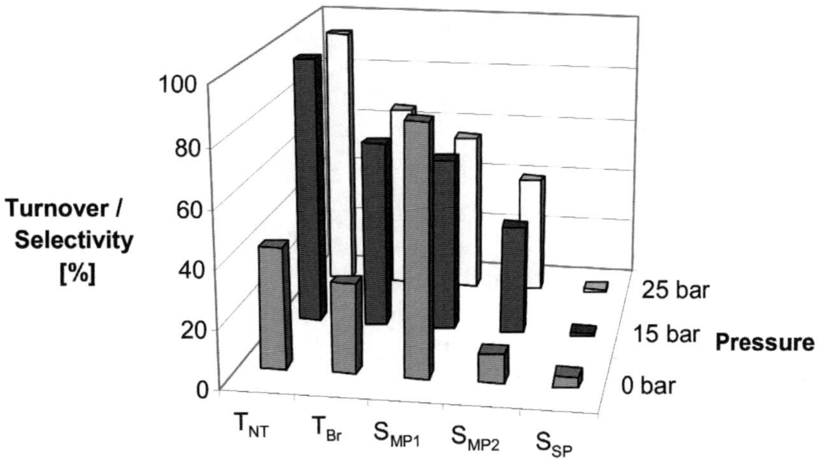

Figure 3. Turnover and selectivity for the bromination of *meta*-nitrotoluene for different pressures at a molar ratio bromine to nitrotoluene of 2 at a temperature of 200 °C. Abbreviations: T_{NT}: turnover nitrotoluene, T_{Br}: turnover bromine, S_{MP1}: selectivity benzyl bromide, S_{MP2}: selctivity benzal bromide and S_{SP}: selectivity side products.

References

[1] P. Löb, H. Löwe, V. Hessel, J. Fluorine Chem. 125, 11 (2004) 1677-1694.

[2] K. Jähnisch, M. Baerns, V. Hessel, W. Ehrfeld, V. Haverkamp, H. Löwe, Ch. Wille, A. Guber, J. Fluorine Chem. 105 (2000) 117-228.

[3] R. D. Chambers, D. Holling, R. C. H. Spink, G. Sandford, Lab Chip 1 (2001) 132-137.

[4] N. de Mas, A. Günther, M. A. Schmidt, K. F. Jensen, Ind. Eng. Chem. Res. 42 (4) (2003) 698-710.

[5] H. Ehrich, D. Linke, K. Morgenschweis, M. Baerns, K. Jähnisch, Chimia 56 (2002) 647-653.

[6] P. Löb, V. Hessel, H. Löwe, Letters in Organic Chemistry, to be submitted.

[7] V. Hessel, S. Hardt, H. Löwe, Chemical Micro Process Engineering – Fundamentals, Modelling and Reactions, Wiley-VCH, Weinheim, 2004.

[8] K. Mazanek, diploma thesis, Europa Fachhochschule Fresenius, Idstein, Germany, 2004.

135a Novel Glass Micro-reactor for Water Gas Shift Reaction Characterization

Mayuresh V Kothare, Miltiadis K Hatalis, Samrat Mukherjee

Introduction:

Glass micro reactors are often overlooked in preference to the ubiquitous silicon micro-reactors. Glass is cheaper and hence a more commercially viable material for mass production of micro reactor. With some adaptation glass micro reactors can have the same capabilities as the ones made of silicon. Furthermore transparency of glass has a variety of advantages including visual understanding of the system and possibility of doing spectrometric studies.

A novel glass micro reactor and novel micro-macro fluidic connector made of Teflon has been designed, fabricated and characterized. The fabrication steps were simple, quick and economical. Moreover, no special facility such as clean room was required for the fabrication. The reactor is able to withstand 260oC and 240 psi without any perceived problems.

The fabricated device consists of a vaporizer, heater, mixer and reactor. Commercial water gas shift (WGS) catalyst from ICI-Synetix (83-3M) was ground and sieved to 50-75 microns and fluidized into the reactor. The packed bed reactor, thus formed, was used for WGS reaction of water and carbon monoxide between 180-240 oC. The reactor was designed for CO flow of 0.15 moles/hr. The reaction can be stated as:

H2O + CO <--> H2 + CO2

Motivation:

It is envisioned that in future portable micro chemical systems (plant-on-a-chip) would be universally used. One application for portable MCS is a fuel cell and fuel processor combination that would replace lithium ion batteries used today to power portable devices. Stored methanol would be reformed to hydrogen which would be fed into a fuel cell that would convert chemical energy into electric energy. Reformation of methanol produces hydrogen and carbon dioxide, but also carbon monoxide (CO) as a by product. CO poisons the fuel cell and hence needs to be removed from the gas. Water Gas Shift Reaction leads to removal of CO and formation of more useful hydrogen. Combined with a membrane, hydrogen can be effectively purified. [1]

Water Gas Shift reaction has been widely used and characterized. Unfortunately, various mechanisms, reaction paths and reaction rates have been reported in literature. The expression fits specific experimental data and experimental setups, but vary widely with reactor size and reaction conditions. For example rate expression for a large WGS reactor operating at high temperature and pressure with larger catalyst pellets would be different from micro reactors. In the present study, commercial WGS reactor has been characterized for use in a glass micro reactor.

Fabrication:

The reactor was fabricated using 2 borosilicate glass wafers 4 inch in diameter and 1.75 mm thick. The channels were machined in wafer A, by a spinning drill bit that was restored over the surface. Glass wool was placed in suitable places to retain the catalyst powder. The second wafer B was then bonded to the wafer A using high temperature epoxy from Cotronics, to form closed channels. The two wafers were aligned by drilling through holes in the two wafers, passing metal wires through the holes and then sliding the wafers along the wires. In wafer B, holes were drilled in for fluid flow and macro-micro fluidic connectors were constructed. On wafer B around the hole, on the free surface, a donut shaped groove, 0.5 mm deep, was etched. Teflon tube, 9/64 inch OD was placed into the groove. High temperature epoxy was then poured around the tube and cured, bonding the tube to the wafer. The groove prevented the epoxy from flowing into the hole. Barb-to-luer connectors was procured from Harvard Apparatus. The barb side was pushed into the Teflon tube while the luer side was used to connect to the flow-controller, syringe pump and the mass spectrometer.

Characterization:

A variety of tests were done to characterize the unit. The reactor and connectors were tested for their ultimate pressure by pumping in water with the outlet closed. The unit was able to withstand 240 psi at which pressure the syringe in the syringe pump broke leading us to abandon higher pressure tests. Clearly 240 psi pressure is not a limitation of the reactor unit. The connector was subjected to pull-out-test to analyze its strength. The unit was able to withstand 25 N which again is the limitation of our testing device. The pressure drop across the packed bed was characterized and was found to conform to Ergun equation.

Set-up:

The unit is heated on a hot plate. Eventually, this can be replaced by heating lines. Carbon Monoxide from a gas cylinder was metered into the reactor using a flow controller. Water was fed in from a second port using a syringe pump. The water was evaporated into steam in situ. Steam and CO were heated and mixed together and then fed into the packed bed reactor. The catalyst was retained using glass wool.

Reaction Characterization:

The reactor was operated under a variety of conditions. CO flow was varied between 0.05 moles/hr to 0.115 moles/hr. Water-to-CO molar flow ratio was varied between 0.5 to 1.4. The temperature was varied between 185 oC to 240 oC. The reaction rate constant was evaluated based on these results.

Summary:

A glass micro reactor was designed and fabricated, and found to work extremely well. The reactor was characterized for its strength and durability. Water Gas Shift Reaction was characterized using a commercial catalyst. Reaction rates were found to vary slightly with time.

Reference

1. Mukherjee, S., M. K. Hatalis and M. V. Kothare. Hybrid Micro Membrane Micro Reactor for in situ PURE Hydrogen Production In the 7th International Conference on Microreaction Technology (IMRET 7), Lausanne, Switzerland, September, 2003.

135b Microprocess Technology for Fischer-Tropsch Gas-To-Liquids

James Cao, John Hu, Terry Mazanec, Yong Wang

Fischer-Tropsch synthesis has fast surface reaction kinetics. However, the two conventional FT reactors, fixed bed and slurry phase, are limited by slow reaction rates due to either mass transport limitations or heat transport limitations. The limited heat transfer together with the fast surface reaction kinetics result in hot spots in fixed-bed catalytic reactors, which favor the methanation reaction; these reactors are operated at lower GHSV and large recycle streams to minimize temperature excursions in the catalyst bed. Because of the restricted operational parameters to minimize methanation, the performances of conventional reactors are not improved even with more active Fischer-Tropsch synthesis catalysts. More importantly, conventional processes are highly dependent on the economy of scale, and may not be economically applicable for stranded gas conversion or biomass derived syngas conversion.

A microchannel reactor offers the advantages of both the slurry and the fixed bed reactors. First, a microchannel reactor has much higher surface area/volume ratios for enhanced heat transfer so that it can be operated at close to optimum conditions while minimizing hot spots. Second, the catalyst is fixed within a microchannel reactor so that there is no need for catalyst separation from the product waxes and water. Successful deployment of the microchannel reactor relies on the development of a highly active and selective catalyst. Here we report the development of a highly active Fischer-Tropsch catalyst and a unique engineered structure, selected to minimize mass transfer resistance. The catalyst performance was evaluated in a microchannel testing reactor.

135c Microreactor for Endothermic Heterogeneous Catalysis Reaction

Christophe Pijolat, Claude de Bellefon, Cyril Delattre, Marilyne Roumanie, Patrick Pouteau, Valérie Meille

Medical diagnostics ha been one of the first research area to work on "Lab-On-A-Chip" and to develop such Microsystems and in particularly DNA chips. Nowadays "Lab-On-A-Chip" attracts the attention of both heterogeneous and homogeneous catalysis field. One of the major interest comes from the miniaturized size and its related very high surface on volume ratio which leads to improve mass and heat transfer but also the safety in regards of industrial application. The aim of this work is to study and to realize an autonomous catalytic microreactor for hydrogen release from methylcyclohexane. For this reaction of dehydrogenation, a classical catalyst is platinum supported on alumina. Consequently, the general objectives of this work are : (i) to develop a microreactor with its heaters, sensors…, (ii) to deposit catalysts in the microreactor, and (iii) to study the catalytic conversion of this system.

The chip used has been developed by the CEA-LETI. It comprises several channels etched in silicon substrate with a classic deep reactive ion etching process. These channels of 100mm depth are structured with pillars to increase the surface and thus improve the overall reaction rate compare to a channel without pillars. The structured microreactor is molecularly bonded with a silicon cover to form an airtight device. The reaction studied is extremely endothermic with a heat of reaction of 204kJ/mol. Consequently it is necessary to include some heaters to have an autonomous microreactor. Microheaters have been fabricated with thick film resistances of platinum deposited using screen-printing method. Tests of microheaters are realized using an infrared thermographic camera which gives temperature cartographies. It enables the observation of temperature homogeneity on the silicon substrate.

Final packaging is achieved by connecting microtubes to the chips. The microtubes consists in a polyimide-glass capillary of 365mm OD and 100mm ID stuck inside the channel entrance with UV-sensitive glue. Such connection is called "low temperature connection" because of its limited working temperature around 150 C. One challenging point was to have the entrance/exit at 150 C whereas the reaction zone is at 370 C. Some simulations with FEMLAB software have been performed leading to different design. The first idea was to make some notches to increase the surface contact with air but the thermal loss was not sufficient (about 30 C). A second idea was to evacuate the heat using a block in Al or Cu. The maximum thermal loss was 200 C with a block looking like a thermal heat sink. This solution was difficult to realise because of the dimension of such pieces. The only solution to get a thermal loss of 220 C was to design a block in which a liquid flow cools the microreactor entrance/exit. Only the entrance and exit are laid on this block and the reaction zone is fixed at 370 C with the screen-printed heaters.

The catalyst is currently inserted in a closed microreactor in liquid phase with a syringe pump. A film of boehmite suspension is deposited, dried on a hotplate under air, and then calcinated at 600 C. After this washcoating step, the microreactor is impregnated with a toluenic platinum solution, then dried and calcinated at 400 C. The resulting catalyst layer is about 1mm.

The reaction of methylcyclohexane dehydrogenation has been firstly carried on in a macro-reactor using pieces of 1mm^2 silicon wafer with two types of catalyst : sputtered platinum and platinum on supported alumina. For these catalytic studies, a specific testing bench has been developed to test such pieces of silicon wafer. This bench have gas analysers (mass spectrometer, gas chromatography, sensors) to follow the hydrogen production. The catalytic results obtained in the macro-reactor show that the sputtered Pt catalyst does not seem adapted for the dehydrogenation of methylcyclohexane contrary to the conventional Pt/Al2O3 catalyst. According to this result, only the conventional catalyst will be tested in the final microsystem.

To test this microreactor a specific test bench is developed. It comprises a syringe pump, a system to evaporate the methylcyclohexane, the cooling block to isolate the microreactor entrance/exit , a flowmeter to monitor the hydrogen flow during the reaction. Only the gas circuit (N2, O2, H2) is in common with the macro-reactor. All this work on testing the microreactor is under work and more data will be presented.

Multichannel microstructured reactor for hydrogen production based on filamentous catalysts

Chrystèle Horny, Lioubov Kiwi-Minsker and Albert Renken
Swiss Federal Institute of Technology, EPFL-LGRC, CH-1015 Lausanne, Switzerland

Introduction

Hydrogen for fuel cell application can be produced autothermally by the combination of endothermic steam-reforming and exothermic oxidation of methanol, which is called the Oxidative Steam-Reforming of Methanol (OSRM):

$$4CH_3OH + 3H_2O + 0.5O_2 \rightarrow 4CO_2 + 11H_2 \qquad \Delta_r H^{573K} \cong 0 kJmol^{-1}$$

Since oxidation is faster than steam-reforming, these reactions carried out in a traditional reactor with randomly packed catalytic beds lead to a temperature profile in the axial direction. High conversion near the reactor entrance results in a hot-spot that can damage the catalyst. The steam-reforming being relatively slow, a cold-spot is observed in the second part of the reactor.

The novel concept of a microstructured reactor proposed in this study is based on catalytically active metal filaments placed in parallel to the reactor walls. The channels for gas flow are formed by the interspaces between the strings and have an equivalent hydraulic diameter in the same range of magnitude as the diameter of the filaments.

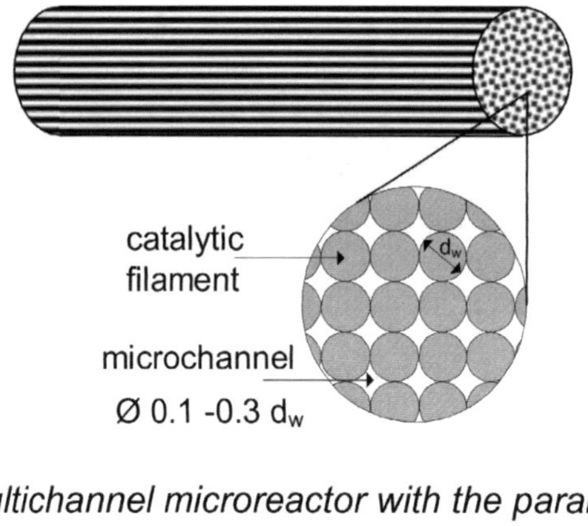

Figure 1. Multichannel microreactor with the parallel filaments.

Microstructured reactors are characterized by their small size: the radius channel, in the micrometer range, leads to a laminar flow and thus in low pressure drop. Furthermore, the radial diffusion time is short, this results in a flat radial concentration profile and consequently narrow residence time distribution [1].

The use of metallic wires with high thermal conductivity suppresses the pronounced axial temperature profile. Indeed it was shown [2] that heat conduction in the support of structured metallic catalyst can effectively remove the energy generated by a strongly exothermic reaction. In this way, the heat necessary for the steam-reforming is provided by the oxidation via the metallic catalyst. Hence, the dynamic behaviour is greatly improved.

Brass is chosen as support material for its chemical composition, similar to the $Cu/ZnO/Al_2O_3$ catalyst generally used for the steam-reforming of methanol. In addition, brass has a very high thermal conductivity (120 W/m.K).

The catalysts were prepared by treating brass wires in a mixed powder of $Al/Al_2O_3/NH_4Cl$, followed by a leaching to increase the specific surface area. A screening was carried out for methanol steam-reforming. The most active and selective catalyst demonstrates a selectivity of carbon dioxide towards carbon monoxide of 99% for 65% methanol conversion [3].

Oxidative steam-reforming of methanol

In the presence of oxygen, treated brass wires show a decline of activity with time. To prevent deactivation, metals are incorporated into the catalysts by impregnation (Cr) or precipitation (Cr, Zr, Ce) methods. The activity of these doped catalysts for the oxidative steam-reforming of methanol is shown in Figure 2.

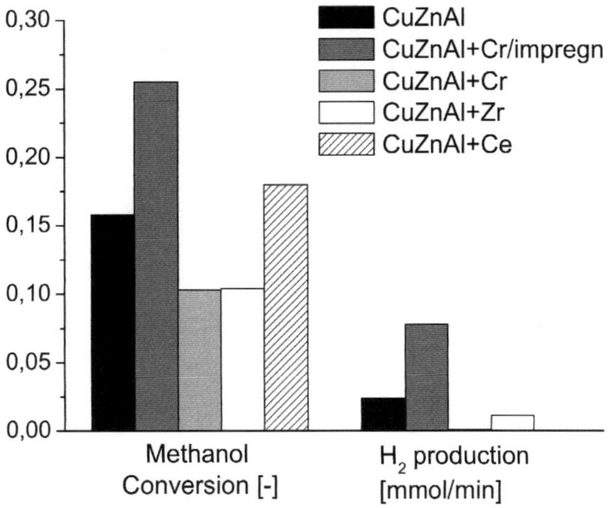

Figure 2. Methanol conversion and hydrogen production for doped brass wires (~1.3g). P = 1.3bar, F = 100NmL/min (5% methanol, 5%water, 0.95% oxygen, argon as diluent), T = 538K.

Chromium incorporated by impregnation method provides the best catalyst for OSRM. Moreover, no deactivation was observed at 523K for 10h. The later is then used to measure the temperature profile along the axial direction under various conditions. The comparison packed bed reactor filled with an industrial catalyst is shown in Figure 3.

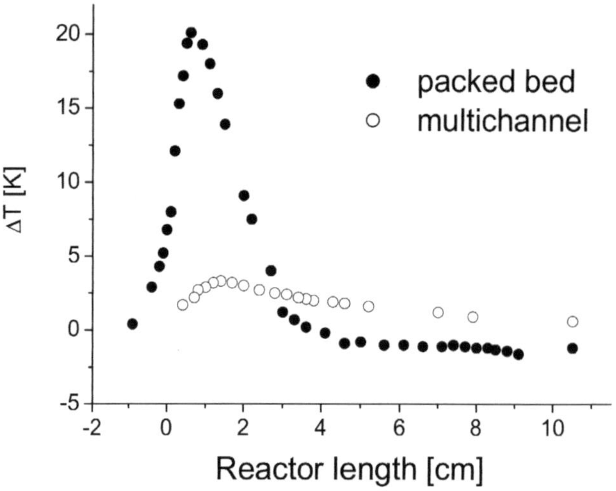

Figure 3. Temperature profile for packed bed and microstructured catalysts. P = 1.3bar, F = 100mL/min (17% methanol, 17% water,2% oxygen, argon as diluent), methanol conversion =0.42.

The catalytic packed bed leads to a hot spot of nearly 20K and the temperature in the main part of the reactor is lower than the initial value. On the other hand, the microstructure allows almost isothermal conditions in the entire reactor: the maximum temperature increase is 3K and never drops below the entrance value. This demonstrates that heat is efficiently transferred along the filamentous catalysts from the oxidation part to the exit.

Conclusions

Catalytic activity of the doped catalysts and kinetic study for OSRM as well as characterization of the wires (XPS, TPR, TPO, SEM) will be presented in details.

Acknowledgements

The authors gratefully acknowledge the financial support of the Swiss National Foundation and the Max-Buchner-Forschungsstifftung. We thank Süd-Chemie for providing the catalyst.

Literature

[1] Rouge, A., B. Spoetzl, et al. (2001) "Microchannel reactors for fast periodic operation: the catalytic dehydration of isopropanol." Chem. Eng. Sci. **56**(4): 1419-1427

[2] Tronconi, E., G. Groppi. (2000) "A study on the thermal behaviour of structured plate-type catalysts with metallic support for gas/solid exothermic reactions." Chem. Eng. Sci. **55**: 6021-6036

[3] Horny, C., L. Kiwi-Minsker, and A. Renken. (2004) "Micro-structured string reactor for autothermal production of hydrogen." Chemical Engineering Journal. **101**: 3-9

135e Novel Microreactor Design for Balancing Heat and Mass Transfer

Benjamin Q Roberts, David L King, Jamelyn D Holladay, James M Davis, Kriston P Brooks, Robert A Dagle, Ward E TeGrotenhuis

The water-gas shift (WGS) reaction is a member of the class of reversible, exothermic reactors where reaction kinetics are faster at high temperature but conversion is favored at lower temperature. As a consequence, removing heat as the reaction proceeds through the use of microreactors can substantially reduce the amount of catalyst needed and the size of the reactor. Integrating heat exchange into microchannel reactors not only makes it possible to remove heat of reaction but also progressively reduce the reaction temperature along the reaction pathway. The inlet of the reactor can be maintained at higher temperature to initially favor kinetics, while the outlet end is kept at lower temperature to favor conversion. Appropriate management of the temperature profile can substantially reduce the amount of catalyst required, the residence time, and the size of the reactor, resulting in a greater degree of process intensification than what is achievable through adiabatic or even isothermal operation in a microreactor.

A novel design is presented that provides an effective method for balancing heat and mass transfer within a microreactor. The WGS reaction is only mildly exothermic, so the standard approach of interleaving planar layers of reaction channels and heat exchange channels tends to provide too much heat transfer. This problem becomes accentuated when trying to design for a desired temperature profile. A unique microreactor geometry has been developed that effectively decouples the areas for heat and mass transfer. Not only does the geometry provide versatility in reaction engineering and reactor design, but results in a much higher fraction of the volume being occupied by engineered catalyst, which is typically only 20-25% in an interleaved microreactor. A reactor will be presented that contains over 50% by volume of engineered catalyst structure. This WGS reactor was designed to achieve 90% CO conversion in a simulated reformate, based on steam reforming of a gasoline stimulant at a 3:1 steam to carbon ratio, at a flow rate that would support 2 kW electrical output from a PEM fuel cell. Actual performance exceeded 2.5 kW equivalent electrical output at 90% conversion. The full-scale reactor for a fuel processor producing reformate for a 50 kW PEM fuel cell is projected to be less than 3.5 liters. The reactor geometry and design are presented along with experimental results and a size comparison with a conventional 2-stage adiabatic reactor system.

135f The Effects of Microreactor Geometry on the Performance of Anodized Alumina Microreactors

Ed Seebauer, Rich Masel, Zheng Ni

The objective of this paper is to determine how microreactor geometry affects performance. In particular we consider how the characteristics of a posted microreactor and a channel microreactor are different for a test reaction: the production of hydrogen from ammonia. The experiments include kinetics measurements, using a recycle reactor to obtain an empirical expression for the rate of reaction under the experimental conditions, and rate measurements to verify performance. We find that the performance characteristics of the two reactors are quite different. In the range of conditions studied, the conversion in the posted reactor is quite similar to that for an ideal CSTR, while the performance of the channel reactor is quite similar to that for an ideal PFR.

Catalyst Stabilization for Cyclohexene Hydrogenation in a Microstructured Falling Film Reactor

*J. W. Lee, K. K. Yeong, A. Gavriilidis**
Department of Chemical Engineering, University College London, Torrington Place, London, WC1E 7JE, UK

R. Zapf, V. Hessel
Institut für Mikrotechnik Mainz GmbH, 55129 Mainz, Germany

Abstract

The hydrogenation of cyclohexene on palladium catalyst was performed continuously in a falling film microreactor at 25-65°C, 0-40psig hydrogen pressure and 5 vol% cyclohexene in toluene. Organic peroxides contained in the feed can deactivate the catalyst, and for this reason cyclohexene was purified by treatment with alumina. It was observed that the treatment time and the amount of alumina utilized should be above a certain threshold to completely avoid deactivation. In cases where deactivation occurred, air oxidation at 200°C was sufficient to restore catalytic activity.

Keywords
Microstructured Falling Film Reactor, Cyclohexene Hydrogenation, Deactivation, Palladium

1. Introduction

Microreactors have an important role to play in catalyst development and find many applications in industrial synthesis [1]. Various catalytic gas-liquid reactions have been investigated in microreactors. These include nitrobenzene hydrogenation on Pd in a microstructured falling film reactor [2], α-methylstyrene hydrogenation in a Pd or Pt containing mesh microreactor [3], cyclododecatriene hydrogenation on Pd-coated microstructured aluminium wafers [4] and cyclohexene hydrogenation on Pt in a microfabricated packed bed reactor [5]. All the above reactors have excellent mass transfer characteristics and hence can be used to intensify solid-catalyzed gas-liquid hydrogenations.

Cyclohexene hydrogenation has been used as a model reaction for various reactors or catalytic systems [6-7]. The reaction is exothermic with ?H=-117kJ/mol, and fast under mild conditions. Hydrogen is only sparingly soluble in the liquid-phase reactant and product [8]. The order of reaction on Pd catalyst is 0.5-1 with respect to hydrogen and zero with respect to cyclohexene [7, 9-10].

2. Experimental

The falling film microreactor contained a plate in which 64 channels (300 μm wide, 100 μm deep, 65 mm long separated by 100 μm wide wall) were machined in stainless steel using wet chemical etching. A γ-alumina layer was deposited in the microchannels using the slurry/washcoating process [2]. The area surrounding the channels was covered with tape and the channels were covered with an aqueous suspension of γ-alumina powder (20 % solids content, average particle size 3 μm), binder (PVA) and acid (acetic acid or nitric acid). Excess suspension was wiped off and the plate was calcined at 700 °C for 2 h. Then the channel area was covered with 10ml of a palladium (II) nitrate solution (Aldrich, Pd 12-16 % w/w) resulting to Pd loading of ca 1.3 wt % on Al_2O_3 and left to dry in a fume cabinet. The plate was calcined in air at 400 °C for 3 h.

Organic peroxides (ROOH) are generally formed when organic liquids, especially alkenes, are exposed to atmospheric oxygen and sunlight. These peroxides have a detrimental effect on catalytic activity [11]. Hence, purification of cyclohexene prior to reaction is necessary. This can be accomplished by passing the cyclohexene feed through a γ-alumina bed [6-7]. Two methods for purification of cyclohexene were used in this work as shown in fig. 1. In the first method (A), 20 ml cyclohexene was stirred with 10 g γ-alumina spheres (Johnson Matthey, particle size 2-4 mm, surface area 300 m^2/g), unless otherwise specified, in a stirred vessel under nitrogen atmosphere overnight. Subsequently, toluene was added in the vessel, which was always kept under nitrogen atmosphere. In the second method (B), cyclohexene and toluene were first mixed and then the reaction mixture was passed continuously through a packed bed containing 8g of γ-Al_2O_3, located upstream of the reactor, similarly to other researchers [11-12].

A HPLC pump (Knauer K-120, 50 ml pump head) was used to feed the reaction mixture, 5 vol% cyclohexene in toluene. Hydrogen flow into the reactor was controlled using a rotameter. The temperature and pressure of the reactor were controlled with the help of water bath and a micrometering valve at the reactor outlet, respectively. Both liquid and gas were fed cocurrently from top to bottom in the reactor. Product analysis was performed with an Agilent 6890 GC with an autosampler. A HP-INNOWAX capillary column (30m long, 0.25 mm ID, 0.25 μm df, Agilent) was used in conjunction with a FID. Carbon balance typically closed within 10-13 %. A three-way valve connected to the reactor inlet allowed cleaning the catalyst with toluene or purging the reactor with nitrogen prior to shutdown.

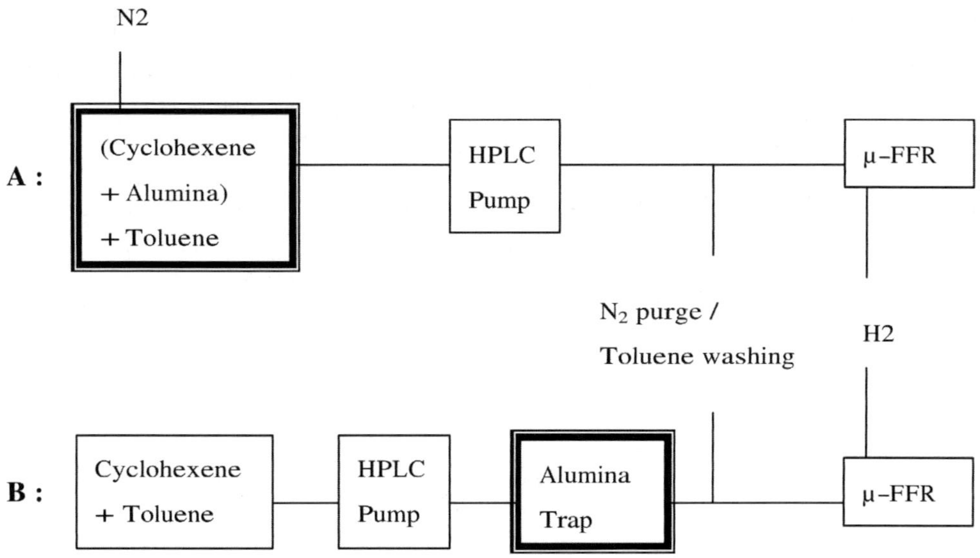

Fig. 1. Schematic of experimental procedures used for cyclohexene purification.

3. Results and discussion

When cyclohexene was not purified, conversion was observed to decrease significantly with time. In order to avoid this catalyst deactivation, the cyclohexene purification methods presented above were evaluated. Fig 2 shows cyclohexene conversion as function of time for various reactant treatment conditions. A, B denote the treatment procedure employed, while 4g, 8g etc indicate the grams of alumina used for the treatment. t1 and t3 denote periods during which the reactor was washed with toluene for 15min, purged with nitrogen for 1h and left overnight isolated by closing all inlet and outlet valves. During t2, washing and purging were carried out as above and subsequently the plate was placed in an oven at 200 °C overnight to regenerate the catalyst. Comparing the first (A-4g) and second (B-8g) runs it is observed that conversion decreased by 2.4% per hour for the first, and 6.1% per hour for the second run. This shows that deactivation is most severe for B-8g, even though it used double the amount of alumina than A-4g. The behaviour is attributed to the short contact time available for method B. This could be rectified by increasing the length of the alumina trap; however this would lead to increase in pressure drop.

After regeneration of catalyst with air oxidation at t2, cyclohexene conversion more than doubled compared with that before the regeneration treatment. The successful regeneration of catalyst indicates that deactivating (organic) compound(s) have been removed from the surface of the catalyst. These most probably are oligomers that are formed by cyclohexene polymerization catalyzed by organic peroxides [12]. Furthermore, for runs A-10g and A-20g for which the feed was treated with 10 and 20g of alumina respectively, no deactivation was observed in contrast with run A-4g. Hence, it is concluded that a minimum amount of alumina is required for efficient prevention of deactivation, which for the feed and alumina utilized is ca. 1 g of alumina for 2 ml of cyclohexene.

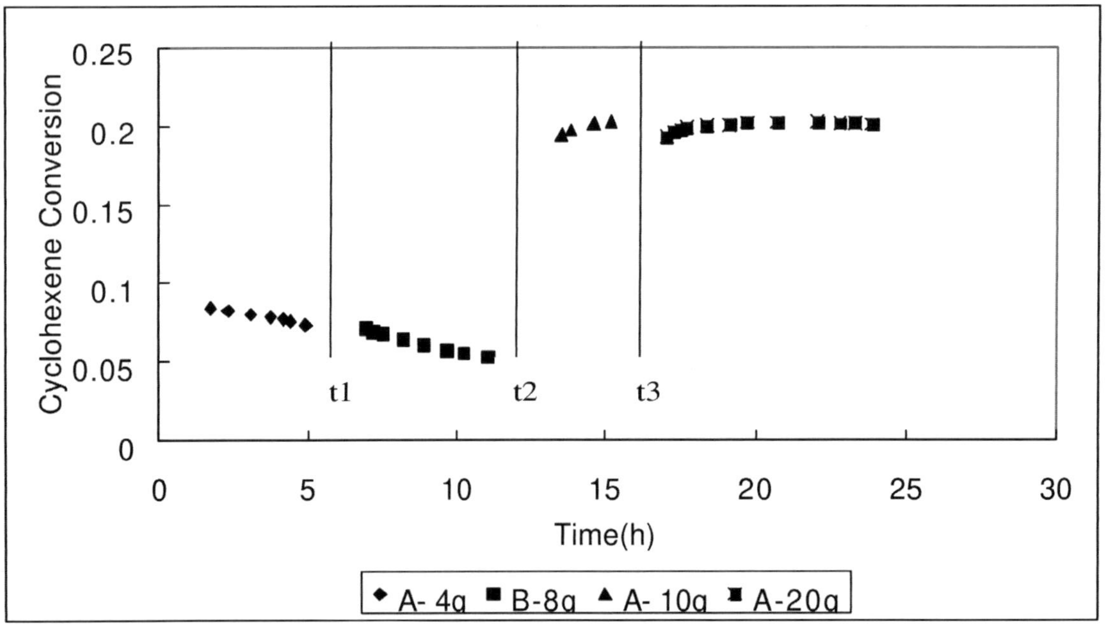

Fig. 2. Cyclohexene conversion as a function of reaction time after purification method A or B. 25 °C, 1bar, liquid flow rate 0.3ml/min (A and B are methods of cyclohexene purification, g=alumina weight).

Temperature (°C) at 0.3 ml/min and 20 psig	45	55	65
Conversion	0.45	0.54	0.6
Reactant Flow Rate (ml/min) at 25 °C and 0 psig	0.3	0.5	1
Conversion	0.17	0.096	0.046
Hydrogen Pressure (psig) at 0.3 ml/min and 25 °C	20	30	40
Conversion	0.28	0.31	0.34

Table 1. Cyclohexene conversion at various pressures, temperatures and flowrates.

A few experiments were performed to investigate the effect of temperature, pressure and flowrate on reactor performance. As shown in Table 1, increase in temperature brought about significant increase in conversion, up to 60% (at 20 psig; higher pressure was utilized to reduce evaporation). Increase of pressure resulted to increase in conversion, but not to the extent than one would expect for completely mass transfer controlled reaction or first order kinetics. Evidently, by increasing the reaction mixture flowrate lower conversion was obtained, due to both decrease of residence time and increase of liquid film thickness.

4. Conclusions

Cyclohexene hydrogenation on Pd in a microstructured falling film reactor was carried out. By pretreating the cyclohexene feed overnight with sufficient amount of alumina, deactivation of catalyst was avoided. In cases, where deactivation was observed, calcination at 200°C was sufficient to restore activity. Conversion of 60% was obtained at 65 °C and 20psig.

References

1. A. de Mello, R. Wootton, "But what is it good for? Applications of microreactor technology for the fine chemical industry", *Lab on a Chip*, 2: 7N–13N (2002)
2. K.K. Yeong, A. Gavriilidis, R. Zapf, V. Hessel, "Catalyst preparation and deactivation issues for nitrobenzene hydrogenation in a microstructured falling film reactor", *Catalysis Today*, 81, 641-651 (2003)
3. R. Abdallah, V. Meille, J. Shaw, D. Wenn, C. de Bellefon, "Gas-liquid and gas-liquid-solid catalysis in a mesh microreactor", *Chem. Commun.*, 4, 372-373 (2004)
4. R. Födisch, D. Hönicke, Y. Xu, B. Platzer, "Liquid phase hydrogenation of p-nitrotoluene in microchannel reactors", in Proceedings of the 5th International Conference on Microreaction Technology, Strasbourg, 470-478 (2001)
5. M. W. Losey, M. A. Schmidt, K. F. Jensen, "Microfabricated multiphase packed-bed reactors: characterization of mass transfer and reactions", *Ind. Eng. Chem. Res.*, 40, 2555-2562 (2001)
6. S.D. Jackson, G.J. Kelly, S.R. Watson, R. Gulickx, "Cycloalkene hydrogenation over palladium catalysts", *Applied Catalysis A: General,* 187, 161-168 (1999)
7. R.J. Modon, J.P. O'Connell, M. Boudart, "Catalytic hydrogenation of cyclohexene", *AIChE Journal*, 24, 904-911 (1978)
8. P.C. Watson, M.P. Harold, "Dynamic effects of vaporization with exothermic reaction in a porous catalytic pellet", *AIChE Journal*, 3, 989-1006 (1993)

9. E.E. Gonzo, M. Boudart, Catalytic hydrogenation of cyclohexene 3. Gas-phase and liquid-phase reaction on supported palladium, *Journal of Catalysis*, 52, 462-471, 1978
10. T.J. Lee, D.H. Han, Y.G. Kim, "Catalytic hydrogenation on supported Pt catalysts; I. Vapor phase kinetics of cyclohexene hydrogenation", Yeungnam University Report, 13(1), 233-241 (1985)
11. M.C. Clark, B. Subramaniam, "1-hexene isomerization on a Pt/γ-Al_2O_3 catalyst: The dramatic effects of feed peroxides on catalyst activity", *Chem. Eng. Sci.*, 51, 2369-2377 (1996)
12. V. Arunajatesan, B. Subramaniam, K.W. Hutchenson, F.E. Herkes, "Fixed-bed hydrogenation of organic compounds in supercritical carbon dioxide", *Chem. Eng. Sci.*, 56 (4), 1363-1369 (2001)

Decomposition of Carbon Dioxide in a Micro Plasma Reactor

Shinsuke MORI[1], *Aguru YAMAMOTO*[2] *and Masaaki SUZUKI*[1]

[1]*Department of Chemical Engineering, Tokyo Institute of Technology, 2-12-1, O-okayama, Meguro-ku, Tokyo 152-8552, Japan*

[2]*The Research Association of Micro Chemical Process Technology (MCPT), 2-12-1, O-okayama, Meguro-ku, Tokyo 152-8552, Japan*

Abstract

Decomposition of carbon dioxide in a plasma reactor is investigated experimentally by using capillary discharge tubes with diameters of 0.5, 1.0, and 3.0 mm over the pressure range from 19 to 152 Torr. The decomposition degree increases with reducing discharge tube diameter and reducing gas pressure and very high decomposition degree over 95 % has been obtained easily without formation of solid carbons. We have also performed the oxidation of carbon monoxide in the micro plasma reactor and those results have been compared with CO_2 decomposition results. For He/CO_2 = 4/1 and $He/CO/O_2$ = 4/1/0.5 reactant gas mixtures, almost same gas compositions with those for CO_2 decomposition experiments are obtained as reaction products. The importance of various parameters for decomposing CO_2 is discussed.

1. Introduction

There has been a continuous interest in the CO_2 decomposition in the plasma reactor and a large number of studies in various discharge systems have been reported [1-5]. In those studies, it has been recognized that, in the steady-state of electrical discharge of CO_2, a significant amount of CO_2 equilibrates with CO even if the input energy and gas residence time are increased and it is difficult to decompose CO_2 completely into CO and O_2 by the conventional plasma reactors [5]. In our previous study, we have investigated the effects of downscaling of discharge tube diameter on CO_2 decomposition and the maximum CO_2 decomposition degree of about 70 % has been obtained by using smaller size discharge tube [6]. In our subsequent experiments, however, we have found that decomposition degree over 95 % has been obtained easily by micro plasma reactor with reducing gas pressure. Here, we present the experimental results of CO_2 decomposition in micro plasma reactor with various discharge pressures and those results are compared with experimental results of oxidation of carbon monoxide in same micro plasma reactors. The most effective parameter for decomposing CO_2 is deduced from the experimental results.

2. Experimental

Figure 1 shows schematic diagram of micro plasma reactor. The micro plasma reactor is a Pyrex glass tube of 50-mm length, 6-mm outer diameter, and 500, 1000, 3000-μm inner diameters, which is connected to the gas inlet and outlet line by 1/4 inch stainless steel union tees. In order to eliminate the influence of the shape and size variety of electrodes on discharge characteristics, the same stainless steel union tees are used as hollow electrodes for the different inner diameter discharge tubes. Between those two electrodes, a DC voltage (Max. 15 kV) is applied and Helium/CO_2 plasma is generated in the capillary tube. Helium gas is used as carrier gas to stabilize the microplasma. The conditions for discharge were as follows: discharge pressure was 19 to 152 Torr, discharge voltage was 1.2 to 5.3 kV, discharge current was 0.3 to 20.0 mA, the mole ratio of helium to CO_2 was 1 to 4, and total gas flow rate was 0.25 to 9.0 cm^3/min (at 101.3 kPa, 293.15K). Under such experimental conditions, the discharge voltage was measured by the digital multi-meter and/or digital oscilloscope and gas compositions of reaction products were measured by gas chromatography.

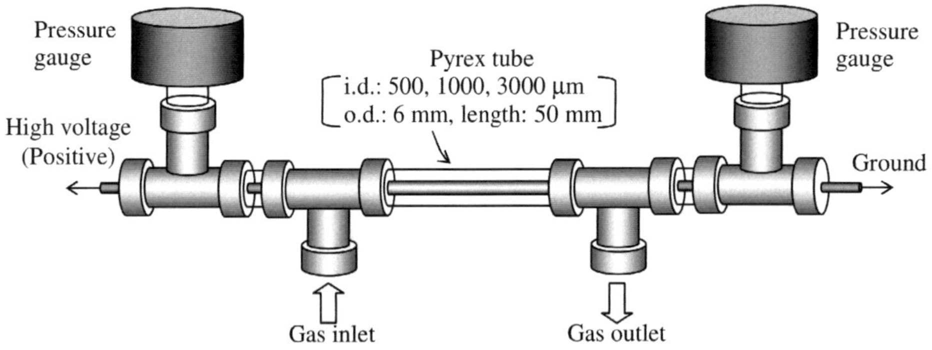

Fig. 1 Schematic diagram of micro plasma reactor

3. Results and Discussion

Figure 2 shows the experimental results of (a) discharge voltage and (b) CO_2 decomposition degree for d = 1 mm discharge tube as a function of discharge current with various discharge pressure. Although discharge voltage decreases with decreasing pressure, CO_2 decomposition degree increases with decreasing pressure. The maximum decomposition degree about 97 % has been obtained when the pressure is 19 Torr.

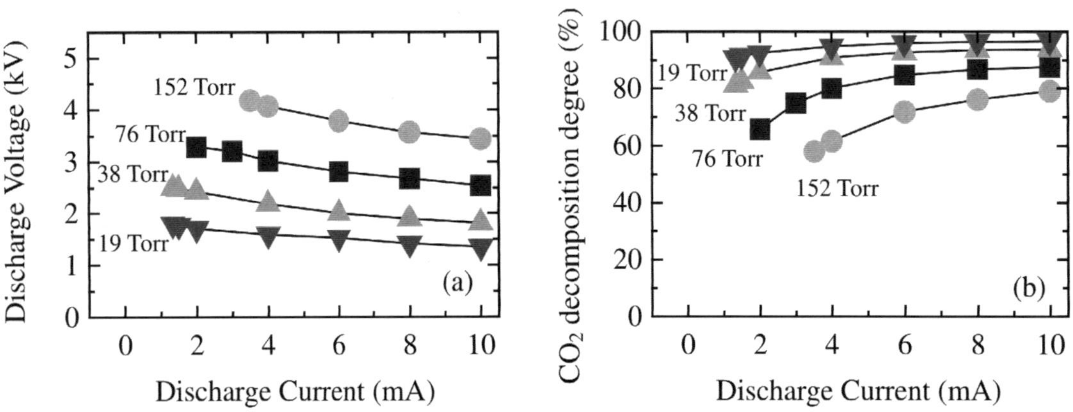

Fig. 2 Experimental results of (a) discharge voltage and (b) CO_2 decomposition degree for d = 1 mm discharge tube as a function of discharge current with various discharge pressure.

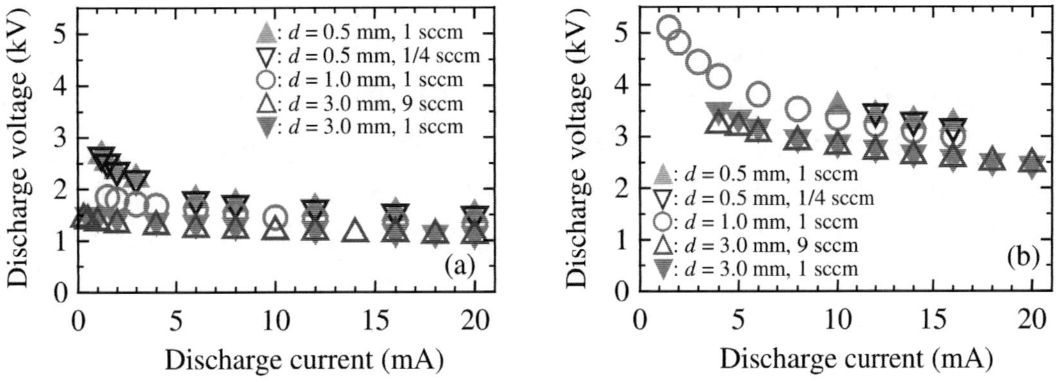

Fig. 3 Current-voltage characteristics of micro plasma reactors: (a) 19 Torr and (b) 152 Torr.

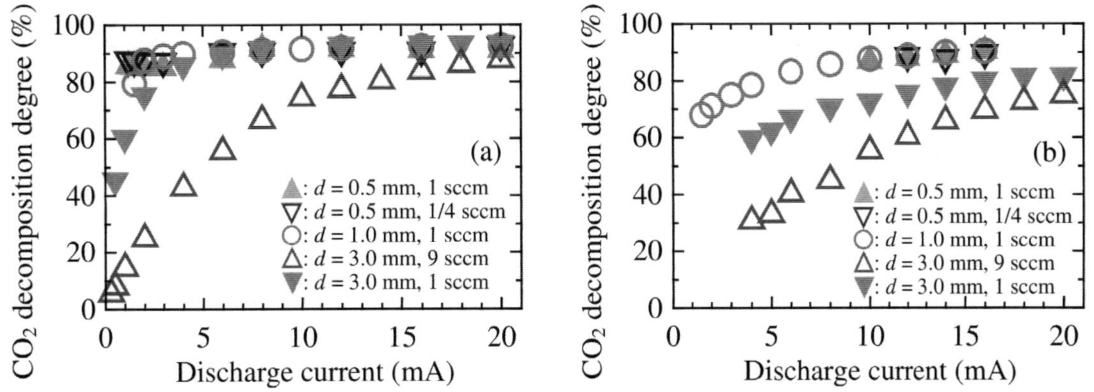

Fig. 4 CO_2 decomposition degree as a function of discharge current: (a) 19 Torr and (b) 152 Torr.

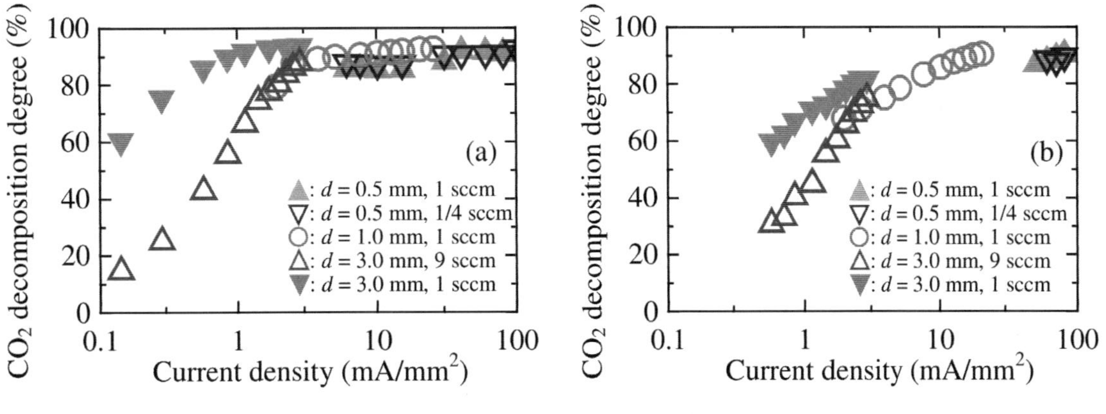

Fig. 5 CO_2 decomposition degree as a function of current density: (a) 19 Torr and (b) 152 Torr.

Experimental results for different discharge tube diameters are shown in figures 3-5. Figure 3 shows the current-voltage characteristics of the plasma reactor for the different inner diameter *d*. For both (a) 19 Torr and (b) 152 Torr, discharge voltage increases slightly as diameter of discharge tube decreases. Figure 4 shows CO_2 decomposition degree as a function of discharge current. As shown in Fig. 4(a), the CO_2 decomposition degree at 19 Torr for $d = 3$ mm is lower than that for $d = 1$ mm and 0.5 mm when comparing it with same residence time of 8.8 ms ($d = 3$ mm, 9 sccm; $d = 1$ mm, 1 sccm, $d = 0.5$ mm, 1/4 sccm). However, as comparing those data with same flow rate of 1 sccm, almost same decomposition degree is obtained for different discharge tube diameters. On the other hand, as shown in Fig. 4(b), the CO_2 decomposition degree at 152 Torr for $d = 3$ mm is always lower than that for $d = 1$ mm and 0.5 mm when comparing it with both same flow rate (1sccm) and same residence time of 70 ms ($d = 3$ mm, 9 sccm; $d = 1$ mm, 1 sccm, $d = 0.5$ mm, 1/4 sccm). Figure 5 shows CO_2 decomposition degree as a function of current density. In this case, data for same residence time in the discharge tube ($d = 3$ mm, 9 sccm; $d = 1$ mm, 1 sccm, $d = 0.5$ mm, 1/4 sccm) are plotted on one curve. From these results we can deduce that increasing current density is most effective to increase the decomposition degree although decomposition degree is increased by increasing residence time and decreasing gas pressure. In Figure 6, the experimental results of CO_2 decomposition and CO oxidation for $d = 1$ mm discharge tube are compared. Figure 6(a) shows experimental results for discharge voltage and figure 6(b) shows $CO/(CO_2 + CO)$ of the reaction products. For both $He/CO_2 = 4/1$ and $He/CO/O_2 = 4/1/0.5$ reactant gas mixtures, almost same current-voltage characteristics and gas compositions as reaction products are obtained. As can be seen in Fig. 6(b), the value of $CO/(CO_2 + CO)$ keeps high and we have not confirmed the formation of solid carbon on the discharge tube wall through our experiments. Hence it can be said that this micro plasma reactor is highly selective to decompose CO_2 into CO.

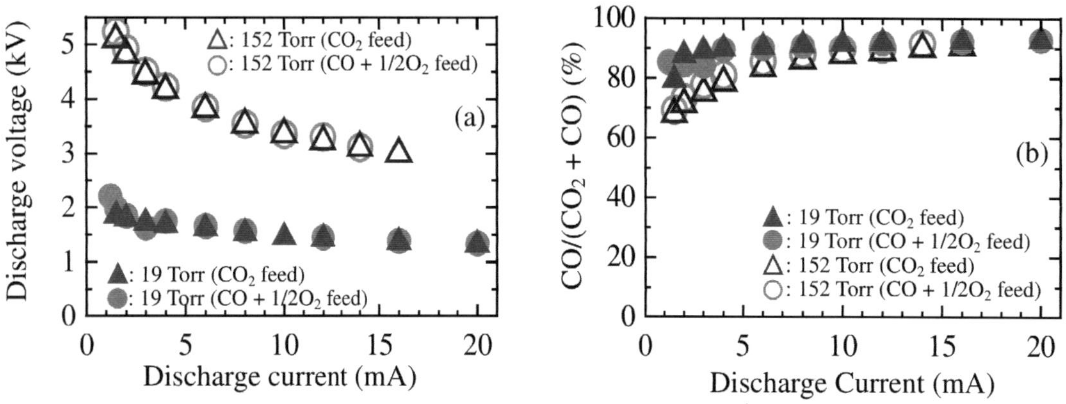

Fig. 6 Comparison of experimental results for CO_2 decomposition and CO oxidation in a micro plasma reactor: (a) discharge voltage and (b) $CO/(CO_2 + CO)$ for $d = 1$ mm discharge tube as a function of discharge current with various discharge pressure.

4. Conclusion

Decomposition of carbon dioxide in a plasma reactor is investigated experimentally by using capillary discharge tubes for various discharge pressure. The decomposition degree increases with reducing discharge tube diameter and reducing gas pressure and the maximum decomposition degree of 97 % has been obtained for d = 1 mm inner diameter discharge tube and p = 19 Torr discharge pressure. We have also performed the oxidation of carbon monoxide in the micro plasma reactor and almost same gas compositions with those for CO_2 decomposition experiments are obtained as reaction products. Through our experiments, formation of solid carbon has not confirmed. Consequently, by using this micro plasma reactor, one can easily convert CO_2 into CO with both high conversion and high selectivity.

Acknowledgements

This work was partially supported by the New Energy and Industrial Technology Development Organization (NEDO) of the Ministry of Economy, Trade and Industry, Japan.

References

[1] S.N. Andreev, V.V. Zakharov, V.N. Ochkin and S.Yu. Savinov, *Spectrochimica Acta Part A,* **60**, 3361-3369 (2004).

[2] R. Li, Y. Yamaguchi, S. Yin, Q. Tang and T. Sato, *Solid State Ionics,* **172**, 235-238 (2004).

[3] S.L. Brock, M. Marquez, S.L. Suib, Y. Hayashi and H. Matsumoto, *J. Catalysis,* **180**, 225-233 (1998).

[4] D. K. Otorbaev, *Chem. Phys.,* **196**, 543-550 (1995).

[5] L.C. Brown and A.T. Bell, Ind. Eng. Chem., Fundam., **13** (3), pp. 203-210 (1974).

[6] S. Mori, S. Akaji and M. Suzuki, Proc. 7th International Conference on Microreaction Technology (IMRET7), Lausanne, Switzerland, Sep. 7-10, pp. 148-150, 2003.

Fuel to hydrogen - An overview over fuel conversion activities at the Institute for Micro Process Engineering

P. Pfeifer, K. Haas-Santo, O. Görke, L. Bohn, K. Schubert

Institute for Micro Process Engineering (IMVT), Forschungszentrum Karlsruhe GmbH, Hermann-von-Helmholtz Platz 1, 76344 Eggenstein-Leopoldshafen, Germany, www.fzk.de/imvt-en

Prepared for Presentation at 2005 Spring National Meeting / IMRET 8 / April 10-14 / Energy Generation Systems

Unpublished

Introduction

Fuel Cells are of great interest for mobile and on-site production of electricity because of legal requirements considering emission control or the saving of resources. Higher efficiency, lower noise emission and negligible amounts of pollutants can be reached by the use of fuel cells. However, big problems for the hydrogen supply are the small energy density and the heavy storage systems for H_2. An alternative could be the on-site production of hydrogen. Good dynamics and a high efficiency are demands for the chemical reactor/reformer and the heat exchangers in the system. Therefore, microsystem technology could be used because high heat and mass transfer coefficients can decrease system size and weight as well as energy loss.

An overview over the developments of the Institute for Micro Process Engineering (IMVT) considering hydrogen production by steam, oxidative or autothermal reforming of methane, methanol, propane and gasoline will be given in the present contribution. Aspects like CO reduction for low temperature fuel cells applying water gas shift, selective oxidation and selective methanation as well as heat generation will also be mentioned in the overview. Data will be given for capacity of the hydrogen generation systems tested by the Institute of Micro Process Engineering and partners so far. The feasibility of mass production will be highlighted.

Overview

Research on hydrogen generation using microstructured components started at IMVT in the year 1997 with oxidative reforming of methane and steam reforming of methanol.

The work on ***oxidative reforming of methane*** was funded by the German Federal Ministry of Education and Research (BMBF) under the acronym ***"MIKREAK"*** [1]. As results about catalyst coatings for metallic microsystems were not well established at that time and oxidative reforming was known to possess fast kinetics it was decided to fabricate a microstructured honeycomb out of pure active metal. Together with the project partner ACA Berlin it was decided to choose Rhodium as the substrate material, having high activity for reforming and high heat conductivity. High heat conductivity of the metal should lead to distribution of heat from total oxidation at the reactor inlet to regions of CO-shift at the reactor end, because the reactors, i.e. microstructured honeycombs, were designed without cooling passage. A method based on micro wire EDM (electro discharge machining) was developed to fabricate the microchannels with a minimum channel width of 60 µm and depth of up to 140 µm into Rhodium. A microstructured cross section of 5.5 x 5.6 mm^2 with up to 1152 channels and a channel length of up to 20 mm was formed by a stack of 200 µm thick foils bonded by electron beam welding (EB) or diffusion bonding. As supposed, even at operating pressures of up to 1.2 MPa with oxygen and up to 2.0 MPa with air, the catalyst was not found to be adversely affected by the reaction heat caused by the total oxidation taking place at the catalyst inlet. The high thermal conductivity of rhodium contributed to a good heat distribution in the flow direction (ΔT measured with a pyrometer was 50 – 120 K), which was reflected by significantly increased conversions, CO and H_2 selectivities compared to those obtained by using a stack of commercial Pt/Rh gauzes. At 1100°C a hydrogen selectivity of up to 88 % was reached at a GHSV of $0.195*10^6$ h^{-1} STP with air and a fuel to oxygen ratio of CH_4/O_2 = 1.75. The weight of the microstructured body was 11.5 g and the pressure drop was in the range of 0.025 MPa.

The ***steam reforming of methanol*** was done together ***with Mannesmann AG*** to build up a complete hydrogen system suitable for dynamic hydrogen generation for automotive fuel cells. Before switching to other fuels like gasoline the main parts of the system, reformer, catalytic burner and selective oxidation reactor were already developed at IMVT. The intermediate state of the reformer demonstrator has been published recently [2]. The PdZn alloy catalyst used was developed on the basis of a washcoating procedure with nanoparticles described in more detail elsewhere [3]. This coating applicable to already bonded microreactors was used for two different reformer foil materials, an aluminum alloy (ALMg3) and copper. To a total number of e.g. 48 microstructured copper foils with channels of 100 x 150 $µm^2$, the empty channels yielding a reaction volume of 19.3 cm^3, 13.2 g of catalyst was applied (20 µm thick layer). The necessary heat for reaction was delivered by heat cartridges within the reformer body. The demonstrators were tested under steady state and dynamic operation. At 310°C a maximum conversion of methanol of 90 % was reached when feeding 70 g/h methanol with a molar water/methanol ratio of 1:2. This is enough for a 200 W fuel cell system assuming a total efficiency of 42 %. Scale up effects, which reduce body to foil weight ratio to approximately 20 %, result in a value of 0.7 kg/kW of electric power for the reformer. Dynamic experiments with the demonstrators highlighted that the hydrogen demand during load changes and temperature changes was met as fast as in other already well optimised systems although the weight of was not optimised in the microsystems.

Other concepts of methane and methanol conversion to hydrogen were funded later by the BMBF in the project ***"Hydrogen Generation in Microsystems"***. There steam reforming of methane and micro fixed beds for reforming of methanol were subject of research then.

For the ***steam reforming of methane*** a final demonstrator for a 500 W fuel cell system was fabricated and tested together with the project partner Technical University of Munich (TUM) in steady state and dynamic operation. For heat supply methane was burnt catalytically in a second passage in counter current flow. The reactor weighs 1.2 kg, has a complete length of 95 mm, a width of 22 mm and an height of 60 mm. The reformer passage was coated with a $NiAl_2O_4$-catalyst, the burner passage was impregnated with Pt-solution. The use of the microstructured reformer enables the production of up to 2 l/min hydrogen (STP). The reactor has been operated autonomously without external heating and achieved quite high methane conversions in excess of 70% at high loads. Methane conversions of 90 % were achieved at lower loads. The temperature could be well controlled combusting a mixture of methane and hydrogen in co-flow with the reforming reaction. Start-up tests showed that a constant product composition is achieved at less than 20 s after admittance of methane [4].

For ***methanol steam reforming*** different reactors consisting of two passages were developed. The first tests were performed at the Swiss Federal Institute of Technology Lausanne (EPFL) with different heating concepts in steady state and dynamic operation [5]. A cramped two-passage reactor was used to couple two reactions: steam reforming of methanol, producing hydrogen, and total oxidation of methanol, producing heat. Different catalysts were used: Commercially available copper based catalyst for the steam reforming and a cobalt based catalyst for total oxidation. Special methods were developed to coat the catalysts on the wall of the micro-channels. The exothermal and endothermal were coupled in the developed two-passage reactor under co-current conditions. The axial temperature profiles never exceeded 2.5 K. For the oxidation reaction complete conversion of methanol (>99%) at temperatures higher than 250 °C was observed. For the steam reforming, the hydrogen and CO_2 selectivity is higher than 96% for methanol conversion up to 90%. The final reactor with an output of hydrogen sufficient for a 500 W fuel cell system was tested at the ZAE Bayern [6]. To avoid too high temperatures in the reformer, the heat source and the reformer were separated. The reformer had two alternating passages for the heat transfer fluid and the reforming reaction. The reaction passage was formed as a microstructured fixed bed. This concept allows to profit from the advantages of microstructured reactors but also to have a large catalyst amount in a low reactor volume. Under static operation conditions methanol conversions of 96 % and H_2 production rates of over 6 l/min STP were reached at comparatively low temperatures. Dynamic tests proved a start-up time of only 18 s and a very low sensitivity of the product composition to load changes.

The ***oxidative as well as autothermal reforming of propane / butane***, i.e. energy supply by oxidation in equilibrium with heat loss and heat demand of reforming, was started together ***with SINTEF and the Norwegian University for Science and Technology (NTNU)*** in Norway by applying the rhodium honeycombs to propane experiments. Due to progress in fabrication of high temperature alloys and catalyst coating technology it was possible to switch to rhodium impregnated microstructured honeycombs out of FeCrAlloy©. The substrate material having high aluminum content segregates aluminum oxide at elevated temperatures (annealing) under air conditions. From the reaction experiments it is clear that this oxide layer provides enough surface area for the desired reaction. Still using the same channel geometries than in rhodium honeycombs it was possible to increase conversion and selectivity considerably by applying the high temperature material. Rhodium and nickel impregnation were compared and rhodium was found to be superior in conversion [7]. Temperature measurements were performed within the honeycomb and compared with foam systems at SINTEF. Results showed that the reduced temperature gradients within the honeycombs were the reason for superior long term stability of the catalyst [8]. The weight of the honeycombs can be reduced by use of the high temperature alloy by 36 %. The hydrogen selectivity for oxidative reforming is already as high as 60 % at 1000°C with a C/O ratio of 0.8 and a residence time of 12.6 ms (STP conditions).

Most of the work done on ***gasoline reforming*** systems has been done in the BMBF project called ***"Micromotive"*** together with DaimlerChrysler and is described in more detail in another presentation at the 8th IMRET [9]. The main objective of this work is to evaluate the mass fabrication possibilities of microstructured components for a hydrogen generation system with a fuel cell power of 5 kW. Different microstructured heat exchangers for evaporation, heat recovery and condensation as well as a reactor for the system are built. Mass production is feasible with today's knowledge, integrating microembossing, microetching and soldering.

Necessary components for ***gas cleaning*** and heat generation have been developed and are still objective of further work. The gas cleaning includes selective methanation, selective oxidation and water gas shift.

Methanation in microreactors is a good possibility to reduce the CO content in a reforming gas mixture in presence of CO_2 [10]. The reactor used by the IMVT consists of 27 foils (length = 78 mm, width = 20 mm, thickness = 230 µm) with 17 channels (width = 600 µm, height = 150 µm). The experiments were performed with a gas mixture of 1 vol.% CO, 1 vol.% O_2, 25 vol.% H_2 (N_2: balance). At a temperature of 250 °C a CH_4 space time yield of 9 mol $l^{-1} \cdot h^{-1}$ was reached. At temperatures up to 250 °C the CH_4 space time yield by CO methanation is higher than by CO_2 methanation. If CO (in a mixture of CO and CO_2) has to be converted over a Ru/SiO_2-catalyst by methanation a sufficient amount of O_2 has to be added and temperature has to be controlled precisely. Because of the dimension and the heat transfer coefficients of the microchannel reactor the latter demand can be met very easily. Temperature ranges can be controlled exactly, which is important to maximize the ratio of CO to CO_2 methanation rate. A microchannel reactor is an excellent tool for studying the reaction network of methanation of CO in presence of oxygen, CO_2 and hydrogen without heat transfer limitation.

Also the ***water gas shift reaction*** (WGS) and the ***selective oxidation of CO*** (Selox) are good possibilities to reduce the CO content in reforming gas mixtures [11]. At small average residence times of less than 30 ms in a relatively small microreactor (structured body: length 20 mm, width = 20 mm, height = 2.5 mm), the water gas shift reaction at 250°C – 300°C using a Ru/ZrO_2 catalyst allowed to reduce the CO content by more than 95%. So it is possible to reach a very high H_2 space time yield of 720 mol$\cdot l^{-1} \cdot h^{-1}$.

By means of the selox reaction, which was performed in a microreactor with same geometrical dimensions like the WGS reaction, a residual CO content of 1 vol.% could be oxidized to CO_2 with a conversion of more than 99% when using a CuO/CeO_2 catalyst at less than 150°C and an average residence time of 14 ms. However, CO-selectivity amounted to

20% for the selox reaction only. That corresponds to a CO_2 space time yield of 75 $mol \cdot l^{-1} \cdot h^{-1}$. The selox catalysts prepared by the IMVT, allow the temperature range of selective CO oxidation to be extended considerably. The usable temperature range is increased from 60°C – 120°C (conventional reactors) to 60°C – 180°C. Because of the small dimension and the high heat transfer coefficients of a microreactor system, temperatures can be reached very fast. Additional temperature ranges can be controlled exactly, which is important to maximize the CO-selectivity.

First results in collaboration with the Max Planck Institut für Kohlenforschung Mühlheim about ***heat generation from hydrogen combustion*** on platinum impregnated aluminum oxide layers prepared by a CVD technique in cross flow devices have been published early in 2000 [12]. A different approach concerning the catalyst preparation based on platinum impregnated aluminum oxide obtained by anodic oxidation of aluminum was presented on the 4th IMRET [13]. At the 5th IMRET further experiments have been presented for hydrogen combustion on sol-gel supported platinum catalysts highlighting the modularity of IMVT's microstructured heat exchangers, mixers and cross flow reactors. The reaction product water was condensed in a subsequent heat exchanger and prior to the reactor a mixer was connected for security reason [14]. In all three cases about 1 kW of heat was generated in approximately 1 cm^3 of structured body. To reduce the thermal stress in the microstructured cross flow devices further development was done and a new design was found which is capable to prevent hot spots [15]. The overall maximum heat transfer is reduced there by a factor of about 10 but temperature distribution is excellent and cold start properties remained nearly constant.

Outlook

The results obtained for hydrogen generation by the use of micro technology are promising in terms of temperature control, catalytic conversion and mass fabrication. However, challenges like quality control of catalytic coatings and improved dynamic operation and strategies therefore have to be faced when a new fuel has to be evaluated for hydrogen generation. Work on some specific tasks has been started already [16]. In the future it might be useful to evaluate biofuels. Work on diesel fuel reforming is already under development.

Literature

[1] Mayer et al., Ind. Eng. Chem. Res. 2001, 40, 3475-3483
[2] Pfeifer et al., Chem. Eng. Res. Des. 2003, 81, A7, 709-720
[3] Pfeifer et al., Appl. Cat. 2004, 270, 165-175
[4] Cremers et al., Proc 7th IMRET 2003, 202
[5] Reuse et al., Chem. Eng. J. 2004, 101, 133-141
[6] Cremers et al., Proc 7th IMRET 2003, 167
[7] Aartun et al., Chem. Eng. J. 2004, 101, 93-99
[8] Aartun et al., 2005, IMRET 8 Energy Systems
[9] Pfeifer et al., 2005, IMRET 8 Poster Paper
[10] Görke et al., 2005, IMRET 8 Poster Paper
[11] Görke et al., Appl. Cat. 2004, 263, 11-18
[12] Janicke et al., J. Cat. 2000, 191, 282–293
[13] Wunsch et al., Proc. 4th IMRET 2000, 481-487
[14] Haas-Santo et al., Proc. 5th IMRET 2001, 313-321
[15] Pfeifer et al., Chemie Ingenieur Technik 2004, 76, 607-613
[16] Pfeifer et al., AIChE J. 2004, 50, 418-425

136b Design Principles Of Multifunctional Microdevices

Dionisios Vlachos, Soumitra Deshmukh

Introduction

The demand for longer lasting or light-weight batteries for portable devices, such as laptops and cellular phones, is increasing. The high energy density of hydrocarbon fuels, compared to existing Li-based batteries, can be harnessed through proton exchange membrane (PEM) fuel cells [1] to potentially meet this increasing demand. Hydrogen production in portable devices usually requires integration of endothermic and exothermic reactions. However, the small scale of these devices imposes challenges on optimal coupling of reactions.

Our work focuses on designing integrated microdevices for the production of hydrogen for portable applications. Specifically, we study using computational fluid dynamics (CFD) a spatially segregated multifunctional microdevice with combustion and reforming reactions occurring in alternate channels separated by walls. Hydrogen production is achieved via ammonia decomposition on Ru [2, 3] to eliminate the need for downstream processing associated with CO removal. The required heat is supplied via homogeneous propane combustion [4] to generate high temperatures necessary for achieving high conversions at short contact times that are necessary for compact devices. Finally, hierarchical models are developed to map operation regimes with low computational effort. It is shown that integrated microdevices are feasible to operate but temperatures and conversions delimit operation in a relatively narrow window.

Materials and Methods

Key issues affecting integrated microdevice design are discussed using results from two dimensional CFD simulations of microdevices. An adaptive mesh with up to 200,000 nodes is used to resolve the gradients in the reaction zone and improve the solution accuracy. One-step reduced rate expressions are used to describe the reaction chemistries. Kinetics of homogeneous propane combustion is taken from [5], whereas that of ammonia decomposition kinetics is taken from our modeling of ammonia decomposition on Ru. In particular, a full microkinetic model was proposed and validated against microreactor experimental data and finally an easy-to-use reduced rate expression was developed using a novel computer-aided model reduction methodology [2]. Steady state solutions of the problem were sought using a finite volume method. Various parametric studies were carried out using natural parameter continuation to understand the principles and operating conditions governing the operation of integrated microdevices.

Due to computational intensity of CFD simulations, hierarchical models are also developed. Specifically, a model comprising of two coupled continuous stirred tank reactors (CSTRs) is proposed. Combustion and reforming reactions are carried out in the two CSTRs with same volumes and catalyst loadings as in the CFD simulations. A simple heat transfer coefficient, obtained from the CFD simulations, is used to describe the thermal coupling between the two CSTRs based on the Newton's law of cooling. Arc-length continuation is used to trace the operating window and the results are compared to those obtained from CFD simulations.

Results and Discussion

We first demonstrate self-sustained operation of these integrated devices. The heat recirculation through the wall separating the two reaction zones is responsible for igniting the cold incoming propane/air and ammonia streams. Furthermore, it has been found that these integrated devices are more robust than single stand-alone microburners because the magnitude and even the sign of the overall heat transfer coefficient significantly varies with location. Various parametric studies involving different flow rates and materials conductivity are performed in co- and counter-current flow configurations to determine the best operating strategy. Different performance measures including device temperatures, conversions, and net hydrogen produced are evaluated to arrive at optimal configurations and operating strategies.

It is found that the optimal flow configuration depends on the materials selection. Specifically, for highly conductive materials, the two flow configurations are found to be practically equivalent. This result departs from the currently known optimal strategies where the counter-current configuration is the best from a heat exchange point of view [6], whereas the co-current configuration gives lowest temperatures due to overlapping of reaction zones [7, 8]. However, for low to intermediate conductivity materials, the co-current configuration does exhibit lower device temperatures and temperature gradients than the counter-current configuration. At the same time, the conversions achieved and power exchanged are also lower in the co-current configuration.

It is found that using properly balanced flow rates, the co-current operation significantly expands the range of useable materials compared to the counter-current operation. Finally, operation maps for coupled combustor/reformer are constructed, and recommendations for engineering integrated microchemical devices are made.

References

1. Hu, J., Wang, Y., Van der Wiel, D., Chin, C., Palo, D., Rozmiarek, R., Dagle, R., Cao, J., Holladay, J., and Baker, E. Chem. Eng. J. 93, 55 (2003).

2. Deshmukh, S.R., Mhadeshwar, A.B., and Vlachos, D.G. Ind. Eng. Chem. Res. 43, 2986 (2004).

3. Mhadeshwar, A.B., Kitchin, J.R., Barteau, M.A., and Vlachos, D.G. Catal. Lett. 96, 13 (2004).

4. Norton, D.G. and Vlachos, D.G. Combust. Flame 138, 97 (2004).

5. Westbrook, C.K. and Dryer, F.L. Comb. Sci. Tech. 27, 31 (1981).

6. Kern, D.Q. in "Process Heat Transfer" McGraw Hill, New York, 1950.

7. Zanfir, M. and Gavriilidis, A. Chem. Eng. Res. Des. 82, 252 (2004).

8. Kolios, G., Frauhammer, J., and Eigenberger, G. Chem. Eng. Sci. 57, 1505 (2002).

136c Transport Phenomena Applicable to Microchemical Systems

Khaled A Alfadhel, Mayuresh V Kothare

Microchemical systems have gained considerable interest in many areas in engineering and technology. These microchemical systems have gone beyond prototypes and laboratory setups to real life implementation. They are widely used in biological applications as micro total analysis systems (μtas), and in low-to-medium power generation as micro fuel cells [Jensen 1999].

To assess and optimize the performance and functionality of these microchemical systems, adequate models that incorporate the unique phenomena pertinent to microfluidic flows in microchemical processes have to be developed. Fluid dynamics models based on partial differential equations will become more complicated [Alfadhel and Kothare 2004-1] and perhaps unsolvable [Lockerby and Reese 2000] when one considers secondary phenomena such as velocity-slip and temperature-jump on the physical walls of the microchannels. Continuum approaches based on Navier-Stokes equations, which are valid only under quasi-equilibrium conditions, are applicable when Knudsen number (Kn) < 0.1 although the continuum assumption upholds above that limit [Gad-el-Hak 2003]. The no-velocity-slip/no-temperature-jump boundary conditions require a more stringent criterion [Arkilic et al. 1997, Jie et al 2000]. In this work, analytic expressions for the pressure gradient and the velocity profiles in rectangular microchannels incorporating velocity-slip will be derived. Error in estimating pressure gradient from the continuum correlations as opposed to our slip-based model can exceed 50% for higher Kn [Alfadhel and Kothare 2004-1].

Mass transfer models involve a higher level of complexity when written for microchemical processes. It is not only that rigorous models for this class of systems do not exist in the literature; the modification of the boundary conditions to relax the existing models to capture the physics of microchemical systems remains an area of on-going research [Alfadhel and Kothare 2004-2]. In this work, we will propose an alternative modeling approach to the multicomponent mass balance problem in membrane microreactors to generate meaningful concentration profiles from which, the sensitivity of the system to variations in physical parameters will be assessed, and conclusions about the optimal design, operability and performance will be drawn.

Microfluidic and mass transfer models are necessary to develop heat transfer models. The general energy equation for multicomponent systems includes terms that are often ignored when dealing with large-scale processes. These terms are: the diffusion thermo effects (Dufour Effect), viscous heating, compressibility, and the possibility of having other body forces different from gravity.

In this work, the energy equation applicable to microfluidic flows in microchannels will be derived from the classical energy equation in its rigorous form. The significance of each term in the derived equation will be carefully addressed. The final form applicable to microchemical systems, which incorporates temperature jumps on the walls of the microchannel, will be analytically solved under some simplifying assumptions. The effect of scale reduction (higher Kn) on the dimensionless viscous dissipation function we develop in this work will be compared to the one from Morini et al. [Morini et al. 1998], which was developed for the case when $Kn = 0$. We will show the effect of changing the tangential momentum and thermal accommodation coefficients (the effects of temperature jump and velocity slip) on Nusselt number (Nu). Finally, we will investigate the influence of permeation and reaction on the heating performance of the device.

Our future study will address the problem of coupled mass and heat fluxes, and will detail a methodology for the evaluation of the temperature and concentration gradients in multicomponent microchemical systems. Due to this coupling, the effect of temperature difference on mass fluxes when there are no concentration gradients (Soret effect) will be investigated.

Acknowledgments

Partial financial support for this research from the Government of the State of Kuwait, the U.S. National Science Foundation ("XYZ-on-Chip" initiative), and the Pittsburgh Digital Greenhouse is gratefully acknowledged.

References

K. A. Alfadhel and M. V. Kothare. Microfluidic Modeling and Simulation of Flow in Membrane Microreactors, *Chemical Engineering Science*, October 2004 (Accepted for publication).

K. A. Alfadhel and M. V. Kothare. Modeling of Multicomponent Concentration Profiles in Membrane Microreactors, *Industrial & Engineering Chemistry Research*, September 2004 (Submitted).

E. B. Arkilic, M. A. Schmidt, and K. S. Breuer. Gaseous slip flow in long microchannels, *Journal of Microelectromechanical Systems*, **6** (2): 167-177, 1997.

A. J. Franz, M.A. Schmidt and K. F. Jensen, Palladium Membrane Microreactors, *In The 3rd International Conference on Microreaction Technology*. DECHEMA, April 18-21, 1999.

M. Gad-el-Hak. Comments on "Critical view on new results in micro-fluid mechanics", *International Journal of Heat and Mass Transfer*, **46**: 3941-3945, 2003.

K. F. Jensen. Microchemical systems: Status, challenges and opportunities, *AIChE Journal*, **45** (10): 2051-2054, 1999.

D. Jie, X. Diao, K. B. Cheong, and L. K. Yong. Navier-Stokes simulations of gas flow in micro devices, *J. Micromech. Microeng.*, **10**: 372-379, 2000.

D. A. Lockerby and J. M. Reese. High-resolution Burnett simulations of micro Couette flow and heat transfer, *Journal of Computational Physics*, **188**: 333-347, 2000.

G. L. Morini, M. Spiga, and P. Tartarini. Laminar viscous dissipation in rectangular ducts, *Int. Comm. Heat Mass Transfer*, **25** (4): 551-560, 1998.

A. S. Pattekar and M. V. Kothare. A Microreactor for Hydrogen Production in Micro-Fuel Cell Applications, *Journal of Microelectromechanical Systems*, **13** (1): 7-18, 2004.

136d Fabrication and Characterization of Integrated Ceramic Microreactors for High-Temperature Hydrogen Production

Dong-Pyo Kim, Inkyung Sung, Michael Mitchell, Nfn Christian, Paul Kenis

The area of microreactors has undergone rapid growth over the past few years. The advantages of using microreactors over macroscale reactors make them useful for various applications, including on-site synthesis of hazardous chemicals, catalyst development and optimization, and the reforming of fuels such as methanol and hydrocarbons for hydrogen production for fuel cells. For the reforming of hydrocarbon fuels, the challenges are to avoid coking by operating at temperatures greater than 800 C, achieving high conversion in a small reactor volume, and minimizing the pressure drop across the reactor. A suitable microreactor, to meet these challenges, must be compatible with high temperature operations, have a high surface area per unit volume, and have a high porosity with open, interconnected pores.

We have successfully synthesized inverted beaded porous monoliths as catalyst support structures made from SiC and SiCN ceramic materials in order to meet the specified requirements. Both SiC and SiCN porous structures are stable to temperatures greater than 900 C and have geometric surface areas between 105 and 108 m2/m3. We will show the fabrication of these structures, their impregnation with catalyst, and their integration into high-temperature-compatible alumina reactor housings. We will show the characterization of SiC and SiCN porous structures to confirm the thermal stability of these structures at temperatures greater than 900°C under oxidizing environments, to confirm their promise for high temperature reforming of hydrocarbon fuels. We will show preliminary conversion data for the decomposition of ammonia on Ru-impregnated inverted beaded structures inside a test fixture up to 500 C and we hope to show conversion data from an all ceramic microreactor at temperatures as high as 1000 C.

Temperature profiles and residence time effects in metallic microchanneled reactor for hydrogen production

I. Aartun[1], B. Silberova[1], H. Venvik[1], P. Pfeifer[2], O. Görke[2], K. Schubert[2], A. Holmen[1]

[1]Department of Chemical Engineering, Norwegian University of Science and Technology (NTNU), NO-7491 Trondheim, Norway.
[2]Institut für Mikroverfahrenstechnik, Forschungszentrum Karlsruhe GmbH, Postfach 3640, D-76021 Karlsruhe, Germany.

Prepared for Presentation at IMRET 8 - 8th International Conference on Microreaction Technology
10-14 April 2005
Energy Generation Systems

Unpublished

Introduction

Large-scale steam or autothermal reforming is presently the most energy efficient and economical way for hydrogen production. For small-scale hydrogen production, catalytic partial oxidation (POX) and oxidative steam reforming (OSR) represent interesting routes, since oxidation reactions are fast and allow for high throughput and small reactors. Metallic microreactors have very high surface-to-volume ratios and favourable heat transfer properties. Narrow residence time distribution, negligible mass transfer limitations and safety in handling of explosive gas mixtures make them particularly interesting in these applications.

In this study, the performance and characteristics of rhodium impregnated Fecralloy microreactors for hydrogen production from propane by POX or OSR are investigated. Effects of temperature, pressure and residence time are discussed. A comparison to previous work on Rh-impregnated alumina foams was also made [1-4].

Experimental

The Fecralloy microreactors were manufactured at Institute for Micro Process Engineering (IMVT) at Forschungszentrum Karlsruhe. Fecralloy foils with micromachined channels of approximately 120x130 μm^2 cross section and 20 mm length were stacked and electron beam welded into compact honeycomb reactors. These reactors were oxidized at high temperatures to form a thin alumina layer in the channels prior to impregnation by an aqueous $RhCl_3$ solution. The Rh/Al_2O_3/Fecralloy monoliths were reduced in situ at 800 °C for 3 h in 10 % H_2 in N_2 [5].

The experiments were carried out in a set-up as described elsewhere [1, 5] at near to atmospheric pressure and furnace temperatures up to 950 °C. The residence time was varied from 6.3 to 31.6 ms by varying the reactant flow between 400 and 2000 Nml/min, keeping the reactant composition constant. The propane concentration was kept constant with C/O ratios of 0.8 and 0.5 for POX and OSR, respectively. An O_2/N_2 ratio as in air was applied for POX. A H_2O/O_2 ratio of 2 was used in the OSR experiments, with N_2 to balance. The dry product gas composition was analyzed by means of a MicroGC (Agilent G2891A).

Results and discussion

Temperature profiles obtained along the central axis of the microreactors clearly demonstrate the metallic structure's ability to reduce temperature gradients, especially when compared to temperature profiles obtained for Rh-impregnated alumina foams. When the furnace temperature is increased beyond 700 °C, the gas phase in front of the microchanneled reactor ignites, leading to lower syngas selectivity and more formation of hydrocarbon by-products [3, 4, 6].

Although the pressure drop over the microchanneled reactor is negligible, increasing the flow leads to higher pressure in the system, mainly due to a downstream filtering unit. To rule out that pressure is responsible for any conversion and selectivity effects, an experiment with changing pressure in the same range as observed during residence time experiments was conducted, while the residence time was kept constant. Only minimal effects on the conversion and selectivities were observed, both in POX and in OSR [6].

Lowering the residence time results in a slight decrease in syngas selectivity in OSR until ~ 20 ms, where the selectivity to both H_2 and CO experiences a minimum. In POX, no evident effects of the residence time are observed in this region. A further reduction of the residence time leads to a noticeable rise in the H_2 and CO selectivity, slightly more pronounced in OSR than POX. In addition, less hydrocarbon by-products are observed at these short residence times, and one may speculate if the flow is so high that gas phase ignition is quenched [3, 6].

Experiments over freshly impregnated alumina foams show higher hydrogen selectivity than over the impregnated microreactors, for POX as well as for OSR. The H_2 and CO selectivity during POX experiments decreases very little with time-on-stream [2] but more severely by temperature cycling [4]. This deactivation is even more pronounced when steam is added to the reactant mixture (OSR), and could be ascribed to sintering of the Rh particles [1, 4]. The microchanneled Fecralloy reactors have been cycled repeatedly without decrease in conversion or selectivity [3, 6]. Coke formation was not observed, neither in the foams nor in the microreactors.

Acknowledgements

The financial support from The Research Council of Norway is gratefully acknowledged.

References

[1] B. Silberova, H.J. Venvik and A. Holmen: *Production of hydrogen by short contact time partial oxidation and oxidative steam reforming of propane*, Catal. Today, In Press (2005).
[2] B. Silberova, H.J. Venvik, J. Walmsley and A. Holmen: *Small-scale Hydrogen Production from Propane*, Catal. Today, In Press (2005).
[3] I. Aartun, B. Silberova, H.J. Venvik, P. Pfeifer, O. Görke, K. Schubert and A. Holmen: *Hydrogen production from propane in Rh-impregnated metallic microchanneled reactors and alumina foams*, Catal. Today, Submitted.
[4] B. Silberova, I. Aartun, H.J. Venvik, J.C. Walmsley and A. Holmen: *Effects of catalyst loading in partial oxidation and oxidative steam reforming of propane over Rh-impregnated alumina foams*, In preparation.
[5] I. Aartun, T. Gjervan, H. Venvik, O. Gorke, P. Pfeifer, M. Fathi, A. Holmen and K. Schubert: *Catalytic conversion of propane to hydrogen in microstructured reactors*, Chem. Eng. J. 101 (2004) 93.
[6] I. Aartun, H.J. Venvik, P. Pfeifer, O. Görke, K. Schubert and A. Holmen: *Temperature profiles and residence time effects during catalytic partial oxidation and oxidative steam reforming of propane in metallic microchanneled reactors*, Catal. Today, Submitted.

136f Microreactors for The Production Of Hydrogen From Ammonia

Ed Seebauer, Jason Ganley, Rich Masel

This paper describes the development of a structured aluminum-anodized alumina microreactor that exhibits high catalytic activity for the decomposition of anhydrous ammonia to nitrogen and hydrogen at moderate temperatures. Modifications such as adjustments to the geometry of the microreactor features, surface area enhancement of the anodized catalyst support, choice of Ru precursor, and application of a catalyst promoter are each shown to affect the reactor performance. The reactor converts 99% of ammonia at 600° C into the equivalent of 60 W of hydrogen. These numbers are sufficient for to merit serious consideration for use in practical applications, particularly in mobile devices fueled by hydrogen where monolithic structures are advantageous.

136g Engineered Catalyst for Steam Reforming of Methane and Diesel Fuel in a Microchannel Reformer

Jeffrey Harrison, Patricia Irving, Quentin Ming, Ying Peng

A proprietary bimetallic precious metal steam reforming catalyst was designed and tailored to fit a microchannel reactor developed and manufactured by InnovaTek. The development work on this high performance steam reforming catalyst was driven by the need for a reliable and compact fuel processor for portable fuel cell applications in the size range of 0.5-5 kW. The engineered catalyst allows for high heat and mass transfer and low pressure drop simultaneously in a micro-channel reactor.

The engineered catalyst consists of a high temperature resistance alloy (HTRA) support and bimetallic active components. An appropriate heat treatment procedure can substantially improve the adhesion of active components coatings to the support by forming a micrometer-sized metal oxide structure at the interface. Thin strips of the catalyst and support were produced in this way for insertion into the catalytic channels of the micro-channel reactor. The thickness of the engineered catalyst is less than 20 micrometers. The BET surface area of the final engineered catalyst was 7.33 m2 per gram.

The engineered catalyst was tested in a single channel reactor, which simulated the exact steam reforming channel dimension in our scaled-up multi-microchannel reformer. Pressure drop in the both the single reactor and scaled-up system was less than 3 psi. The single channel reactor was heated up in an electrical furnace. In the scaled-up microchannel reformer, heat was supplied to the endothermic catalytic channels from cross-flowing heat exchange micro-channels heated through combustion exhaust. Both methane and diesel fuels were tested with the engineered catalyst. The steam to carbon ratio (S/C) of steam methane reforming was 2.5. Scanning electron microscopy photos of the engineered catalyst before and after 40 hours of steam methane reforming showed no sign of microstructure change. Nearly complete methane conversion was achieved at a gas hourly space velocity (GHSV) of 100,000 hr-1 and an exit catalyst temperature of 850 C. Steam methane reforming yielded a product stream consisting of 74% H2, 16.5% CO and 9.5% CO2 .

Diesel fuel with an ultra-low sulfur level (~5ppm) was also tested with the engineered catalyst. The diesel fuel was evaporated and mixed with steam prior to the single channel reactor entrance using InnovaTek's proprietary fuel injector. The S/C ratio was controlled at 5 for diesel steam reforming. The product concentrations in the reformate flow were 70% H2, 13.5 % CO and 16.5% CO2. No methane was detected in the reformate stream.

The single channel test results of our steam reforming catalyst matched well with the results obtained in our scaled-up multi-microchannel reformer. Since the major difference between the two reactors was the heat source, we conclude that the engineered catalyst, sandwiched between the cross-flow micro-channel heat exchangers in the scaled-up reactor design, experienced no heat transfer limitations to the steam reforming reaction.

Spatially resolved monitoring of aqueous CdS nanoparticle synthesis in a microreactor

Thomas L. Sounart[1], *Jessica E. Bickel, David R. Tallant, Carolyn M. Matzke, James A. Voigt, and Terry A. Michalske*
Sandia National Laboratories, P.O. Box 5800, Albuquerque, NM 87185

Introduction

Nanomaterials, including nanoparticles and quantum dots, nanowires, and nanostructured films, are expected to have wide ranging applications such as advanced computing, chemical and biological analysis/detection, drug delivery/discovery, tissue engineering, catalysis, solid-state lighting, and energy conversion and storage [1, 2, 3]. However, fulfilling this potential requires reproducible synthesis of uniform nanostructures in large quantities, and current laboratory synthesis techniques have not been successful upon scale-up. Large-scale production even of simple nanoparticles does not exist because yields are low and the performance and reproducibility of the products are economically unacceptable [4]. Nanoparticles are typically produced in batch processes, where many advances have been made in the last two decades [1, 2]. However, in batch processes, nucleation and growth occur under poorly controlled and unknown local conditions, particularly during reaction initiation and termination. Improvements in batch synthesis will be difficult, due to inherent problems with control of thermal, chemical, and fluid transport, and poor in-situ monitoring and feedback mechanisms for obtaining critical information on the reaction pathways.

Microfluidic reactor systems afford microscale control over process variables, not accessible to macroscale batch chemistry. In microreactors, laminar flow and high heat transfer rates can be exploited to exercise unparalleled control over fluid, mass, and energy transport. In addition to providing better control of reactor conditions, microfluidic systems provide a unique platform for investigation of fundamental reaction processes, which is critical to understanding how to control nanostructured materials on a production scale. Several researchers have recently explored the use of simple continuous-flow microcapillary and microfabricated reactors to gain more control of nanoparticle synthesis and to illucidate the growth mechanisms and kinetics [5]. While these previous microfluidic studies on nanoparticle synthesis have analyzed the product stream of a microfluidic reactor, an advantage of microfluidic reactors that has not yet been exploited is the ability to spatially resolve nucleation and growth in the reactor during synthesis. Microfluidic reactor systems can be made planar and transparent, so entire reactors can be observed with a single optical instrument. In this work, we have examined the synthesis of cysteine-capped CdS quantum dot nanocrystals (CdS-cys) between two interdiffusing reagent streams in a continuous flow microfluidic reactor. Using a two-feed stream system, an aqueous solution containing $CdSO_4$ was continuously fed into one side of a rectangular microchannel while a solution containing Na_2S was fed into the other (Fig. 1). The laminar flow of the impinging streams allowed for controlled diffusional mixing of the reacting cadmium and sulfide ions at the boundary between the two solutions, while the capping agent was present in one or both of the solutions in excess. Using spatially resolved fluorescence imaging and spectroscopy of the microreactor at various reactant concentrations and flow rates, we have acquired kinetic and mechanistic data on the CdS-cys nanoparticle nucleation and growth that would be difficult to achieve in a batch reactor. Furthermore, the photoluminescence of the CdS-cys particles grown under these microfluidic conditions differs from those grown in batch reactors, which is not surprising given the unique conditions of growth between laminar reagent streams.

[1]Corresponding author, tlsouna@sandia.gov, (505) 845-0946

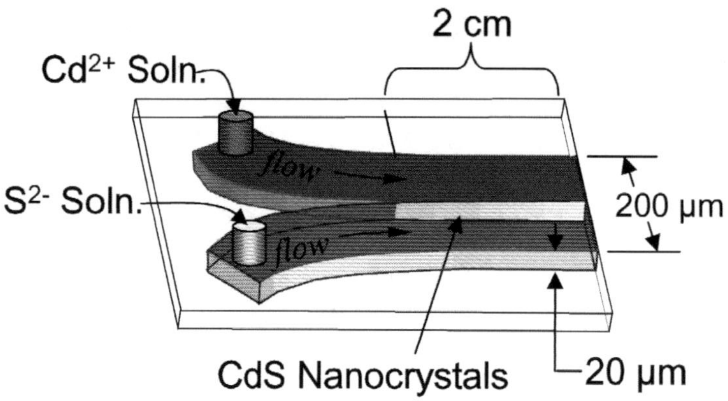

Figure 1: Sketch of continuous flow microfluidic reactor for observation of of CdS synthesis.

Materials and Methods

An established method for batch preparation of CdS-Cys bioNCs [6] was adapted for synthesis in our continuous-flow microfluidic reactor. Aqueous $CdSO_4$ and Na_2S solutions were prepared as described in [6], but the concentrations of reagents in the two solutions were adjusted so that when equal volumes of each reagent were mixed, the final concentrations of all reagents were equal to that in the final batch. The final ratios of cadmium to sulfide and cysteine were 1:1 and 1:4, respectively, and cysteine was included in both solutions. A simple "y-shaped" Si/glass microfluidic reactor was designed such that the two reagent streams could be pumped through two inlet channels (100 μm wide) and react at ambient temperature at the diffusional mixing boundary between the two flowing streams in a 2 cm long straight reaction channel (200 μm wide and 20 μm deep). Photoluminescence (PL) spectra and intensity distributions were acquired, respectively, using an argon laser/spectrometer system, and image analysis of photomicrographs from an epifluorescence microscope (360 nm excitation). The PL images and spectra are sampled as an average over the depth, but variations along the depth are minimized by diffusional mixing across the small depth dimension.

Results and Discussion

The CdS-Cys PL intensity profiles measured across the width of the fluorescent zone are symmetric and Gaussian (Fig. 2a). Further, the PL spectra sampled at several positions across the width of the plume at a residence time of 5 seconds are essentially identical (data not shown), which suggests that all the NCs reflected in the PL spectra are grown at the same reagent concentration ratios because of the known sensitivity of the PL to these variables [6]. This further implies that essentially all reagents are consumed at the boundary between the two reagent streams, thus preventing significant interdiffusion, which would result in large variations in the reagent concentration ratios (high Cd/S on one side and low Cd/S on the other), and thus asymmetry in the fluorescent plume. We also expect that if the NCs grow only at the boundary, the width of the Gaussian zone will reflect diffusion of the nanoparticles, rather than the reacting ions, which have an order of magnitude higher diffusivity. Gaussian curves were fit to the measured intensity profiles to yield diffusivities based on the average life of the particles at each axial position. The estimated diffusivities are within a factor of 2 of the measured diffusivities of CdS-Cys NCs produced in batch reactions, and about an order of magnitude smaller than the diffusivities of the reactants (Fig. 2b).

Figure 3a shows the normalized CdS-Cys PL spectra measured at the boundary between the two

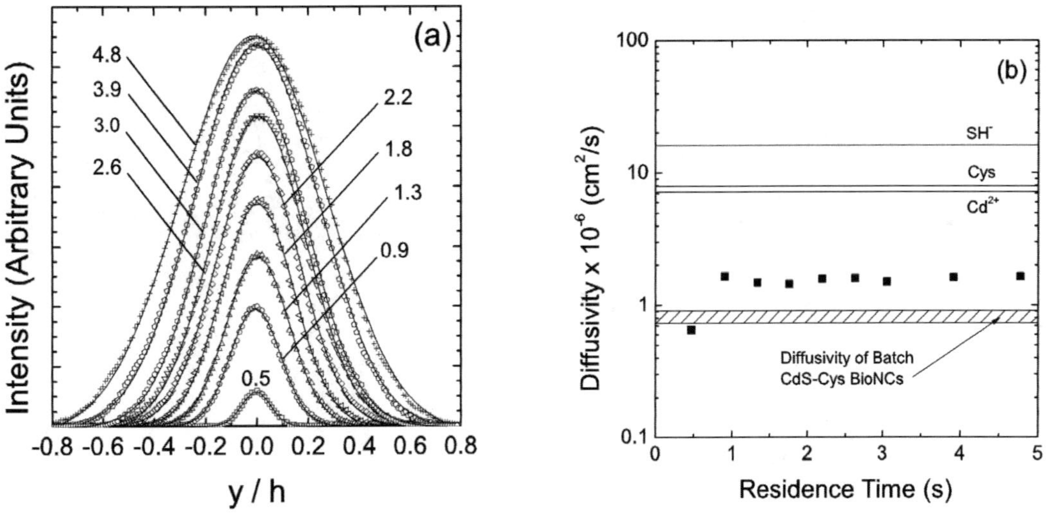

Figure 2: (a) Cross-sectional PL intensity profiles at selected residence times. The symbols are measured intensity data, and the solid lines are Gaussian fits to the measured data. (b) Diffusivity calculated from each Gaussian fit compared to measured or literature diffusivities of reactants and products. Flowrate = 700 nL/min.

flowing streams, at selected residence times. The spectra are similar to the PL spectra presented previously for the analogous batch synthesis [6], but the PL spectra in that work peak at about 480 nm, and the shape of the distribution is slightly different. This difference is not surprising given the unique reaction conditions of this microfluidic configuration, as discussed above. Also, there is a significant shift in the distribution of the PL along the length of the reaction zone. At 0.3 s, the PL spectrum is bimodal with the peak at 528 nm, and a smaller peak at 450 nm, suggesting two sources (or groups of sources) of emission. The relative contribution of shorter wavelength PL decreases with longer residence times. The kinetics of this bimodal shift are presented in Fig. 3b. The contribution to the integrated intensity from wavelengths less than 500 nm shifts from 32% to 18% between 0.3 and 5 s of residence time. The mean and peak wavelength of the PL spectra shift by 23 and 12 nm, respectively (inset), so the red shift results both from a shift of the whole spectrum and the diminishing contribution of the higher energy, lower wavelength mode. The change in PL occurs rapidly in the first few seconds and approaches an asymptotic state with little variation in either the intensity or spectral distribution of the PL after 5 s.

Extrapolating the kinetic curves in Figs. 3b to zero time suggests that the NCs might initially luminesce entirely by the higher energy mode. However, we do not observe a continuous shift in the entire emission spectrum with time as would be expected if the observed change in fluorescence reflected a band gap shift from continuous growth along the reactor length. These data reflect two discrete modes of fluorescence, which could result from two classes of particles or two distinct modes of emission from the same particles—e.g., the higher energy peak from band gap fluorescence and the lower from surface traps [7]. So why would the PL exhibit a binary shift in fluorescence during nucleation and growth, rather than a continuous shift in the spectrum? The batch chromatography fractions in Fig. 4 of [6], which exhibit a broadband emission that is nearly independent of particle size but sensitive to sulfide

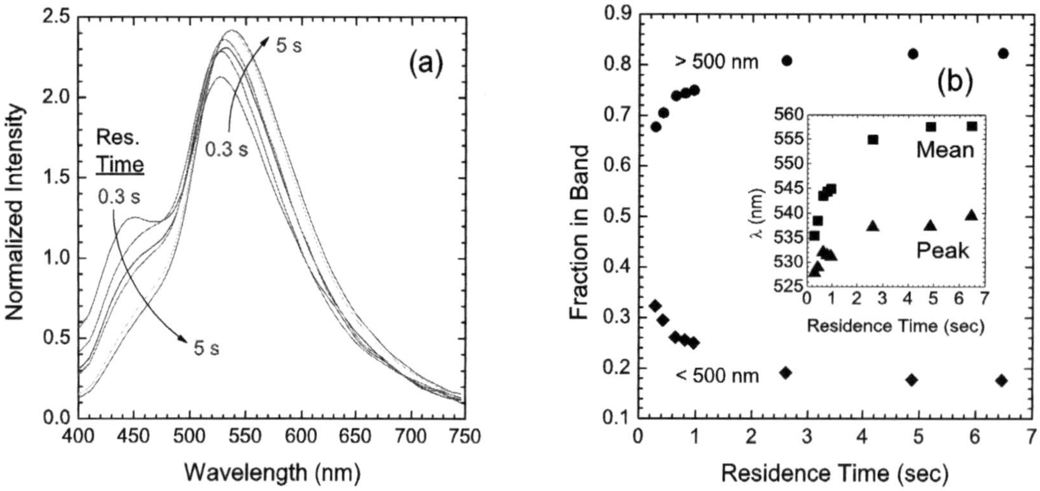

Figure 3: (a) CdS-Cys PL spectra, normalized by the integrated intensity, at selected residence times. (b) Kinetics of PL red-shift. Fraction of CdS fluorescence greater than or less than 500 nm obtained by integration of the emission spectra. Flowrates = 700 nL/min, 4 μL/min.

concentration, show that photon emission in capped CdS-Cys particles is dominated by recombination through surface traps. Therfore we do not expect a continuous shift in the spectrum during growth, as would be expected for an NC dominated by bandgap emission. We expect rather a discrete shift from surface states in the uncapped NCs to surface states in the cysteine-capped NCs. Thus, the binary shift in fluorescence during nucleation and growth along the boundary may reflect the change in the surface as the CdS particles are capped with cysteine.

Conclusion

We have demonstrated the first spatially resolved on-chip investigation of NC growth during synthesis in a microfluidic reactor. Results show a binary shift in the particle emission spectrum from a higher (2.8 eV) to lower (2.3 eV) energy emission peak within the first second of residence time. This binary shift is consistent with a shift in NC surface traps from the uncapped to cysteine-capped state. A Gaussian spatial distribution of the fluorescence intensity is observed across the channel width, and this fluorescent reaction plume grows in width and intensity along length of the channel. The integrated area of the fluorescence intensity profile increases linearly with residence time until sidewall effects become significant after about 3 seconds. Also, the particle photoluminescence emission spectrum is symmetric about the channel centerline. Several reaction models and kinetics are being tested against the spatially and spectrally resolved signals. The elegant results defy explanation by the simplest reaction mechanisms, and the constant growth rate is the most difficult feature explain. The analysis shows that homogeneous reaction and particle nucleation are diffusion limited and occur only at the boundary between the two laminar streams, while growth and/or capping occur on a slower (seconds) time scale. Spatially resolving nanocrystal synthesis at the boundary between two laminar reagent streams provides direct insight to the rapid processes that occur during crystallization in microfluidic mixers, and is a useful tool for investigating nucleation and growth kinetics and mechanisms. Although the analysis is more complicated, mechanisms that would be difficult to distinguish in a batch kinetic study can be differentiated in this configuration because of the added information from analysis of the product spatial distribution. This is particularly useful for quantum dot synthesis because the particles are fluorescent, and their transport can be differentiated from precursor transport on the basis of a significant shift in diffusivity. Our results also have broad implications for microreaction technology in general because most reactions that are initiated upon contact in micromixers begin by diffusional mixing between laminar reagent streams. In fact the study of this configuration may also help to understand and control batch syntheses because it approximates the instantaneous processes occurring at the length scale of one turbulent eddy.

Acknowledgements

Sandia is a multiprogram laboratory operated by Sandia Corp., a Lockheed-Martin Company, for the U.S. Department of Energy under Contract No. DE-AC04-94AL85000. We thank Diana Moore for assistance with reagent preparation, and Regina Simpson for optical instrumentation support and data collection.

References

[1] Murray, C.B.; Kagan, C.R.; Bawendi, M.G. *Annu. Rev. Mater. Sci.* **2000**, *30*, 545–610.

[2] Trindade, T.; O'Brien, P.; Pickett, N.L. *Chem. Mater.* **2001**, *13*, 3842–3858.

[3] Parak, W.J.; Gerion, D.; Pellegrino, T.; Zanchet, D.; Micheel, C.; Williams, S.C.; Boudreau, R.; Le Gros, M.A.; Larabell, C.A.; Alivisatos, A.P. *Nanotechnology* **2003**, *14*, R15–R17.

[4] *Chemical Industry R&D Roadmap for Nanomaterials by Design: From Fundamentals to Function* **2003**, Chemical Industry Vision2020 Technology Partnership; www.ChemicalVision2020.org.

[5] DeMello, J.; DeMello, A. *Lab on a Chip* **2004**, *4*, 11N–15N.

[6] Bae, W.; Abdullah, R.; Mehra, R.K. *Chemosphere* **1998**, *37*, 363–385.

[7] Resch, U.; Eychm uller, A.; Haase, M.; Weller, H. *Langmuir* **1992**, *8*, 2215–2218.

137b Micro Fluidic Synthesis of Metallic Nanoparticles

Challa Kumar, Josef Hormes, Yujun Song

One of the key challenges in synthesis of nanoparticles is to precisely control nanocrystal growth parameters in order to produce nanoparticles of defined size, shape and crystal structure which control both physical and chemical properties. Micro reactor technologies present a unique opportunity for creating nanomaterials with tailored characteristics as demonstrated by recent publications. 1 Additionally, micro reactor technologies are cost-effective and environmental friendly. We have recently fabricated a continuous flow polymeric micro reactor using a negative photo resist SU-8 on a 8×8 cm PEEK (polyetherether ketone) substrate by a standard UV lithography. 2 The micro reactor was used to synthesize palladium nanoparticles of controlled sizes. 3 In order to generalize the micro reactor based synthetic approach we have recently synthesized Cu and Co nanoparticles using the same reactor.

In order to develop a micro reactor process for wet chemical synthesis of Cu nanoparticles, the mechanistic aspects of the formation of sulfobetaine-stabilized copper nanoparticle were investigated in the batch process using in situ XANES (X-ray absorption near edge structure), UV-vis spectroscopy and reaction calorimetry. It was found that two types of Cu+ complex (a linear and a tetra-coordinated) intermediates are formed when the tetragonal Cu2+ complex was reduced to metallic copper. The linear Cu+ complex changes to a tetra-coordinated complex when the ratio of reducing agent to CuCl2 is more than one. The stability of the linear Cu+ complex is sensitive to the addition rate of the reducing agent and the concentration of the stabilizer. The micro fluidic reactor process was developed using the stable and soluble Cu(I) complex as the starting material. The micro reactor process provided Cu nanoparticles with narrower size distribution and a smaller size compared to those obtained by conventional batch processes. The Cu nanoparticles from the micro reactor were also found to be more stable to oxidation.

Cobalt nanoparticles were prepared in the micro fluidic reactor by the reduction of CoCl2 in tetrahydrofuran (THF) using lithium hydrotriethylborate (LiBH(C2H5)3) as a reducing agent and 3-(N,N-dimethyldodecylammonia)-propanesulfonate (SB12) as a stabilizer. We have been able to control the size, size distribution and the crystal structure by controlling the flow rates, reaction times, reaction temperatures and the quenching procedures. A high flow rate (0.9ml/min) of reactants followed by quick quenching of the reaction generated the fcc phase and a low flow rate (0.084ml/min) followed by quick quenching provided the hcp phase. However, a low flow rate(0.084ml/min) with slow quenching resulted in the formation of -phase.

The particle size, size distribution of the Cu and Co nanoparticles was determined by TEM. The crystal structure was analyzed using SAED, XRD and XANES. The interaction between the metallic core and the ligands was studied using UV/vis and FT-IR spectroscopy. The magnetic properties of the Co nanoparticles were investigated by SQUID.

1. E. M. Chan, R. A. Mathies and A. P. Alivisatos. Nano Letters 2003, 3(2), 199-20.

2. Y. Song, Challa S.S.R.Kumar, and Josef Hormes, Journal of Nanoscience and nanotechnology, 4(7), 788-793, 2004.

3. Y. Song, Challa S.S.R.Kumar, and Josef Hormes, Journal of micromechanics and microengineering, 2004, 4, 932-940.

Novel Synthesis of Fullerene Nanowhiskers in a Microchannel Reactor

*Sung-Ho Lee[a,] *, Kun'ichi Miyazawa[b], Sohei Matsumoto[a], Ryutaro Maeda[a]*

[a] *National Institute of Advanced Industrial Science and Technology, 1-2 Namiki, Tsukuba, Ibaraki 305-8564, Japan*

[b] *National Institute for Materials Science, 1-1 Namiki Tsukuba, Ibaraki 305-0047, Japan*

In this work, we mention the fullerene nanowhisker(C_{60} nanowhiskers; C_{60} NWs) synthesis using a microchannel reactor of Y-type micromixer. For the preparation of a silicon microchannel reactor, we fabricated the silicon microchannel reactor using deep reactive ion etching and bonded anodically a silicon channel device with a pyrex glass. We connected the silicon tube to the inlet/outlet of the fabricated microchannel reactor for delivery of chemicals. We delivered isopropyl alcohol and fullerene mixed with toluene to inlets of the microchannel reactor using a 10 ml glass syringe connected to a syringe pump. We obtained the reaction interfaces of the isopropyl alcohol and fullerenes mixed with toluene as shown in Fig. 1. We observed that the C_{60} nanowhiskers grew to the vertical direction of the reaction interfaces.

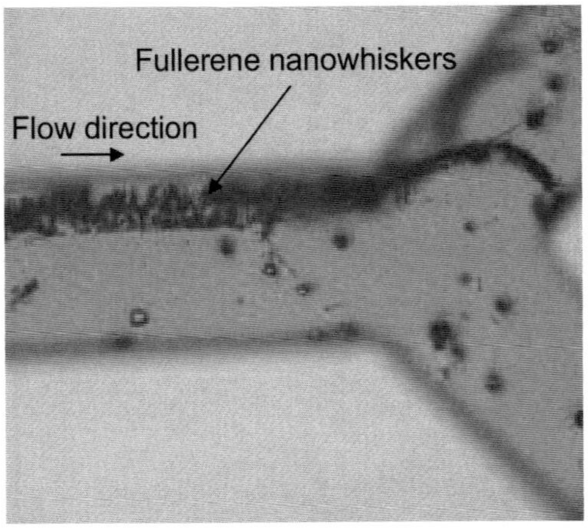

Fig.1. Liquid-liquid interfaces in a micro channel reactor by isopropyl alcohol and fullerenes mixed with a toluene.

In Fig. 2, obtained C_{60} nanowhiskers showed the length of a 200 micrometers and thickness of a 2 micrometers. These C_{60} nanowhiskers obtained using the silicon microchannel reactor show the spectra peaks of 527, 576, and 1183cm^{-1} in the FTIR measurement and lattice parameter of about 1.411nm (FCC crystal structure) and (111) inter-plane distance of 0.81nm with HRTEM image and SAEDP image. These values were almost same to the 1.417nm of a C_{60} nanowhiskers molecular crystal(in air and atmosphere conditions). We believe that C_{60} nanowhiskers synthesized by a microchannel reactor will have many properties that differ from the glass flask method.

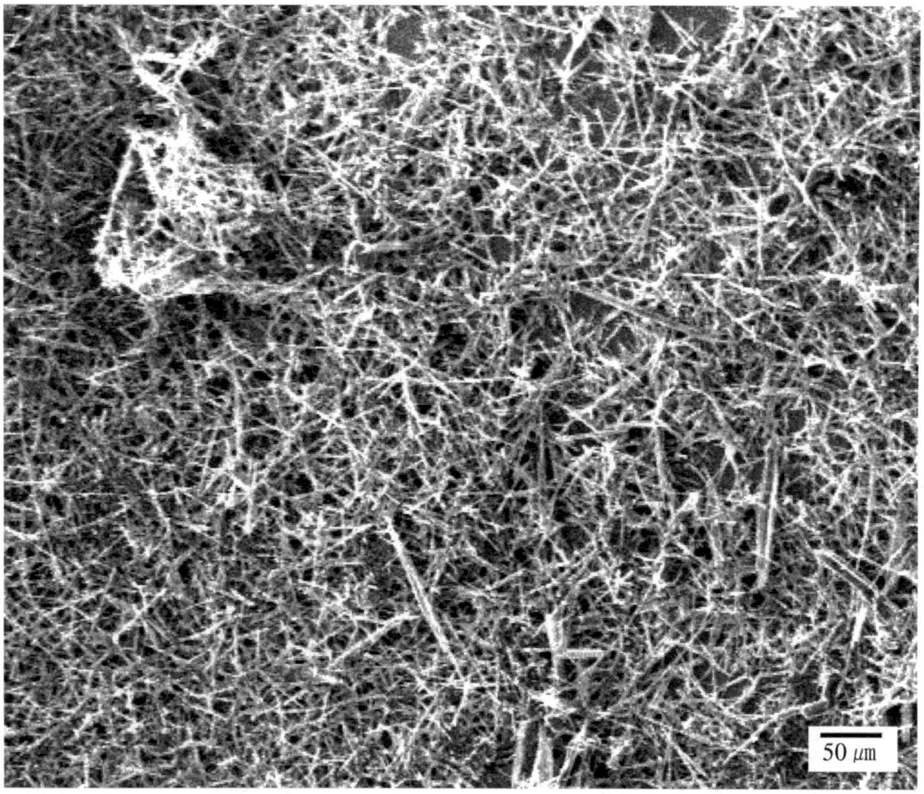

Fig.2. SEM micrograph of C_{60} nanowhiskers obtained by liquid-liquid interfacial precipitation method in the micro channel reactor.

As a synthesis method of the C_{60} nanowhiskers, a micro channel reactor showed that it will be a promising synthesis method for future.

Acknowledgements

This research has been partially supported by Cooperative System for Supporting Priority Research of Japan Science and Technology (JST).

137d High Surface Area Mixed Oxide Supports for High Temperature Catalysis in Micro-reactors

Rich Masel, Vaidyanathan Subramanian, Zheng Ni

The lifetime of many high temperature microreactors is often limited by surface area loss in catalyst washcoats. This paper reports the synthesis of TiO2 -Al2O3 mixed oxide nanoparticles that are useful as high temperature catalyst supports in microreactors. The porous oxide nanoparticles are prepared by employing sol-gel synthesized TiO2 and Degussa Alumina. The mixed oxide properties are evaluated following prolonged exposure to high temperature (900C, 9 hrs) using conventional surface analysis techniques. Preliminary results from XRPD show anatase phase of TiO2. Method of calcination plays an important role in the final available surface area. A 1:1 weight ratio of the mixed oxide provides surface area=74m2/gm. This is the highest surface area reported to date on a support that had been annealed to 900C. These results suggest that mixed oxide nanoparticles will be excellent washcoats for high temperature microreactors.

137e Design and Fabrication of Infiltratable Multiscale Catalytic Cellular Structures into Microchannel

Haibiao Chen, Woo Lee

Multiscale cellular structures containing periodically interconnected open cells of 10 to 20µm were synthesized in the 100 µm x 500 µm microchannel of a silicon-based microcreactor by developing a fabrication procedure based on microsphere templating, sol-gel, and nanoparticle infiltration. Polystyrene microspheres were selectively infiltrated and self-assembled into the microchannel from suspension. The polystyrene microspheres were partially sintered to create the inverse template of a desired cellular structure. SiO_2 precursor sol or nanoparticle (20 nm) suspension was subsequently infiltrated into the interstices of the template. Upon drying and sintering, the solid silica skeleton was formed and the microspheres were removed. The resulting cellular structure was infiltrated with a 1 mm catalyst layer of Pd/SiO_2 by sol-gel. The thin-film layer contained pores in the range of 1.2-1.4 nm and Pd particles of 3 nm. Cell size, cell interconnectivity and skeleton density were mainly considered for overall procedural refinement, as these structural variables are expected to be important in addressing the mechanical stability, pressure drop, and mass transfer resistance issues associated with designing this type of catalytic cellular structures for microreactor applications.

Evaluation of Mixing Efficiency of Microdevices by Preparation of Polymer Nano-particles

Taisuke Maki,[1] Yoshikage Omukai,[1] Hideharu Nagasawa,[2] and Kazuhiro Mae[1]
[1] *Department of Chemical Engineering, Kyoto University, Kyotodaigaku-Katsura, Nishikyo-ku, Kyoto 615-8510 Japan*
[2] *Kyoto Laboratory, The Research Association of Micro Chemical Process Technology, Kyotodaigaku-Katsura, Nishikyo-ku, Kyoto 615-8510 Japan*

Abstract

Polymer particles were successfully prepared by immersion precipitation using micromixer and the effects of microchannel size of the micromixer, flow rate of fluid and solvent on the formation of particle were investigated. Polymer dispersion was obtained by micromixing an aqueous surfactant solution with a solution of the polystyrene in organic solvent. Some kinds of micromixers, IMM, YM-1 and YM-2 were used. The microchannel size of IMM, YM-1 and YM-2 was 40μm, 400μm and 400μm, respectively. A microdevice with the double pipe structure, which microchannel size was about 1500μm, was also used. An aqueous solution containing diisooctyl sodium sulfosoccinate (AOT) was used as nonsolvent. As solvent, tetrahydrofuran (THF) with low mutual affinity with water, N,N-dimethylacetoamide (DMAc) with middle affinity and N,N-dimethylformamide (DMF) with high affinity were used. Using all microdevices, polystyrene particle was formed without clogging of microchannel. The size of particle was dependent on the width of the microchannel of the micromixer. In the cases of THF and DMAc with the low affinity with water, the particle size was increased with an increase of total flow rate. This result suggested that the segment size of water and polymer solution also affected the particle size. The particle size was possible to control in the range from 150nm to 10 μm by changing the micromixer, the total flow rate. It can be expected that this method of polymer particles will be used for the evaluation of the mixing performance of micromixers.

Introduction

The mixing speed and method of reactant greatly influences product yields and selectivity for very fast reactions. When mixing is mainly driven by molecular diffusion, reducing the scale is essential for fast mixing, since mixing time is proportional to the square of its length. In many micromixers (Ehrfeld et al., 1999; Hassel et al., 2003; Lowe et al., 2000), the reactant flow is split into many laminations to shorten mixing time. Using these micromixers, many researchers performed experiments and reported that higher conversion and product yields were obtained than by the conventional batch reactor. (Bayer et al, 2000; Kurummaradt et al., 2000)

As mentioned above, the micromixing is a key operation to achieve the rapid productions and the high products yield and also to control product selectivity for fast reactions. In addition, the micromixing seems to be a promising approach to achieve the rapid production of high performance materials such as dispersion of various pigments, nano-particles, and various polymers due to the rapid and homogeneous mixing in microspace. Nano-particles are receiving interesting attention because of their distinct physical and chemical properties. It is expected that nano-particles will be applied to many new and high-functional devices. Considering such an application of the nano-particles, it is important to develop a manufacturing process for the nano-particles with narrow size distribution. To date, various chemical methods have been applied for the production of the nano-particles with narrow size distribution such as micro emulsion method (Motte et al., 1992; Hirai et al., 1993), hot soap method (Murray et al., 1993), and electrospray pyrolysis (Lengoro et al., 2000). However, it is difficult to control the reaction precisely in a conventional macroreactor because of the non-uniformity of the residence time, mixing and so on. To overcome this, it is expected that microdevices can be applied to the production of nano-particles with narrow size distribution, because microdevices provide a uniform reaction field in which the reaction condition is precisely controlled. Generally, it is essential to design precise nuclei formation and aggregation processes for high quality particles. The way to form the nuclei is to keep a proper super-saturation state by controlling the gradient of temperature, concentration of reactant and so on. From this viewpoint, a microchannel is very attractive for easily forming the reaction conditions in continuous production. Many studies for the production of nano-particles have been reported using microdevices and the advantages of microreactor for the nano-particle production were claimed (Edel et al., 2002; Wagner et al., 2004; Takagi et al., 2004).

In this study, we prepared polystyrene particles successfully by immersion precipitation in several micromixers and examined the effects of the mixer type, the flow condition and solvent type on the size distribution of particle, and then clarified that the size distribution of particles produced was applicable to the evaluation of the mixing performance of micromixers.

Materials and Experimental

Micro mixers and a microdevice used

Three types of micromixers, IMM (CPC-Systems, Gmbh), YM-1 and YM-2, were used. IMM mixer was a single mixer, which consisted of the mixing element based on a layer containing 18 sinusoidal shaped fluid channels for each fluid with a width of 40 μm. Both mixers of YM-1 and YM-2 (Mae et al., 2003 and 2004), were made of stainless steel, and had 11-stages structure of split-recombination of fluids. The mixing element of YM-1 consisted of two inflow holes and two outflow holes of 400 μm fabricated by micro machining. The lowest stage has 11 mix elements and the number of element diminished at a time to one toward the upper stage. In each element, two kinds of liquids entered from two inflow holes were split into two streams, respectively, and then were recombined each other at two outflow holes, finally send to the inflow holes in upper stage and met with the stream next to element. Repeating these motions from the bottom to the top, two liquids were finally mixed as one stream at the exit of mixer. YM-2 is also composed by the same element but the number of elements, which number was 36, was constant in each stage. Two liquids were respectively distributed to the 1st stage through the inlet sheet, and splitting and recombination of the flow were repeated in each element up to the 11th stage, and the flows were finally gathered as to flow out of the center exit of the top outlet sheet.

A micro device with the double pipe structure (Takagi et al., 2003) was also used. This device was developed to control the condition for the nuclei formation process for high quality fine particles. Two liquids were flowed in the inner and outer tubes, respectively, and maintained an annular and laminar flow of separated phase to create a micro space by outer fluid wall, which size was about 1500μm.

Materials

Polystyrene (Aldrich, Ltd., Mw: ca. 225,000) as a polymer material, tetrahydrofuran (THF), N,N-dimethylacetoamide (DMAc), N,N-dimethylformamide (DMF) (Nacalai tesque, Ltd.) as solvents and diisooctyl sodium sulfosoccinate (AOT) (Wako Pure Chemical Industries, Ltd.) as a surfactant were used.

Preparation and analysis of polymer particles

Polystyrene particle was prepared by mixing of 0.5 wt% of polystyrene solution and 0.05wt% of AOT aqueous solution. Polymer solution and AOT aqueous solution were supplied to the microdevices using two sets of micro syringe pumps (IC3210, Kd Scientific Co.) at 293 K. The volume ratio of polymer solution to nonsolvent was adjusted to 0.14. Total flow rate of polymer solution and nonsolvent changed in the range of 1.0 to 5.0 mL/min for IMM single mixer, 1.0 to 50.0 mL/min for YM-1, 2.0 to 90.0 mL/min for YM-2. When the microdevice with double pipe structure was used, total flow rate was 40 mL/min. Polystyrene particles were also produced using a 500 mL batch vessel with a stirrer for comparison. For comparison, titania particles were also conventionally produced by a batch reactor under the same condition.

Thus prepared particles were observed with a scanning electron microscope (SEM, JEOL JSM-6340F). To prepare SEM samples, a droplet of the colloidal suspension of the product was applied to a mica substrate and subsequently dried by a vacuum desiccator for 12 h at 303K. The particle size distribution was analyzed by the laser diffraction method (HORIBA Co. Ltd., LA-920).

Results and Discussion

Performance for polymer particles production by micromixers

First, we examined whether micromixers, IMM, YM-1 and YM-2 had a potential to produce polymer particles by immersion precipitation or not. As typical examples, Figure 1 shows the SEM photograph of the polystyrene particles prepared by YM-1. From this SEM photograph, spherical polystyrene particles of less than 1μm in diameter could be produced by YM-1. Figure 2 shows the photographs of polystyrene suspension prepared by mixing THF solution including 0.5wt% polystyrene and AOT aqueous solution as a nonsolvent with changing the total flow rate using YM-1. When the flow rate was low, the clogging in microchannel was observed and the polystyrene particle was randomly formed. But, the fine polystyrene particle successfully formed without clogging by adhesion of precipitated particles to wall surfaces in the micromixer at high flow rate. The same results were obtained by the experiment using other micromixer, IMM and YM-2. When the total flow rate was 50 mL/min, contact time was about 0.3 s for YM-1 micromixer. From these results it was confirmed that micromixers have a potential to produce polymer particle suspension

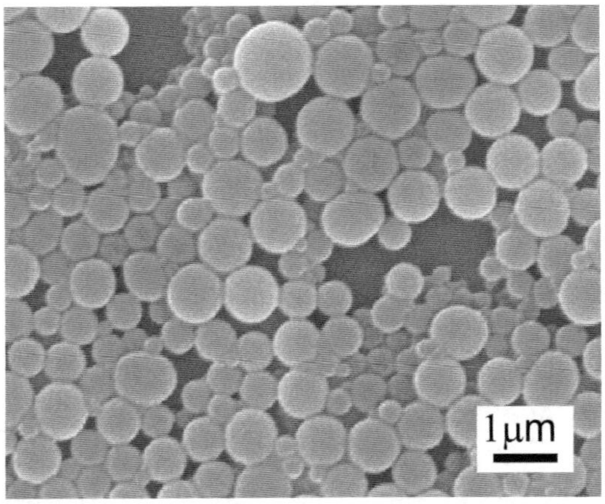

Figure 1. SEM photographs of the polystyrene particles produced by YM-1.

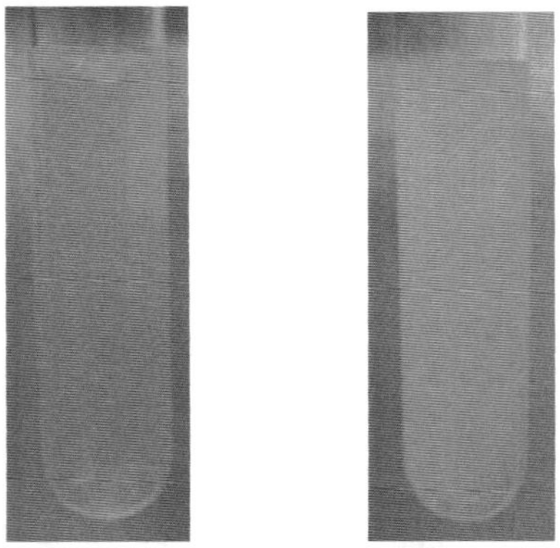

(a)0.1mL/min **(b)50mL/min**

Figure 2. Photograph of polystyrene suspension produced at various flow rates by YM-1

continuously at short contact time.

Comparison of size distribution of polystyrene particles produced by several microdevices and a batch-mixing vessel

Next, we examined the size distribution of polystyrene particles produced by micromixers. Figure 3 shows the volume density distribution of the particle size produced by mixing THF solution including 0.5 wt% polystyrene and AOT aqueous solution by using some kinds of micromixers, the microdevice with double pipe structure and a conventional batch mixing. Since THF has the low affinity with water, the rate of precipitation was expected to be low. The flow rate was maximum value for each micromixer carried out in the experiment. From Fig.3 a bimodal distribution having two peaks of 0.5 and 30 μm was obtained by in a batch-mixing vessel, but the narrow size distributions were obtained by the micromixers. This was probably due to the regular and rapid mixing in micromixers at short contact time. The peak size of the distribution by IMM, YM-1, YM-2 and the double pipe device was 0.14, 0.37, 0.98 and 1.76 μm, respectively. This results shows that the narrower microchannel size, the smaller particles. Although YM-1 and YM-2 have the microchannel of same width, the size distribution of droplets produced by YM-2 shifted toward large size. At that time, the total flow rate of fluid in YM-2 was 90 mL/min. Since the number of elements at the outlet of YM-2 was 36, which meant the area of cross section at outlet was 36 times larger than that in YM-1. Comparing the flow rates between YM-1 and YM-2, the apparent velocity of fluid in YM-2 is ca. one tenth smaller than that in YM-1. This was the reason for the large size distribution of the particles produced using YM-2. We could not perform the experiment at high flow rate using YM-2 because of the capacity of the pump. These results suggest that the velocity in the mixing zone as well as the microchannel width effects on the particle size.

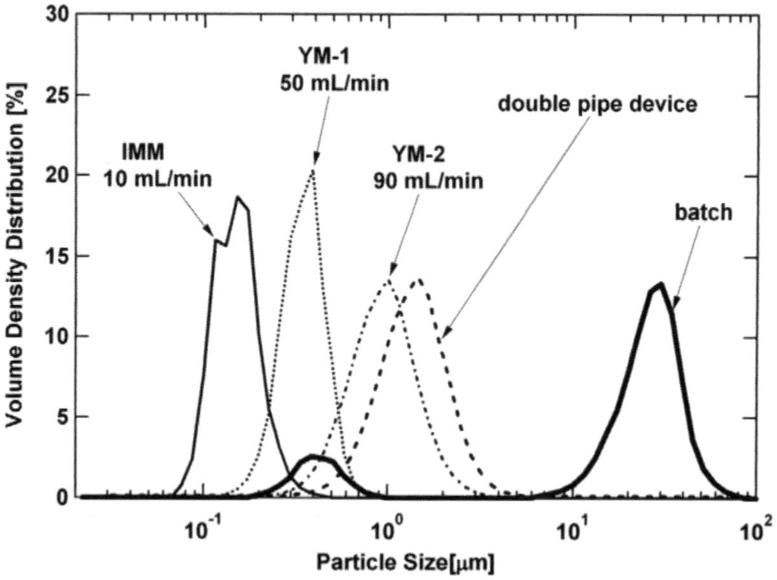

Figure 3. Comparison of the size distribution of polystyrene particle produced by mixing among three micromixers, a microdevice with double pipe structure and a batch mixing vessel.

The size distribution of particles produced by the microdevice with double pipe structure was also narrow, but the particle size was larger than that by micromixers. This is because this device has large size microchannel of ca. 1500μm. In our recently work, it was clarified that the grain of TiO_2 and $CaCO_3$ particles was uniformly grown near the interface between two liquids in the microdevice and that this microdevice was valid for synthesis of a mono-modal nano-particle by controlling the nucleation and aggregation processes. This result shows that this microdevice was also useful for the preparation of polymer particles by the immerse precipitation method if the affinity between solvent water is low.

Effect of the flow rate on the particle size distribution

Figure 4 shows the effect of the total flow rate on the volume density distribution of the polystyrene particle. For all micromixers, the particle size distribution became narrow and the peak particle size became small of the distribution decreased with an increase in the total flow rate. As shown in Figs.4 (a)(b), the distributions for IMM and YM-1 were mono-modal in all range of the flow rate. On the other hand, the distributions for YM-2 became to be bimodal at low flow rate. This means that the mixing between various sizes of solvent and water segments occurred in YM-2 at low flow rate, as the batch-mixing vessel.

The peak size of the particles increased from 0.14 to 0.54 μm for IMM, from 0.35 to 1.1 μm for YM-1, from 0.93 to 11 μm, with decreasing in the total flow rate. Since the production capacity of YM-1 is more than 50 ton/year, YM-1 mixer can be applied to an efficient and flexible device for the polymer production of various particle sizes but narrow size distribution only by change of the flow rate. These results strongly suggest that the flow rate is a crucial factor to control the particle size and to produce mono-disperse polymer particles by using the micromixer.

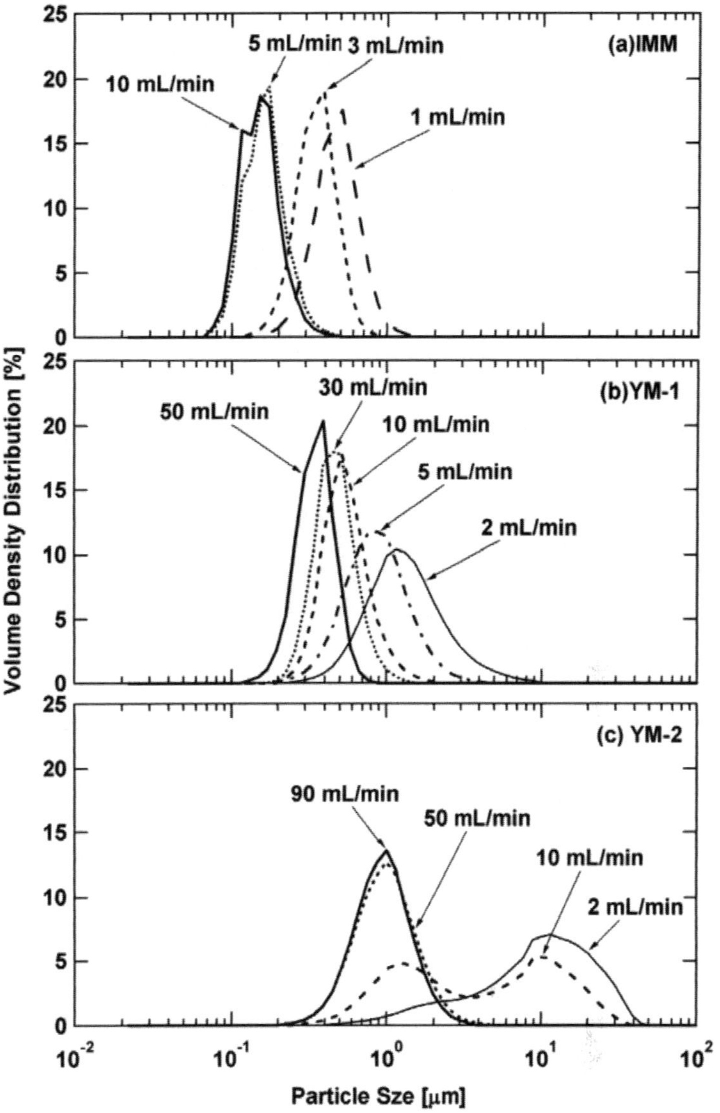

Figure 4. Effect of the total flow rate on the distribution of particle produced by (a)IMM, (b)YM-1 and (c)YM-2

Effect of solvent for the polystyrene solution on the particle size

It was found that the proposed method had a possibility to control the size of polymer particles. To clarify the validity of the method, we examined the effect of solvent type on the particle size. It is well known the rate of immersion precipitation is dependent on the affinity between solvent and nonsolvent. The difference in the solubility parameter is well used as an index of the miscibility between solvents. The solubility parameter of the solvent, δ, was calculated by

$$\delta^2 = \delta_d^{\ 2} + \delta_p^{\ 2} + \delta_h^{\ 2} \qquad (1) \quad \text{(Hansen et al., 1971)}$$

where δ_d, δ_p, and δ_h represent the solubility parameters deriving form dispersion forces, polar forces, and hydrogen bonding, respectively. The miscibility is characterized by the distance

Table 1. Solubility parameter of solvents used

	δ_d	δ_p	δ_h	δ	R_{12}
THF	**16.8**	**5.7**	**8.0**	**19.5**	**36.0**
DMAc	**16.8**	**11.5**	**10.2**	**22.8**	**32.6**
DMF	**17.4**	**13.7**	**11.3**	**24.9**	**31.4**

between two solvents coordinates

$$R_{12} = [4(\delta_{1d} - \delta_{2d})^2 + (\delta_{1p} - \delta_{2p})^2 + (\delta_{1h} - \delta_{2h})^2]^{1/2} \quad (2)$$

The smaller R_{12} value is, the higher the affinity between two solvents is. Table 1 lists the solubility parameter of used solvents and R_{12} value. As shown in Table 1, the affinity with water increased in order of THF, DMAc, DMF.

Figure 5 shows the effect of solvent type for the polystyrene solution on the volume density distribution of the polystyrene particles produced by IMM. As shown in Figs. 5(a) and 4(b), the peak size of the particles decreased with an increase in the total flow rate using THF and DMAc as solvents. This indicates that the particle size decreased with increasing in the total flow rate when the solvent with the low or middle mutual affinity with nonsolvent was used. On the other hand, as shown in Fig. 5(c), the peak of size distribution for DMF did not change by increasing in the total flow rate This means that the particle size did not change, when the solvent with the high affinity with nonsolvent was used. Namely, the immersion precipitation rate was much higher than the mixing rate of the solvent and water. These results means that the size of particle dependent on the flux of solvent into water from the polymer solution and can be controlled when the mixing rate, which is controlled by the segment size of fluid and the flow rate, is higher than the immersion precipitation rate in the micromixer.

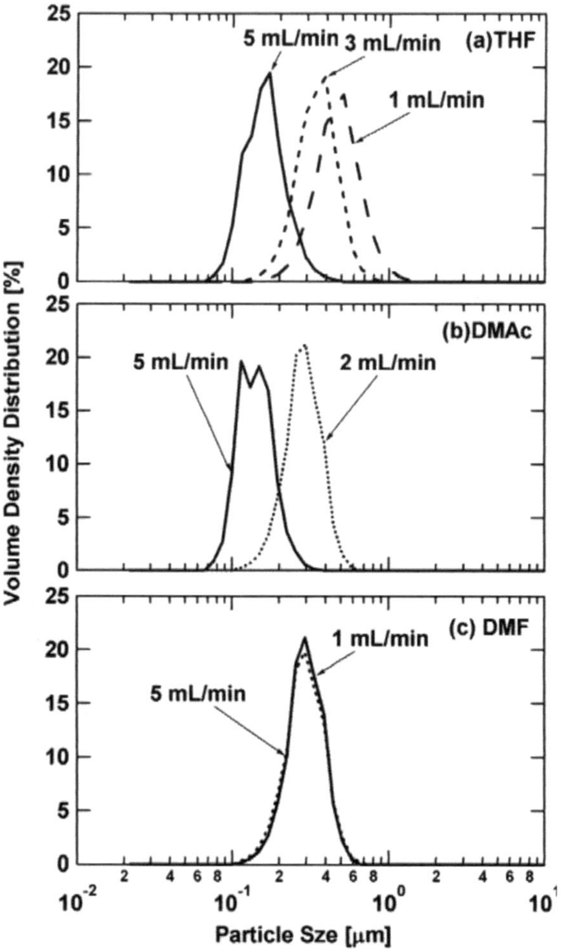

Figure 5. Effect of solvent type for the polystyrene solution on the volume density distribution of the polystyrene particle produced by IMM

Presumed mechanism of the formation of polymer particles by the immersion precipitation method in micromixers

Summarizing the above discussion, the mechanism for the formation of polystyrene particles in the micromixer was presumed. The formation of polystyrene particles by immersion precipitation method consists of two processes: one is the formation of the primary particles via the nucleation, and the other is the growth of the primary particles. The nucleation occurs after the polymer concentration was supersaturated by exchange of the solvent in polymer solution to nonsolvent. Thus, the precipitation was occurred slowly, resulting that the particle size was dependent on the size of the mixing segment when the mutual affinity between solvent and nonsolvent was low. Since the size of the mixing segments of two fluids were determined by the width of the microchannel or the flow conditions, the particle size could be controlled by the type of the micromixer and the total flow rate. On the contrary, the affinity between solvent and water is high, the precipitation was rapidly occurred after the heterogeneous nucleation, resulting that the particle size was independent of the flow condition.

Application for the evaluation method of the mixing performance of micromixers

As mentioned above, it seems that the peak size of the particle reveals the size of the mixing segment in micromixers. In addition, the distribution seems to reveal the uniformity of the segment size. Therefore, we found that the preparation method of polymer particles shown here was available to evaluate the mixing performance of micromixers. Since THF is the lowest affinity with water, the mixing efficiency clearly reflects to the particle size. Figure 6 shows that the effect of the total flow rate on the peak value and the coefficient of variation of the size distribution of particles produced from THF solution. For the single micromixer, IMM, the particle size decrease with the increase in the total flow rate, but the variation coefficient was almost constant. This means that only a possibility to be mixed two fluids increased with the increase of the flow rate keeping a same size of the mixing segment, because IMM has uniform microchannels of single stage. On the other hand, for YM-1 and YM-2, the variation coefficient also decreased with the increase of the flow rate. This was because the mixing occurred based on repeating the split and recombination of fluids in YM-1 and YM-2. (Mae et al., 2003) To clarify these results, we compared the Reynolds number and the peak size of the particle among three micromixers as listed in Table 2. The segment size of the fluid in IMM mixer is several times smaller than that of YM-1 and YM-2. The peak size of the particle was small for the small segment size of fluid at the same Reynolds number. On the other hand, the same peak size of the particle was obtained at the same segment size and Reynolds number between YM-1 and YM-2. From these results, it was found that the desired particle size could be controlled by two factors, the design of the segment size of fluid and a proper Reynolds number. To turn the viewpoint, we can estimate the mixing performance of the mixer in

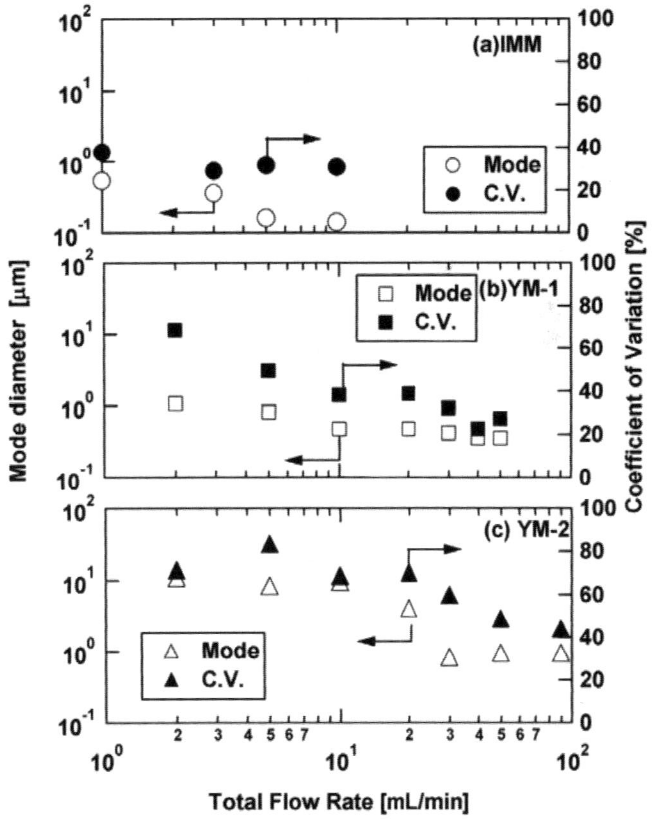

Figure 6. Effect of the total flow rate on the peak size and the coefficient of variation of the distribution of particles produced from THF solution.

accordance with the factors of the segment size of fluid in mixer and Reynolds number by using this precipitation method of polymer. Thus, it was clarified that the proposed method was valid for the evaluation of the mixing performance of micromixers. In addition, this evaluation method is very simple, because its procedure was only that dilute polymer solution and water containing a little surfactant are poured into the micromixer, following the measurement of the size distribution of polystyrene particles, which are stable and do not aggregate in the suspension.

Table 2. Comparison of Reynolds number and the peak size of the particle among three micromixers

IMM	Re_{max} [-]	210	420	1000	2100	
	D_p [μm]	0.54	0.36	0.16	0.14	
YM-1	Re_{max} [-]	61	140	280	840	1400
	D_p [μm]	1.1	0.82	0.48	0.52	0.34
YM-2	Re_{max} [-]	9	20	200	360	
	D_p [μm]	11	9.4	0.94	0.89	

D_p : Peak size of the particle

Conclusions

We tried to prepare polystyrene particles successfully by immersion precipitation in several micromixers. Using all microdevices, polystyrene particles were formed without clogging of the microchannel. The particle size and its distribution were varied with the microchannel size of the micromixer, the solvent type of polystyrene solution, the total flow rate of fluid. We successfully controlled the particles size from 150 nm to 10 μm by changing micromixer and the total flow rate when the affinity between the solvent of polystyrene solution and water as a nonsolvent. It can be expected that this method of polymer particles will be used for the evaluation of the mixing performance of micromixers.

Acknowledge

The research was conducted in the aid of the Micro Chemical Plant Technology Union (MCPT) for the Project of Micro-Chemical Technology for Production, Analysis, and Measurement Systems financially supported by NEDO. We appreciate to MCPT and NEDO for their assistance.

Reference

Bayer, T., D. Pysall and O. Washen, "Micro mixing effects in continuous radical polimerization," *Proc. of IMRET 3*, 165 (1999)

Edel, J.B., R. Fortt, J.C. deMello, "Microfluidic routes to the controlled production of nanoparticles," *Chem. Commun.*, 1136-1137(2002)

Ehrfeld, W., K. Golbig V. Hessel., H. Lowe and T. Richter, "Characterization of mixing in micromixer by a test reaction: single mixing units and mixer arroys," *Ind. Eng. Chem. Res.*, 38, 1075-1082 (1999)

Hansen, C. and A. Beebpwer; Solubility parameters in Kirk-Othmer, Encyclopedia of Chemical Technology, 2nd Ed., pp. 889, Interscience, New York (1971)

Hessel, V., S. Hardt, H. Lowe, F. Shonfeld, "Laminar mixing in different interdigital micromixers," *AIChE J.*, 49, 566-577 (2003)

Hirai, T., H. Sato, I. Komasawa, Mechanism of formation of titanium dioxide ultrafine particles in reverse micelles by hydrolysis of titanium tetrabutoxide," *Ind. Eng. Chem. Res.* , 32, 3014-3019 (1993)

Kurummradt, H., U. Kopp and J. Stoldt, "Experiences with use of microreactor in organic synthesis," *Proc. of IMRET 3*, 181 (1999)

Lenggoro, I.W., K. Okuyama, J. Fernández, "Preparation of ZnS nanoparticles by electrospray pyrolysis," *J. Aerosol Sci.*, 31, 121-136(2000)

Lowe, H., W. Ehrfeld, V. Hessel, T. Richter and J. Schiewe, "Micromixing technology," *Proc. of IMRET 4*,5-9 (2000)

Mae, K., T. Maki, I. Hasegawa, U. Eto, Y. Mizutani and N. Honda "Development of Micromixer for Rapid Production of Soap Free Emulsion," *Proc. of IMRET 7*, 120-122(2003)

Mae, K., T. Maki, I. Hasegawa, U. Eto, Y. Mizutani and N. Honda "Development of a new micromixer based on split/recombination for mass production and its application to soap free emulsifier," *Chem. Eng. J.*, 101, 31-38(2004)

Takagi, M., T. Maki and K. Mae, "A New Micro Device Assembled with Same Axle Dual Pipe for Production of Various Fine," *Proc. of IMRET 7*, 266-268(2003)

Takagi, M., T. Maki, M. Miyahara, K. Mae, "Production of titania particles by using a new microreactor assembled with same axle dual pipe," *Chem. Eng. J.*, 101, 269-276(2004)

Motte, L., C. Petit, L. Boulanger, "Synthesis of cadmium sulfide in situ in cadmium bis(ethyl-2-hexyl) sulfosuccinate reverse micelle: polydispersity and photochemical reaction," *Langmuir*, 8 1049-1053(1992)

Murray, C.B., D.J. Norris, M.G. Bawendi, "Synthesis and characterization of nearly monodisperse CdE (E = S, Se, Te) semiconductor nanocrystallites," *J. Am. Chem. Soc.*, 115, 8706-8715(1993)

Wagner, J., T. Kirner, G. Mayer, J. Albert, J. M. Kohler, "Generation of metal Nanoparticles in a microchannel reactor," *Chem. Eng. J.*, 101, 251-260(2004)

137g A Microfabricated Segmented-Flow Reactor for the Synthesis of CdSe Quantum Dots

Axel Guenther, Brian K H Yen, Klavs F Jensen, Martin A Schmidt, Moungi G Bawendi

We present CdSe quantum dot (QD) synthesis on a chip using multiple temperature zones and taking advantage of the mixing and residence time characteristics associated with segmented gas-liquid flows. QDs have been extensively studied for use in a variety of applications including fluorescent labels, electroluminescent devices, and lasers. QDs (2-10 nm diameter) are typically prepared in a batch process where the precursors are rapidly injected into a heated flask containing a mixture of solvents and coordinating ligands. The optical properties and average size of QDs synthesized in the batch process can depend strongly on factors which are difficult to control such as injection process, local temperature and concentration fluctuations, rate of stirring, and rate of cooling. Single-phase flow reactors in which precursor solutions are delivered into a heated reaction section were the first continuous processes to address some of these issues. However, a single-phase flow reactor has limitations in terms of the residence time distribution (RTD) and slow mixing by diffusion. We have therefore realized a gas-liquid segmented flow reactor which allows for rapid and controlled mixing of precursors as well as narrow RTDs. In contrast to the case of single-phase flow, recirculation within the liquid segments provides a mechanism to exchange fluid elements located near the channel walls with those in the center. This recirculation has the dual of effect of narrowing the RTD and improving mixing of the reagents - factors which have a strong influence on the ultimate size and size distribution of QDs produced.

The silicon reactor accommodates a 1 m long reaction channel (hydraulic diameter ~400um) and two shallow side channels for collecting reaction aliquots. The entire device is protected through a thermal oxide, and then capped and sealed with an anodically bonded pyrex wafer. Two temperature zones are maintained on the reactor chip, a heated reaction region (>260^oC) and a quenching region (~60^oC). The Cd and Se precursor solutions are delivered separately at room temperature into the heated chip. An inert gas stream is introduced further downstream to form a segmented gas-liquid flow, thereby rapidly heating and mixing the precursors and initiating the reaction. The reaction is stopped when the fluids enter the cooled outlet region of the device. The rapid mixing of precursors and narrow RTDs realized by the design make it possible to prepare high quality samples. Under conditions for a typical synthesis the gas and liquid segments are very uniform, and the resulting QD samples possess narrow size distributions as indicated by absorbance and photoluminescence (PL) spectra. For instance, the band-edge and several higher transitions are clearly resolved in the absorbance spectrum, and PL line widths are narrow (FWHM <30 nm).

Progress Towards Chip-Based High-Throughput Dendrimer Synthesis

C.-H. Chang+, S.-H. Liu+, Y. Tennico, J.T. Rundel*, V. T. Remcho*, E. Blackwell#, T. Tseng# and B.K. Paul#*

+Department of Chemical Engineering
**Department of Chemistry*
#Department of Industrial and Manufacturing Engineering
Oregon State University, Corvallis, Oregon 97331 USA

Abstract

A critical barrier to the routine use of nanomaterials is the tedious, expensive means of their synthesis. It is expected that the field of nanoproduction can be advanced using the improved process control made available within highly-parallel, process-intensified microsystems. Microsystems technology has the capacity to transform current batch nanoproduction practices into continuous processes with rapid, uniform mixing and precise temperature control. Advantages of microsystems technology include eliminating air contact thereby minimizing contamination and improving yield; minimizing the environmental impact of nanoproduction including solvent free mixing, integrated separation techniques and reagent recycling; and the possibility of synthesizing nanomaterials in the required volumes at the point of use eliminating the need to store and transport potentially hazardous materials. A key requirement of this microsystems technology is the need for highly-parallel, high-yield reaction technology and high-throughput extraction technology to minimize excess reagent and defective product to further improve yields in downstream reactions. In this paper progress is reported for continuous mixing, reaction and extraction of PAMAM dendrimer products for implementation in microsystems. In particular we explore the effects of interdigital micromixing on the PAMAM reaction. In addition, we explore microextraction methods via the surface immobilization of monolithic sorbents in microfabricated fluidic pathways. Progress toward the implementation of the microsystem within a composite polymer/elastomer microlamination architecture capable of pneumatic valve integration is reported. The fabrication architecture bears the added advantage of providing an economical pathway to "numbering up" for high-throughput production.

Introduction

Dendritic macromolecules, or dendrimers, are highly-branched, nanometer-sized molecules with fascinatingly symmetrical fractal morphologies (Figure 1). Dendrimers consist of a core-unit, branching units, and end groups located on their peripheries. Their dendritic architecture presents great potential for a wide variety of applications including drug delivery, MRI imaging[i,ii,iii], gene transfer techniques, antimicrobial agents[iv], environmental clean-up and DNA microarray analysis[v] among others.

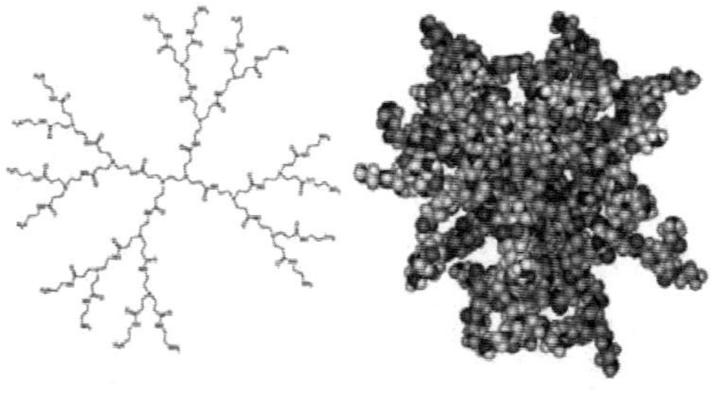

Figure 1. Generalized structure of a dendrimer showing its branched architecture and 3-D space-filling model of a dendrimer.

In many of these applications, limitations for deployment of these technologies are the production costs for realizing dendrimers. This paper will report recent findings for a microreactor-based dendrimer production methods aimed to address this problem. In particular, this paper reports progress in the synthesis, separations, valving and detection necessary to implement this system.

Progress in High-rate Synthesis of Lower Generation Dendrimers

To explore the potential benefits of a microreactor for dendrimer synthesis, we have investigated EDA-core PAMAM synthesis using a continuous flow microreactor. Being the first reported dendrimer, polyamidoamine (PAMAM) is probably the most studied dendrimer today. In 1985 and 1986, Tomalia et al.[vi] described the preparation of PAMAMs by the divergent approach. The synthesis involves *in situ* branch cell construction in step-wise, iterative stages around a desired core (e.g. ammonia or ethylenediamine (EDA)) to produce defined core-shell structures. Take the synthesis of EDA cored PAMAM as an example, the typical synthesis takes 24 hr for half generations and 96 hr for full generations to complete the reaction.[vii] The half generations mean that the terminal functionality is ester and the full generations with terminal functionality of amine. Fig. 2 shows the structure of a generation-2 of EDA–cored PAMAM.

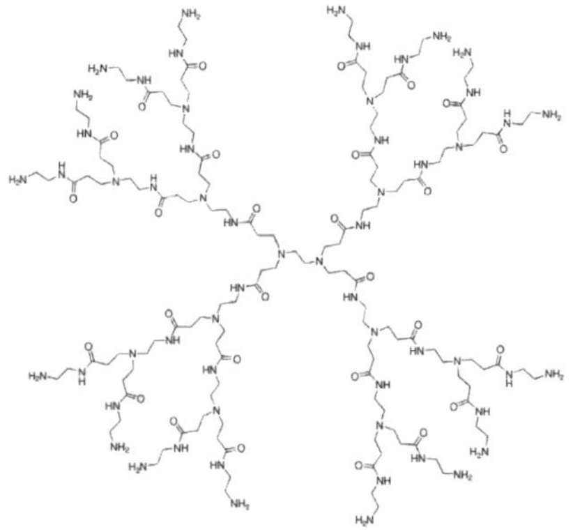

Fig.2. Structure of generation 2 of EDA-cored PAMAM

Microreactors offer several new opportunities to suppress the competing side reactions and maximize the purity for dendrimer synthesis. These include uniform and precise temperature control and low moisture permeability to avoid water content. Most importantly, the key is to increase the conversion rate of the alkylation/amidation reaction sequence through enhancement of effective mixing between reactants. It is extremely difficult to achieve uniform mixing at the micrometer scale in a short time by a traditional mixing apparatus. Micromixers offer features which cannot be easily achieved by macroscopic devices.

A schematic diagram of the microreactor is given in Fig. 3.a. The reactants were delivered to an interdigital micromixer (Figure 3.b) through two syringe pumps. The interdigital micromixer (SSIMM from Institut für Mikrotechnik Mainz, Germany) consists of an array of interdigital microchannels patterned in the center of a channel on a substrate made of thermally grown silicon dioxide. The mixing element is housed within a stainless steel container.

A methanol solution of precursor and a methanol solution of reagent, either EDA (for synthesis of full generations) or methyl acrylate (for synthesis of half generations), are fed into the mixing element through the two micromixer's inlets respectively by means of syringe pumps at room temperature. Once the solution streams are introduced into the micromixer, each stream is divided by the micromixer into many thin substreams, which leave the channels perpendicular to the direction of the feed flows yielding a mixed solution at the outlet of the micromixer. The mixed solution passes through the outlet of the micromixer and the tube connected with the outlet with an estimate mean residence time of 1.08 seconds. Consequently, the adduct solution is collected, solvent is removed by a rotary evaporator, and the trace residue of labile reactants is further removed by a vacuum pump (0.1mm Hg, 40°C).

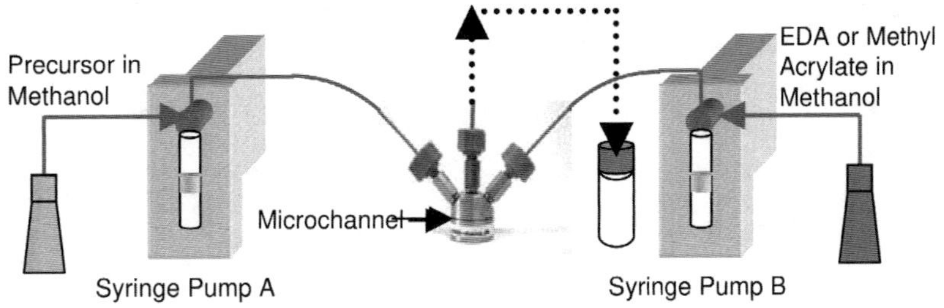

a

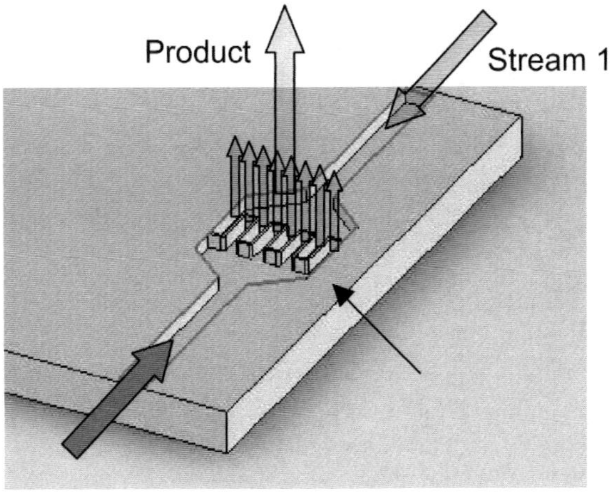

b

Fig. 3. a. A schematic diagram of the continuous flow microreactor;

b. Concept of an interdigital micromixer.

The first two reactions of EDA-cored PAMAM were conducted using this continuous flow microreactor. Starting from the synthesis of the first generation G-0.5 from EDA and methyl acrylate in a solution of methanol, G-0.5 was prepared with 99% yield without observing any side products according to NMR and mass spectrometry. Sequentially, generation G0.0 was synthesized from the starting materials of G-0.5 and an excess amount of EDA with a yield of 98% and without observing any side product. Figures 4.a and 4.b show the NMR spectra of generation G-0.5 synthesized by flask and the microreactor, respectively. The NMR results indicated no side products were produced from the microreactor. On the other hand, the NMR spectrum from the batch reactor shows some side products (shoulders appeared at peak 3.67ppm in Fig. 3.a) even after 3 days of vigorous mixing. The side reaction could be caused by intra-dendrimeric cyclization. NMR spectra were recorded in deuterochloroform, with a Brook nuclear magnetic resonance spectrometer (300MHz).

This microreactor demonstrated several advantages over the conventional batch process. The conventional reaction requires cooling and stirring system, dropwise addition of reagents at the beginning to conduct the released heat and inert gas protection; while all reactions conducted using micromixer were continuous flow at ambient temperature without the need of inert atmosphere. In addition, several experiments were conducted using the micromixer and the results showed good reproducibility. The most attractive advantage is that the residence time for the micromixer approach is 1 second versus 3 days (for half generation) and 96 hours (for full generation) in a conventional batch reaction. In addition, by achieving the high purity of the low generations PAMAM, further syntheses of higher generations are made easier.

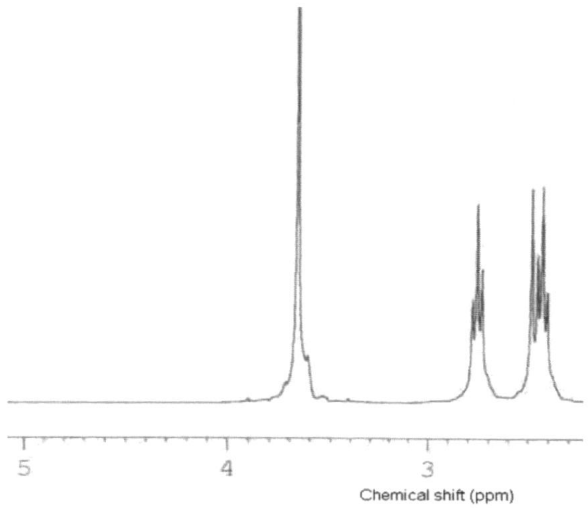

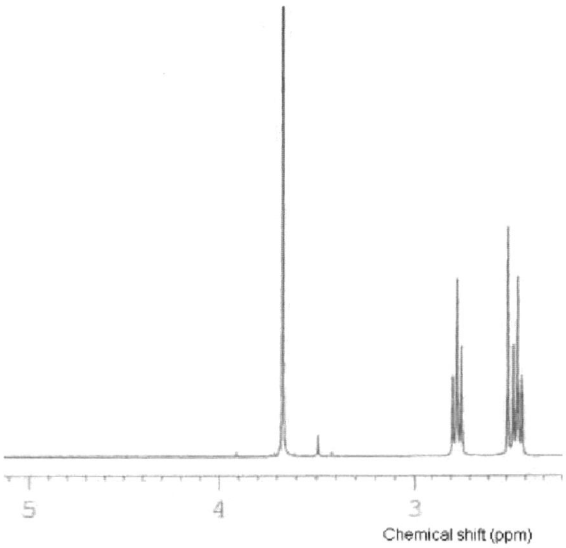

a. b.

Fig. 3. NMR spectrum of PAMAM G-0.5 a. synthesized by flask for 3 days, and b synthesized by the microreactor.

On-line Analysis of Microreactor-Produced Dendrimers

The function of the on-line analysis subsystem is to separate the dendrimer from structurally similar side products and residual reactants. This process will facilitate collection of the dendrimer as the desired product, elimination of side products from the reaction stream, and recycling of reactants. The process will be conducted in a continuous or quasi-continuous operation depending on the chosen method of separation.

For initial exploration of separation methods, microfluidic channels have been realized by hot embossing poly(carbonate) (PC) using an embossing tool produced by Ni-electroforming on top of a stainless steel (SS) substrate. To produce the tool, a film (~50 μm) of the negative photoresist SU-8 is spun onto a polished SS substrate, crosslinked by UV-exposure through a transparency photomask and developed. Ni features are electroformed into the thick resist mold and the SU-8 is ashed away. The resulting SS-Ni master is used to hot emboss the microfluidic features into a PC lamina.

The final form of the chip consists of three distinct layers: a PC substrate containing the microfluidic features; a silicon dioxide, SiO_2, layer to enable bonding and to functionalize the microfluidic channel; and a poly(dimethylsiloxane) (PDMS) slab to cap the microfluidic channels. The PC substrate, as fabricated in previous steps, is sputtered with a thin film (~500 nm) of SiO_2. A PDMS slab (thickness ~1 mm) is cut to size and punched with holes to form fluidic reservoirs for the microfluidic channels. The SiO_2 and PDMS surfaces are then irreversibly bonded after exposure to an oxygen plasma and direct contact for several hours.

Two methods are being developed off-chip in silica capillaries for the separation of the dendrimer from the side products and residual reactants. The chosen method will be incorporated into the separation channel in the microfluidic chip. A capillary zone electrophoresis, CZE, method (typical results shown in Figure 4) would simply utilize the open SiO_2 channel.[viii] A capillary electrochromatography, CEC, method (typical results shown in Figure 5) would rely on a porous monolithic polymer anchored to the SiO_2 walls and filling the channel.[ix] The CZE method could possibly allow the system to operate in a continuous mode. The CEC method would likely be used in a pulsed or quasi-continuous mode.

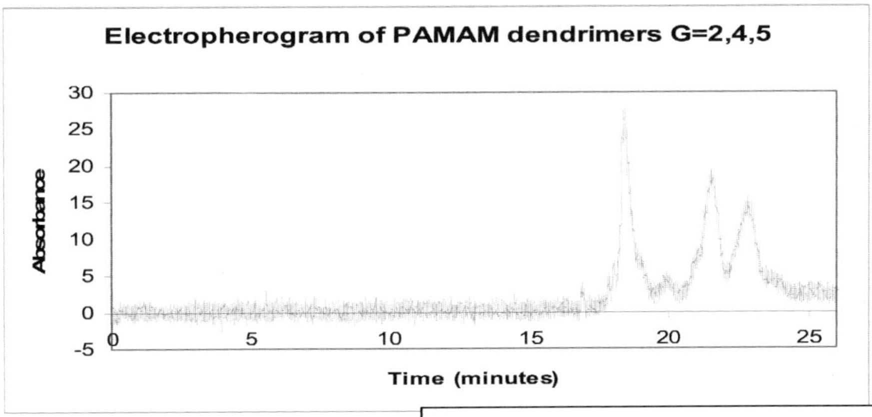

Figure 4. CZE separations of various generations of EDA-core PAMAM (Polyamidoamine) dendrimers (generation 2, 4, and 5) in phosphate buffer solution.

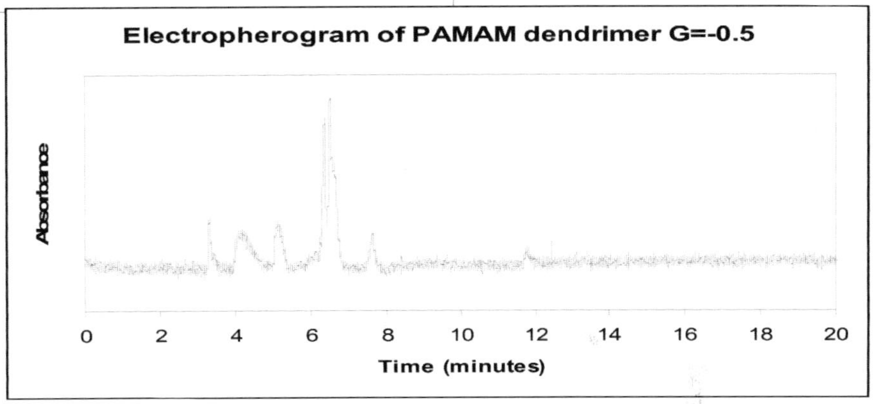

Figure 5. Separations of dendrimer generation -0.5 from side products and residual reactants using dendrimer-templated monolithic sorbents

With a CEC method the eluted dendrimer and precursors will be detected in-channel following separation using indirect laser-induced fluorescence. This method relies on an eluent solution matrix containing a uniform concentration of a background fluorophore. The eluate, the dendrimer or precursors, will cause a decrease in concentration of this background fluorophore resulting in a decrease of detected signal. This decrease in signal can provide quantitative analyis of concentration of the eluate or simply serve as a photogate for valve timing. With a CZE method the same detection method would be used to simply indicate the presence or absence of a separation stream.

The detection system is outlined in Figure 6. The laser diode will be end-coupled with a waveguide at the edge of the assembled chip. The photomultiplier tube (PMT) will be place directly below the detection window in the microfluidic channel.

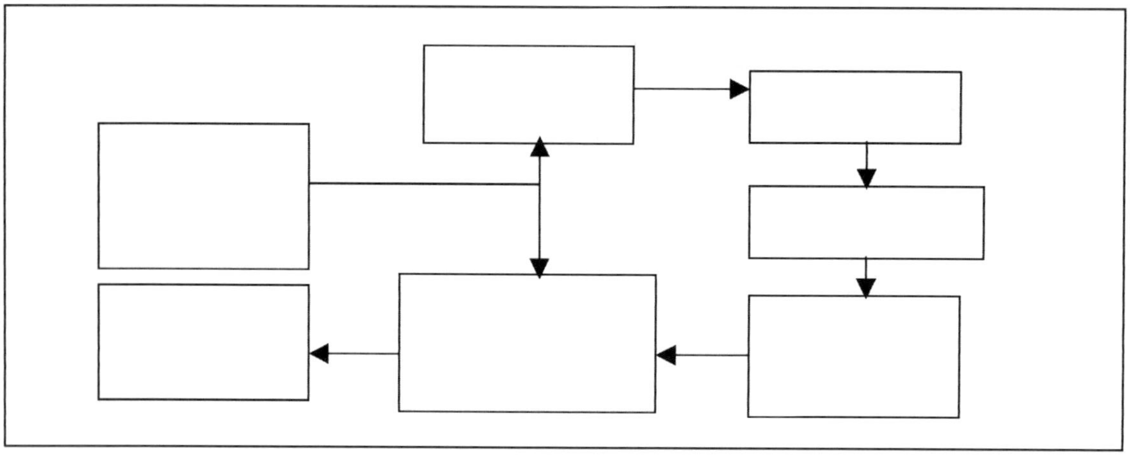

Figure 6. System diagram for a chip-based indirect laser-induced fluorescence detection apparatus.

Valve Integration and Packaging

A unique architecture and packaging method has been developed involving the ultrasonic welding (USW) of PDMS membranes between polycarbonate channel substrates. Advantages include the ability to integrate microfluidic valves in an inexpensive microfluidic device. This architecture will provide for greater 3D spatial intensification of the microsystem by using high modulus films for the fluid channels instead of previous work that used bulk low modulus PDMS for both fluid channels and valve membranes.[x] Prior work performed with silicon and glass devices to integrate elastomer membranes for greater 3D intensification do not take advantage of the cost effective nature of polymers.[xi,xii] This 3D intensification will become more important in numbering up the microreactor using microchannel arrays.

The architecture involves embossing both sunken microchannel and raised energy director features into PC. A PDMS membrane is put between two PC laminae. Holes and slots within the PDMS layer permit contact of energy directors through the membrane with the underlying PC lamina (see cross-section in Figure 7a). Layers are pressed together and welded thereby compressing the elastomeric layer forming a conformal seal between PC laminae (see cross-section in Figure 7b). Figure 8 shows a cross section of 100 micron wide x 50 micron deep channels sealed by a 270 micron thick PDMS layer that has been compressed 43 microns by ultrasonic welding. Future efforts will work to implement the microreactor and analysis platforms within this architecture for device integration and packaging.

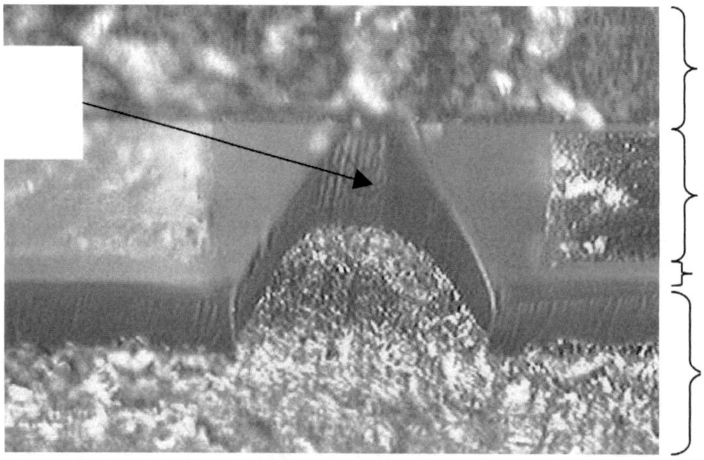

a.

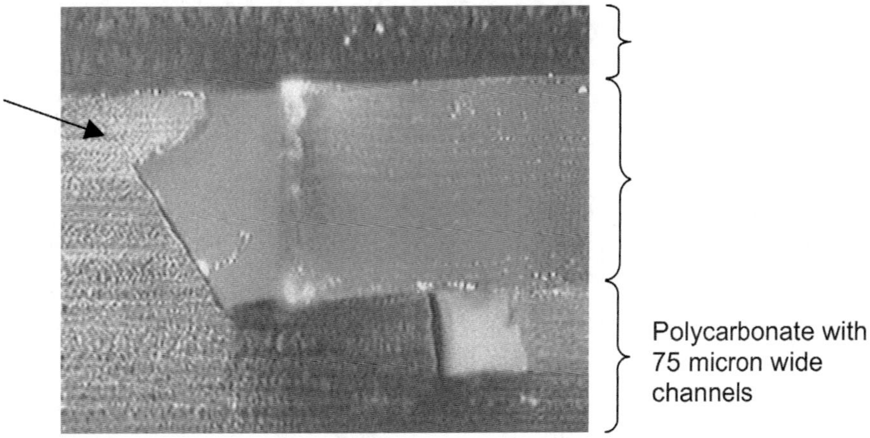

b.

Figure 7. a. Test device before ultrasonic welding; b. ultrasonically welded test device with PDMS layer compressed

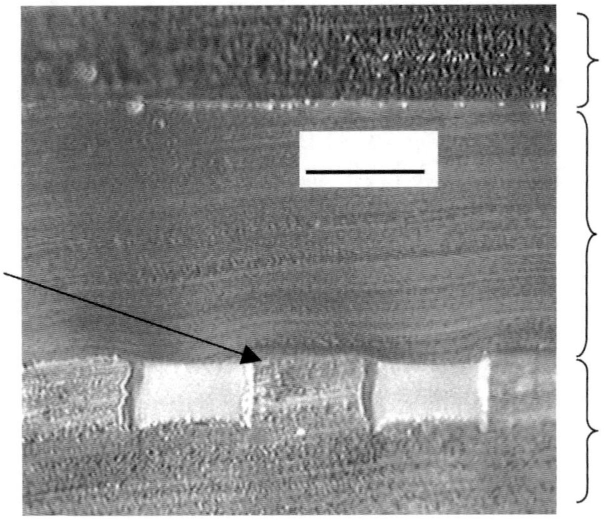

Figure 8. Array of microchannels sealed by compressed PDMS

Summary

Based on the successful syntheses of the first two generations of EDA-cored PAMAM dendrimer using an interdigital micromixer, the synthesis of higher generations are underway. Attributes of this synthesis are fast continuous flow, high yield, high selectivity and most importantly, the ability to number-up for industrial production scale. Progress has also been made toward the development of a microfluidic system for separating and differentiating between the dendrimer products, side products and residual reactants. Finally, a platform for packaging these technologies together with integrated microfluidic valves has been conceived and validated. Future work will emphasize integration of these technologies into a single microsystem for producing dendrimers.

Acknowledgements

This work is sponsored by the W. M. Keck Foundation.

References

i. Wu, C.; Breshbiel, M. W. : Kozack, R. W.; Gansow, O. A. Biorganic and Medicinal Chemistry letters, 1994, 4(3), 449.
ii. Margerum, L. D.; Campion, B. K.; Koo, M.; Shargill, N.; Lai, J; Marumoto, A.; Sontum, P. C. J. Alloys Comp. 1997, 249 (1-2), 185.
iii. Kim, Yoonkyung; Zimmerman, Steven C. Curr.Opin.Chem. Biol. 1998, 2(6), 733-742.
iv. Balogh, L.; Bielinska, A.; Eichman, J.D.; Valluzzi, R. and others. Dendrimer Nanocomposites in Medicine. The University of Michigan Center for Biologic Nanotechnology, Tuff Biotechnology Center and Department of Radiation Oncology at The University of Michigan.
v. Le Berre, Nucleic Acids Res. 2003, 31 (16) 88.
vi. D.A. Tomalia, H. Baker, J. Dewald M. Hall, G. Kallos, S. Martin, J. Roeck, J. Ryder, and P. Smith, *Macromolecules,* 19(9), 2466-2468, 1986.
vii. J. M. J. Frechet, D.A. Tomalia, Dendrimers and other dendritic polymers. John Wiley & Sons Ltd. (2001).
viii. A. Ebber, M. Vaher, J. Peterson, M. Lopp. J. Chromatogr. A, **2002**, 949, 351–358.
ix . A. Doneanu, G.S.Chirica, V.T. Remcho. J. Sep. Sci. 2002, 25, 10-16
x. Unger, M. A., H.P. Chou, T. Thorsen, A. Scherer, S.R. Quake, “Monolithic Microfabricated Valves and Pumps by Multilayer Soft Lithography.” Science, v 288, April 7, 2000, pp. 113-116
xi. Baechi, D., J. Dual, “A High Density Microchannel Network with Integrated Valves and Photodiodes.” *Sensors and Actuators A,* v 95, 2002, pp 77-83.
xii. Grover, W., A.M. Skelly, C.N. Liu, E.T. Lagally, R.A. Mathies, “Monolithic membrane valves and diaphragm pumps for practical large scale integration into glass microfluidic devices.” *Sensors and Actuators B,* v 89, 2003, pp 315-323.

138a Solids Technology using Microstructured Devices

Dirk Kirschneck, Maria Kober, Rolf Marr, Walter Linhart

The aim of solids technology is the production of disperse solid products. Disperse substances require to have specific characteristics in terms of production and application. The handling and production of solid substances is a main task in chemical industry because a vast majority of products are solid substances or at least pass through a disperse state in the course of their production.

The application of microstructured devices offers new opportunities for the production of solid chemicals (Khan et al. 2003, Donnet et al. 2002). Microstructured devices usually contain structures below 1000 microns. Their main advantage is their high surface to volume ratio which leeds to a large specific surface area (Ehrfeld et al. 2000). In terms of solids technology microstructured devices can be used to achieve specific application-related product properties. However, the handling of solid substances in microstructured devices is always challenging.

In the current paper the usability of different micromixing concepts for the industrial production of well-defined particles in the submicron area will be discussed. The micro mixers examined in this study were the Starlaminator and the Caterpillar Micro Mixer which were all designed by IMM (Institut fuer Mikrotechnik Mainz). The Starlaminator is based on the micromixing concept of flow multilamination whereas the Caterpillar Micro Mixer uses a concept known in literature as 'split and recombine'. The solid products produced in the micro mixers were compared to those obtained from conventional stirred batch reactors. Special attention was given to the influence of surface-active agents on the precipitation. Product properties, i.e., particle size and particle size distribution and process conditions were observed in order to characterize the precipitation process. Particle size analysis was performed using laser diffractometry.

Khan, S. A.; Jensen, K. F.; Microchemical systems for colloidal materials synthesis and separation, Proc. Int. Conf. On Microreaction Technol., IMRET 7, Lausanne, p. 62, Sept. 7-10, 2003.

Donnet, M.; Bowen, P.; Jongen, N.; Lemaitre, J.; Hofmann, H.; Schreiner, A.; Jones, A. G.; Schenk, R.; Hofmann, C.; De Carlo, S.; Successful scale-up from millilitre batch optimisation to a small scale continuous production using the Segmented Flow Tubular Reactor. Example of calcium carbonate precipitaion. Chemical Engineering Transactions,3, 1353-1358, 2002.

Ehrfeld, W.; Hessel, V.; Loewe, H.; Microreactors - New Technology for Modern Chemistry, Wiley-VCH, Weinheim, 2000.

138b Living Cationic Polymerization Based on "Cation Pool" Method Using Microsystem

Aiichiro Nagaki, Jun-ichi Yoshida, Kohsuke Kawamura, Mitsuo Sawamoto, Seiji Suga, Tsuyoshi Ando

The cation pool of an N-acyliminium ion was found to serve as an effective initiator of cationic polymerization of vinyl ethers in a microsystem consisting of two micromixers and a micro tube reactor. The polymerization led to very narrow molecular weight distribution (Mw/Mn = 1.14). The molecular weight (Mn) increased linearly with an increase of the amount of the monomer. The carbocationic polymer end was effectively trapped by allyltrimethylsilane. The present observations illustrate the potential of microsystems, in conjunction with the cation pool, to effect polymerization in a highly controlled manner without the deceleration inherent in the dynamic equilibrium between active and dormant species.

138c Swern Oxidation Reactions Using a Microreactor

Hiroyuki Miyata, Jun-ichi Yoshida, Kazuhiro Mae, Kikuo Ataka, Tatsuya Kawaguchi

Microreactors (micro-structured chemical reactors) exhibit numerous practical advantages including safety, easy modulation, and numbering-up for industrial production when compared with conventional macro-scale batch reactors. It is also advantageous that highly exothermic reactions can be conducted on the basis of efficient mass and heat transfer. It should be noteworthy that microreactors also enable precise control of reactive intermediates to facilitate highly selective reactions that are difficult to be achieved by conventional reactors. In this paper we wish to report that the Swern oxidation can be accomplished using a microreactor at temperatures (-20 ^oC ~ 20 ^oC) much higher than those for conventional macro batch reactors (-50 ^oC or below).

We examined the Swern oxidation using a micro flow system consisting of multilamination type micromixers. Oxidation of various alcohols took place smoothly to give the corresponding carbonyl compounds in good yields and selectivities at -20 ^oC ~ 20 ^oC. A dramatic effect of the micro flow system seems to be attributed to precise temperature control, extremely fast and efficient mixing, and optimum residence time.

The present observations demonstrate a striking example of the effectiveness of micro flow system for the reaction involving highly unstable intermediates.

Kinetic Investigations on Selective Acetylation of Nucleophilic Groups of Cysteine in a Micromixer-Microchannel Reactor Module

[1]P. Pfeifer, [1]K. Schubert, [2]M.A. Liauw, [3]J. Herzog, [3]G. Emig,[4]T. Messerschmitt, [4]A. Aigner

[1]Institute for Micro Process Engineering (IMVT), Forschungszentrum Karlsruhe GmbH, Hermann-von-Helmholtz Platz 1, 76344 Eggenstein-Leopoldshafen, Germany, www.fzk.de/imvt-en

[2]ITMC, RWTH Aachen, Worringer Weg 1, 52074 Aachen, Germany

[3]LTC1, University Erlangen-Nürnberg, Egerlandstr. 3, 91058 Erlangen, Germany

[4]Noveon Pharma GmbH & Co. KG, Rosenheimer Str. 43, 83064 Raubling, Germany

Prepared for Presentation at 2005 Spring National Meeting / IMRET 8 / April 10-14 / Single Reactors

Unpublished

Introduction

Applying micro reaction technology instead of stirred tank reactors for fast liquid-liquid reactions is a promising approach in process intensification. Diazotation and acetylation reactions are well known examples for fast reactions. During batch processing mixing and heat exchange are crucial parameters for yield, reaction time and selectivities because acetylation of amino- and thiol- groups can be controlled thermodynamically or kinetically, respectively.

In the present study, the selective acetylation of cysteine was carried out in a micromixer-microchannel reactor module. Conventionally, the process is carried out in three steps. The first process step is the neutralisation of the cysteine chloride with soda lye (equation 1) yielding -59 kJ/mol.

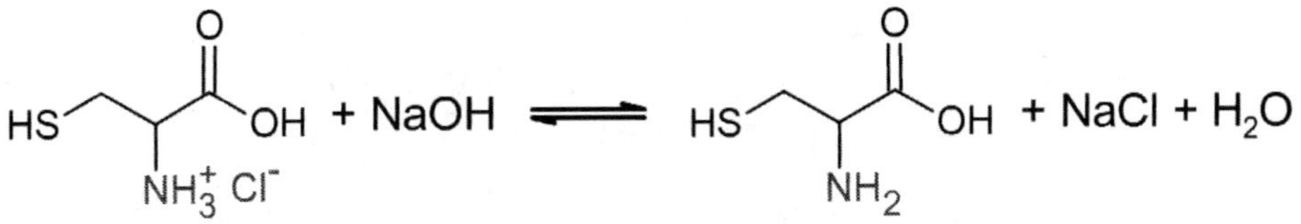

(1)

Cysteine Chloride Cysteine (↓)

The second process step, the reaction of cysteine with acetic anhydride (equation 2), is also exothermic (-172 kJ/mol). To compensate for acetic acid formed during reaction, additional soda lye has to be added.

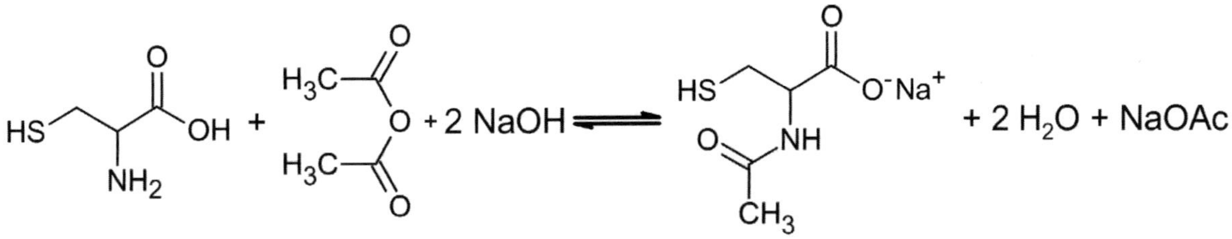

(2)

Anhydride (Ac_2O)

The third step is the addition of hydrochloric acid to precipitate the desired product N-acetyl-L-cysteine (NAC) (equation 3).

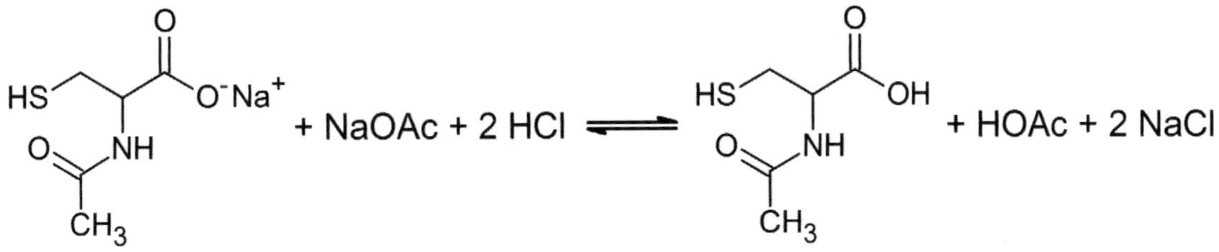

(3)

NAC (↓)

In order to avoid the formation of the suspension in the first step, i.e. precipitation of free amino acid at the isoelectric point, the first two steps were combined for the micro reactor study. If the third step was carried out, it was performed with the collected product afterwards. However, most of the samples were already analysed in the liquid state at a pH = 1.8 - 3.

Experimental

The overall flow in the micromixer-microchannel module was 0.8 kg/h. The three different inlet flows into the module were 33% NaOH, pure anhydride and 3.05 mol/l aqueous solution of cysteine. Temperatures in the reactor were set to 35, 50 and 80°C, respectively. The variation in molar ratios was NaOH/cysteine = 2 – 3 and anhydride/cysteine = 0.52 – 1.05.

An exchangeable three-component cyclone mixer and reaction channels were integrated in the micro reactor module. The reactor was made out of Hastelloy (2.4602) for corrosion reasons. The design of the mixer was varied during the study (three mixer types).

The residence time in the mixer was lower than that in the reaction zone (3.1 – 17 s) by orders of magnitude to ensure low conversion in the mixing chamber. The reaction zone itself consisted of up to 11 channels and was divided into three different cooling/heating zones which allowed to examine temperature profiles. The scheme and a photo of the reactor module are shown in Figure 1.

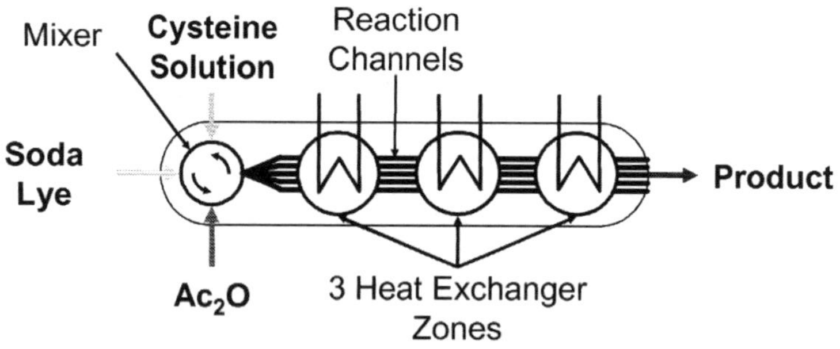

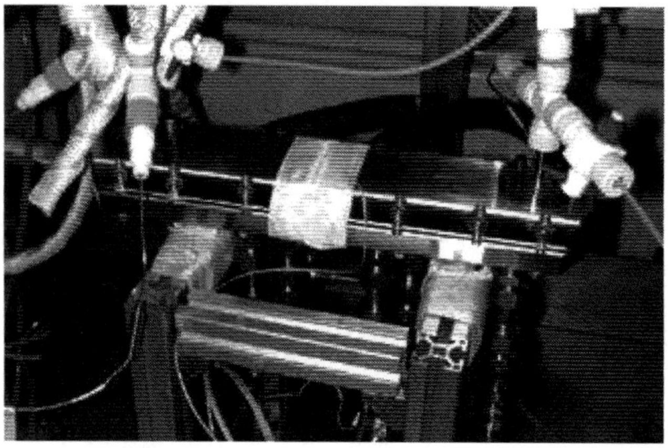

Figure 1: Scheme and photograph of the reactor module for the acetylation of cysteine.

In this study chemical analysis was done via potentiometric titration, i.e. determination of SH-groups with $HgCl_2$, and high pressure liquid chromatography (HPLC). The product concentration was calculated by subtracting the total amount of byproduct and cysteine from the total amount of SH-components.

Results and Discussion

Influences of mixing and reaction parameters were investigated and considered under the aspect of different concentration profiles and temperatures in batch and micro reactors.

The major aspect is that the concentrations of soda lye and anhydride are extremely high at the reactor entrance. While the pH value is nearly constant in a semi-batch process for reaction step 2, the pH value is dependent on the mixing quality, i.e. has a spatial dependence and can be lower or higher than for an optimum ratio of nucleophilic amino groups for acetylation. This and other pH effects on the results can be highlighted by the scheme in Figure 2 showing all byproducts as well as main and byproduct routes.

The second aspect, the temperature, which is preferably elevated in the micro reactor to enhance kinetics to obtain short residence times and small reactor volumes, is also discussed with respect to the scheme.

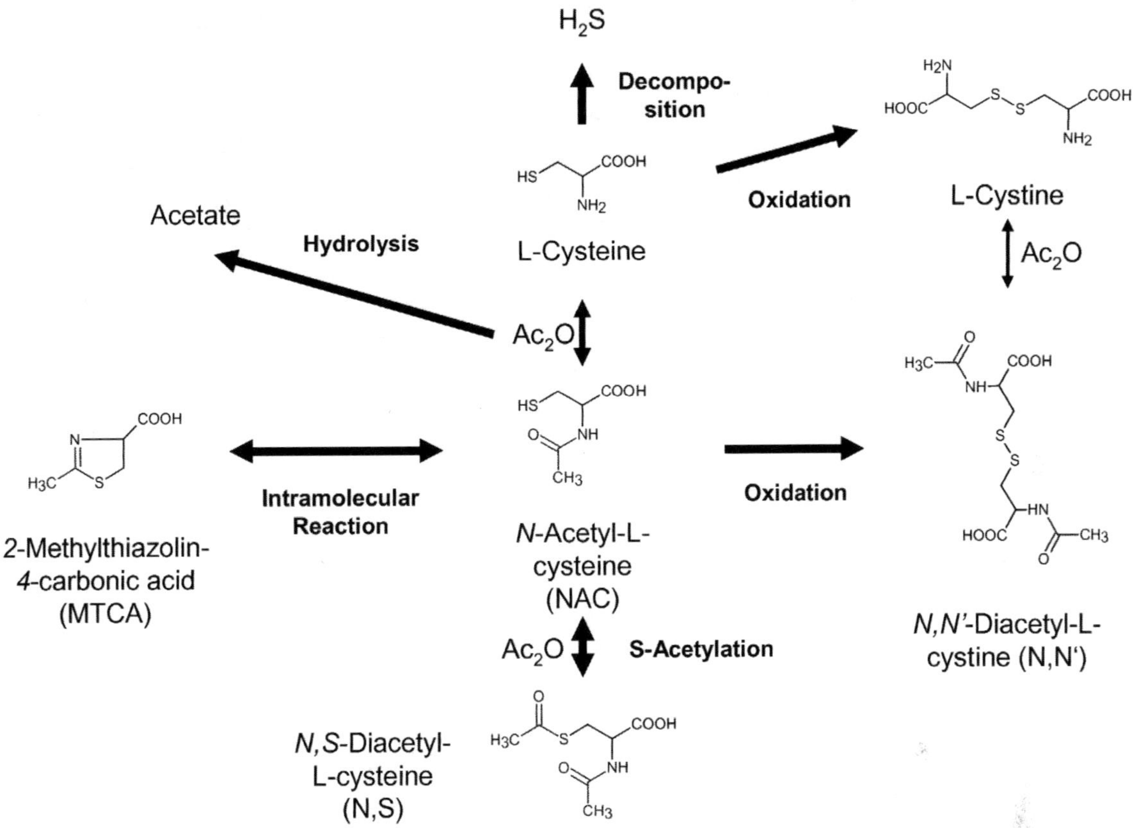

Figure 2: Scheme of all byproducts and reaction pathways (main reaction shaded).

Decomposition reactions are increasing at high pH values and high temperatures. However, this effect in one of the first experiments, i.e. adding only NaOH to cysteine, this effect was found to be negligible at the reaction conditions studied.

Oxidation reactions of cysteine and the main product NAC also are accelerated by increasing temperature and pH. In our experiments, with extremely short residence times (< 20 s), oxidation products were not observed in considerable amount neither. Slightly elevated concentrations of oxidation products in the micro reactor were found upon using aged soda lye. Laboratory experiments with batch processing confirmed this observation.

The intramolecular reaction of ring closure to form MTCA is favoured by elevated temperature at low pH. The reaction to MTCA can be reversed by precipitation and dilution of NAC. This byproduct thus is not of relevance for the comparison of batch and micro processing.

The byproducts most interesting for studying the kinetics of cysteine acetylation were the N,S-diacetyl-L-cysteine (N,S) and acetic acid. The acetylation of both nucleophilic groups, the amino- and thiol- group, appears to proceed extremely fast. Further as competitive reaction, the hydrolysis of the anhydride to acetic acid had to be considered. Both reactions, the competitive hydrolysis and the competitive or subsequent thiol-acetylation, were expected to accelerate with increasing pH and temperature. For the hydrolysis the influence of temperature and pH was confirmed.
However for the N,S formation the trend dependent on temperature observed was quite contrary. The concentration of N,S decreased with increasing temperature. Tests run at varying residence times and varying mixing efficiencies by changing the mixer inlays gave to the following results: N,S is formed predominantly at high pH, as it occurs at the entrance of the mixer. The concentration of N,S is already lower at the exit of the reactor. The acetyl group appears to be transferred from the S-acetyl- to the amino- group of un-reacted cysteine, when comparing the mole equivalents of anhydride added to the acetylation partners isolated. This is in agreement with the results obtained when using only one half equivalent of anhydride which yielded less N,S byproduct and high conversion of cysteine.

In order to validate the conversion rate in the reactor module versus the final conversion succeeding in the collection container, the reaction mixture was diluted immediately, when leaving the reactor module. The residence time in the micro reactor for achieving a maximum conversion of 95% was only 17 s at 35°C. A maximum yield of 87 % NAC at 92 % conversion was obtained at 50° C while suppressing N,S formation. A further raise of temperature did not increase yield, but led to additional unknown byproducts, as observed at 80°C.

Taking in consideration the observed influence of anhydride concentration on the formation of N,S , it was obvious that the N,S formation could be used as an indicator of the mixing efficiency of the micromixer. From simulation and mixer characterization via the Villermaux reaction system ($H_2BO_3^-$ / IO_3^- and H^+, see also [1]) it was known that the mixer parameter variation done for the three mixer inlays influenced mainly the anhydride mixing, since this component was determining the mixer design. From the reaction experiments run with the optimized mixer inlay type 3 a higher N,S concentration was observed (Figure 3 shows anhydride concentration for mixer 3 and the N,S selectivity for the different mixers at the same reaction conditions), which illustrates that the selectivity of the competitive acetylation is considerably dependent on the mixing conditions.

In neither of the experiments (including mixer variation) precipitation, i.e. blocking of channels was observed demonstrating that mixing in the cyclone is extremely fast even at high viscosities (33 % NaOH).

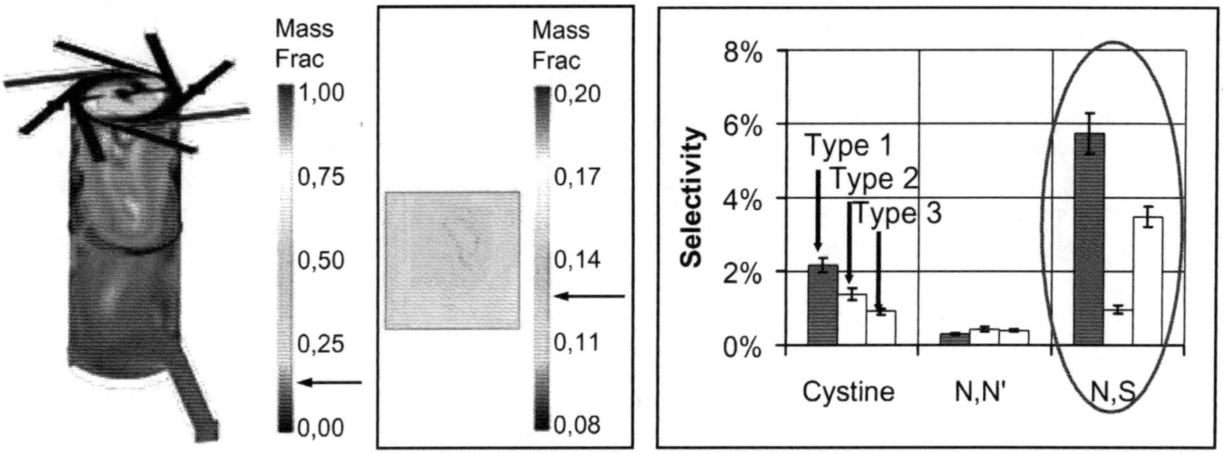

Figure 3: Anhydride concentration in mixer type 3 (left), at the outlet of mixer type 3 (middle), and byproduct selectivities for the different mixers at the same reaction conditions (right).

Literature

[1] R.A. Rakoczy et al., 7th World Congress of Chem. Eng., 10-14 July 2005, accepted.

138e Engineering of Depletion Boundary Layers in Electrochemical Microreactors; Application to cofactor regeneration

Paul Kenis, Seong Kee Yoon

Taking into account the characteristics and transport phenomena of the microscale is a common feature of all research in microchemical systems. The inherently short distances lead to the potential for short residence times and fast heat and mass transfer compared to the macroscale. Moreover, high surface to volume ratios can be utilized in performing heterogeneous catalytic reactions. The occurrence of laminar flow and the capability to create laminar flows consisting of multiple streams each containing different chemistries is the key microscale characteristic being utilized in this work. The dimensions and operating conditions of the proposed electrochemical microreactors are such that fluid flow is pressure driven and characterized by a Reynolds number, Re, less than 10. Two or more different liquid streams introduced into the same channel at Re $<\sim 2100$, will proceed to flow laminarly in parallel as the viscous effects dominate over the inertial effects. The only remaining mechanism of mixing is slow diffusion across the mutual liquid-liquid interface between the two streams transverse to the direction of flow.

In this work, electrochemical conversions are being explored in a microreactor in which electrodes line opposing, inside walls of a microfluidic channel. Pressure-driven laminar flow allows for two streams of different composition to be guided over each of these electrodes respectively, without the need for a physical barrier. The two mass transfer related physicochemical phenomena that determine performance for this novel microreactor concept are (i) slow mixing by interdiffusion of the two stream at the liquid-liquid interface and (ii) the formation of reaction-induced depletion boundary layers on the electrodes. The width of the diffusional mixing zone at the liquid-liquid interface can be controlled precisely by adjustment of the flow rates, while the shape is dictated by the Peclet number: a so-called hourglass profile at high Pe yet a rectangular profile at low Pe. In the microreactors studied here, the width of the mixing zones is typically narrow (<15 µm.), barely forming an hourglass shape. The effect of this diffusional broadening on the performance of the microreactors can thus be neglected.

For optimal performance (i.e. high conversion rates), control of the concentration boundary layer is more important. Minimization of the boundary layer thickness will create a steeper concentration gradient and thus drive a larger flux of reactants through the depletion region to the electrode. Knowledge of theoretical depletion region concentration profiles for different residence times enables prediction of concentration profile for any position in the system. Scaling analysis indicates that by operating at high Peclet and Schmidt numbers (i.e. at Pe $\gg 1$ and Sc $\gg 1$) the concentration boundary layer is pressed close against the wall while simultaneously preventing extensive inter-diffusion of the two streams at the liquid-liquid interface. Scaling analysis also indicates that reducing the dimensions of the channel and/or increasing the flow rate will decrease the width of the depletion region. In addition, reactant stream can be focused closely to the electrode by control of the ratio of the volumetric flow rates of the cathode and anode streams.

Once the limits of minimizing the thickness of the boundary layer by changes in operation conditions have been exhausted, one can only search for engineering solutions to increase the performance of the electrochemical microreactor. First, one can modify the microreactor design to have multiple inlets or outlets to add fresh reactants (multiple inlets) or remove depleted boundary layer (multiple outlets) respectively. Another way to increase mass transfer to the electrode is the introduction of a chaotic micromixer, as described by Stroock et al. Also the use of this micromixer concept to continuously replenish the concentration boundary layers on the electrodes of microreactors will be discussed.

The above-described concepts of depletion boundary layer engineering will be applied and tested in an electrochemical microreactor for enzyme cofactor regeneration. To date efficient regeneration of cofactors such as NADH and NADPH has been challenging due to unfavorable reaction equilibria involved in regeneration. Use of a multistream laminar flow-based microreactors allows for efficient cofactor regeneration as well as actual electro-enzymatic conversions of achiral substrates into chiral products, as will be shown.

138f New Functions for Microfluidic Components by Using Micro Metal Injection Molding (µMIM)

Astrid Rota, Carsten Harms, Marko Hoffmann, Michael Schlüter, Natalie Salk, Norbert Räbiger, Torben Seemann

Microfluidics and microreaction technology is an upcoming field in a variety of industrial branches. As this technique gets implemented in the daily business of industrial companies the cost effectiveness of the production process starts to become an issue

The micro metal injection molding (µMIM) process combines the freedom of design with a broad spectrum of materials (metals, metal alloys and polymers). Catalytic- as well as bio compatible materials have already been structured using this technology. Furthermore µMIM shows the potential of a cost effective series production technique that can be adapted to produce microstructured parts for various applications.

As case example a microfluidic structure was applied for water purification. The reactor surface was coated with an active photocatalytic (TiO_2) modified nano-layer using dip coating. Subsequently it was covered with a glass plate which was adhered with a 2 component adhesive.

For the determination of hydrodynamics and mixing performance measurements with µ-PIV and laser scanning confocal microscopy have been performed. For the investigations of water purification and sanitization the reactor was first fed with a water stream contaminated by a special dye to optimize the hydrodynamic parameters and afterwards with a water stream contaminated by Escherichia coli to investigate the degradation of germs. This was done by illuminating the TiO_2 coated reactor with UV light.

The results indicate that this new type of reactor has an exceptional high potential to eliminate germs in drinking water. The manufacturing process as well as results of mixing performance and germ elimination will be presented.

Keywords: micro metal injection molding (µMIM), microfluidics, microreactor, photocatalysis

138g Advantages of Microsystems Technology in Lead Discovery

Maria J.V.Chapela and Stephanie Y. F. Wong

The highly competitive world of drug market forces pharmaceutical companies to an expensive search for suitable leads. Although industrialization and high-throughput implementation of automation has increased the individual productivity, this is still insufficient to identify the best lead. In addition, the current methodology of lead discovery results in logistic problems of reagent consumption, sample storage and waste production.

Microfluidics and Microsystems technologies are regarded as one of the more promising new disruptive techniques for carrying out ultra-high throughput chemical synthesis that is compatible with highly miniaturized modern screening techniques. The aim is to develop a continuous flow system to perform multiple functions as synthesis, detection, separation and screening all integrated in a microfabricated device eliminating storage and supply chain issues and exploiting the advantages of combining microfluidics and electronic components.

In this work, we present some of the results obtained with continuous flow microreactors and we will discuss the use of various automated systems. The products of the reactions on these systems were identified and quantified using an integrated post-chip analysis and detection system.

In addition, we will present our latest developments towards multi-step synthesis protocols and coupling with screening assays.

American Institute of Chemical Engineers
3 Park Avenue
New York, NY 10016-5991

ISBN: 978-1-60423-468-8